UNDERSTANDING ELECTRICITY AND WIRING DIAGRAMS FOR HVAC/R

UNDERSTANDING ELECTRICITY AND WIRING DIAGRAMS FOR HVAC/R

AIR CONDITIONING AND REFRIGERATION INSTITUTE

and

Robert Chatenever
Oxnard College

Prentice Hall
Upper Saddle River, New Jersey *Columbus, Ohio*

Library of Congress Cataloging-in-Publication Data

Chatenever, Robert.
Understanding electricity and wiring diagrams for HVAC/R / Air Conditioning and Refrigeration Institute and Robert Chatenever.
p. cm.
ISBN 0-13-517897-5
1. Air conditioning—Electric equipment. 2. Heating—Equipment and supplies. 3. Electric wiring, Interior. I. Air-Conditioning and Refrigeration Institute. II. Titl
TK4035.A35C43 2000
697.9'3—dc21 99-18026
CIP

Editor: Ed Francis
Production Editor: Christine M. Buckendahl
Production Coordinator: Lisa Garboski, bookworks
Design Coordinator: Karrie Converse-Jones
Text Designer: Meryl Poweski
Production Manager: Patricia A. Tonneman
Illustrations: Academy Artworks, Inc.
Marketing Manager: Chris Bracken

This book was set in Meridien by The Clarinda Company, and was printed and bound by R. R. Donnelley & Sons Company. The cover was printed by Phoenix Color Corp.

Printed in the United States of America

10 9 8 7 6 5 4 3 2 1

ISBN: 0-13-517897-5

Prentice-Hall International (UK) Limited, *London*
Prentice-Hall of Australia Pty. Limited, *Sydney*
Prentice-Hall of Canada, Inc., *Toronto*
Prentice-Hall Hispanoamericana, S. A., *Mexico*
Prentice-Hall of India Private Limited, *New Delhi*
Prentice-Hall of Japan, Inc., *Tokyo*
Prentice-Hall (Singapore) Pte. Ltd., *Singapore*
Editora Prentice-Hall do Brasil, Ltda., *Rio de Janiero*

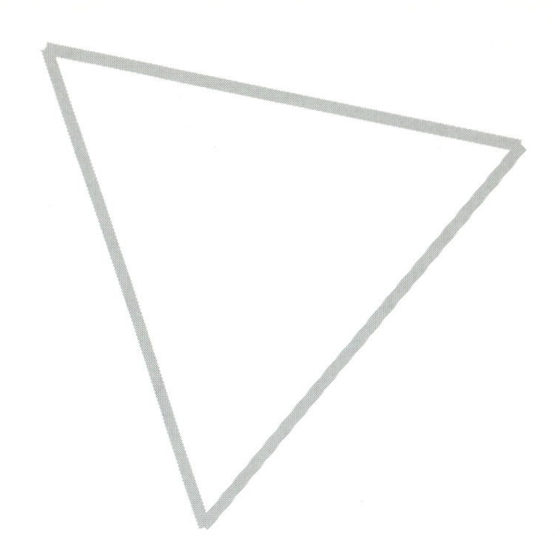

PREFACE

This text is aimed at the service technician and the vocational student studying to become a service technician. Towards serving this audience, this text contains a number of features, many of which are unique in texts of this type. Some of those features are as follows:

1. The text begins with general electricity and electrical circuits and moves quickly into electrical wiring diagrams. The bulk of the text is devoted to explaining wiring diagrams and the new devices that are encountered with each new wiring diagram.

2. Questions and exercises are presented throughout the text immediately following the applicable wiring diagram description. These questions are designed to challenge students' understanding of concepts, rather than to challenge their ability to recall miscellaneous facts or parrot back passages in the text.

3. Mathematical presentations, which are helpful to some students and confusing to others, are separated into identified sections (Advanced Concepts) so that the instructor or the student can easily choose whether to use them or not.

4. The text points out and identifies common pitfalls and mistakes made by new technicians, which could only be known by an author who has experienced many of the pitfalls firsthand (no modesty here!).

5. Helpful hints are identified with a ☛ symbol. These are not just book learning, but techniques actually used by service technicians.

6. The various electrical devices are introduced one or two at a time, as each is encountered in a circuit. This transforms the discussion of each device from an academic exercise to one with an application readily at hand.

7. Explanations of electronics are handled by describing the inputs and the outputs, without attempting explanations at the nuclear level of how the diodes, IC, and triacs work or are put together. This is appropriate because the service technicians who use this text will not be the ones who design or repair these "black boxes."

8. The theory of "How Motors Work" is segregated into its own chapter (Chapter 12). This entire chapter can be deleted by those instructors who feel that service technicians are never called upon to do tasks for which motor theory is required. They can spend more time on motor applications, dealing with the nuts and bolts of starting relays, motor troubleshooting, and circuit testing. Alternately, this chapter section can be used as a starting point, providing a different approach to explaining current, transformers, inductance, and so on.

9. Jargon, which is common within the industry, is identified within the text and defined in the Glossary.

There is no way that one author can know about all the unique wiring diagrams, methods, technician short-cuts, and potential pitfalls. However, this text represents a revolutionary and unique start in that direction. To those manufacturers whose equipment is not presented here, the author invites you to submit your most common wiring diagrams for inclusion in the next revision. To those technicians who have developed better methods and would like to share them with others, the author invites your contributions as well.

This text is not designed to be a single text to be used for a complete course in refrigeration and air conditioning. This text presumes a prior knowledge of the refrigeration circuit and the function of basic electrical refrigeration components such as compressors, condenser fans, and evaporator fans.

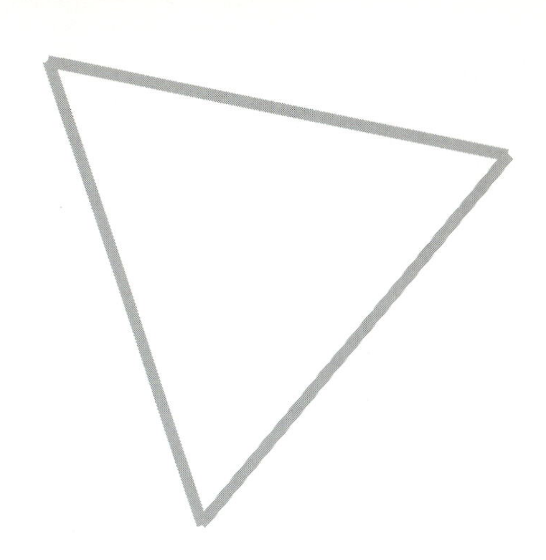

BRIEF CONTENTS

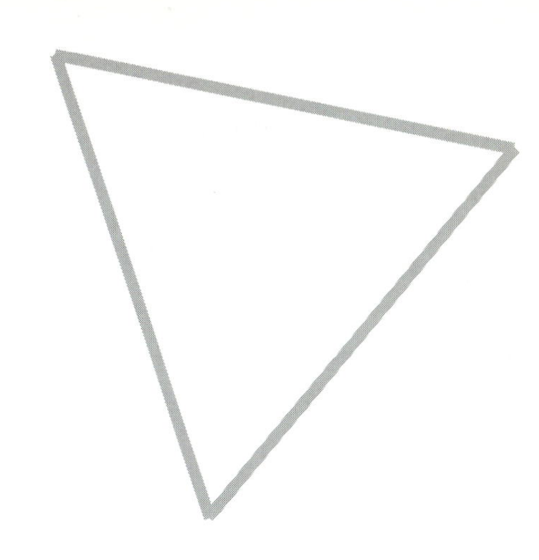

Contents

CHAPTER EIGHT COMMERCIAL SYSTEMS 109

CHAPTER NINE MOTOR APPLICATIONS 130

CHAPTER TEN POWER WIRING 150

CHAPTER ELEVEN TESTING AND REPLACING MOTORS AND START RELAYS 161

CHAPTER TWELVE HOW MOTORS WORK 170

CHAPTER EIGHTEEN HEAT PUMPS 231

CHAPTER NINETEEN ICE MAKERS 246

CHAPTER TWENTY MISCELLANEOUS DEVICES AND ACCESSORIES 277

CHAPTER TWENTY-ONE WIRING TECHNIQUES 287

CHAPTER TWENTY-TWO DDC CONTROLS 292

GLOSSARY AND ABBREVIATIONS 300

INDEX 307

UNDERSTANDING ELECTRICITY AND WIRING DIAGRAMS FOR HVAC/R

Electrical Concepts

Electricity is a mystery to many people, and it's easy to understand why. You can't see it, but you know it's there when it lights up a bulb, runs a motor, or gives you a shock. What is electricity, and how does it work? The easiest way to understand electricity is to make an analogy with something that you can easily visualize such as water flowing through pipes. In many ways, electricity flowing through wires behaves very similarly to water flow through pipes.

WATER FLOW THROUGH PIPES

Consider the water system shown in figure 1–1. A storage tank full of water is connected to a pipe that contains a small propeller. The water will flow from the tank through a pipe under pressure, through the propeller, and then into a pipe that is at atmospheric pressure (for our purposes, we will describe atmospheric pressure as zero pressure). As the water flows through the propeller, it causes the propeller to turn. The flow of the water across the propeller results in a pressure drop across the propeller, and the mechanical rotation of the propeller can be converted to useful work.

Why does the water flow through the pipe and the propeller? It flows because there is a higher pressure at one end of the pipe than there is at the other. If there is a path available for the water to flow from the higher pressure to the lower pressure, it will. If the pipe that is now open to atmosphere were piped back to the tank as shown in figure 1–2, there would be no flow because there would be no pressure difference between one end of the pipe and the other which would cause any flow to happen.

In figure 1–3, we have duplicated the piping system in figure 1–1, except that we have replaced the propeller with a different propeller. The propeller in figure 1–3 has more blades, and the blades have more pitch. It is harder to push water through the propeller in figure 1–3 than the propeller in figure 1–1. As you can see in figure 1–3, the pressure before the propeller and after the propeller are the same as in figure 1–1, but the amount of flow is less, because of the higher resistance propeller. From figures 1–1 and 1–3, we can draw the following conclusions:

1. Flow and pressure are two completely different properties in a water circuit. The circuit in figure 1–3 has the same pressures as in figure 1–1, but the flow rate is different. Flow of water is measured in gallons per minute (gpm), while water pressure is measured in pounds per square inch (psi).

2. When water is flowing through a pipe under pressure, and passes through a resistance like the propeller, downstream from the propeller we will have the same amount of flow as we had before the propeller, even though the pressure of the water has

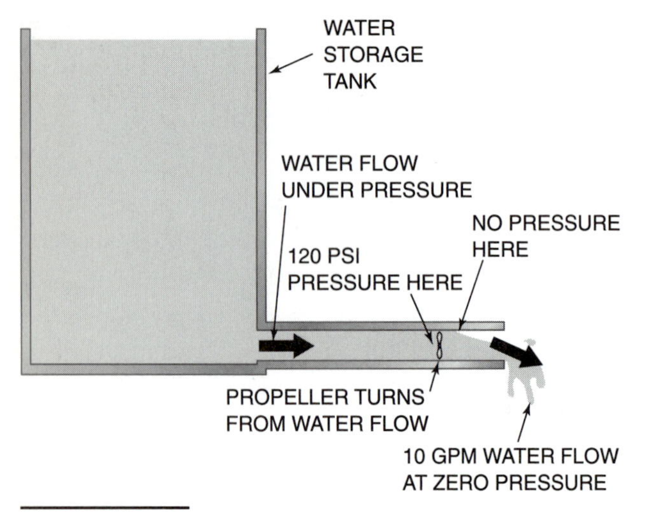

Figure 1–1 Propeller turns because there is flow resulting from a pressure difference.

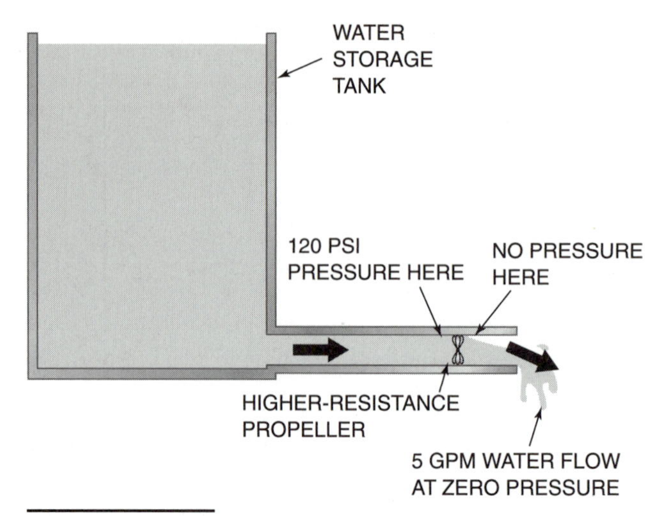

Figure 1–3 Higher resistance = less flow.

been reduced to zero. It doesn't matter that the pipe in front of the propeller is full and under pressure, and that the pipe behind the propeller is only partially filled and under zero pressure. The flow rates are the same. Simply stated,

GOZINS = GOZOUTS

That is, whatever flow rate goes into the propeller, also goes out.

3. The amount of flow that goes through the propeller depends upon the resistance of the propeller to flow. The higher the resistance of the propeller to flow, the lower the flow rate of water will be.

4. Changing the resistance of the propeller changes the flow rate of water, but it does not change the pressure drop across the propeller. In this circuit, the propeller, regardless of its physical design, will have a pressure drop across it equal to the difference in pressure between the ends of the pipe.

Figure 1–2 When there is no pressure difference, there is no flow.

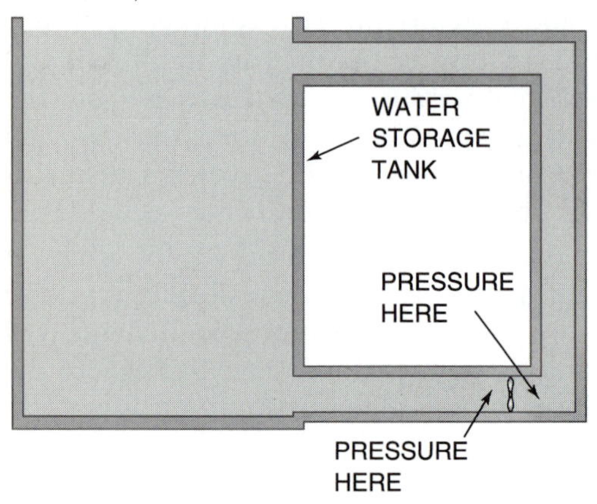

5. If the level of water in the tank were increased, increasing the pressure difference across the propeller, the flow rate of water across the propeller would increase.

ELECTRICAL FLOW THROUGH WIRES

Obviously, electricity is not water. Then, what is electricity?

To answer the question, we must understand something about the structure of matter. All matter is made up of atoms. Different atoms make up all of the naturally occurring elements in the universe. There are atoms of copper, aluminum, oxygen, carbon, and over 100 other elements. Each of these elements' atoms consists of some electrons that revolve around a nucleus of protons and neutrons, very similar to the planets revolving around the sun

Figure 1–4 Atomic structure.

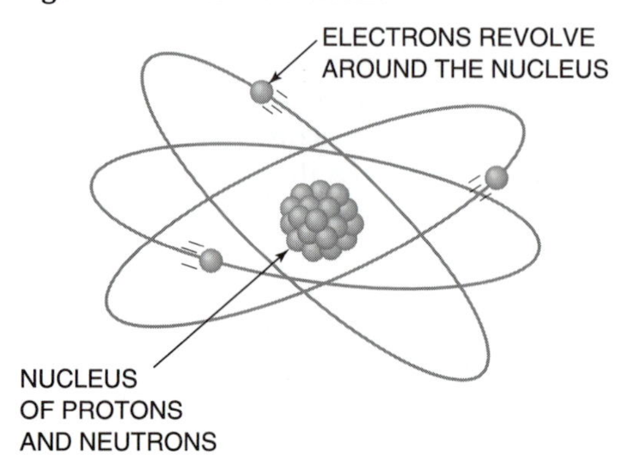

Figure 1–5 Pushing on ball #1 will cause all of the balls to move through the tube.

(figure 1–4). Electrons, protons, and neutrons are simply very small particles of matter. Electrons carry a negative magnetic charge, and protons carry a positive magnetic charge.

For some materials such as copper, aluminum, silver, and gold, if a magnetic force is applied, an electron in orbit around one nucleus can be easily knocked out of its orbit into the next atom. This negatively charged electron then repels an electron in the adjacent atom, causing it to be knocked out of orbit into the next atom in line. The result is a "flow" of electrons through the wire. The action of the electrons, each pushing on an adjacent electron, is similar to the ping-pong balls inside the tube shown in figure 1–5. The balls are analogous to the electrons, while the tube is analogous to a wire. When pressure is applied to one end of the tube, it pushes on the first ball, which, in turn pushes on the second ball, and so on until the last ball moves. Although not strictly accurate, it is convenient to think of electricity as very small particles (electrons) flowing through a wire.

Compare the circuit in figure 1–6 with the water system in figure 1–1. A circuit is a system of wires that provides a path through which electrons can flow. At the left end of the wire, we have a source of electrons under pressure, such as from a battery. A wire carries electrons under pressure to the light bulb. The wire on the right side of the light bulb is connected to a rod that is buried in the ground, which is at zero electrical pressure. The light bulb consists of a very thin wire, and it is difficult for the electrons to move through such a thin wire. This is similar to the propeller in figure 1–1. However, if there is a higher electrical pressure in the wire on the left side of the bulb than there is on the right side of the bulb, electrons will be forced to flow through the filament in the bulb. And, as with the propeller in figure 1–3, if the filament is made thinner (more restrictive), then the flow rate of electrons will be reduced.

All of the observations that we made for the water system apply equally well to the electrical circuit. If the resistance in the circuit is increased, the flow rate of electrons will be reduced. If the available pressure difference across the resistance is increased, the flow rate of electrons will be increased.

Following are some differences between the water circuits and the electrical circuits described above:

1. In the water circuit, the flow rate was measured in gpm. In the electrical circuit, the flow rate of electrons is measured in amperes, usually abbreviated to **amps**. One amp is a flow rate of 6,280,000,000,000,000,000 electrons per second. The symbol used for **amps** is either "A" or "I." To give you some points of reference, a 100-watt light bulb would consume slightly less than one amp, a household refrigerator would consume 4 to 6 amps, and an electric hair blower/dryer would consume almost 15 amps.

2. In the water circuit, the water pressure was measured in psi. In an electrical circuit, the electrical pressure is measured in **volts.** The symbol used to represent voltage is "V." Sometimes, electrical pressure is called electromotive force. In that case, the symbol used is "E."

Figure 1–6 Electrons under pressure flow through the light bulb and into the ground.

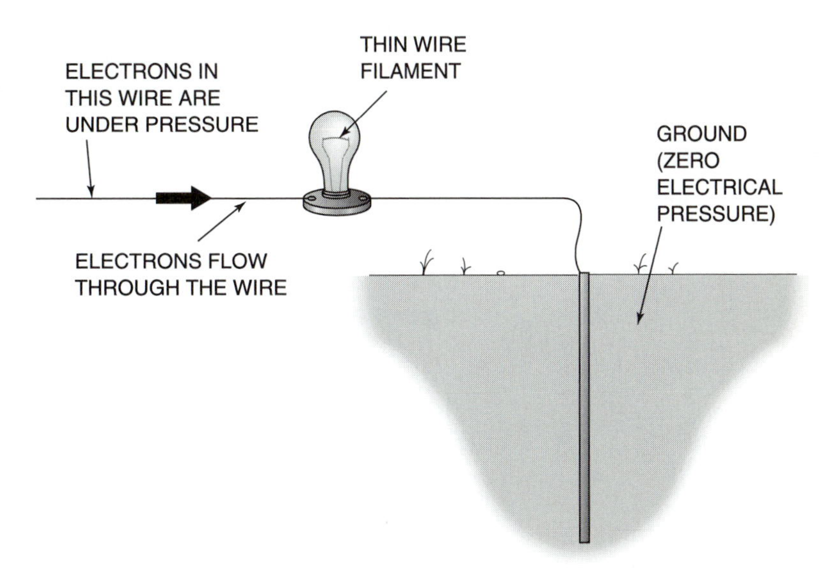

3. In the water circuit, we don't have a commonly used method of measuring the resistance of the propeller. In an electrical circuit, the resistance of the device in the circuit is measured in **ohms.** The symbol used for resistance is "R." The symbol used for ohms, which is the measurement of how much resistance there is to the flow of electricity, is Ω (the Greek letter omega).

CONDUCTORS AND INSULATORS

Every different element is comprised of atoms that consist of a different number of electrons, protons, and neutrons. The different atomic structure of each element is responsible for giving each element its unique physical properties. For some elements, the electrons are easily knocked out of their orbit into the next atom. Electrons flow easily through these materials, so they are called **conductors.** For other elements, it is very difficult to knock electrons loose. Electricity will not flow through these materials, so they are called **insulators.** Most metals are good conductors. Conductors in electrical circuits are usually made from copper, or sometimes aluminum. Rubber, plastic, and glass are commonly used as insulators.

Conductors have very low resistance. Insulators have very high (almost infinite) resistance. An ordinary wire consists of a conductor and an insulator (figure 1–7). The diameter of the conductor is determined by how many amps it is designed to carry. A greater diameter conductor can carry more amps than a smaller diameter conductor. On the other hand, the thickness of the insulation is determined by the voltage that is to be carried in the conductor. The insulation must be thick enough to keep the electrons contained within the wire.

Figure 1–7 Insulated wire.

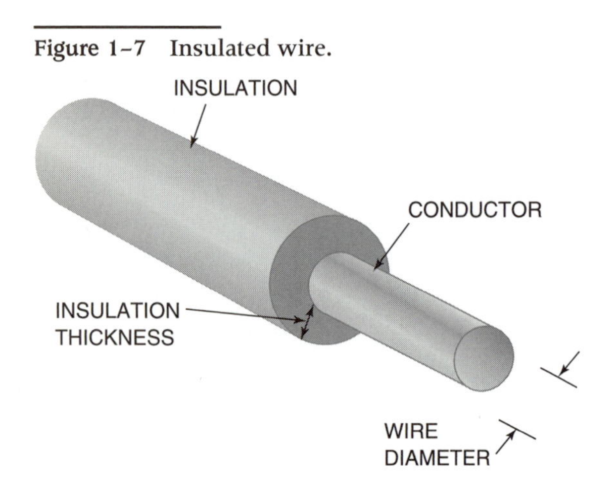

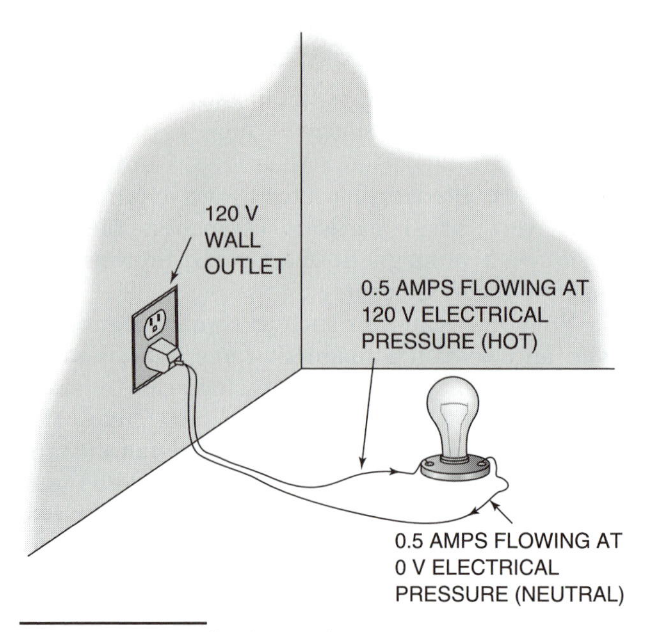

Figure 1–8 Simple electrical circuit.

A SIMPLE ELECTRICAL CIRCUIT

Figure 1–8 shows a simple electrical circuit, a lamp plugged into a wall outlet. The power company provides electrons under a pressure of 115 to 120V at one of the outlet slots (figure 1–9). [Note: the values of 115V and 120V are used interchangeably in this text. Where calculations are involved, 120V will be used.] The other outlet slot is at zero electrical pressure. Electrons flow from the pressurized outlet slot, through a wire under pressure, to the light bulb. As the electrons flow through the filament, the electrical pressure is dissipated. The amps that leave the light bulb are at zero pressure, and they flow back to the zero pressure slot at the outlet. In residential electrical service, the pressurized slot at the outlet is called **Hot,** and the zero pressure slot is called **Neutral.** When electrons flow from a wire at some electrical pressure **(potential),** through a load to another wire at a lower potential, it is called a **complete circuit.** If the path is interrupted, it is called an **open circuit.** If there is an unintended path that allows electrons to flow from a wire at one pressure to another wire at another

Figure 1–9 115V duplex outlet.

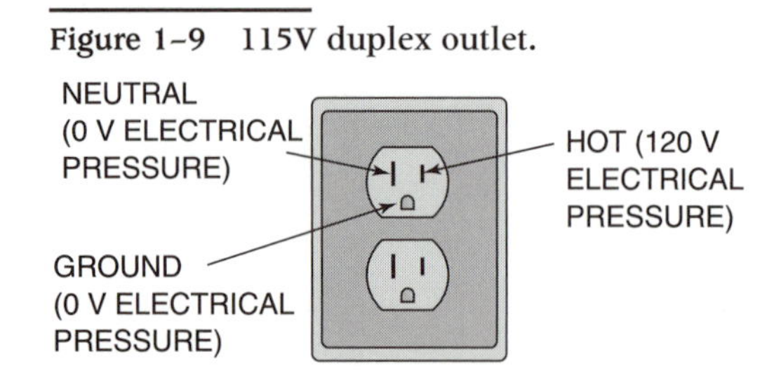

lower pressure *without passing through a load,* it is called a **short circuit,** and it will cause a fuse or a circuit breaker to open. If the short circuit has been caused inside a load such as a motor, that load is said to be **shorted.** When there is an unintended connection that allows electrons to flow from a current carrying wire into the casing of a device, it is called a **grounded circuit,** and the device that failed is **grounded.**

COMMON, GROUND, NEUTRAL

Some technicians use the terms **common** or **ground** interchangeably with Neutral. "Common" and "ground" are terms that should not be used when referring to "Neutral." Common is used to indicate an electrical point that is a part of two different devices. For example, the switch shown in figure 1–10a can make contact from terminal A to terminal B, or it can make contact from terminal A to terminal C. Terminal A would be called the common terminal, because it is part of two different switching circuits, A-B and A-C. Similarly, if a motor has two windings inside that are connected together at one end (figure 1–10b), that point would be called the common terminal on the motor. The reason that some technicians refer to "Neutral" as "common" is understandable, because many different loads are usually connected to the Neutral wire.

Some technicians use the term "ground" instead of "Neutral" because they are at the same electrical pressure—zero. However, a ground wire and a Neutral wire are quite different. A ground wire (green, or bare) is provided for safety purposes, and ordinarily does not carry any current. However, a Neutral wire (white) carries current whenever the circuit is operating.

Figure 1–10 Two "commons."

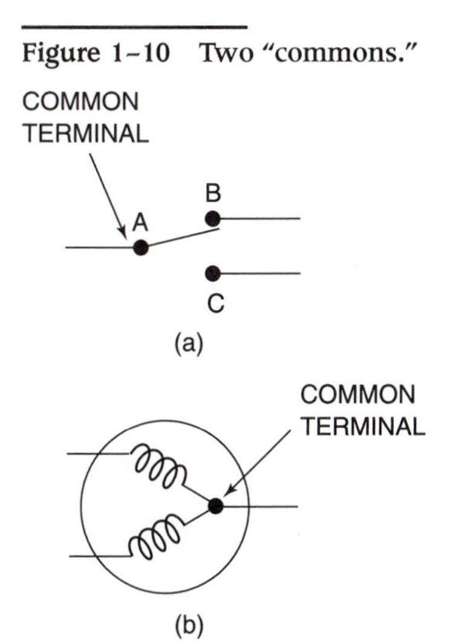

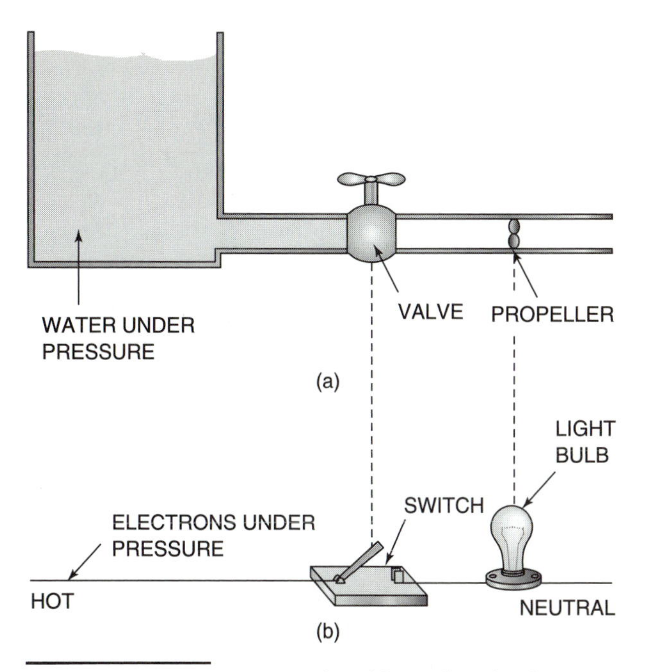

Figure 1–11 An open switch is like a closed valve.

SWITCHES

In figure 1–11a, a valve has been added to the previously described water circuit, and in figure 1–11b, a **switch** has been added to the electrical circuit. When the valve is open, there is flow of water through the propeller. This is the same as the operation of a closed switch in the electrical circuit. When the water valve is closed, the flow stops. In an electrical circuit, the flow of electrons can be stopped by opening the switch. Do not get confused by the terminology. An *open valve* is equivalent in function to a *closed switch,* and a *closed valve* is equivalent to an *open switch.*

The switch and the valve can be thought of as the "traffic cops" for flow. When a switch is closed, it acts just like a wire. It has no effect in the circuit, and amps will flow freely through the switch without any loss in electrical pressure. However, when the switch is open, the electrical pressure that is supplied to one side of the switch cannot pressurize the downstream side of the switch. In this case, the switch merely stops the flow of electrons. A common mistake made by students is to think that one side of the switch is Hot, and the other side is Neutral. This is not necessarily the case. Figure 1–12 shows how the pressure in the wires connected to the switch changes, depending on whether the switch is open or closed. Figure 1–12a shows that with one side of the switch connected to Hot, and with the switch open, the other side of the switch will be at zero pressure (Neutral pressure). However, when the switch is

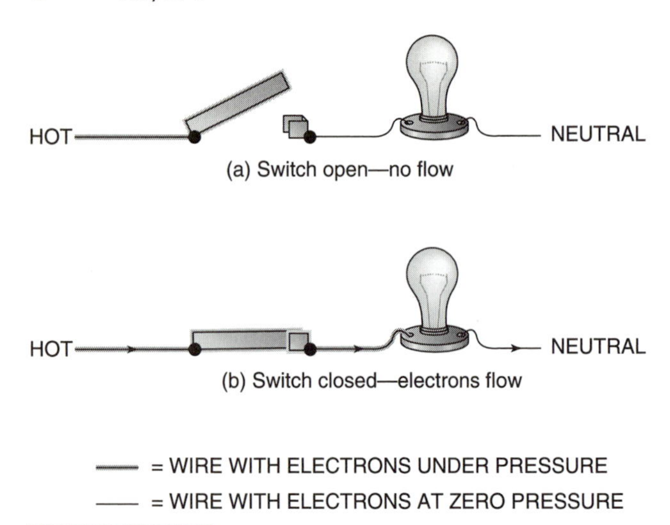

Figure 1–12 Electrical pressure change in a circuit.

closed (Figure 1–12b), the wire connecting the switch to the load becomes the same electrical pressure as Hot. The wire between the switch and the load is commonly referred to as a **switched Hot.** Sometimes it's at the same pressure as Hot, and sometimes it's at the same pressure as Neutral.

LADDER DIAGRAMS

Figure 1–13a shows how a switch would be wired into the previous light bulb circuit. This type of diagram is called a **pictorial wiring diagram.** It shows the physical locations of the devices, and it shows what each device looks like. Figure 1–13b shows the same circuit, but in a schematic wiring diagram called a **ladder diagram.** The power supply is shown as two vertical lines. The circuits are then drawn between the two vertical lines. When there are many different circuits, they look like the rungs on a ladder. A ladder diagram is very easy to read, but it does not tell you anything about the physical locations of the devices, or their physical locations relative to each other. Also, a ladder diagram uses symbols to represent various devices, rather than pictorial representations of the devices. Figure 1–14 shows some of the symbols that are used in wiring diagrams. They will be explained later, as each is encountered in a wiring diagram. Note that not all manufacturers use exactly the same symbols for each device. The symbols given here are commonly used, but you will see others.

EXERCISE:
Redraw the pictorial diagram shown in figure 1–15 as a ladder diagram. Label all the wire colors. In this exercise, and in all future exercises, when you are asked to draw wiring, make all lines straight, and either vertical or horizontal.

SWITCHES IN SERIES

Figure 1–16 shows a circuit with two switches in **series** with a load. When we say that electrical devices are connected in series, we mean that the flow of electrons must all pass through one device, and then they pass through the next device, one after the other. When switches are in series, both switches must be closed in order for the load to operate. If ei-

Figure 1–13 (a) Pictorial diagram. (b) Ladder diagram.

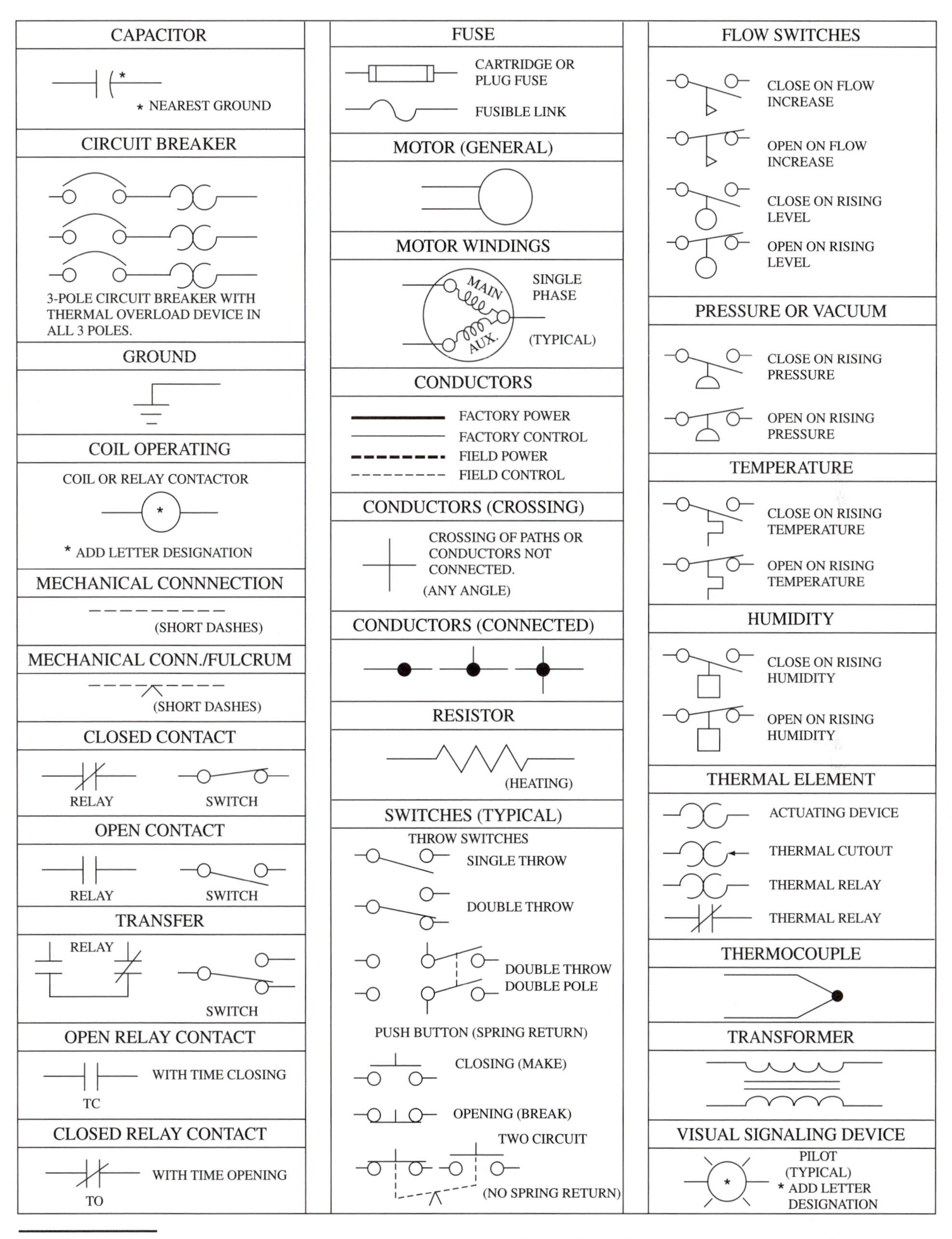

Figure 1–14 **Ladder diagram with wiring symbols.** *(Reprinted from Refrigeration and Air Conditioning, 3rd Ed. by Air Conditioning and Refrigeration Institute, copyright © 1998, Prentice Hall, Inc. Reprinted by permission.)*

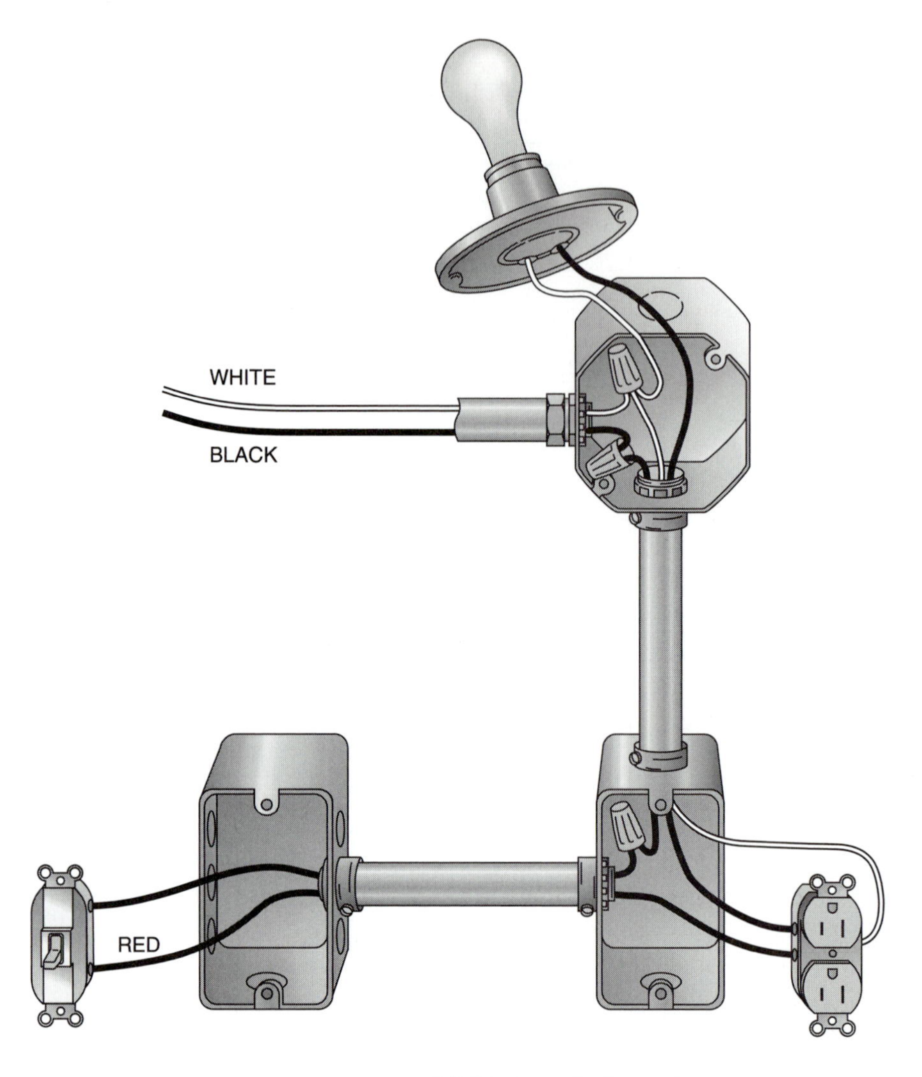

Figure 1–15 Exercise.

ther one of the switches opens, the load will become de-energized.

SWITCHES IN PARALLEL

Figure 1–17 shows a circuit with two switches in **parallel**, controlling a load. With either of the switches closed (or with both switches closed), the load will operate. Compare this with switches in series, in which both switches needed to be closed in order for the load to operate.

Figure 1–16 Switches in series.

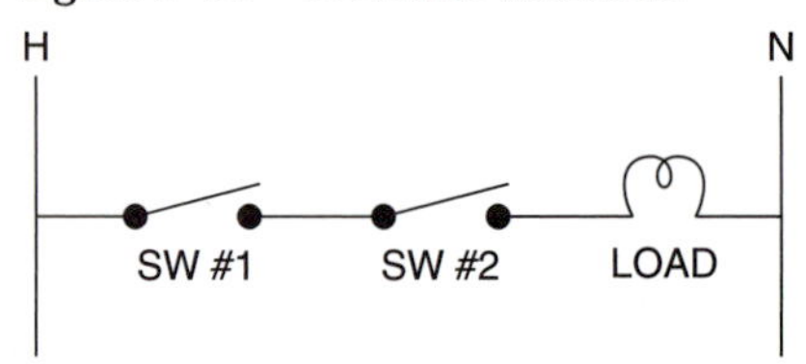

EXERCISE:
Redraw the components in figure 1–18. Add wiring so that the motor will run when either switch 1 or switch 2 is closed. Redraw the components again. Add wiring so that the motor will run only when both switches are closed.

SWITCH TYPES

A switch is described by the number of moving parts **(poles)**, and the number of positions in

Figure 1–17 Switches in parallel.

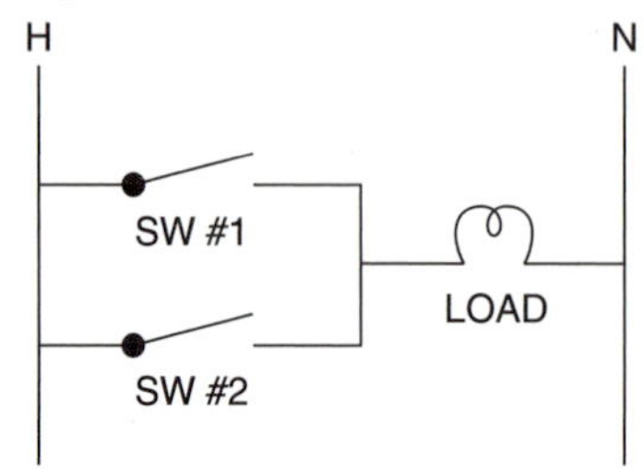

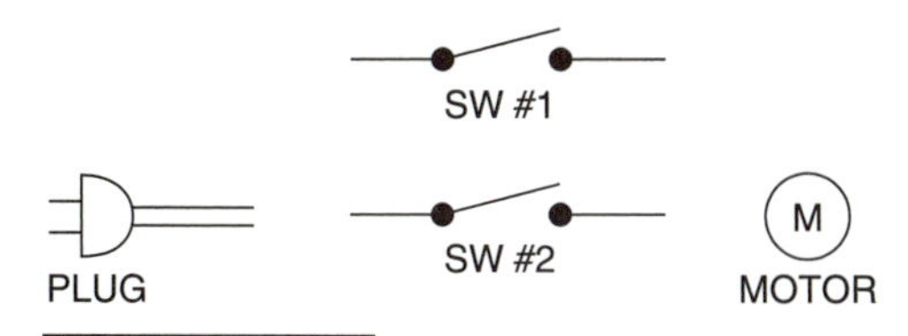

Figure 1–18 Exercise.

which a circuit can be completed (**throws**). Several types of switches are shown in figure 1–19. Illustration (a) shows a single-pole single-throw switch, abbreviated **SPST**. This switch is either open, or it is closed. When it is closed, bulb A is energized.

Illustration (b) shows a single-pole double-throw switch (**SPDT**). It can make contact in two different circuits, but not at the same time. Depending upon its position, either bulb A or bulb B will be lit. Terminal 1 is a part of the circuit in either switch position. We say that terminal 1 is common to both circuits, and it is called the **common** terminal. This is not to be confused with the terminology used by some technicians who refer to "Neutral" as "common." As you can see from figure 1–19b, it is perfectly acceptable to have a common terminal on a switch connected to the "Hot" side of the circuit.

Illustration (c) shows a double-pole single-throw switch (**DPST**). It can make contact in two different circuits. The dotted line that connects the poles indicates that they are mechanically **ganged** together and must operate simultaneously. Bulb A and bulb B will either both be energized or both be de-energized.

Illustration (d) shows a double-pole double-throw switch (**DPDT**). It is simply two single-pole double-throw switches that are ganged to operate together.

There are also triple-pole and triple-throw switches.

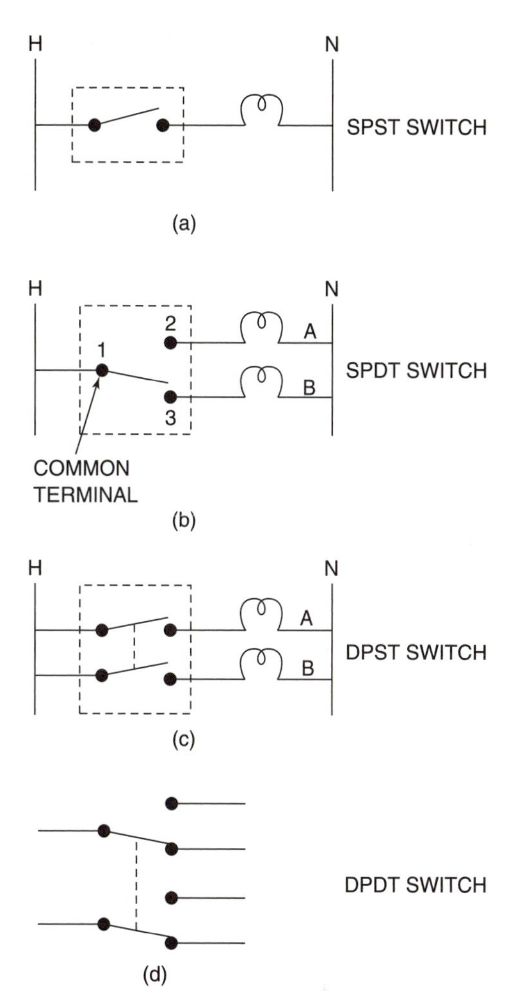

Figure 1–19 Switch types.

EXERCISE:
Redraw the components in figure 1–20. Add wiring so that one of the lights is always on, but never both at the same time.

EXERCISE:
Redraw the components in figure 1–21. Add wiring so that when the water level in the sump rises, the sump pump will start and the alarm will sound. The level switch makes R-B and breaks R-W on a rise in level.

EXERCISE (OPTIONAL):
Redraw the components in figure 1–22. Add wiring so that the light can be turned on or off

Figure 1–20 Exercise.

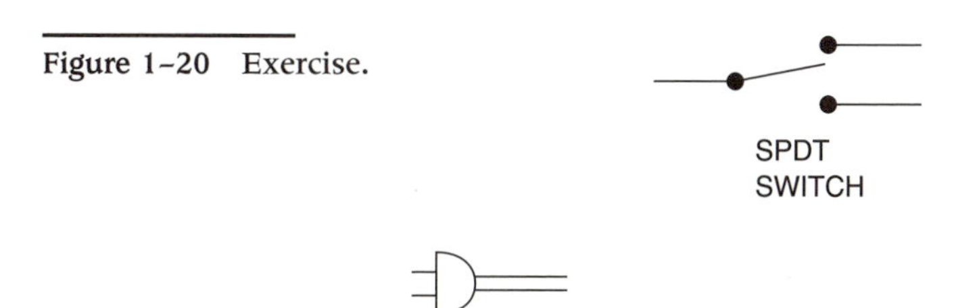

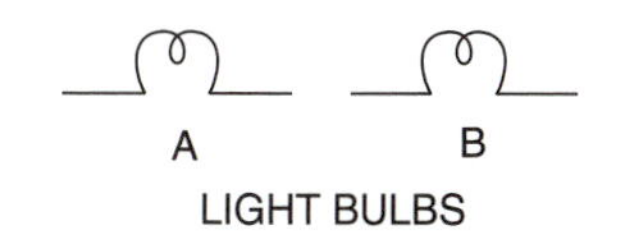

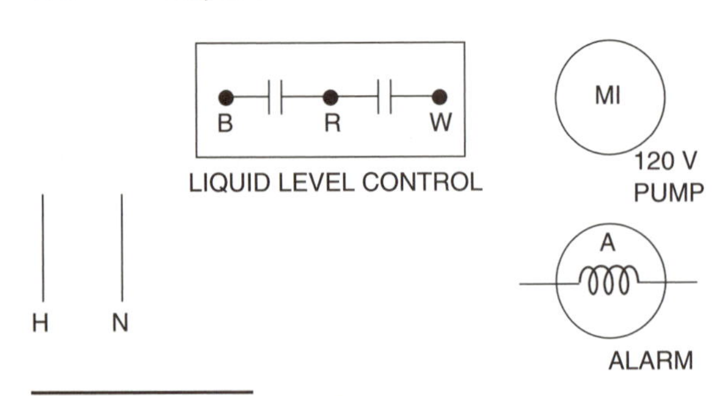

Figure 1–21 Exercise.

from either switch. This is an application that is commonly used in a residence to allow the hall light to be turned from either end of the hallway, and then shut off from the other end.

AUTOMATIC SWITCHES

The switches described in the previous section can be manually operated or automatically operated. For example, switches that automatically close ("make") or open ("break") in response to a change in temperature are called **thermostats** ("thermo-" indicates that it is temperature operated). One example of how a change in temperature can be used to operate a switch is shown in figure 1–23. The temperature in a space is sensed by the bulb, which is filled with a fluid that will expand and contract as temperature increases or decreases. The expansion and contraction of the fluid causes the bellows to move. The bellows is then attached to any type of switch, as shown in figure 1–24.

A second method of automatically operating a switch is by the use of a **bimetal element** (figure 1–25). A bimetal element consists of two dissimilar metals with very different coefficients of expansion. When the temperature around a bimetal element increases, one metal will grow at a faster rate than the other. The result will be that the bimetal will bend. This mechanical motion can then be attached to a switch, as described for the bellows-operated thermostat. The bimetal element can be fabricated into coils, spirals, or "U" shapes to permit them to be used in confined spaces.

Thermostats are also known by other names that are more descriptive of what they do. The following are examples:

High Limit Switch: A thermostat that opens if the sensed temperature gets abnormally high.

Low Limit Switch: A thermostat that opens if the sensed temperature gets abnormally low.

Thermal Overload: A thermostat that senses the temperature of the windings inside a motor, and opens if they get too hot.

Heating Thermostat: A room thermostat that opens on a rise in temperature.

Cooling Thermostat: A room thermostat that closes on a rise in temperature.

Aquastat: A thermostat that senses water temperature.

Cold control: A thermostat used in a small refrigerated box such as a residential refrigerator.

Freezestat: A thermostat that opens if it senses a temperature below 32°F.

There are many other types of automatic switches. Those that are operated by a change in pressure are called **pressurestats.** They work on the same principle as the bellows-operated thermostat, except that the sensed pressure is connected directly to the bellows (figure 1–26). A pressure switch can also sense air pressure (figure 1–27). Pressurestats also are known by other names, depending on their function:

High Pressure Cut-Out: A safety device that senses pressure on the high pressure side of the refrigeration system and opens if it becomes abnormally high.

Low Pressure Cut-Out: A safety device that senses either refrigerant pressure on the low side, water pressure to an ice machine, or gas pressure to a furnace, and opens if the pressure drops too low.

Other automatic switches respond to changes in humidity (**humidistat**), flow (**flow switch**), or

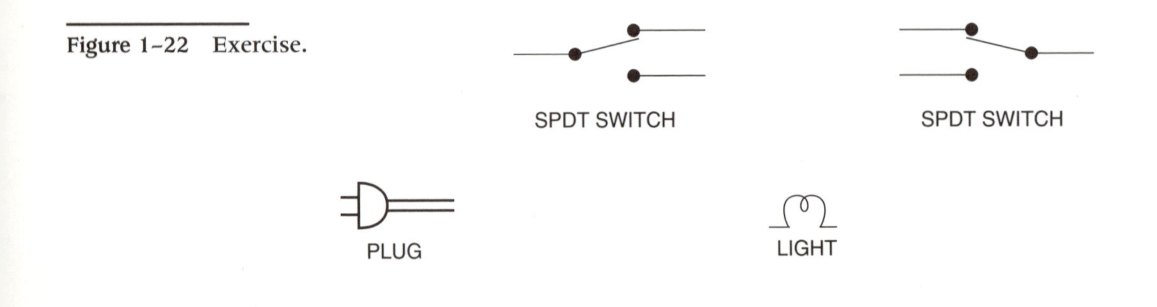

Figure 1–22 Exercise.

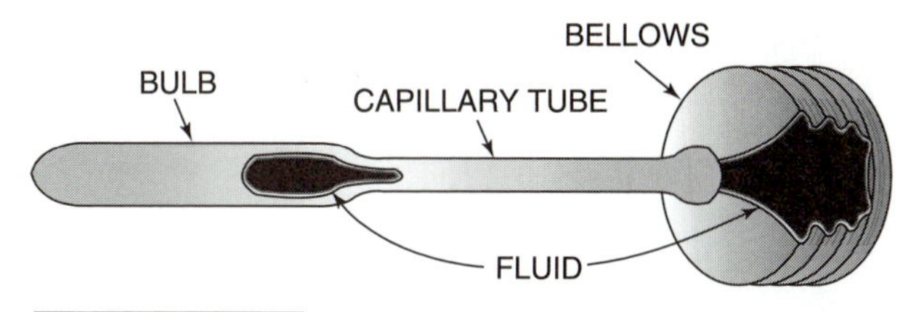

Figure 1–23 Temperature-operated bellows.

position. The term *flow switch* is normally reserved for a switch that senses the flow of water (figure 1–28). A flow switch that senses the flow of air is usually called a **sail switch** (figure 1–29).

Switches that sense position are either **door switches** (figure 1–30) or **microswitches** (figure 1–31). These both are types of momentary switches. A **momentary switch** is one that is held open (or closes) by an internal spring, but when the button or lever on the switch is pressed, the switch position changes. When the pressure on the button or lever is released, the internal spring causes the internal switch to return to its normal position.

A microswitch is also sometimes called a **position switch** or an **end switch.** For example, there may be an actuator that opens a damper, and we want to make sure that the damper actually opens before we allow some other control function to happen. We would use an end switch (microswitch) that would be pushed closed by the damper physically pressing against the switch as it approached its fully open position.

Figure 1–24 Temperature-operated switches.

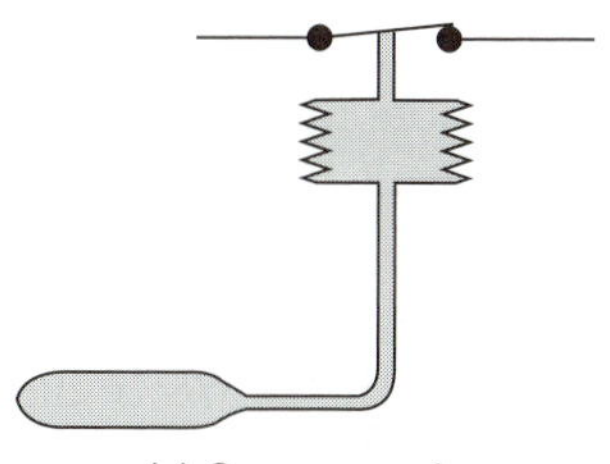

(a) Opens on a rise in temperature.

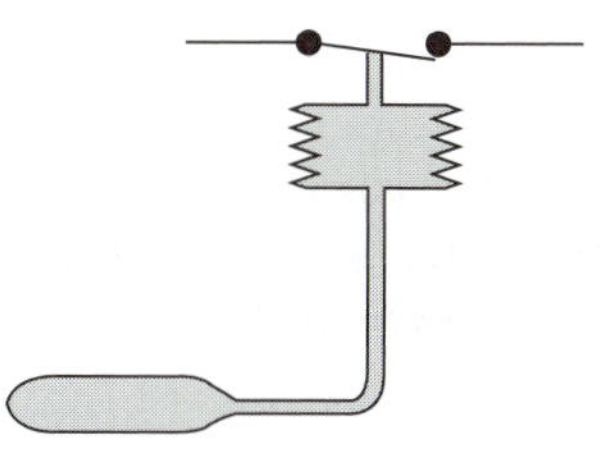

(b) Closes on a rise in temperature.

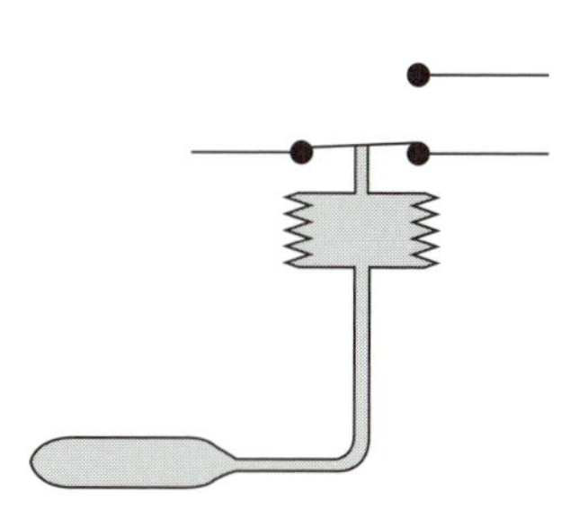

(c) SPDT Bellows-operated thermostat.

METAL ACTUATORS

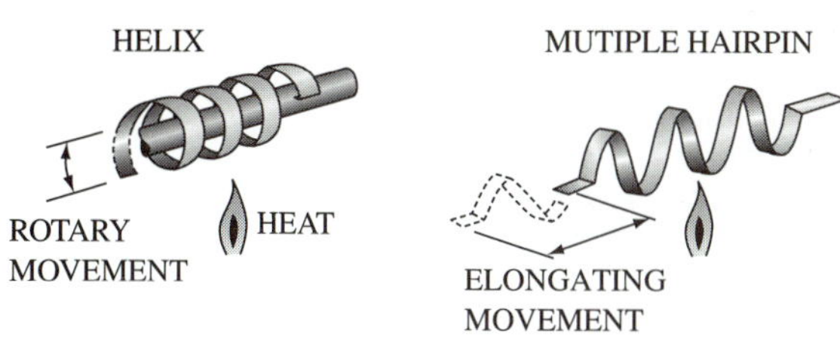

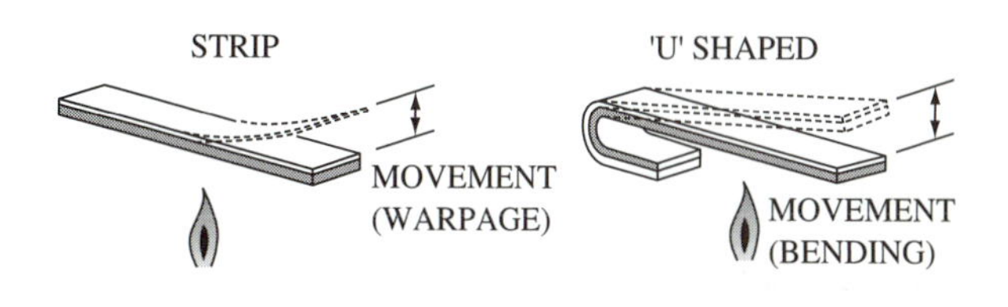

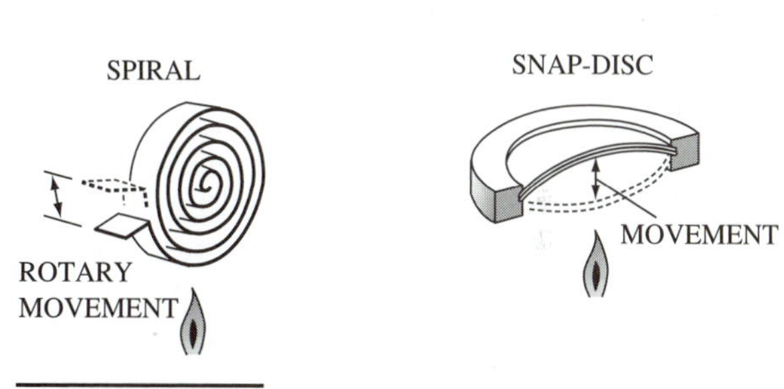

Figure 1–25 Bimetal temperature controls or actuators. *(Reprinted from Refrigeration and Air Conditioning, 3rd Ed. by Air Conditioning and Refrigeration Institute, copyright © 1998, Prentice Hall, Inc. Reprinted by permission.)*

Figure 1–26 Bellows-type pressure switch used as a low-pressure cut-out. (Opens on drop in pressure.)

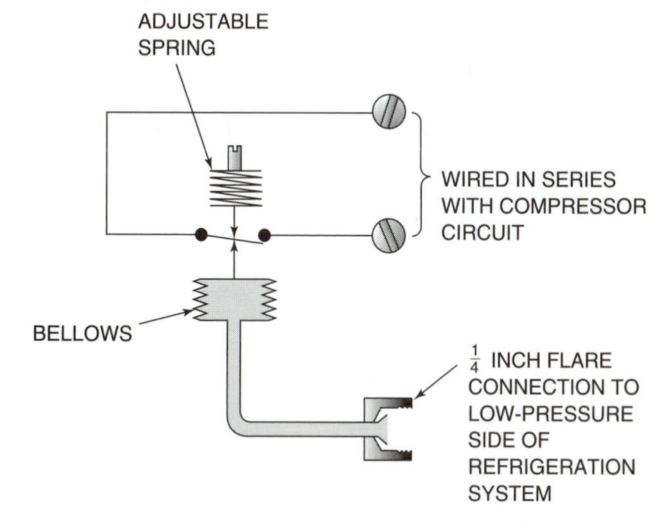

Figure 1–27 Air pressure switch. *(Reprinted from Refrigeration and Air Conditioning. 3rd Ed. by Air Conditioning and Refrigeration Institute, copyright © 1998, Prentice Hall, Inc. Reprinted by permission.)*

Range and Differential

A common misconception about automatic switches is that they have "a set point," that is, one setting where the switch opens or closes, depending on whether the sensed temperature or pressure or other variable is above or below the set point. To illustrate why this cannot be true, consider a lamp controlled by an ordinary photocell. The photocell senses light intensity, and if it gets too dark, it closes a switch to turn on a lamp. With the lamp now on, if there were only "a set point," the light level sensed by the photocell would now be higher than the set point, and the lamp would turn off. This cycle would repeat, turning the lamp on and off as quickly as the photocell could operate, and would obviously not be a usable control scheme.

Instead of just "a set point," there must be a setting where the lamp turns on, and a different setting where the lamp turns off. For example, the photocell switch may be set to close when the sensed lighting level drops to 10 lumens (a measure of lighting intensity), and open when the sensed lighting level rises to 15 lumens. If the light level drops to 10 lumens, the lamp comes on, immediately increasing the light level to 12 lumens. This is not enough to turn off the lamp. However, when the sun comes out, and the light produced by the sun plus the lamp rises to 15 lumens, the lamp turns off. This immediately drops the light intensity to 13 lumens, but the lamp will not turn on again until the sun goes down.

Figure 1–32 shows this operation in graph form. 15 lumens is the **cut-in**, 10 lumens is the **cut-out**, 5 lumens is the **differential.** The **set point** of this switch is 12.5 lumens, midway between the cut-in and the cut-out. The set point may be adjustable between zero lumens and 20 lumens. This is called the **range.**

Figure 1–28 Flow switch. *(Courtesy Johnson Controls, Inc.)*

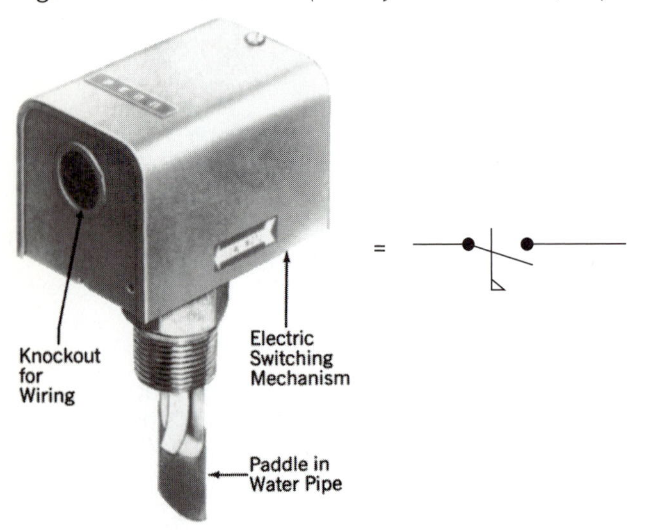

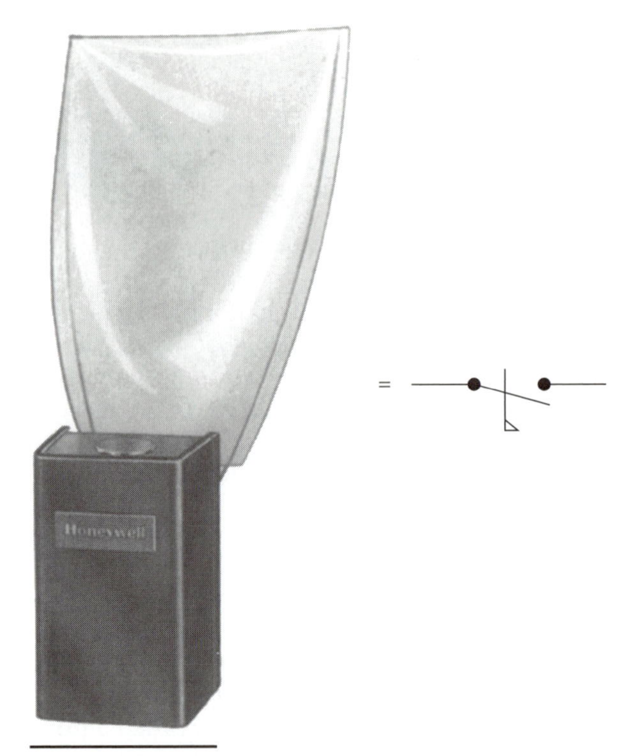

Figure 1–29 Sail switch. Sail switch detects movement of air by use of a sail or paddle attached to arm of a switch. *(Courtesy Honeywell, Inc.)*

> **EXERCISE:**
> On each of the following devices, state which would be higher, the cut-in setting or the cut-out setting:
> **a.** A thermostat used to control a heater.
> **b.** A thermostat used to control an air conditioner.
> **c.** A high pressure cut-out.
> **d.** A low pressure cut-out.

Figure 1–30 Door switch.

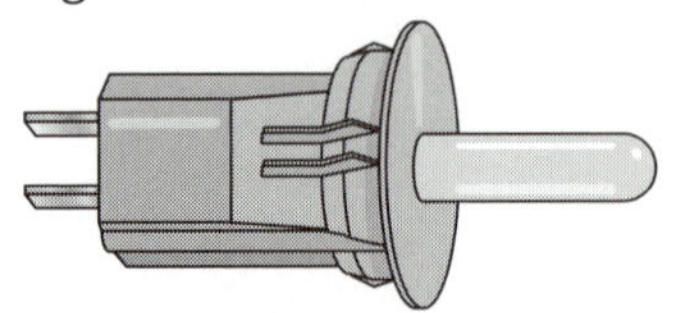

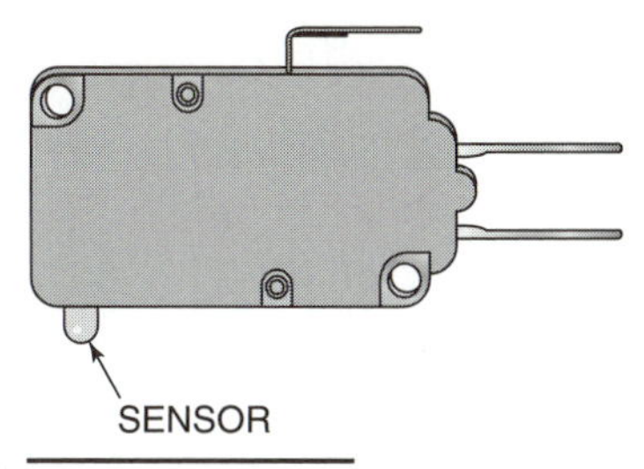

Figure 1–31 Microswitch.

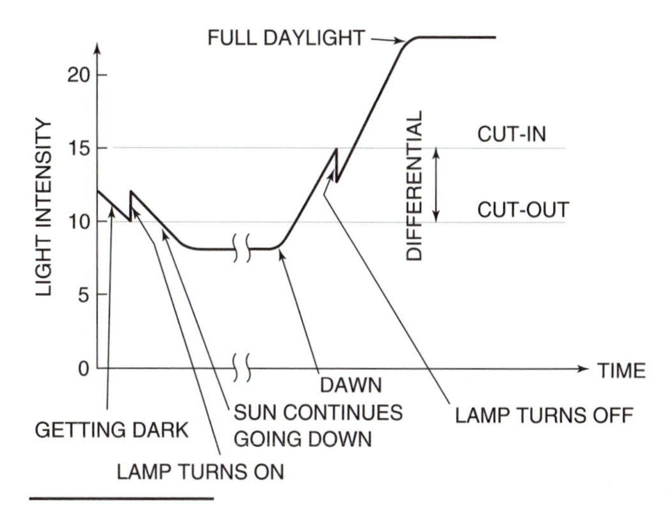

Figure 1–32 Photocell operation.

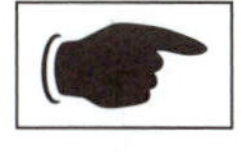

VOLTAGE DROP

When analyzing why an electrical system doesn't work, you will use a volt meter. Usually, the volt meter is combined with an ohm meter in a single instrument called either a **volt ohm meter (VOM)** or a **multi-meter**. Figure 1–33 shows two different types of VOMs.

Analog meters are rapidly being replaced by digital meters due to their accuracy and ruggedness. However, whichever meter you may purchase, the following features are essential:

a. There must be either a fuse or other protection provided to protect the ohm meter if it is mistakenly attached to a live circuit while it is set on ohms.

b. It must be able to distinguish between a reading of 1Ω and 2Ω, and it must be able to distinguish between 10,000Ω and infinite ohms.

All volt meters, regardless of the type, use two probes which can be used to touch various points in the circuit. Each lead senses the electrical pressure in the wire it touches, and the meter reads out the *difference* in the two pressures. Figure 1–34 shows the voltage readings you would get if you measured at various places in the circuit. Note that you get a voltage reading of zero when measuring across the closed switch, even though there is a significant electrical pressure inside the wires. Why, then, do we read zero volts across the switch? The answer lies in remembering that the volt meter reads a pressure difference. Even though there may be 115V of electrical pressure on each side of the switch, the pressure difference between the two pressures is zero.

Figure 1–35 shows the voltage readings you would get in the same circuit, but this time with the switch open. The electrical pressure can only get as far as the switch. But because the switch is open, the pressure can get no further.

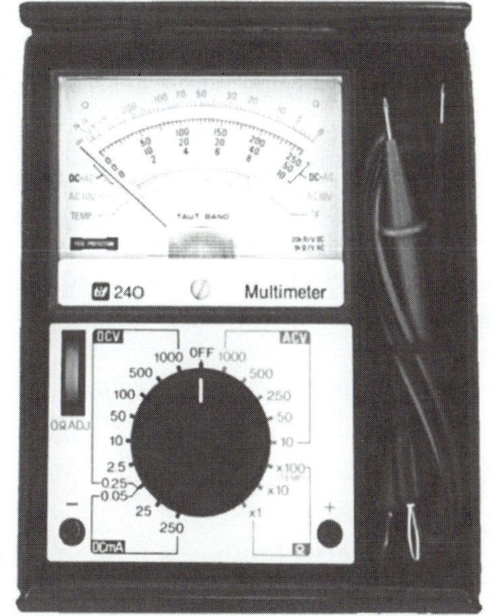

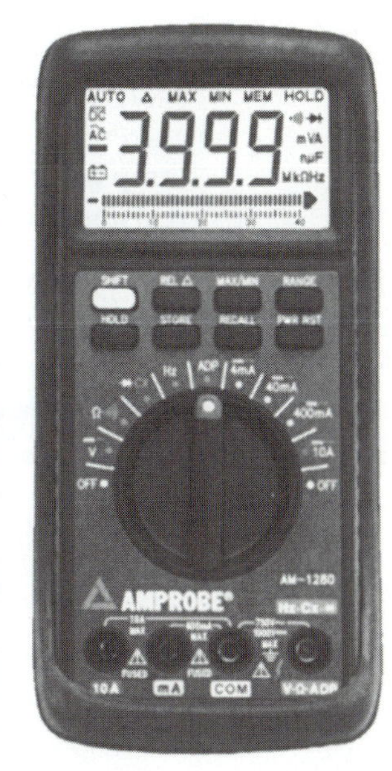

Figure 1–33 Different types of VOMs. (a) Analog VOM. (b) Digital VOM. *(Courtesy Amprobe Instrument®)*

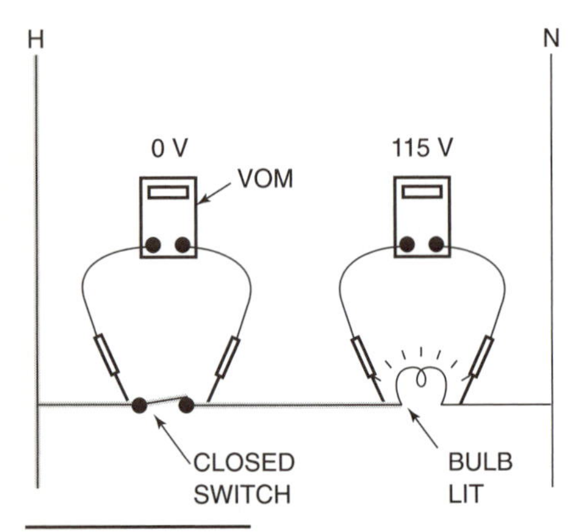

Figure 1–34 Voltage readings with switch closed.

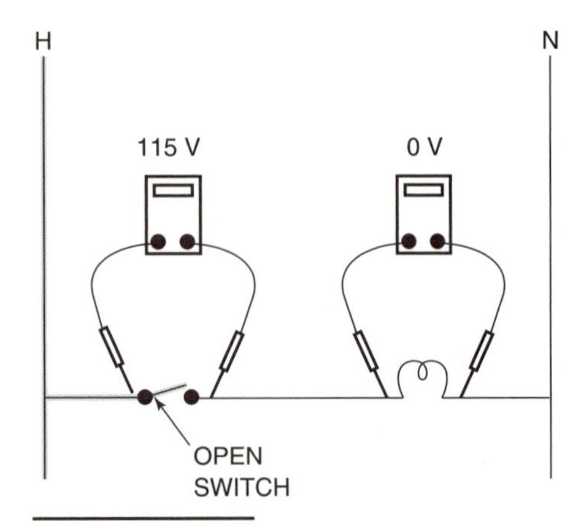

Figure 1–35 Voltage readings with switch open.

QUESTION:
If, in the circuit shown in figure 1–35, the wire labelled H were actually N, and the wire labelled N were actually H, what voltage reading would you get across the switch? The bulb?

In figure 1–34, the volt meter indicated that there was a voltage drop across the load, and in figure 1–35, the VOM indicated that there was a voltage drop across the switch. Each reads the same 115V, but the physical mechanism that causes these two voltage drops is quite different.

In figure 1–34, the reason there is a voltage drop is because as the electrons pass through the light bulb, there is electrical "friction" that shows up as a reduction in electrical pressure. This is commonly called an I × R drop or an **IR drop**, because of the relationship between voltage drop across a load, the current through the load, and the resistance of the load. This relationship is given as:

$$E = I \times R$$

where E = voltage drop across the load (volts)
I = current flow through the load (amps)
R = resistance of the load (ohms)

It is not important that you be able to calculate voltage drops, but it is important to understand the mechanism that causes voltage drops in electrical circuits. An I × R drop is a voltage drop caused by the electrical friction caused by current flowing through a load.

Compare the voltage drop in figure 1–35 with the voltage drop in figure 1–34. In figure 1–35, because of the open switch, there is no flow of electrons, and therefore there can be no electrical friction. Instead, the electrical pressure at the left side of the switch has the potential of pushing electrons through the load as soon as the switch is closed. It is for this reason that the voltage drop across the open switch is called a **potential** voltage drop.

HOW DO WE USE ELECTRICITY?

With few exceptions, devices in electrical circuits can be classified as either **switches** or **loads.** We have already discussed switches. When a switch is open, it has infinite resistance (ohms), and it stops the flow of electrons. When it is closed, it has zero resistance, and it acts just like a wire. It carries the amps without causing any other effect in the circuit.

Loads, however, are the devices that convert the energy of the electrons passing through them into some other form of energy. You are familiar with loads that convert electrical energy into light energy (a light bulb), sound energy (a siren), mechanical energy (a motor), or heat energy (an electric heater).

ADVANCED CONCEPTS

This relationship between voltage drop, current flow, and resistance is known as Ohm's law. It can also be written as

$$I = E/R$$

or

$$R = E/I$$

There are two properties of electricity flowing through a wire that allow these loads to convert electrical energy into other forms of energy. Those properties are:

1. Electricity flowing through a wire will create heat.

2. Electricity flowing through a wire will produce a magnetic field around the wire.

USING ELECTRICITY TO PRODUCE HEAT

The amount of heat that will be produced when electricity passes through a wire depends on the amount of current flowing and the size (diameter) of the wire. As the amps increase through a wire, or as the resistance of a wire is increased, the amount of heat generated will increase. The resistance of a wire is determined by its diameter. The larger the diameter of the wire, the lower its resistance will be (the larger wire has a greater cross section area, making it easier for the electrons to flow through it). Generally, the wiring that is used to connect the devices in a circuit is sized large enough so that its resistance is low enough to neglect. However, when the diameter of the wire is made small, as in a light bulb or an electric heater, then the amount of heat generated increases. A light bulb is nothing more than a very thin wire that gets so hot from the flow of electricity through it that it glows, giving off light. The rate of energy consumption is called **power**, and it is measured in **watts** (W) or kilowatts (1kW = 1000W). Power can be calculated by the formula:

$P = V \times A$

where P = power (watts)
V = voltage drop (volts)
A = current (amps)

For three-phase systems, power is calculated as:

$P = V \times A \times 1.73$

Power can be measured with a **watt meter**. Several types are shown in figure 1–36. In practice, the technician will rarely use a watt meter.

USING ELECTRICITY TO PRODUCE A MAGNETIC FIELD

The relationship between electricity and magnetism was discovered when a physicist found that when a compass needle is brought near a wire carrying current, the needle will deflect. If you were to run a

ADVANCED CONCEPTS

From a previous discussion on the relationship between volts, amps, and ohms, we recall that:

$V = I \times R$

If we substitute the quantity $(I \times R)$ for voltage in the power formula, we get another power formula:

$P = I^2R$,

where P = power (watts)
I = current (amps)
R = resistance (ohms)

EXAMPLE: An electric heater has 15 ohms of resistance, and draws 7 amps. How much power will it produce?

SOLUTION:

$P = I^2R$
$P = (7 \times 7) \times 15$
$P = 735$ watts

If you wanted to know how much heat was being generated by the heater in Btu/hr, you could use the conversion factor of:

1W = 3.4 Btu/hr

Therefore, the heating capacity of the above heater could also be expressed as:

$$735 \text{ watts} \times 3.4 \frac{\text{Btu/hr}}{\text{watt}} = 2500 \text{ watts}$$

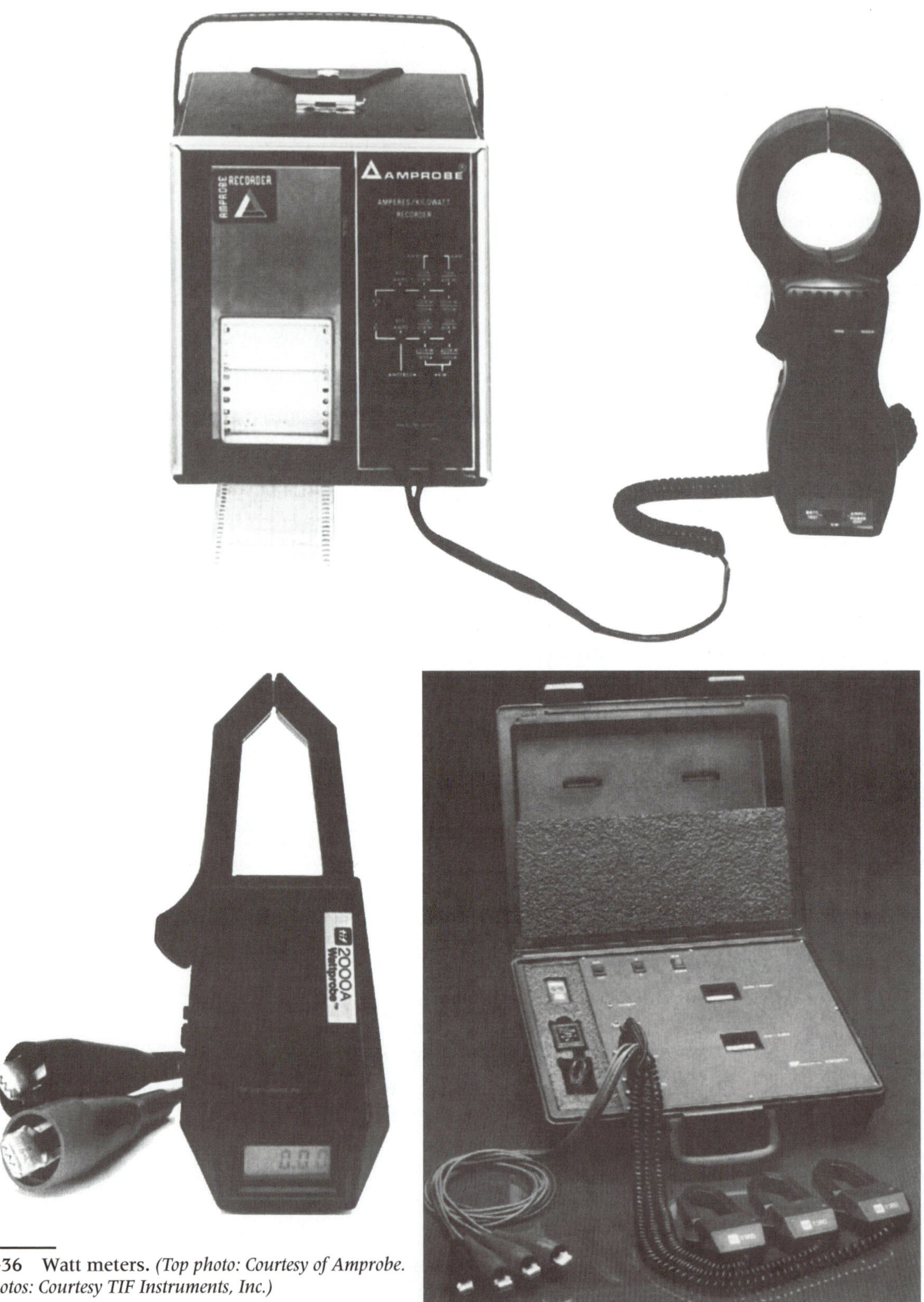

Figure 1–36 Watt meters. *(Top photo: Courtesy of Amprobe. Bottom photos: Courtesy TIF Instruments, Inc.)*

current carrying wire through a piece of paper covered with iron filings (figure 1–37), you would find that the iron filings arrange themselves in the pattern shown. This shows the magnetic field around the wire. If the wire were coiled into a loop (figure 1–38), the magnetic field around the wire would be increased. If several turns of wire are used, as in figure 1–39, the magnetic field is increased further. The magnetic field of each turn of wire adds to the magnetic field of the other turns. If an iron bar is inserted into the coil of wire (figure 1–40), it concentrates the lines of magnetic force and the iron bar

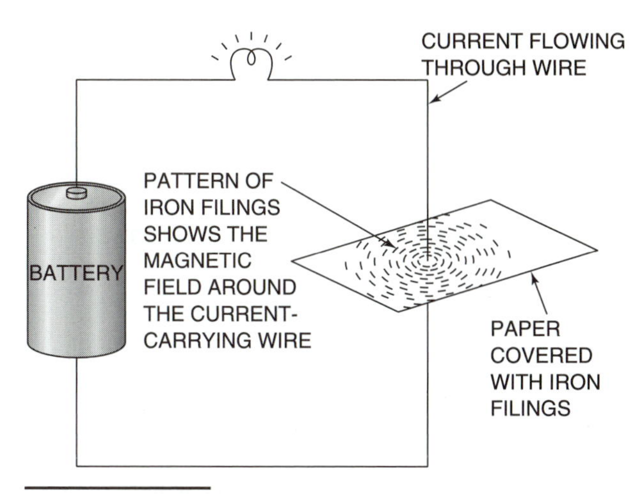

Figure 1–37 Magnetic field around a wire.

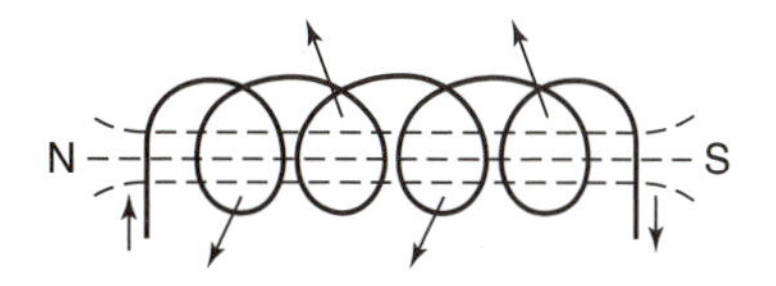

Figure 1–39 Wire coil.

becomes an **electromagnet** whenever there is current flowing through the coil. When the coil of wire is energized, the magnetic field that is produced can be used to cause a valve or a switch to open or close. When used in this way, the coil of wire is called a **solenoid.** The valve is called a **solenoid valve** (figure 1–41a). When the current is allowed to flow through the solenoid coil, it pulls the **armature** of the valve, pulling the valve off its seat. This is called a **normally closed** solenoid valve. The valve can also be arranged physically as in figure 1–41b. When the coil of wire is not energized, the valve is open. When the coil of wire is energized, the armature is pulled into the magnetic field and the valve closes. This is called a **normally open** solenoid valve.

The magnetic field produced by a current-carrying wire can also be used to open or close a switch. The switch and coil together are called a **control relay** (figure 1–42). When switch A is closed, current flows through the control relay coil, pulling the control relay switch closed. The switch in this control relay is **normally open** (closes when the coil is energized).

A control relay can also have a normally closed switch, an SPDT switch, or any other type of switch. The symbols that are used to indicate a normally open switch or a normally closed switch that is operated by a magnetic field are shown in figure 1–43. Note that the term "normal" does not refer to the position in which it spends most of its time. It refers to the position of the switch when the coil is de-energized. In the control relay shown, terminal 4 is called the normally closed terminal, 5 is the normally open terminal, and 2 is the common terminal.

The magnetism produced by a current carrying wire is also the moving force that makes an electric motor run. The way in which magnetism can be used to make the armature in a motor turn is dealt with in a later chapter.

A simple control relay circuit is shown in figure 1–44. When the thermostat closes, it completes a circuit through the coil of the control relay. The magnetic field produced by the coil causes the normally open control relay switch to close, completing the circuit through the motor. You may ask, "Why use a control relay instead of simply having the thermostat wired in line with the motor?" The advantage to using the control relay is that the thermostat may be a light-duty device, with contacts not rated to carry the high amp draw of the motor. By using the control relay, the thermostat contacts will only have to carry the current consumed by the control relay coil (a few tenths of an amp).

EXERCISE:
Using figure 1–44 as a model, draw a ladder diagram in which a light-duty pressure switch is mounted on the tank of an air compressor. The pressure switch is to operate the compressor motor as needed through a control relay. Use correct symbols.

LOADS WIRED IN PARALLEL

Figure 1–45 shows a circuit in which a single switch controls two loads in parallel. When loads are in parallel with each other, they have the following characteristics:

1. There are two different paths that electrons can take to get from one side to the other.

Figure 1–38 One loop of wire.

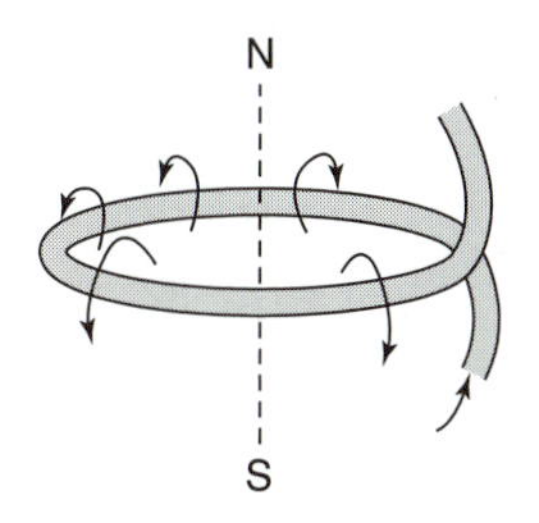

Figure 1–40 Coil with iron bar.

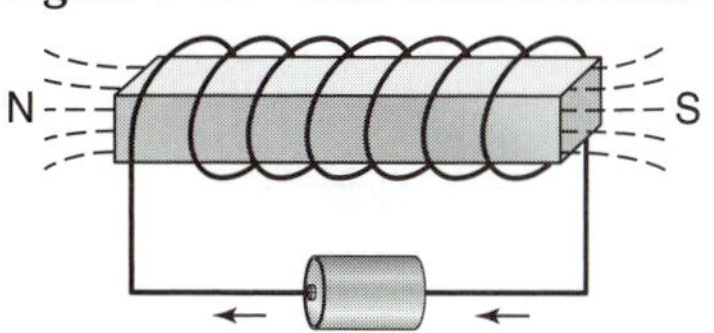

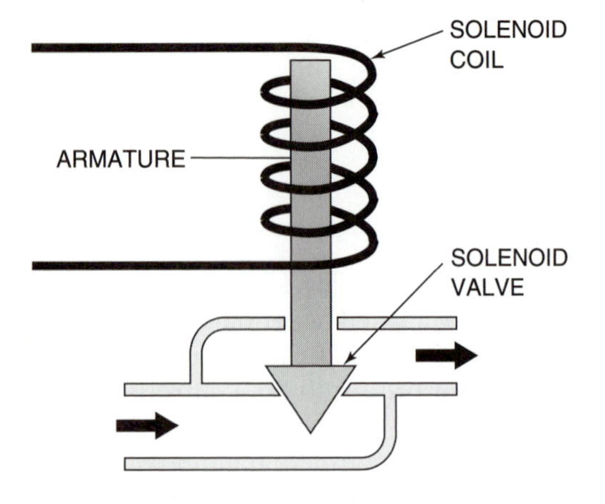

(a) Normally-closed solenoid valve.

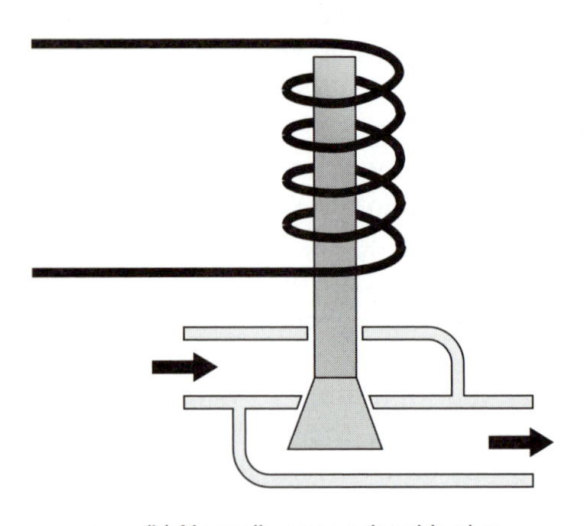

(b) Normally-open solenoid valve.

Figure 1–41 Solenoid valves.

Figure 1–42 Control relay.

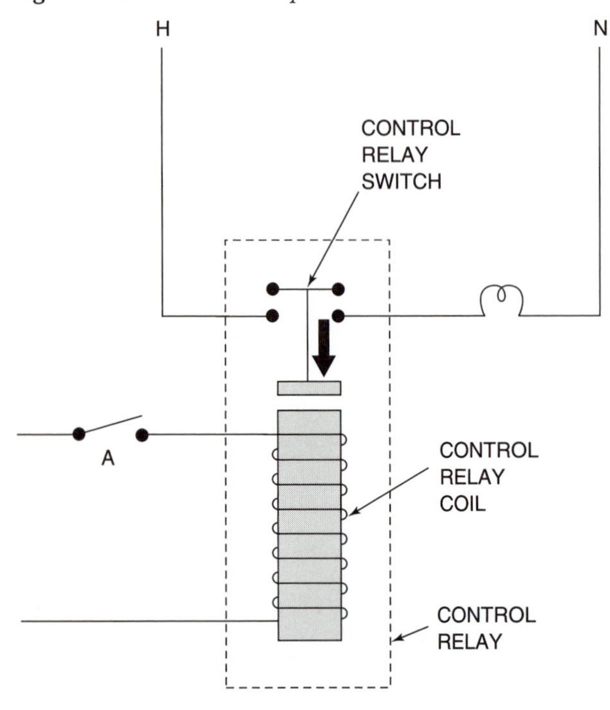

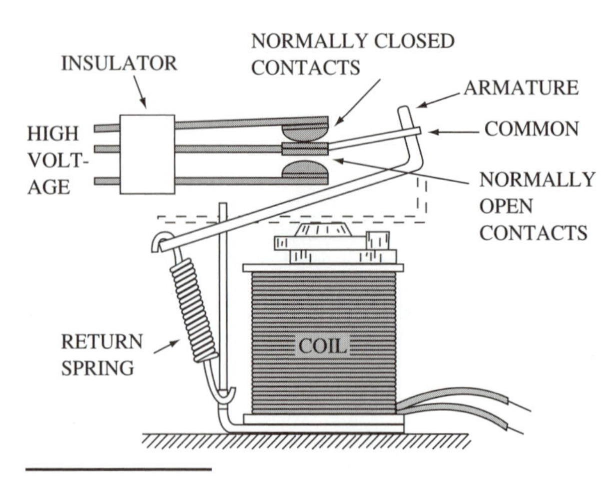

Figure 1–43 Control relay with SPDT switch. *(Reprinted from Refrigeration and Air Conditioning, 3rd Ed. by Air Conditioning and Refrigeration Institute, copyright © 1998, Prentice Hall, Inc. Reprinted by permission.)*

2. Different electrons pass through each load.
3. One side of load #1 is wired to one side of load #2, and the other side of load #1 is wired to the other side of load #2.
4. Each load "sees" the same voltage difference from one side to the other side.
5. The load with the higher resistance will carry the lower amps.
6. The amps flowing through the circuit are equal to the sum of the amps flowing through the individual loads.
7. The overall resistance of the combination of loads in parallel will be lower than the lowest resistance of all the loads wired in parallel.
8. Each load behaves as if it were the only one. The addition of a second load in parallel does not affect the operation of the first load.

It is easy to visualize loads in parallel as holes in a wall that separates two rooms at different pressures (figure 1–46). A 100Ω load can be represented by a hole in the wall that allows flow from the higher pressure room to the lower pressure room. A 200Ω load can be represented by a smaller hole in the wall. Because it is higher resistance than the 100Ω load, it allows less flow to pass than the 100Ω load. However, when the 200Ω hole is *added* to a

Figure 1–44 Control relay circuit.

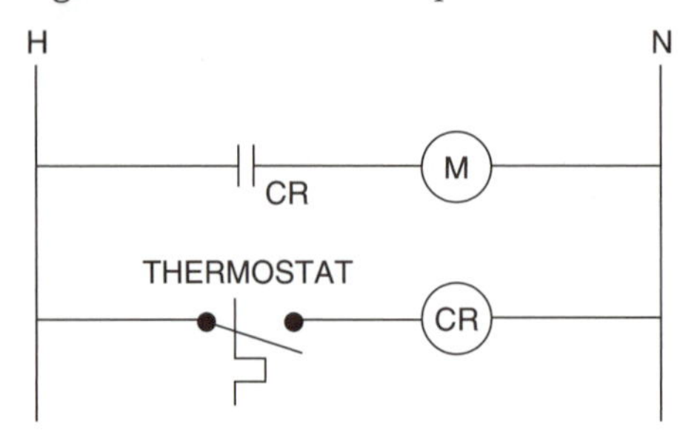

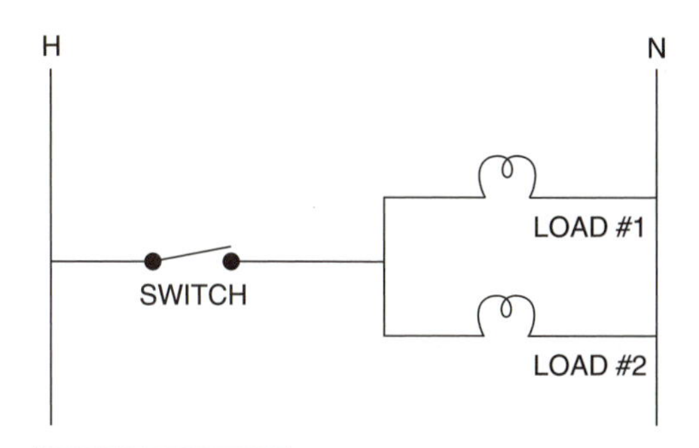

Figure 1–45 A single switch controlling two loads in parallel.

wall that already has a 100Ω hole, the result is that the total flow through the wall increases. Even though the 200Ω hole has higher resistance than the 100Ω hole, both together (in parallel) have a lower resistance than either one by itself, and the total flow through both will be higher than the flow through either one individually.

Where an HVAC/R system has more than one load (as almost all do), the chances are very high that they will be wired in parallel. Each load will "see" the same voltage. (**HVAC/R** is a commonly used abbreviation for heating, ventilating, air conditioning, and refrigeration).

LOADS WIRED IN SERIES

Figure 1–47 shows a circuit in which a single switch controls two loads in series. When loads are in series with each other they have the following characteristics:

1. There is only one path that electrons can take to get from one side of the circuit to the other.

Figure 1–46 Two resistances in parallel are like two holes in the same wall.

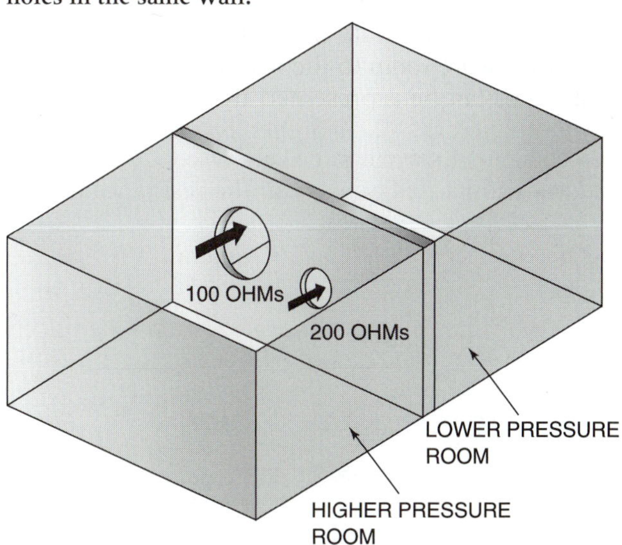

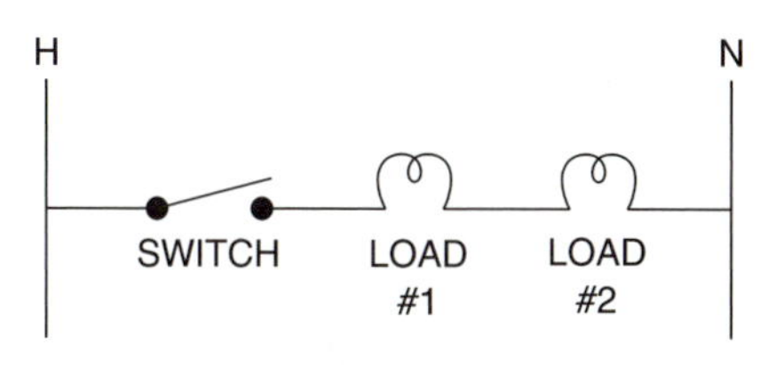

Figure 1–47 A single switch controlling two loads in series.

2. The same electrons pass through each load.

3. Electricity flows into one side of load #1, out of load #1, and then into load #2.

4. Each load "sees" a different voltage difference, depending on the resistance of each load.

5. The load with the higher resistance will have a higher voltage drop across it. The sum of the voltage drops across both loads equals the voltage available from the circuit.

6. The amps flowing through each load are the same and equal to the circuit amps.

7. The overall resistance of the combination of loads in series will be the sum of the individual resistances.

8. The operation of each load will depend upon the resistance of other loads that are wired in series with it.

Loads in series can also be visualized as holes in a wall. This time, the two rooms at different pressure are separated by a corridor. Suppose the wall separating the corridor from room 1 has a 200Ω hole (figure 1–48). There will be flow from the higher pressure room to the other. If another wall with a 100Ω hole is added (figure 1–49), even though the new hole is larger than the existing hole, the over-

Figure 1–48 Flow of air through a hole.

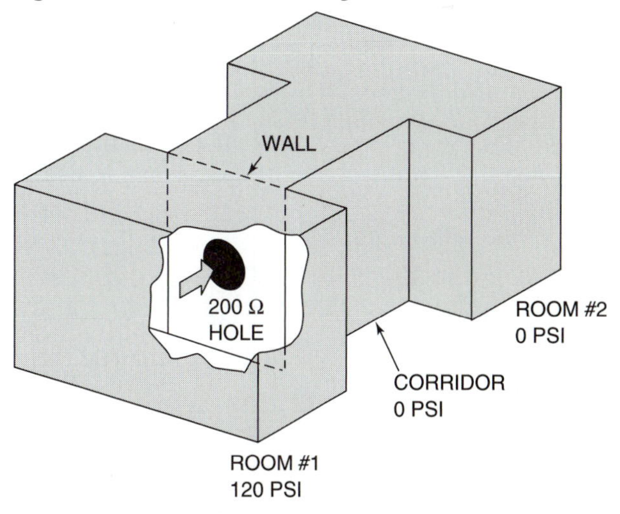

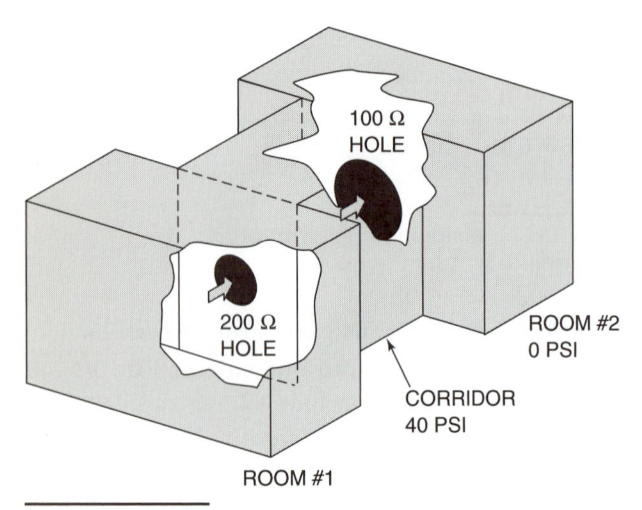

Figure 1–49 Two holes in series.

all resistance to flow increases and the flow rate decreases.

Note that the pressure in the corridor between the two walls is less than the pressure in the high-pressure room, and more than the pressure in the zero-pressure room. If the flow encounters the 200Ω hole first, the pressure between the walls will be 40 psi. If the flow encounters the 100Ω hole first, the pressure between the walls will be 80 psi. But either way, the pressure drop across the 200Ω hole will be twice the pressure drop across the 100Ω hole, and the total pressure drop across both holes will equal the available pressure difference between the two rooms. The schematic representation of resistances in series is shown in figure 1–50. Two resistors can be replaced with a single resistor of the sum value of the first two.

ADVANCED CONCEPTS

In a parallel circuit, the total resistance can be calculated as:

$$\frac{1}{R_T} = \frac{1}{R_1} + \frac{1}{R_2} + \cdots$$

or,

$$R_T = \frac{1}{\frac{1}{R_T} = \frac{1}{R_1} + \frac{1}{R_2} + \cdots}$$

or, for two resistances in parallel,

$$R_T = \frac{R_1 \times R_2}{R_1 + R_2}$$

For a series circuit, the overall circuit resistance can be calculated as:

$$R_T = R_1 + R_2 + \cdots$$

In a parallel circuit, the voltage drop across each load is the same as the circuit voltage. In a series circuit, the voltage drop across each load is proportional to the resistance. For the series circuit shown in figure 1–50, the total resistance is (100Ω + 200Ω) = 300Ω. The total amps flowing in the circuit may be calculated using Ohm's Law:

$$\begin{aligned} I &= E/R \\ &= 120V/300\Omega \\ &= 0.4 \text{ amps} \end{aligned}$$

The voltage drop across each load may then be calculated, because the amps through each load are known (0.4 amps) and the resistance of each load is also known.

Load #1 $\quad \begin{aligned} E &= I \times R \\ &= .4a \times 100\Omega \\ &= 40V \end{aligned}$

and,

Load #2 $\quad \begin{aligned} E &= .4a \times 200\Omega \\ &= 80V \end{aligned}$

Note that the voltage drops across the loads do not depend upon which load comes first.

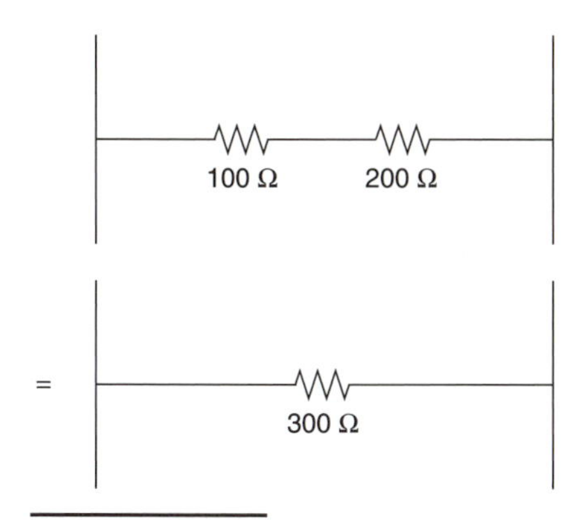

Figure 1–50 Resistances in series add.

EXERCISE:
Redraw the partially completed circuit in figure 1–51, exactly as shown. Add the circuit wiring so that the two loads are wired in parallel. Redraw the partially completed circuit again. Add circuit wiring so that the two loads are wired in series.

EXERCISE:
Wire a 7.5-watt light bulb in parallel with a 100-watt light bulb across a 120V power source. They will both light to full brightness. Unscrew either bulb. The other remains lit. Why?

Wire the same two light bulbs in series across the same 120V power source. Only the 7.5-watt bulb appears lit. Why? Unscrew either bulb. The light goes out. Why?

RESISTORS

Normally, we are only interested in the resistance of loads such as motors, coils, and heaters. However, in electronic circuits and some other special applications, a device called a **resistor** (figure 1–52) is used for the sole purpose of providing a resistance, and providing no other function. You will find resistors on circuit boards, but you will usually not troubleshoot these boards down to the component level. But sometimes you will. For example, if you use your ohm meter incorrectly, you can ruin it by blowing out one of the resistors. If you can find the failed resistor (usually by the burn mark around it), it can be replaced with a resistor of the same resistance value. The resistance value of a resistor can be determined from the colors of the four bands. Each different color stands for a different number, as follows:

Figure 1–51 Exercise.

Plug

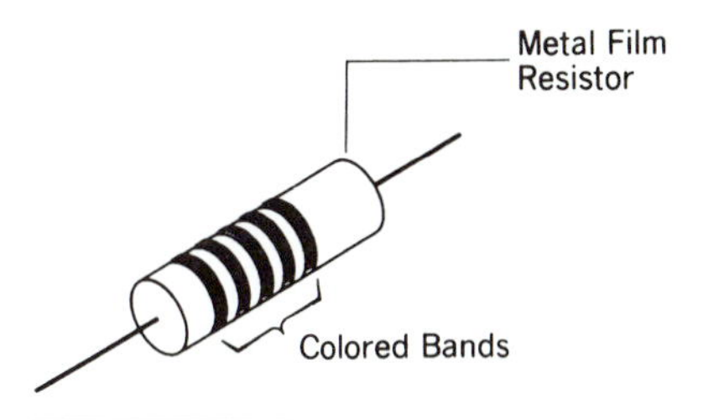

Figure 1–52 Resistor.

Black	= 0
Brown	= 1
Red	= 2
Orange	= 3
Yellow	= 4
Green	= 5
Blue	= 6
Violet	= 7
Gray	= 8
White	= 9

The first colored band (closest to the end of the resistor) is read as the first number. The second band is read as the second number. The third band is read as the number of zero's that must be added to the first two numbers. The fourth band tells you the manufacturing tolerance of the resistor, with the following values:

Gold	= 5%
Silver	= 10%
None	= 20%

The manufacturing tolerance tells you how close the actual resistance of the resistor will be to the value that you read by the colors.

EXAMPLE

A burned out resistor has the colors gray, red, orange, gold. What value resistor will you buy to replace it?

SOLUTION The first band, gray, stands for 8. The second band, red stands for 2. The third band, orange, stands for 3, which means three zeros. The value of the resistor is 82000 ohms. The fourth band, gold, means that the actual resistance is within 5% of the 82000 ohms that you just determined. ■

CREATING ELECTRICITY

Electricity is the effect of moving electrons through a wire. The electrons only move, however, if there is some force that "pushes" the electrons. If the "push" is always in one direction, it will create a one-way flow of electrons. This is called **direct current** (DC). If the "push" constantly reverses its direction, it is called **alternating current** (AC). All building wiring is AC.

DIRECT CURRENT

Electricity can be generated by electrochemical means . . . a battery. A battery will push electrons through a wire and a load as shown in figure 1–53. The lower the resistance of the load, the more electrons will flow. The flow of electrons will continue until the electrochemical reaction happening inside the battery ceases. When the battery is spent, it must be replaced or recharged. The current that flows through the circuit flows in one direction only. This type of current is called **direct current,** and is abbreviated DC. Direct current is of limited interest to the heating, ventilating, air conditioning, and refrigeration technician.

Of greater interest to the HVAC/R technician is **alternating current.** Almost all the compressor and fan motors, controls, and heaters that you encounter in HVAC/R systems will operate on alternating current. There are only a few specialized controls that use direct current.

ALTERNATING CURRENT

Alternating current is somewhat different from the water analogy used earlier in this chapter. In that analogy, there was always a positive pressure of water at the left end of the pipe, and it "pushed" the electrons through the load, to the zero pressure portion of the pipe. However, the Hot wire in an alternating current circuit does not always "push" the electrons through the circuit. It alternately pushes and then pulls. Figure 1–54 illustrates the mechanical analogy. If you were to push on the piston handle, water would flow towards the right, causing the propeller (the load) to turn. If you were to pull on the piston handle, water would flow towards the left, also causing the propeller to turn.

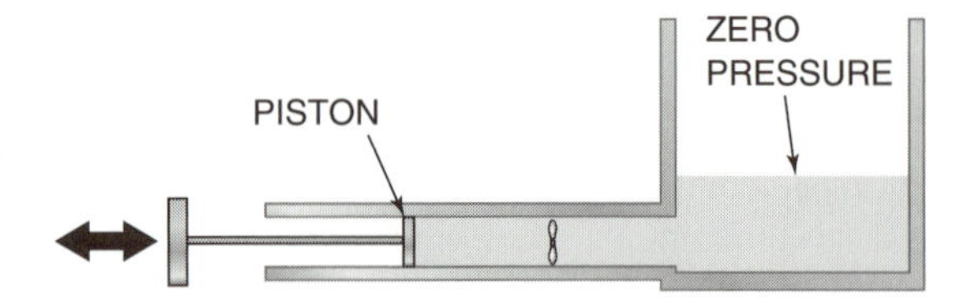

Figure 1–54 Mechanical analogy for alternating current.

In alternating current circuits, the hot wire is sometimes at a pressure higher than zero, and electrons are pushed through the circuit. Then the pressure in the Hot wire becomes less than zero, and electrons are pulled through the circuit. In the United States, the pressure alternates 60 times every second. This is called the **frequency** of the electricity, and it is measured in **Hertz** (abbreviated Hz).

Just as electricity flowing through a wire produces a magnetic field around the wire, electricity can be produced when a wire moves through a magnetic field. Figure 1–55 shows a single coil of wire in a magnetic field. If the wire is turned 1/4 turn in a clockwise direction, the volt meter will register a slight voltage. If the wire is rotated 1/4 turn in a counterclockwise direction, the volt meter will register a voltage in the other direction. When the rotation of the wire is stopped, the voltage mea-

Figure 1–53 Direct current circuit.

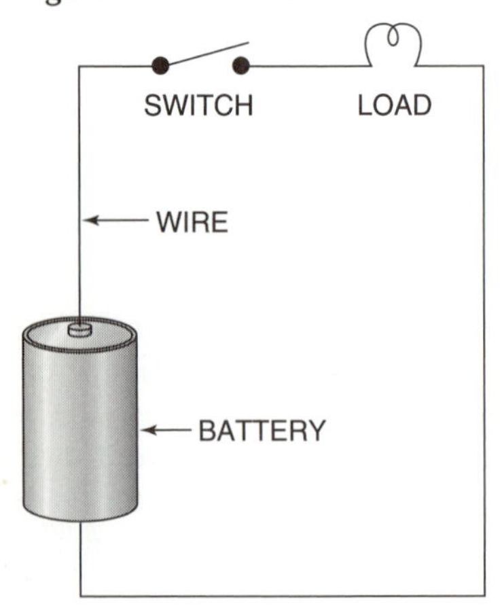

Figure 1–55 Generating a voltage with a moving coil of wire in a magnetic field.

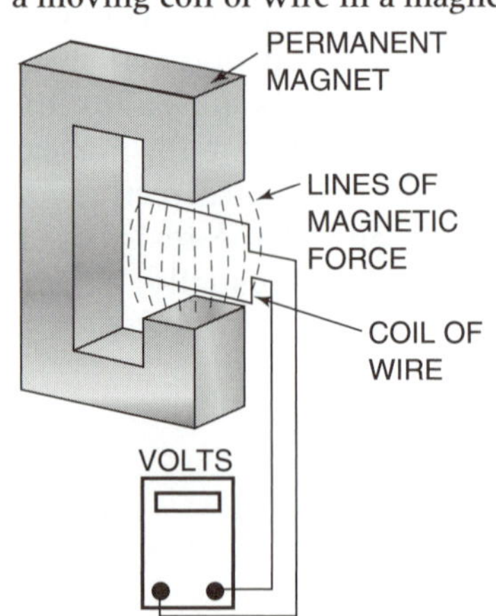

sured drops to zero. If the wire could be rotated continuously (without twisting the wires) each 1/2 rotation, the volt meter would register a voltage in the alternate direction. This is the same principle that is used by electric utility companies. They use large turbines driven by steam (or water for hydroelectric power) to turn coils of wire inside a magnetic field. The driven devices that convert this mechanical motion into electricity are called **generators.**

INDUCTANCE

If you were to measure the current that flows through a coil of wire, and then you measured the resistance of the coil of wire and calculated the current flow (using Ohm's law), you would find that Ohm's law does not accurately predict the correct amperage. Why?

When alternating current flows through a coil of wire, the magnetic field builds and then collapses. The moving magnetic field cuts through the coils of wire and produces the same effect as when a coil of wire is moved through a magnetic field. The moving magnetic field induces a voltage in the coil of wire, and the direction of that voltage is in the opposite direction of the voltage that is being applied to the coil. This voltage is sometimes called a **back EMF.** Therefore, much less current flows through the coil than you would predict using Ohm's law.

Ohm's law will accurately predict the amps that will flow in a DC circuit, or in an AC circuit without coils of wire. But coils of wire in an AC circuit present a special type of resistance to the flow of electricity. This resistive effect is called **inductance.** Its unit of measurement is the Henry. However, we cannot measure inductance on a coil with a VOM. Nor is there any need for the HVAC/R technician to know the inductance of a coil. The only reason you need to know about inductance is to understand why some loads do not appear to follow the rule of Ohm's law.

EXAMPLE

You have measured a resistance on a motor winding of 2 ohms on a refrigerator compressor. It is plugged into a 120V circuit capable of supplying 15 amps. Is the circuit sufficiently large?

SOLUTION: If you used Ohm's law, you might think that the compressor will draw ($120V/2\Omega$), or 60 amps. However, that is not the case. The refrigerator compressor will only draw 4 or 5 amps. You have no way of knowing exactly how much current it will draw, other than relying on the nameplate data on the compressor. ■

Sometimes switches have two different load-carrying capability ratings. A switch might be rated for 18 amps on an inductive circuit, but 20 amps on a resistive circuit. If the switch were being used to control a motor, the inductive circuit rating would be used.

GROUNDING

Whenever electricity flows through a wire, the wire must be insulated. This is to prevent the electrical pressure inside the wire from finding another path, other than the intended path. Wire that is coiled in a motor may appear to be uninsulated, but it is not. It is coated with a thin layer of clear enamel, which individually insulates each turn of wire from the adjacent turns.

Sometimes, however, a wire or its insulation breaks down, and the electrical pressure is allowed to flow into the casing of the unit (figure 1–56). This is called a **ground fault.** In this case, if a person were to touch the unit, there would be a path for the electricity to flow from the faulty wire to the casing through the person and into the ground. The amount of current that would flow through the person would increase dramatically if s/he were holding onto a pipe at the same time. This is because the resistance of the path from the fault through the person would be much lower.

The risk of personal injury from such an occurrence is not acceptable. It only takes a current of approximately 0.1 amps through a person (even less for some persons) to do serious damage, or even cause death. In order to prevent such an occurrence,

Figure 1–56 Safety risk from a ground fault.

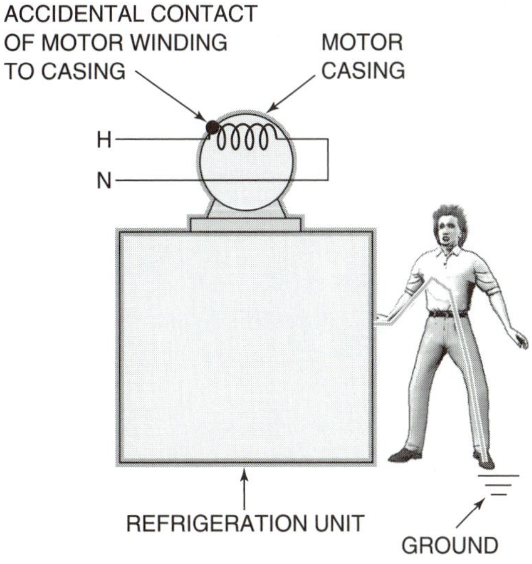

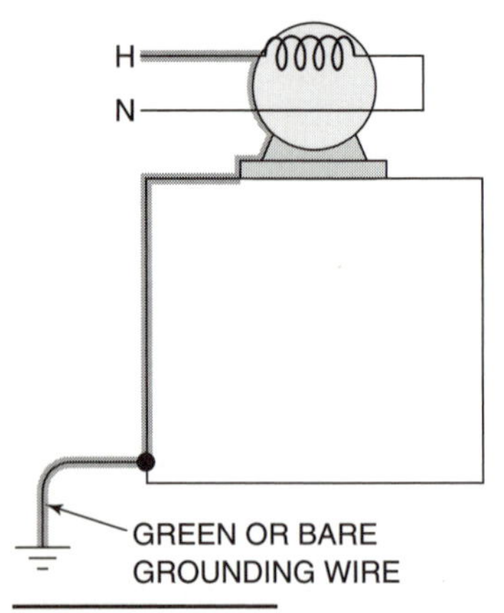

Figure 1–57 Casing ground.

an equipment ground (**casing ground**) is required (figure 1–57). A ground wire is always green or a bare wire. It connects the casing to a rod in the ground that is always at zero electrical potential. If there is unintended leakage of electricity from the wiring into the casing, the casing ground will prevent electrical pressure from building up in the casing. If no electrical pressure can build up in the casing, then there will be no force available to move electrons through an unsuspecting person who touches the casing.

A standard appliance plug has three prongs. Two are flat, being the Hot and the Neutral legs of the power supply. The U-shaped prong is the ground. The only appliances that are manufactured today without ground wires are ones that are **double insulated.** They are designed with plastic casings, so that even if the wiring inside the device breaks and touches the casing, the casing itself is an insulator, and will not become electrically pressurized.

In some locations (especially wet locations such as bathrooms and kitchens), conventional grounded outlets are no longer permitted by Code. A **Ground Fault Current Interrupter** (GFCI) shown in figure 1–58 provides an extra safety feature. It senses if the current being supplied in the Hot leg equals the current being returned in the Neutral leg. If they are at all different, it means that there is some current leakage, and the GFCI will open, stopping the flow of current. The GFCI provides ground fault protection only. It does not provide any protection against a short circuit.

During normal operation, no current flows through the ground wire. It is only provided as a personnel safety precaution in the event of a circuit malfunction. Although a device does not require that the ground wire be connected in order to run, it would be a serious mistake to not connect the casing ground wire.

Some people use the term "ground" when they are referring to a "Neutral" wire. Although in some circumstances (such as automotive wiring), ground and Neutral are the same, they are not the same in building wiring. While both are designed to be at zero pressure, the Neutral wire normally carries current, while the ground wire carries current only in the event of a malfunction.

Figure 1–58 GFCI.

240V (230V) SINGLE-PHASE POWER

The voltage source that you are familiar with in your home is called 120V or 115V single phase. There is a Hot wire, and a Neutral wire. The electrical pressure in the Hot wire alternates from higher than zero to lower than zero. The Neutral wire is always at zero electrical pressure. A graphical representation of the pressure variation in the wiring over time is shown in figure 1–59. While it is convenient to think of the Hot wire as being at a pressure of 115V higher than Neutral, it is really more accurate to say that on the average, the pressure in the Hot wire is 115V *different* from the Neutral wire (sometimes higher and sometimes lower).

230V single-phase is actually a power supply that consists of two Hot wires that are out of phase with each other (see figure 1–60 for a graphical representation). Each Hot is 115V different from zero,

Figure 1–59 Electrical pressure changes in a 115V AC wire.

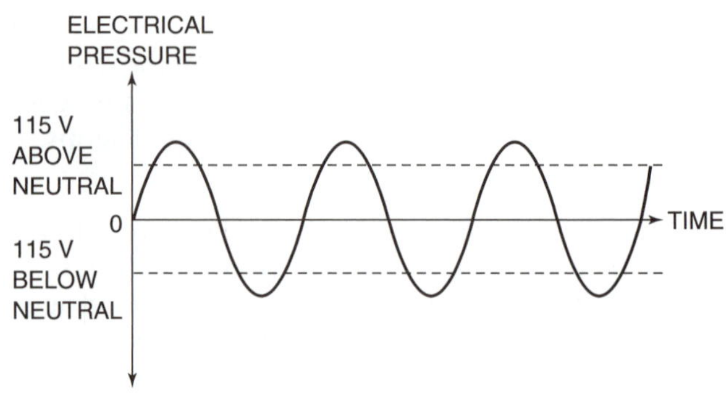

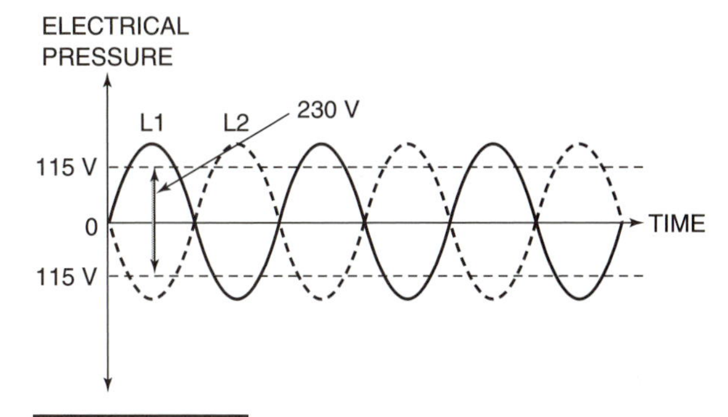

Figure 1–60 Electrical pressure in a pair of 230V 1ϕ wires.

but at any instant, in opposite directions. When one is above zero potential, the other is below. With 115V single-phase circuits, the two wires were identified as "H" and "N" (Hot and Neutral). For 230V single-phase power, the wires are identified as L1 and L2 (Line 1 and Line 2). In practice, either L1 or L2 might be used as a Hot to provide voltage to a 115V load, with the other side of the load being connected to a Neutral.

An analogy for 230V single-phase power would be two lumberjacks (figure 1–61). Each, with his own saw, can push and pull with 115 lbs. of force, while the other end of the saw has no force applied to it. However, if both lumberjacks use the same saw, one pulling while the other pushes, it would produce the same effect as a single lumberjack pushing and pulling with 230 lbs. of force.

Figure 1–61 Mechanical analogy for 230V 1ϕ power.

THREE-PHASE POWER

Very large loads (usually above 5hp), commonly operate on three-phase power, either 230V 3 phase or 460V 3 phase. Figure 1–62 shows the interrelationship between three Hot wires, all operating out of phase. You will note that at any point in time, two of the wires are on one side of the zero line, while the third wire is on the other side of the zero line. The roles of these wires continuously change over time, but at any instant, the two wires on one side of zero will carry a total current equal to the current being carried by the wire on the other side of the line.

POWER MEASUREMENT

Electrical **power** is a measure of the rate of energy consumption being delivered or consumed by a device. An electric motor delivers mechanical energy and is rated in **horsepower.** An electric heater delivers energy in the form of heat. The rate at which the electrical energy must be supplied to each of these two devices is a function of how many amps and at what pressure (voltage) these amps must be supplied. Electrical power is measured in **watts,** or **kilowatts** (1kW = 1000 watts).

ADVANCED CONCEPTS

The reason that motors are manufactured for higher voltages such as 230V is because they are cheaper to produce. Recall that the formula for power is

$$P = E \times I$$

If two motors are rated for the same horsepower but one is designed for 115V and the other is designed for 230V, the 230V motor will only require half as many amps. Wire size is determined by the current carrying requirement, so the 230V motor will actually use *smaller gauge* wire for the windings than the 115V motor, making the cost to manufacture the motor lower.

MORE ADVANCED CONCEPTS

The windings inside a motor have some small resistance. When current passes through the windings, some heat is produced. This heat is wasted energy, because it represents energy input that is not being converted into useful work. The amount of heat generated in the motor windings can be calculated using the equation

$$P = I^2 \times R$$

With the lower current (I) necessary to run an equivalent 230V motor, there will be less heat generated in the motor windings as compared to a 115V motor. As a result, 230V motors are generally slightly more efficient than 115V motors of the same capacity.

MORE ADVANCED CONCEPTS

The meter that the power company places on the side of your house is not a watt meter. It is a watt-hour meter. It doesn't measure power (which is a rate of consuming electrical energy); it measures the consumption of electrical energy. A 100-watt light bulb burning for two hours will consume 200 **watt-hours**, the same amount of energy consumed by a 40-watt light bulb burning for five hours.

POWER FACTOR

The power calculation in the box on page 27 is valid only when the voltage and amperage through the load rise and fall at the same time. However, with motors, the motor windings cut through lines of magnetic force and generate a voltage of their own. The direction of this voltage opposes the voltage being applied to the motor. This causes the voltage and the amperage cycles to be out of phase with each other. That is, the voltage will peak slightly earlier than the amperage. This causes an electrical inefficiency, and the power delivered is actually less than what you would calculate by using the formulas given. This reduced power is called the **apparent power. Power factor** is the number used to describe this discrepancy. It is defined as the ratio of the apparent power to the calculated power as given by one of the given formulas. For resistive loads, the power factor is 1.00. Where inductive loads are involved, the power factor will range between 0.85 and 1.00. Figure 1–63 shows a meter that can be used to read power factor. A service

Figure 1–62 Electrical pressure in 3 wires supplying 3ϕ power.

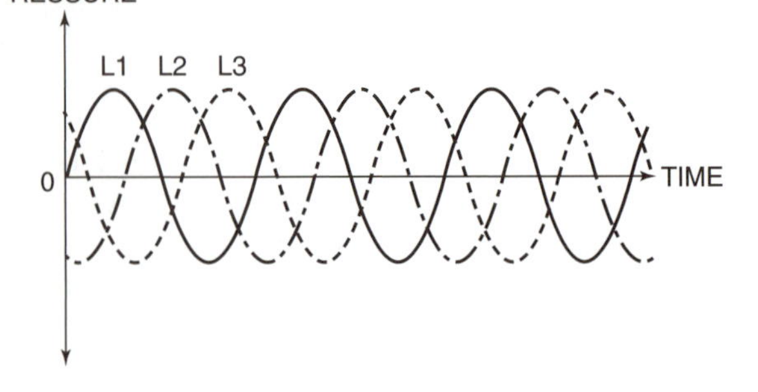

Figure 1–63 Power factor meter. *(Courtesy TIF Instruments, Inc.)*

ADVANCED CONCEPTS

For resistive loads, power consumption is calculated as:

Watts = volts × amps

EXAMPLE: An electric heating element is rated at 2000 watts when operating at 230V. How many amps should it draw?

SOLUTION:

2000W = 230V × amps
amps = 2000/230
amps = 8.7

For three-phase applications, the power consumption is calculated differently:

Watts = volts × amps × 1.73

The 1.73 number is actually the value of $\sqrt{3}$, and it accounts for the fact that power is being delivered by three different legs, but at different times.

technician will rarely (if ever) have to worry about taking a power factor reading.

CAPACITORS

A capacitor (figure 1–64) is a device commonly used in conjunction with a motor. There are two different types. A **start capacitor** is used to give a motor more starting torque. A **run capacitor** is used to improve the running efficiency of a motor. In concept, a start and a run capacitor work the same. Each consists of two plates with a material of high resistance between them. When a voltage is applied to the plates, the voltage source pushes electrons onto one of the plates. That plate, in turn, pushes electrons off the other plate. If the voltage is then removed, the net result is an excess of electrons on one plate, and a shortage of electrons on the other plate. The capacitor, in that condition, is storing electrons. If an electrical path is then provided between the plates, electrons will flow off the plate that has an excess and onto the plate that has a shortage.

Figure 1–64 Start and run capacitors.

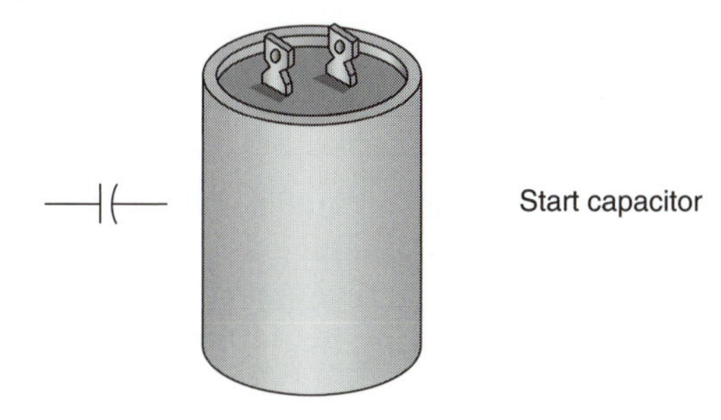

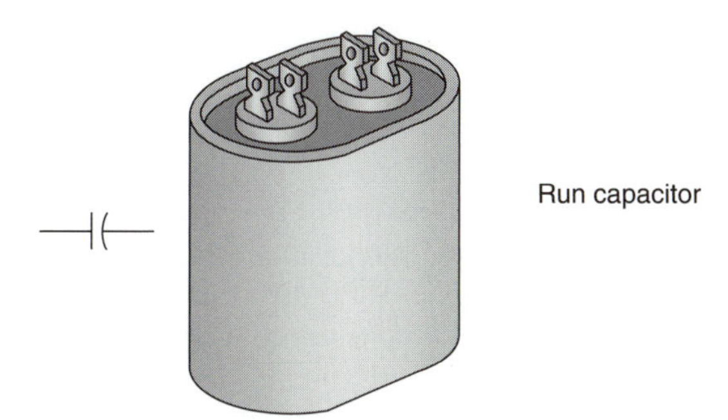

Even though the power has been turned off to a circuit that contains a capacitor, the capacitor may still be storing a charge, and can give the unwary technician quite a shock. Before touching capacitor terminals, with the power off, place a screwdriver across the terminals (shorting the capacitor out) to equalize the electrons on the two plates (figure 1–65). Some manufacturers recommend against this practice, as it can damage the capacitor. They suggest that a resistor be used instead of a screwdriver. Most technicians use a screwdriver anyway.

After the capacitor has been charged, if the applied voltage source is reversed, the electrons will flow in the opposite direction. However, the charged capacitor will "help" push the electrons in that opposite direction because the capacitor wants to discharge itself. The result is that more electrons flow than you would expect just from the voltage source applied. To demonstrate this effect, try the following exercise:

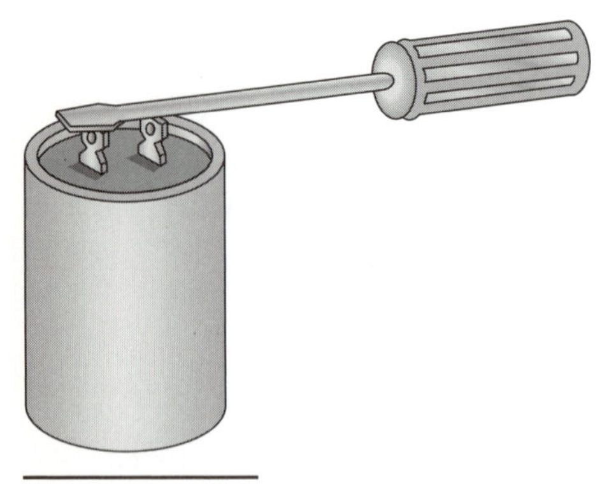

Figure 1-65 Discharging a capacitor.

When an alternating current is applied to a capacitor, the capacitor is continuously charging and then discharging as the direction of the applied voltage changes. The capacitor amplifies the voltage being applied, so that it has the effect of a higher voltage. To help you visualize this effect, think of a small child in a bathtub. If he moves his body slightly, in the right rhythm, first towards his head, and then towards his feet, the small wave of water that he creates will be amplified each cycle until he sloshes the water completely out of the bathtub!

Capacitors have two ratings, voltage and capacitance. The voltage rating is the maximum voltage that can be applied to the capacitor. The capacitance, measured in **microfarads** (abbreviated mf, mfd, or μf), is essentially a measure of how many electrons the capacitor can store.

EXERCISE:

Obtain a capacitor. Discharge it through a screwdriver across the terminals. Using an analog ohm meter set to the highest resistance scale (R×10,000), place the ohm meter probes on the capacitor terminals. Note that the needle swings towards zero and then turns around and moves towards infinity (the symbol for infinite resistance is ∞). When the needle stops moving, the capacitor is charged. Remove the ohm meter probes. Re-attach them, but with the probes on the opposite capacitor terminals. This time, the ohm meter needle will swing much further towards the zero direction than it did the first time. This is because the charge on the capacitor as it discharges is helping the electrons move in the same direction as they are being pushed by the battery in the ohm meter.

ADVANCED CONCEPTS

Capacitors may be connected together in series or in parallel. When connected in series, the voltage rating is the sum of the individual voltage ratings, and the capacitance is less than either of the individual capacitors. The capacitance of two capacitors in series is given by the formula

$$C_T = \frac{C_1 \times C_2}{C_1 + C_2}$$

where

C_T = the total capacitance

C_1 = capacitance of one capacitor

C_2 = capacitance of the other capacitor

Capacitors are almost never found wired in series. However, they are sometimes found wired in parallel. In parallel, the voltage rating of the combination is equal to the lowest voltage rating of any individual capacitor, and the capacitance of the two capacitors in parallel is equal to the sum of the individual capacitors.

CHAPTER TWO

SIMPLE FREEZER CIRCUITS

FREEZER CIRCUIT NO. 1

Figure 2–1 is an interconnection diagram for a freezer. It shows the simplest possible refrigeration circuit. It consists of one switch and one load. The thermostat is located inside the freezer, and it senses the freezer temperature. The compressor is located below the freezer. The wiring for the thermostat, the compressor, and the power supply (plug) all come together in a small box called a **junction box** (sometimes referred to as a J-box). The detailed drawing shows how the wires are connected inside the junction box. Figure 2–2 shows the same wiring, but in a ladder diagram. When the freezer temperature rises above the set point of the thermostat, it will cause the compressor to run. It will continue to run until the thermostat senses that the freezer has cooled to a temperature below the thermostat set point. Then the thermostat will open its contacts, and the compressor will stop running. The thermostat in a small refrigerated box is sometimes called a **cold control.**

Notice the symbol that is used for the thermostat in the ladder diagram. The "squiggle" below the switch tells you that this is an automatic switch that opens and closes its contacts in response to a change in the *temperature* that it senses. Notice also the position of the switch mechanism. It is shown *below* the wire. This tells you the *action* of the thermostat. It closes on a rise in temperature. In other applications (e.g. heating), you will find thermostats that close on a fall in temperature. Figure 2–3 shows the two different symbols for thermostats. The principal of operation of this cold control is that a temperature change can cause the liquid inside a sensing element to expand or contract. Figure 2–4 shows a cold control with a sensing line. The expansion and contraction of this liquid causes changes in the hydraulic pressure inside the sensing line, and this change in pressure can be used to operate a switch using a bellows as shown in figure 2–5.

The temperature in the freezer is not maintained at the exact temperature corresponding to the set point of the thermostat. Instead, it is controlled over a *range* of temperatures. For example, even though the thermostat might be set for −10F, the thermostat contacts might not close until the freezer temperature rises to −8F, and might not open until the refrigerator temperature drops to −12F. In this example, we would say that the thermostat has a **differential** or **range** of 4F. In other terminology, we might say that the switch **makes** at −8F, and **breaks** at −12F.

TROUBLESHOOTING

The only possible customer complaint that could result from this simple circuit is that the compressor does not run. Use the volt meter to measure voltage across the switch and across the compressor motor

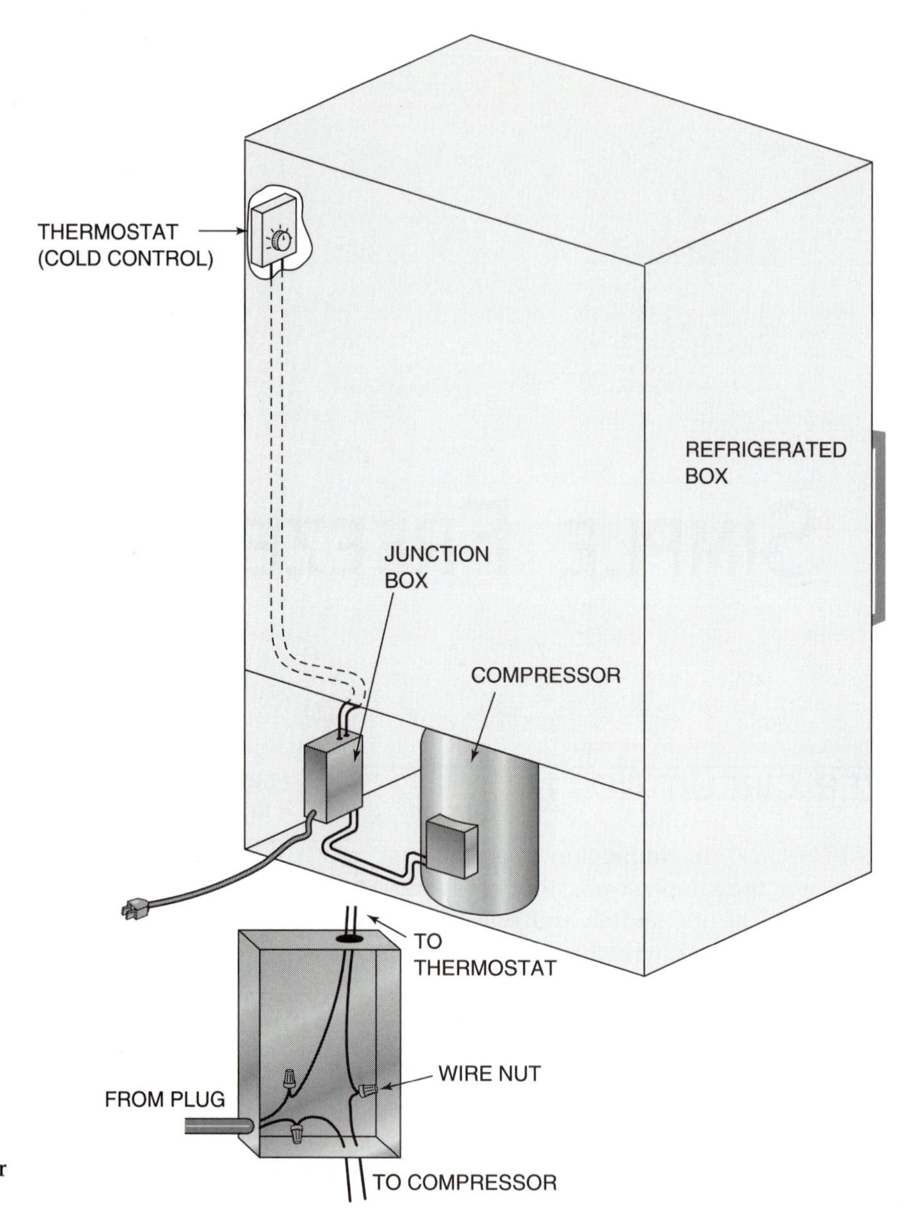

Figure 2–1 Household freezer interconnection diagram.

Figure 2–2 Household freezer ladder diagram.

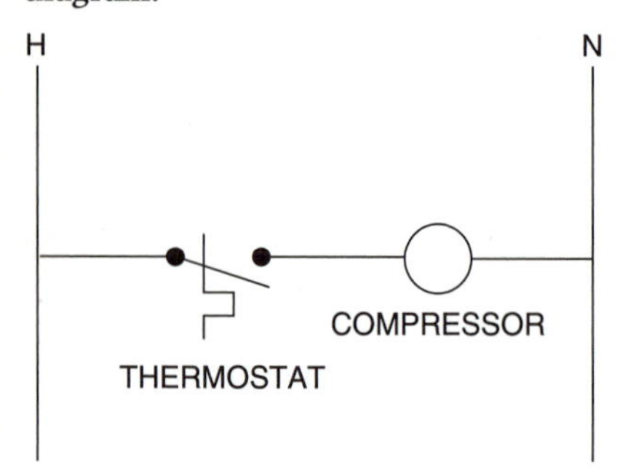

Figure 2–3 Thermostat symbols. (a) Cooling thermostat—closes on a rise in temperature. (b) Heating thermostat—opens on a rise in temperature.

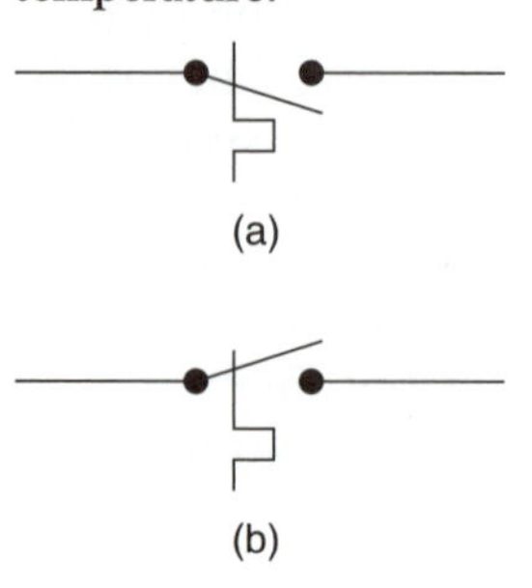

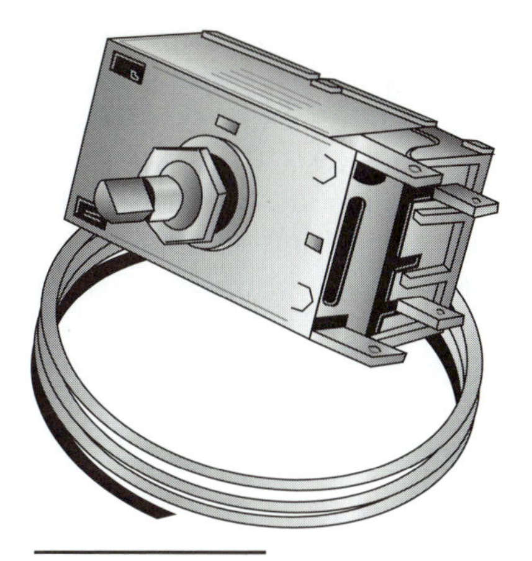

Figure 2–4 Cold control with temperature sensing line.

(Figure 2–6). There is no right or wrong place to start. You should measure whichever is easier to measure first. If you find that there are 115V across the thermostat, then it is open and must be replaced. If you find that there are 115V across the compressor, then it has failed. If you don't find 115V across either, then you must measure the voltage across Hot and Neutral to determine that you probably do not have voltage available to the refrigerator.

Figure 2–5 Switches operated by hydraulic pressure. (a) Closes on a rise in temperature. (b) Opens on a rise in temperature.

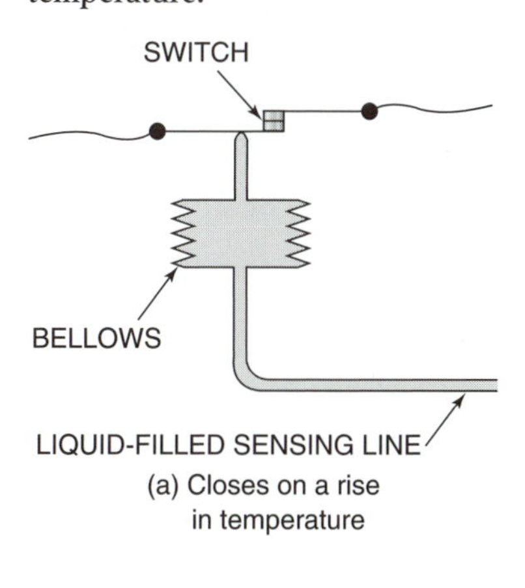

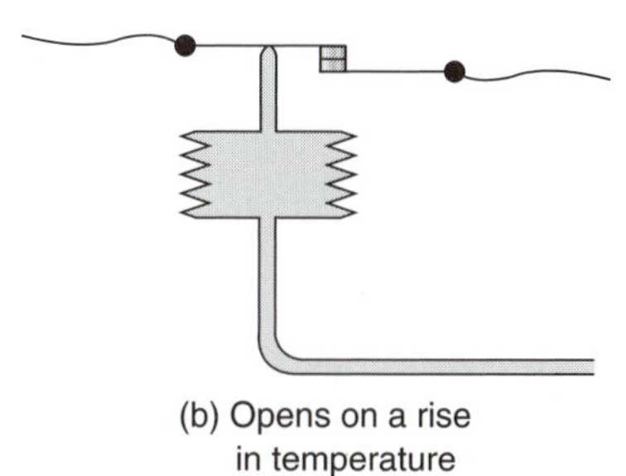

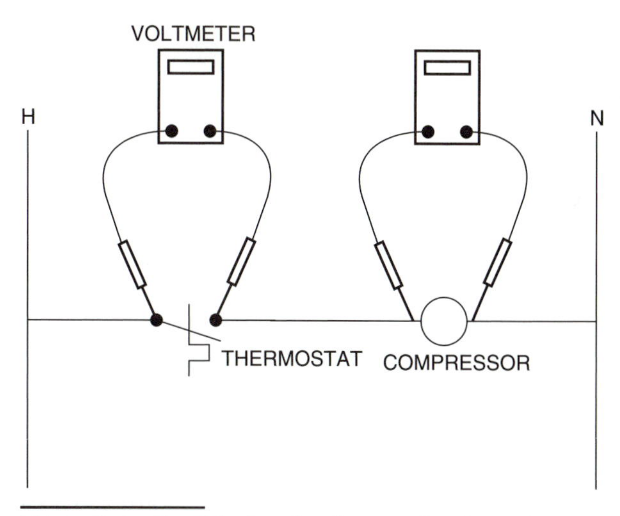

Figure 2–6 Troubleshooting freezer circuit no. 1.

EXERCISE:

1. Draw the correct symbol for each of the following thermostats:
   - **a.** To operate an attic ventilation fan during the summer months.
   - **b.** To operate an electric room heater.
   - **c.** To operate an oven.
   - **d.** To operate a radiator fan in your car.
   - **e.** A freezestat.

FREEZER CIRCUIT NO. 2

In figure 2–7 we have added another device called an **overload**. This type of overload is commonly called a "**Klixon,**" although this is actually jargon for this type of overload. Klixon is actually a registered brand name that has fallen into common use, like Freon, Kleenex, or Amprobe. A picture of this type of overload is shown in figure 2–8).

An overload is a switch that senses the current passing through it, and opens if the current exceeds a predetermined rating. In this case, the overload is

Figure 2–7 Freezer circuit with overload.

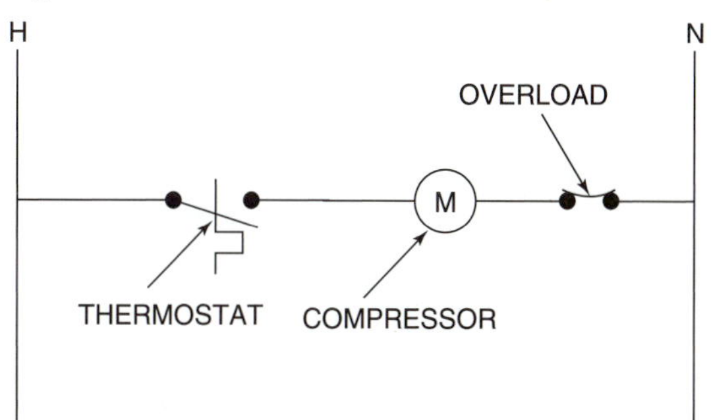

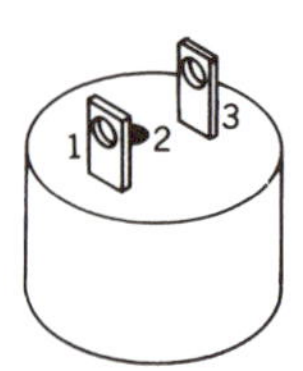

Figure 2–8 Klixon overload.

carrying the same current as the compressor. It must be selected so that if it carries a current that exceeds the maximum allowable current through the compressor, it will open, thereby preventing the compressor motor from burning out. Two observations are in order about the overload:

1. It must be matched to the load that it is protecting. Different compressors can be rated for a wide range of maximum allowable currents. Therefore, you cannot arbitrarily replace an overload with a different one. If the replacement you select is rated for a too-low amperage, it will stop the compressor from operating when it is still within its acceptable operating amperage range. If the replacement you select is rated for a too-high amperage, it may allow the compressor to operate at an amperage draw that exceeds its maximum allowable rating.

2. The overload is a **safety control,** as compared to the thermostat which is an **operating control.** An operating control is a switch that is expected to open and close regularly, to turn the system on and off as required by the area in which temperature is being controlled. However, a safety control is one that will never open unless a malfunction occurs.

The overload relies on two different principles to make it work:

1. The more current that flows through a wire, the more heat will be generated.

2. Different metals expand at different rates when heated. If two metals with different coefficients of expansion are joined together, it is called a **bimetal.** When a bimetal element is heated, the differential expansion causes the bimetal element to bend. A bimetal element is shown in figure 2–9.

Figure 2–9 Bimetal element.

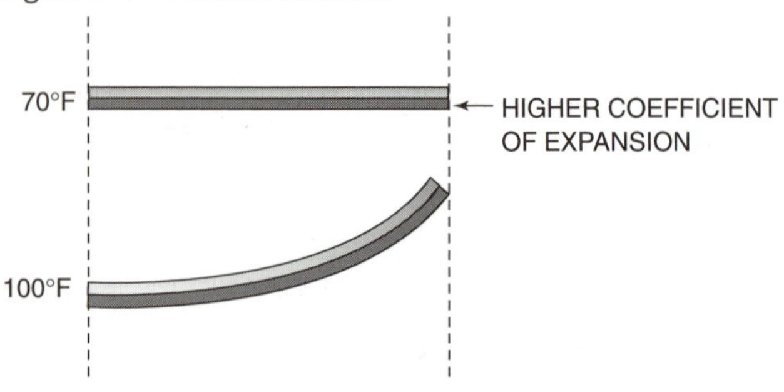

Figure 2–10 shows the internal construction of the Klixon overload. When it is carrying less than its rated amps, the heater wire produces very little heat. But when the current exceeds the rated amp draw, the heat from the heater wire will be sufficient to cause the bimetal element to warp, opening the switch, and de-energizing the compressor. When the heater wire cools, the bimetal will warp back to its original closed position. Each time the overload opens or closes, there will be an audible "pop," similar to the noise made when you push on the bottom of an oil can.

Troubleshooting

If the compressor is not running, either the thermostat or the overload could be open, or the compressor may have failed (we will assume that power is available). Using your volt meter, measure across each of the three components. Whichever has voltage across it is the one that has failed.

If the compressor turns on, runs for a few seconds, turns off, and then repeats the cycle a few minutes later, it is most likely **cycling on the overload.** You will probably hear the overload open and close just before the compressor turns off and on. A reading of 115V across the overload when the compressor is off, and then a 0V reading across the overload when the compressor is on will confirm it. In this event, it may be *either* the overload or the compressor which is at fault. There may be something wrong with the compressor that is causing it to draw too many amps, and the overload is working properly. Or, the compressor may be fine, and the overload is opening at a too-low amperage. In order to determine which is the problem, you must mea-

Figure 2–10 Internal construction of a Klixon overload.

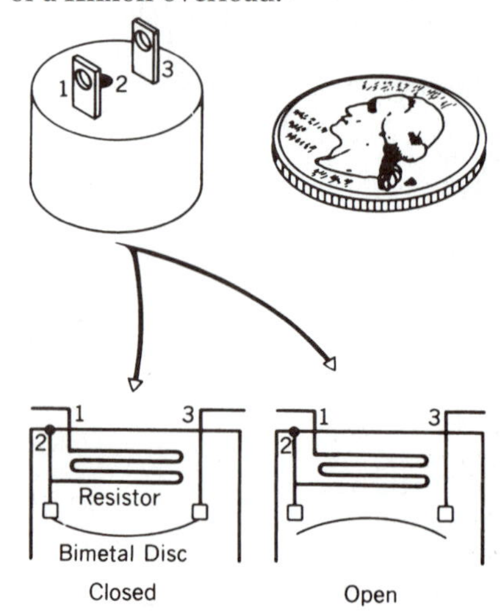

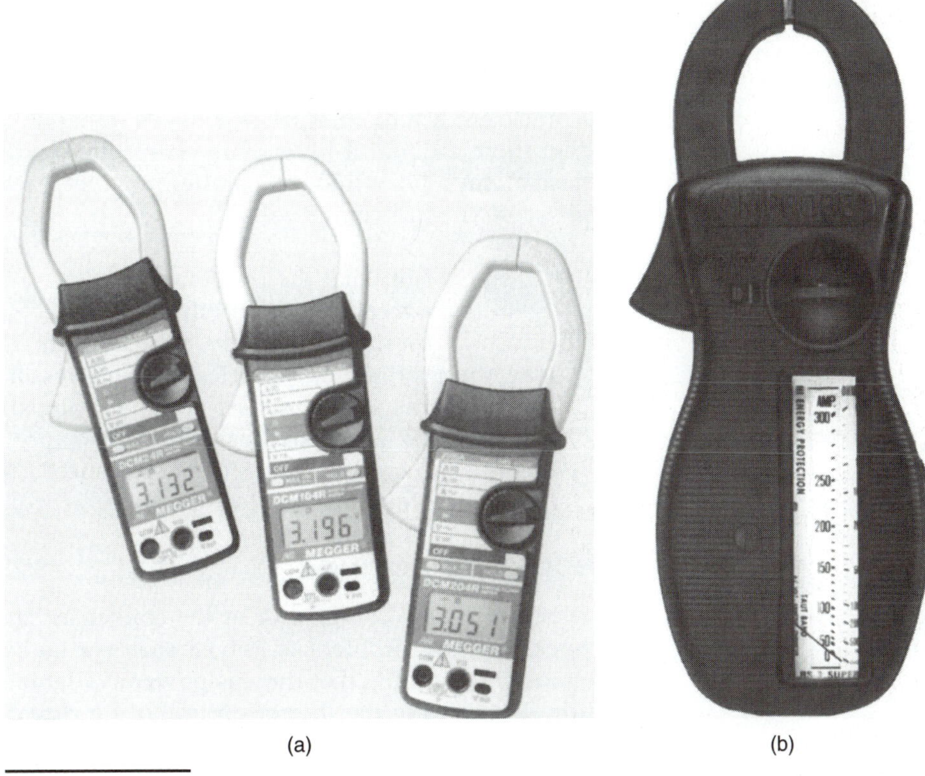

Figure 2–11 Amp meters. (a) Digital amp meter. *(Courtesy of AVO International, manufacturer of Biddle®, Megger®, and Multi-Amp®, products.)* (b) Analog amp meter. *(Courtesy Amprobe)*

sure the amperage when the compressor turns on, and compare it to the amperage rating on the nameplate of the compressor. If your amperage reading exceeds the nameplate rating, the compressor is the problem. However, if the measured amperage is lower than the nameplate rating, then you must replace the overload. An amp reading may be taken by using an amp meter shown in figure 2–11. Many modern VOMs have an accessory amp probe (figure 2–12) available that reads out amps on the VOM. To use it, you simply clamp the jaws of the meter around the wire whose current you wish to mea-

Figure 2–12 Accessory amp probe. *(Courtesy AEMC Instruments.)*

ADVANCED CONCEPTS

A clamp-on amp meter works on the principal of sensing the amount of magnetic field that is generated around a current-carrying wire. On a 120V single-phase circuit, if you clamp around two wires at the same time, one carrying current towards a device and the other carrying current away from the same device, their magnetic fields will cancel each other out, and the amp reading on the meter will be zero. See figure 2–13. If you wind a wire around the jaws of the amp meter, the magnetic fields will add, and the amp reading will equal the actual amps times the number of turns passing through the jaws (figure 2–14). This is particularly useful when trying to read very low amperages that won't accurately register on the lowest amp scale.

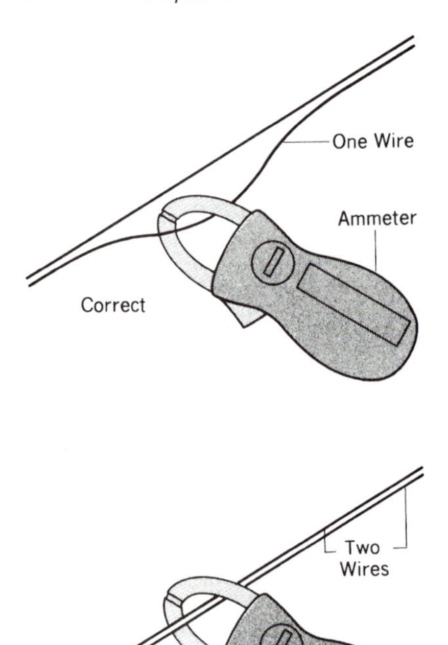

Figure 2–13 Using amp meter.

sure, and plug the probe wires into the appropriate jacks on the VOM.

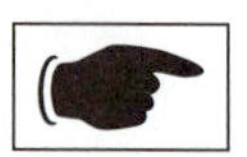

The accessory amp probe has its own batteries, and must be turned on when in use. Technicians commonly forget to turn the accessory amp probe back off, leaving them with a dead instrument the next time it is needed. Some accessory amp probes have a "sleep" mode. The probe will automatically turn itself off when not in use for a few minutes. This is an option well worth looking for when you buy this instrument.

This circuit is an example of switches in series. Some circuits have several safety controls wired in series with the load. If any one of the safety controls detects a problem with current draw, temperature or pressure, that particular switch will open and de-energize the load.

Figure 2–14 Multiplying the amp reading.

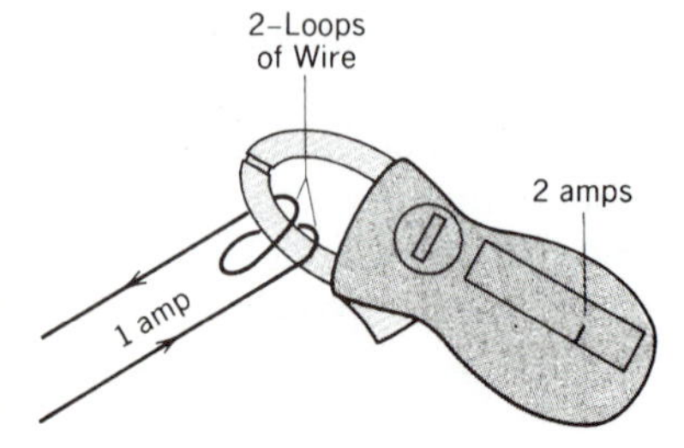

FREEZER CIRCUIT NO. 3

In figure 2–15, we have added a condenser fan motor in parallel with the compressor. When the thermostat closes, it will energize both loads in parallel. Where there are multiple loads in a unit, they will almost always be wired in parallel, so that each load "sees" the full circuit voltage.

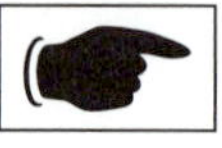

Note that the condenser fan is wired in parallel with both the compressor and overload. In this way, the overload only carries the compressor amps, but not the condenser fan motor amps. If mistakenly wired so that the overload carries the current from both loads, nuisance trips on the overload will result.

Troubleshooting

1. If neither the compressor or the condenser fan is operating, the problem is almost surely a faulty thermostat (assuming that there is power available). While theoretically the thermostat might be closed, and both the compressor and condenser motors might be burned out, it is so unlikely that we would not investigate that possibility first. Also, note that if the overload was open, it would prevent the compressor from running, but it would not prevent the condenser fan from running. Always direct your initial investigation to a device whose failure could cause all of the symptoms you observe. In the great majority of troubleshooting service calls, you will find that there is only one device that has failed. Go to the thermostat first. If it measures 115V across it, it is open.

2. If the compressor is running but the condenser fan is not, the problem must be a failed condenser fan motor. Although an open thermostat can prevent the condenser fan from running, that is not the case in this situation. If the thermostat was open, the compressor could not run. But since the compressor *is* running, the thermostat *must* be closed.

Figure 2–15 Freezer with condenser fan motor.

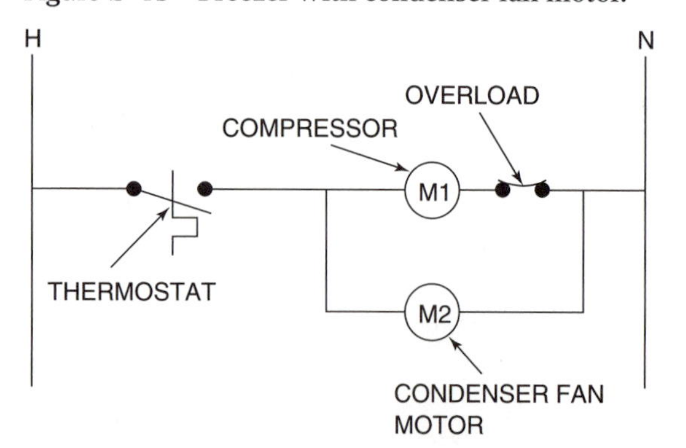

3. If the condenser fan is running but the compressor is not, the problem must be that either the compressor has failed, or the overload is open. Note that an open overload will shut down only the compressor. Measure voltage across the compressor and the overload. Whichever has voltage across it has failed.

FREEZER CIRCUIT NO. 4

In figure 2–16, we have added two more devices, an evaporator fan and another switch. The switch symbol tells us that it is a **momentary switch** (figure 2–17). A momentary switch may have contacts that are **normally open** or **normally closed.** When the movable part of the switch is pressed, the switch will operate. A normally open switch will close, and a normally closed switch will open. When the movable part of the switch is released, the switch contacts return to their normal position. The symbols for a normally open and a normally closed momentary switch are shown in figure 2–18. The momentary switch for the evaporator fan is normally open. When the freezer door closes, it presses on the momentary switch, causing it to close and allowing the evaporator fan to run whenever the cold control closes. A momentary switch that is operated by a door is commonly referred to as a **door switch.**

QUESTION:
The compressor is running while the freezer door is open. What voltage reading would you obtain *across* each of the following devices?
- a. Cold control.
- b. Compressor.
- c. Condenser fan motor.
- d. Door switch.
- e. Evaporator fan motor.

Figure 2–16 Freezer with two fan motors.

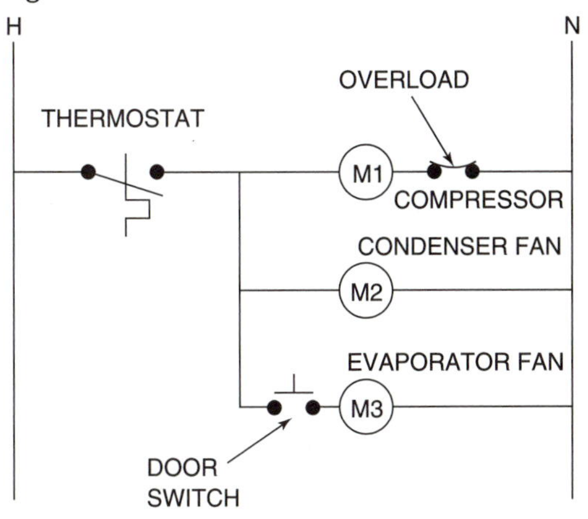

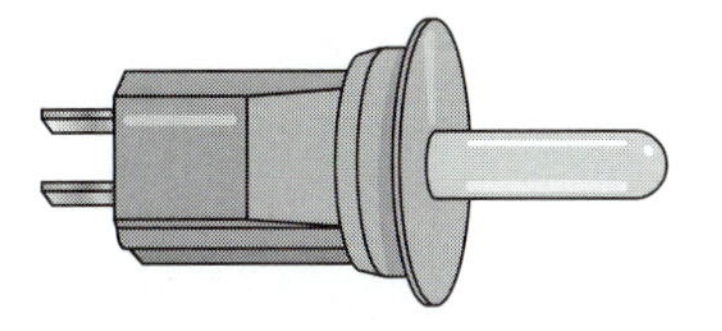

Figure 2–17 Momentary switch.

EXERCISE:
Figure 2–19 shows a partial interconnection diagram for a freezer with a compressor, condenser fan, evaporator fan, overload, door switch, and cold control. Complete the wiring in accordance with the wiring diagram of figure 2–16. For the sake of clarity, draw all wiring as vertical or horizontal lines.

EXERCISE:
For each of the following applications, draw the correct symbol for the momentary switch:
- a. The dome light in your car.
- b. The switch on a flashlight.
- c. The door switch on a microwave oven.

FREEZER CIRCUIT NO. 5

In this circuit, we have added a cabinet light and another momentary switch. Note that this momentary switch is normally closed. When the door is closed, it will press on the switch, opening its contacts, and turning the light off. Some refrigeration systems have two different switches, each operated by the freezer door. However, in this diagram, note that the two momentary switches are connected by a dotted line. This means that they are ganged together, and operate simultaneously. The momentary switch for this circuit looks just like the one in figure 2–20, but instead of two terminals, it has four. One pair of terminals is the normally open fan switch. The other pair of terminals is the normally closed cabinet light switch.

Figure 2–18 Symbols for momentary switches.

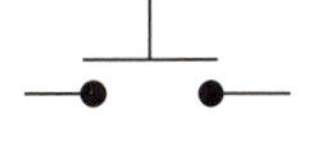

NORMALLY OPEN MOMENTARY SWITCH CLOSES WHEN BUTTON IS PRESSED.

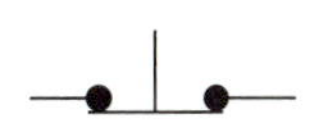

NORMALLY CLOSED MOMENTARY SWITCH OPENS WHEN BUTTON IS PRESSED.

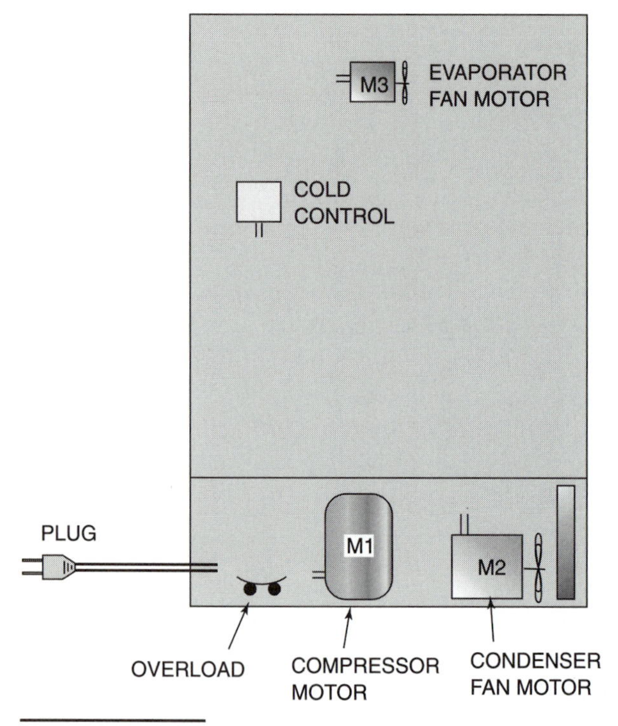

Figure 2–19 Exercise.

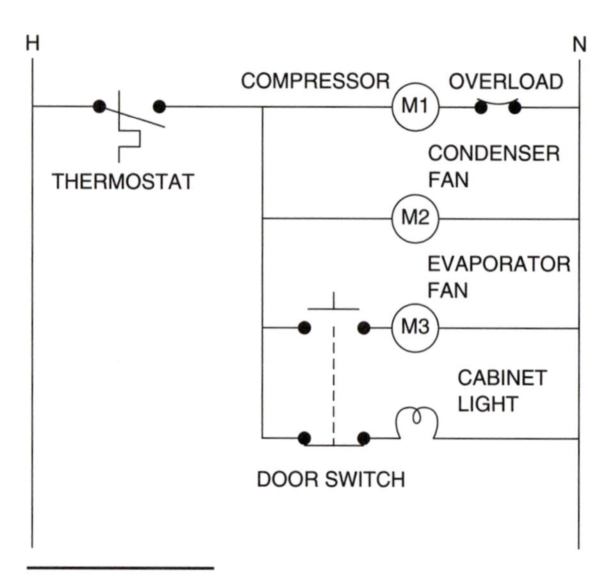

Figure 2–20 Freezer circuit no. 5.

QUESTION:
The compressor is running while the freezer door is open. What voltage reading would you obtain *across* each of the following devices?
 a. Cabinet light.
 b. Momentary door switch for the cabinet light.

QUESTION:
You found a door switch in your truck, but it fell out of the package. It has two terminals. How can you use your ohm meter to determine if this switch works, and whether it is a switch for a fan or for a cabinet light?

FREEZER CIRCUIT NO. 6

A freezer operates with a coil temperature of well below freezing. Moisture that condenses out of the air onto the coil will freeze. If the evaporator is of a finned-tube design, the spaces between the fins will be completely blocked within a few hours unless the coil is periodically defrosted. Figure 2–21 shows the wiring diagram with an **automatic defrost** system. This system consists of three devices:

1. A defrost timer (figure 2–22). It has a time clock inside that moves gears and cams that operate an **SPDT** switch. The switch is in one position for approximately 7½ hours, and in the other position for approximately ½ hour. There is a knob that may be rotated manually to advance the timer, causing the switches to operate.

2. A defrost heater (figure 2–23). It is a heating element similar to the type found in an electric oven. It is mounted directly to the evaporator coil.

3. A defrost termination switch (figure 2–24). It is a thermostat that is fastened to the evaporator coil. It opens when the evaporator is above a temperature of 45°F.

Figure 2–25 shows the arrangement of the electric defrost heaters and defrost termination

Figure 2–21 Freezer with automatic defrost.

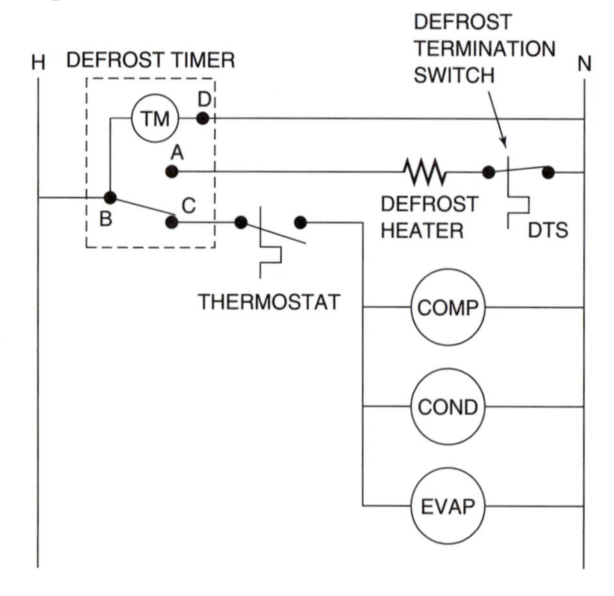

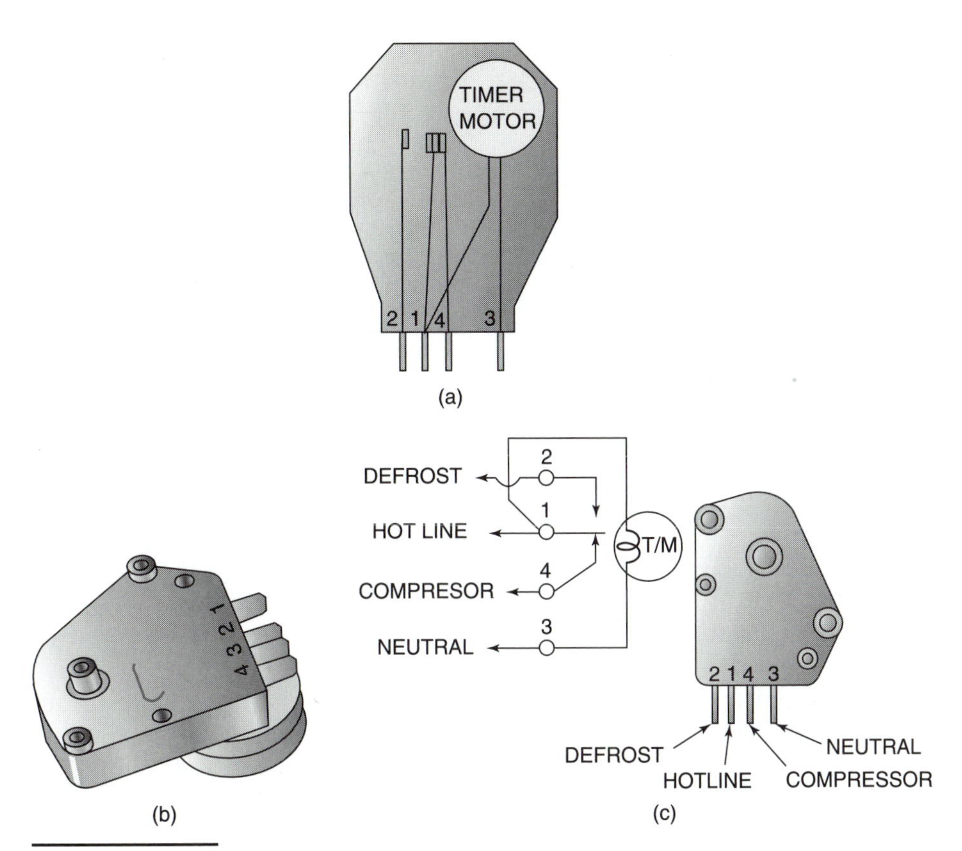

Figure 2–22 Defrost timer.

switch mounted on an evaporator coil of the type that would be found in a walk-in freezer.

EXERCISE:
Obtain a defrost timer. Rotate the manual advance knob in a clockwise direction. You will find that it takes a lot of rotation until you hear a first "click," indicating that the switch has gone into the "defrost" position. Then, after a slight additional rotation, you will hear a second "click," indicating that the switch has gone into the "freeze" position. Use your ohm meter to figure out the internal wiring of the defrost timer. Between which terminals is the timer motor? Which terminals are connected together after the first click? After the second click?

When the switch is in the 7½ hour position (the "freeze" position), the operation of the compressor, condenser, and evaporator is normal. The timer motor is also energized, turning the gears inside the defrost timer. After the timer motor runs for 7½ hours, the switch moves to the other position (the "defrost" position). It will stay there for 1/2 hour. During that time, the compressor, condenser, and evaporator are de-energized. The heater coil is energized, and the ice will melt off the evaporator. [Note: the 7½ hour and 1/2 hour times are just for example. Timers may operate on a six-, eight-, ten-, or twelve-hour frequency, and the defrost time is also variable, depending upon the manufacturer.] When all the ice has melted, the evaporator coil

Figure 2–23 Defrost heater.

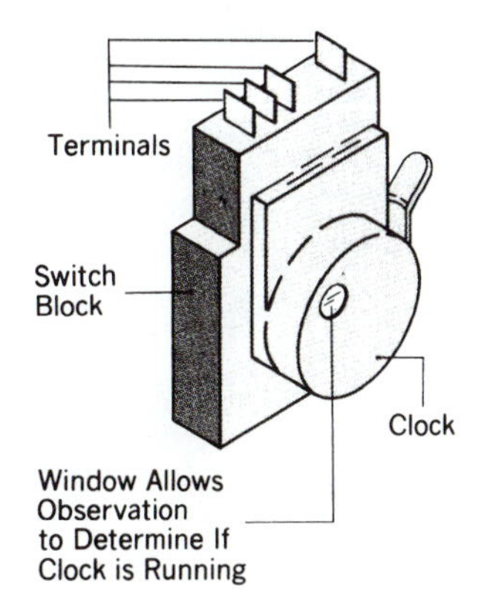

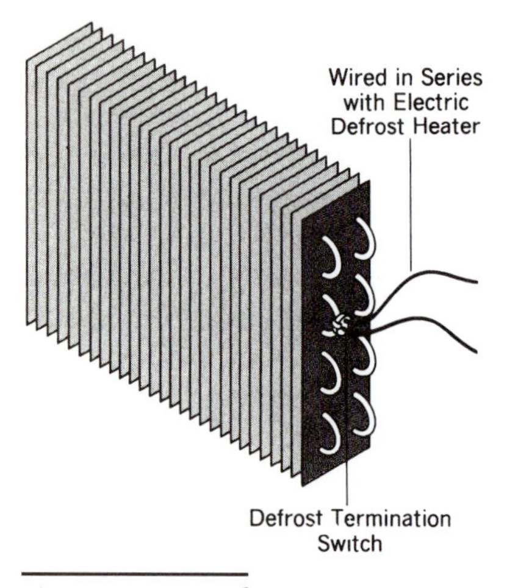

Figure 2–24 Defrost termination switch.

temperature will start to rise above 32°F. When it reaches 45°F, the defrost termination thermostat opens, de-energizing the heater. This prevents the box from overheating when "defrost" has been accomplished, but the timer is still in the "defrost" position. After the defrost termination thermostat opens, only the timer motor is energized. Neither the heater nor the refrigeration circuit is energized until the defrost timer motor switch returns to the "freeze" position.

Figure 2–25 Defrost heaters. *(Courtesy of Witt Division of Ardco.)*

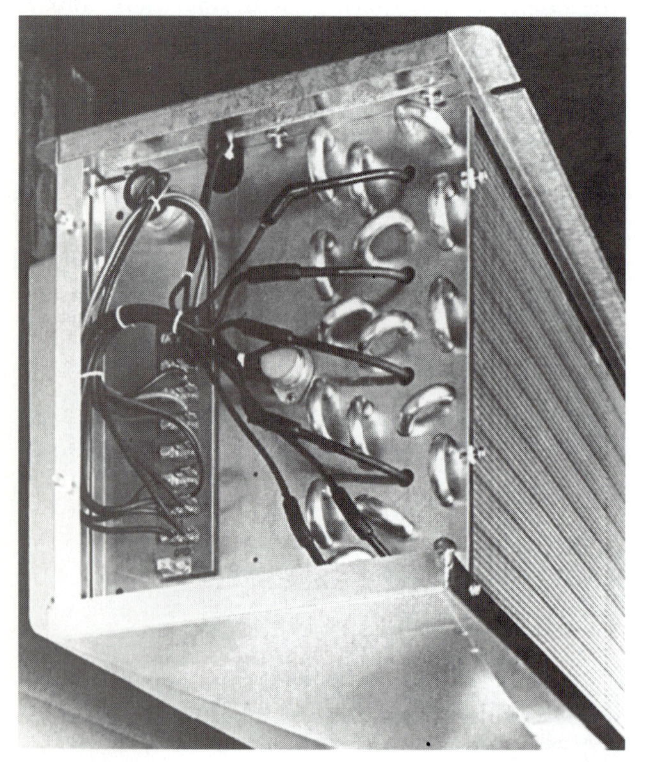

Troubleshooting—Box Is at Room Temperature, Compressor Not Running

Check to see that there is power to the unit (the cabinet light comes on). Rotate the defrost time clock manually. If the compressor comes on, it means that the defrost timer was not switching automatically. Measure voltage at the power input terminals of the timer (see practice pointer below) to make sure that 115V is available to run the timer motor. Then replace the defrost time clock.

The terminal arrangement of most 4-wire defrost timers is the same (figure 2–22a). There are three terminals together and one by itself. Power comes in the center terminal in the group of three (terminal 1). From there it goes into the timer motor and into the SPDT switch. The current that leaves the timer exits at the lonesome terminal (terminal D), which is connected to the other side of the circuit. The current that enters the SPDT switch is switched out to either the "freeze" circuit or the "defrost" circuit. Usually (but not always), the switched terminal closest to the edge of the timer (terminal 2) goes to the defrost circuit, and the switched terminal closest to the lonesome terminal (terminal 4) goes to the compressor circuit. The numbering of the terminals may differ from timer to timer, but the arrangement of terminals is pretty universal.

Troubleshooting—Compressor Running, Evaporator Coil Airflow Is Blocked with Ice

Manually advance the defrost timer until it goes into "defrost."

You can tell when the timer is in the "defrost" mode by listening as you rotate the manual advance knob. You will be able to rotate the manual advance several turns before you hear the first click. Then, a slight additional manual advance will produce a second click. After the first click, the clock is in the "defrost" mode. After the second click, the clock is in "freeze" mode.

Check to see if the evaporator is defrosting. This can be done three ways:

1. Listen. You will be able to hear a "sizzle" as the heater melts the ice.

2. Look at the drain pan that collects the water that runs down a hose from the evaporator. If water is coming out, the defrost is working.

3. Measure the amps through the defrost heater. If amps are flowing, the heater is working.

If the "defrost" is taking place, it means that the defrost heater and defrost termination thermostat are both ok. Change the defrost timer.

There is a potential trap in the above discussion. If the evaporator is concealed, you can be fooled into thinking that the heater is OK because you read amps. Sometimes, several heater elements are wired in parallel. There may be one or more heater elements burned out, but as long as there is at least one functioning heater element left, you will read amps on the defrost cycle. The only way to discover this trap is to inspect the coil as it defrosts to make sure that all of the coil is defrosting.

If, after placing the timer in the "defrost" position, no defrost occurs, there are two likely problems:

1. The defrost timer switch between terminals 2 and 1 is not making contact.

2. The defrost timer switch *is* making contact, but either the heater or the defrost termination switch is open.

[Note: For the following troubleshooting sequence, it is assumed that Hot is connected to terminal 1 and that Neutral is connected to terminal 3. Even if this is opposite from the actual wiring, the same voltage readings described below would be applicable.]

Check voltage between terminals 1 and 3. It should be 115V. This means that electrical pressure is coming into the SPDT switch at terminal 1. Then check for electrical pressure at terminals 2 and 4, by measuring 2–3 and 4–3. There should be electrical pressure available at the terminal going to the defrost circuit. A 115V reading between 2 and 3 would indicate that there is electrical pressure available at terminal 2. If 2–3 and 4–3 both read zero volts, then the SPDT switch is not making contact in the defrost position, and the defrost timer must be replaced.

If you get a 115V reading between 2 and 3, that means that voltage is being supplied to the heater circuit. But we know that the heater circuit is not working. So, either the defrost heater has failed or the defrost termination switch is open. Some disassembly of the unit will be necessary to gain access to the evaporator. Then, measure voltage across the heater and the defrost termination thermostat. Whichever one shows 115V has failed.

FREEZER CIRCUIT NO. 7

This circuit (figure 2–26) is the same as Freezer Circuit No. 6, except that the cold control is in a different place. In this circuit, the timer motor only runs when the cold control is energized. There is an advantage to doing it this way. Instead of going into "defrost" every eight hours, this system will go into "defrost" more frequently during periods of heavy usage and less frequently during periods of light usage. Because this time clock doesn't run all the time, it would probably use a timer that switches the SPDT switch every four or six hours. This would result in a "defrost" taking place probably every eight to ten hours, because it would take that long for the timer motor to get six hours of run time.

The disadvantage to this system is that it will "defrost" at different times each day, compared to regular eight- or twelve-hour intervals, which would start a "defrost" at the same times each day. This could possibly result in a nuisance service call from an owner who notices the refrigeration system not running when the box is above 0°F.

Figure 2–26 Freezer circuit no. 7.

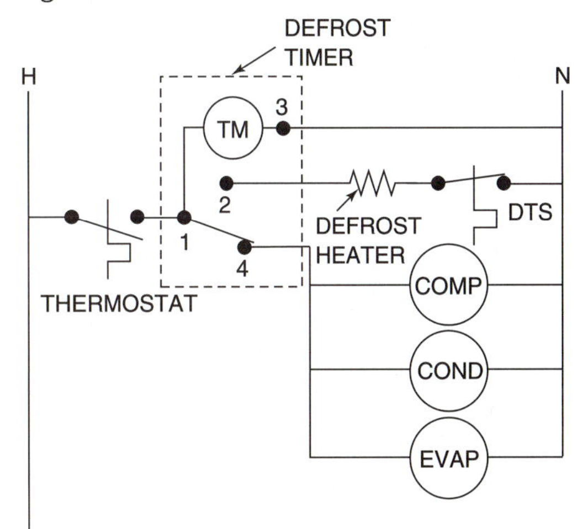

FREEZER CIRCUIT NO. 8

This circuit (figure 2–27) is the same as Freezer Circuit No. 6, except that the evaporator fan has been wired differently. When Freezer Circuit No. 6 returns to the "freeze" cycle, the evaporator coil is warm (45°F), and the box is also warmer than normal. At the beginning of the freeze cycle, the load on the compressor is abnormally high because the evaporator coil and the box are both warm. This could cause the compressor to overload. Freezer Circuit No. 8 addresses this problem by providing a time delay for the evaporator fan. When the defrost timer returns to the "freeze" position, the compressor and condenser fan motors are energized. But the evaporator fan motor does not start because the defrost termination switch is still open. When the compressor and condenser fan run without the evaporator fan, the load on the compressor is reduced. The evaporator temperature (and suction pressure) will drop to a more-normal value relatively quickly. When the defrost termination switch senses that the evaporator coil is cold, it will close, turning on the evaporator fan.

Some technicians who are not familiar with the evaporator fan delay circuit have been fooled. A technician may arrive at a walk-in freezer to find that nothing is running, the box is at room temperature, and the product has been removed for storage elsewhere. After discovering the problem and repairing it, the technician starts up the system. While checking the operation the technician notes that the evaporator fan doesn't run. Rather than tearing into the wiring again, leave it alone for a few minutes. The evaporator fan will probably restart all by itself!

Figure 2–27 Freezer circuit no. 8.

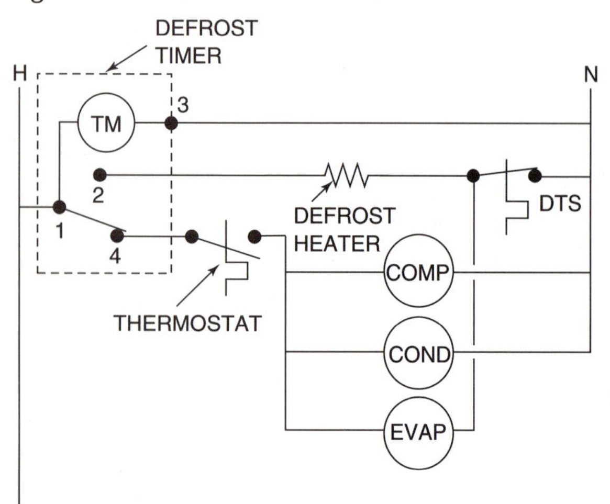

Some older refrigeration systems use a five-wire defrost timer to provide an evaporator time-delay. The extra terminal is wired to the evaporator fan motor only. There is an extra cam inside the defrost timer that allows this extra terminal to become energized a few minutes after the compressor is energized. If you cannot locate a replacement five-wire defrost timer, use a four-wire timer. The wire going to the evaporator fan motor may be placed on the same terminal as the wire going to the compressor.

FREEZER CIRCUIT NO. 9

This circuit (figure 2–28) is similar to Freezer Circuit No. 6, except that the defrost timer has *three* terminals instead of four. Note how the internal wiring of the timer motor has changed. The timer motor is now wired between terminals 2 and 4, and terminal 3 is missing.

When the SPDT switch is made between 1 and 4, the refrigeration components run, as in the previous circuits. But the circuit through the timer motor

Figure 2–28 Freezer circuit no. 9.

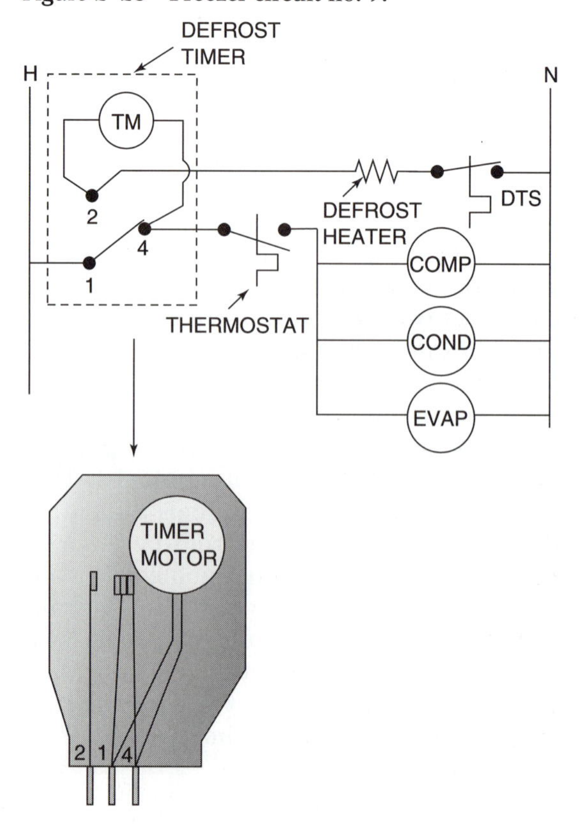

is different. Starting from Hot, current travels to terminal 1, then terminal 4, then to the timer motor, then to terminal 2, then to the defrost heater, then to Neutral. Even though the same amps flow through the timer motor and the heater, the timer motor works and the heater does not. Recall that when loads are wired in series, each takes a portion of the total available voltage drop. The voltage drop across each is proportional to their resistance. The resistance of the timer motor is *much* higher than the resistance of the heater. The timer motor takes maybe 118V, while there is only 2V drop across the heater. The current is sufficient to run the timer motor normally (a few tenths of an amp), but it is insufficient to cause the heater to produce any heat. The heater simply acts as a wire, connecting the downstream side of the timer motor to Neutral.

After a while, the timer motor causes the SPDT switch to switch to the 1–2 position. Then, the circuit through the timer motor is from Hot to terminal 1, then terminal 2, then to the timer motor, then to terminal 4, then to the compressor-condenser-evaporator in parallel, then to Neutral. Again, the compressor-condenser-evaporator combination presents very low resistance compared to the timer motor. The timer motor operates, and the refrigeration system is off. Notice that the timer motor does not care which way the current flows. It always rotates in the same direction, just like any other clock motor.

ELECTRONIC DEFROST TIMER

The defrost timers used in residential and small commercial freezers may all look about the same, but there can be major differences in the internal gears that provide wide variations in the frequency and duration of the "defrost." For example, a system in which the defrost time clock operates only when the compressor operates might use a four-hour frequency, compared with a continuously running time clock defrosting every eight hours. If the compressor runs 50% of the time, each system will "defrost" three times per day. In order to be able to match all of the different frequencies and durations of defrost timers that are used, the service technician would need to carry quite an assortment of defrost timers.

Figure 2–29 shows a new-design replacement defrost timer. Instead of the timing being determined by the internal gears, it is electronic in operation. There are two adjustment knobs provided. One allows setting the defrost duration from 10 to 35 minutes. The other allows setting of the defrost frequency for 4, 6, 8, 10, or 12 hours. With these adjustments, the technician can use this one timer to simulate the operation of any of the common timers.

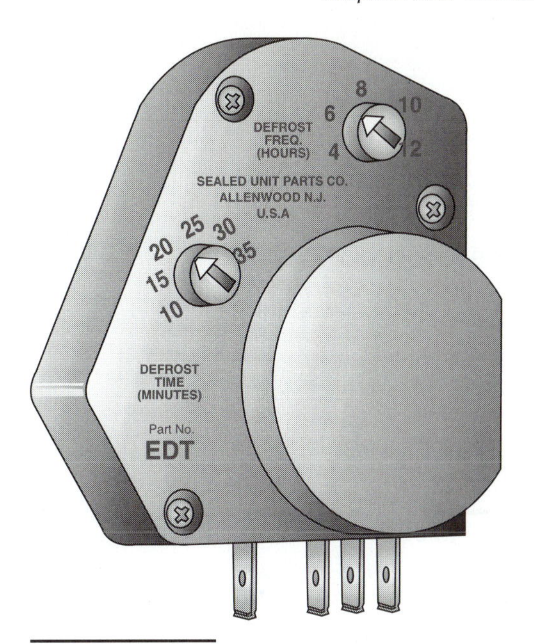

Figure 2–29 Electronic defrost timer.

HEAD PRESSURE CONTROL

Most refrigeration systems and some air conditioning systems must operate during the winter. If they are air-cooled systems, and they are using outside air as the condensing medium, operational problems can result when the outdoor temperature drops too low. The low temperature increases the capacity of the condenser dramatically, and it lowers the head pressure. If the head pressure drops too low, there is insufficient pressure differential between the high pressure and low pressure sides to push sufficient refrigerant through the metering device. **Head pressure control** is any method that is used to reduce the condensing capacity so that the head pressure does not drop too low. One method of head pressure control is called **fan cycling.** Figure 2–30 shows how a pressure switch that senses high side pressure can be used to turn the condenser fan off if the head pressure drops too low. In figure 2–30(a), the condenser fan will not turn on until the head pressure builds to 250psi. As soon as the condenser fan starts, the head pressure may drop to below 250psi. In order to prevent the condenser fan from cycling on and off, the cutout setting is well below the cut-in setting. In figure 2–30(a), once the condenser fan starts, it will

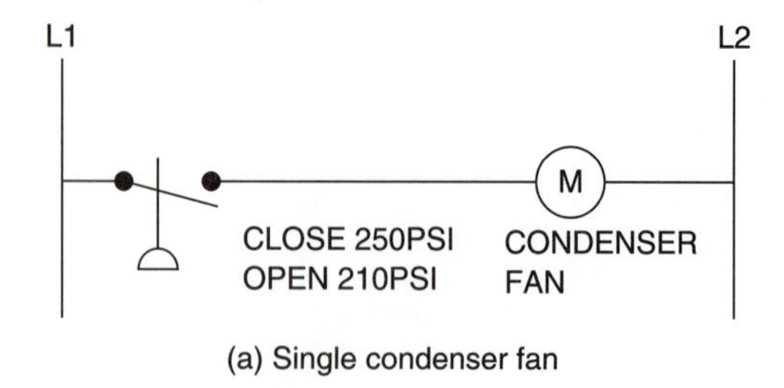

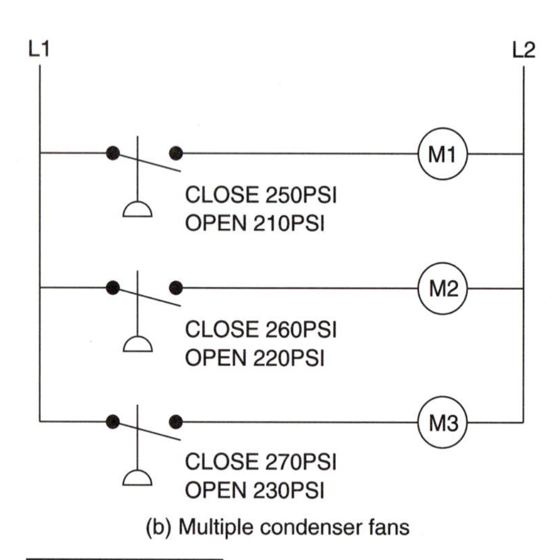

Figure 2–30 Condenser fan cycling. (a) Single condenser fan. (b) Multiple condenser fan.

not turn off again until the head pressure drops below 210psi.

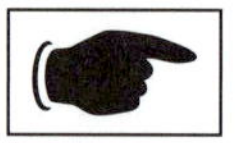

When replacing a head pressure switch, here's how to set the cut-in and cut-out.

1. Estimate the maximum ambient temperature around the condenser coil.
2. Add 30°F to that temperature, and convert the result to the corresponding pressure for the refrigerant in your system. Set the cut-in to that pressure.
3. Determine the lowest acceptable condensing temperature from the manufacturer's data, and convert that to a pressure. Set the cut-out to that pressure. If no manufacturer's data is available, set the cut-out to a pressure that corresponds with 65°F.

For multiple condenser fans as shown in figure 2–30(b), the pressure settings are slightly different for each fan motor. When the head pressure rises to 250psi, M1 is energized. If that is sufficient to keep the head pressure from rising to 260psi, then M1 will be the only condenser fan motor to run. If the ambient temperature later rises, the head pressure will also rise, bringing on M2 when the head pressure reaches 260psi, and M3 when (and if) the head pressure reaches 270psi. When the ambient temperature drops, the head pressure will also fall, turning off the condenser fan motors in the reverse order from which they turned on.

Electronic Head Pressure Control

Head pressure control can also be accomplished by modulating the speed of the condenser fan motor instead of motor cycling. Some electronic fan speed controllers are a combination of a pressure sensor and electronics. The unit is mounted to the top of the receiver or liquid line after the condenser where it will sense the high side pressure. Most will mount on a standard 1/4″ SAE flare connector or Shrader valve. As the sensed pressure drops below the set point of the controller, the electronics wired in series with the condenser fan motor will reduce the motor speed. Generally these speed controllers are limited to use with permanent split capacitor and shaded pole fan motors (the different types of motors are described in Chapters nine and twelve). Figure 2–31(a) shows how this type of electronic speed control is wired into the circuit. The operation of the electronic head pressure controller is shown in figure 2–32(b).

Figure 2–31 Electronic head pressure control.

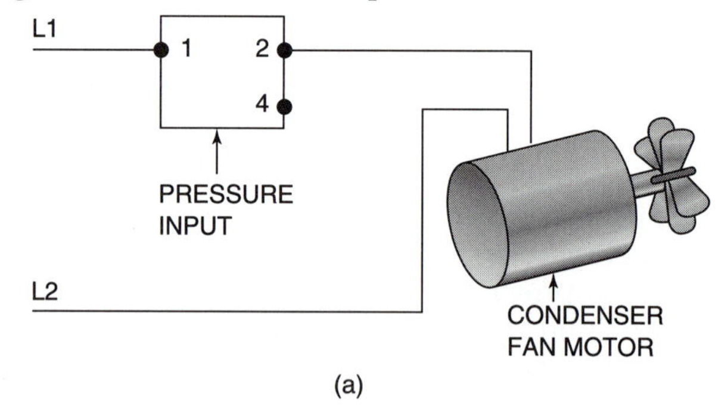

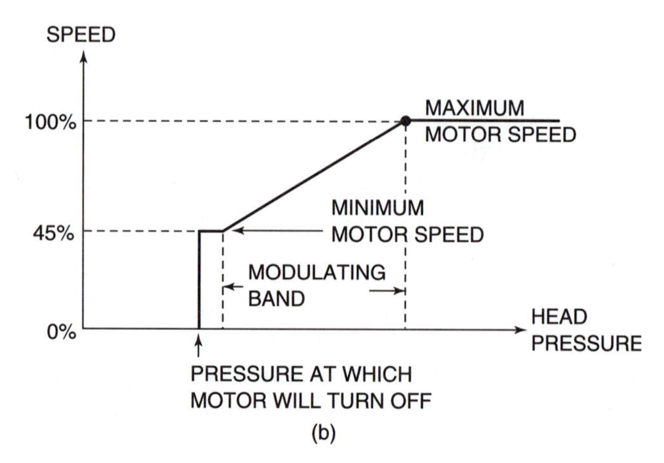

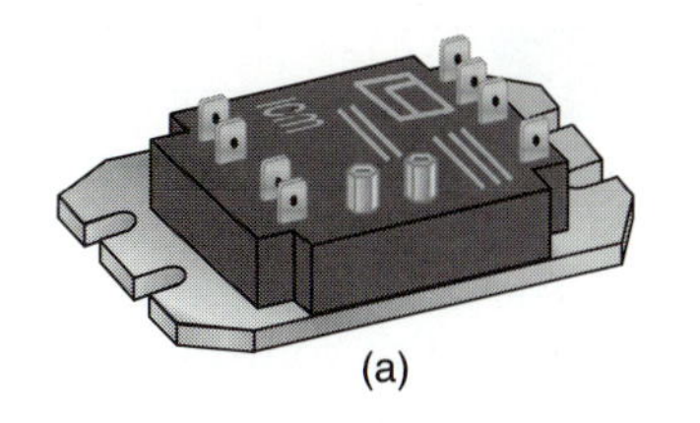
(a)

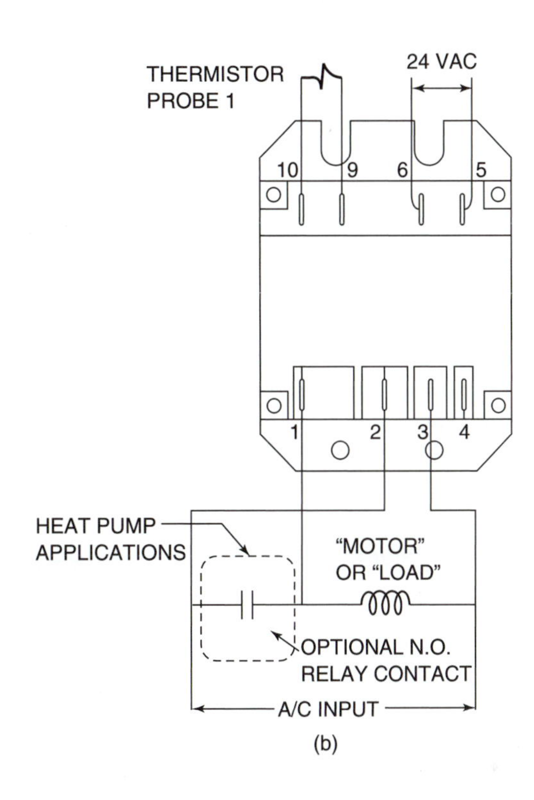

(b)

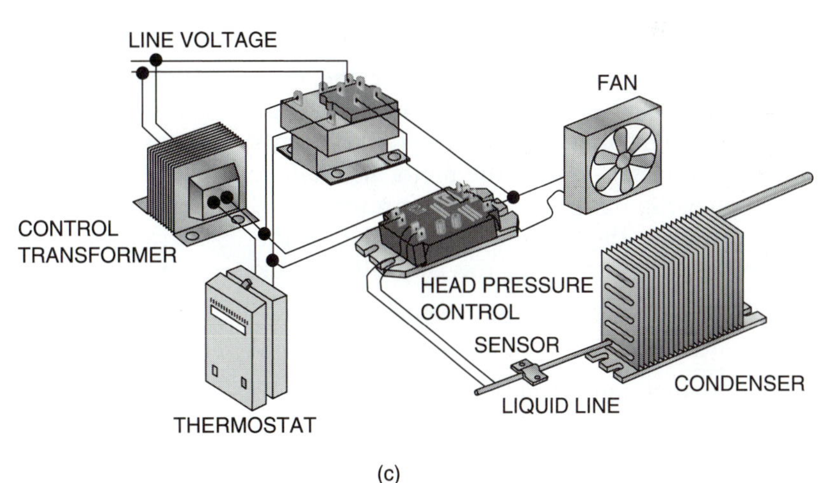

(c)

Figure 2–32 Electronic head pressure control.

Figure 2–31 shows another electronic head pressure control. Instead of sensing pressure, it uses a **thermistor** probe between terminals 9 and 10 to sense the liquid line temperature (which rises and falls to correspond with head pressure). A thermistor is simply an electronic device whose resistance varies with the sensed temperature. When 24V is applied to terminals 5–6, the controller will apply full voltage to the fan motor. After the motor starts (a few seconds), the output from the controller will be determined by the temperature sensed by the thermistor. As the temperature being sensed de-

ADVANCED CONCEPTS

The output of the speed controller in figure 2-32 is not really a reduced output voltage. If the voltage going to the motor were really lower than the rated voltage of the motor, the motor would fail. Actually, the output from this controller is not a reduced voltage, but part of each cycle of output voltage is cut off (see figure 2–33). This has the effect of reducing the motor speed, and if you measured the voltage output of this controller at part load with a standard volt meter, it would appear to be less than line voltage. But by "chopping off" a portion of each time cycle, overheating of the motor is not a problem.

creases, the output voltage decreases. A low temperature cut-off setting is provided to allow the technician to set the minimum condenser fan rpm. If the liquid line temperature falls below that point, the condenser fan will cycle off. There are two adjustments on the controller. One sets the initial full-voltage output time. The other sets the cut-out temperature. The physical wiring of another solid-state speed controller is shown in figure 2–32. Instead of sensing head pressure directly, it uses a thermistor to sense the temperature of the liquid line.

EXERCISE:
Redraw the circuit in figure 2–32(c) as a ladder diagram.

TROUBLESHOOTING

If the condenser fan doesn't run, either the condenser fan has failed or it is not receiving voltage from the speed controller. If correct voltage is available from terminals 1–3, then the motor has failed. If correct voltage is not present from terminals 1–3, it could be because the controller has failed, or the thermistor has failed, making the controller think that it is so cold outside that the condenser fan should be cycled off. The thermistor used on this type of controller is a **negative temperature coefficient** (as sensed temperature increases resistance decreases). Short the thermistor terminals (9–10), simulating a very hot outside air temperature. If the output voltage becomes the same as line voltage, the thermistor has failed.

Figure 2–33 Output signal from electronic speed controller.

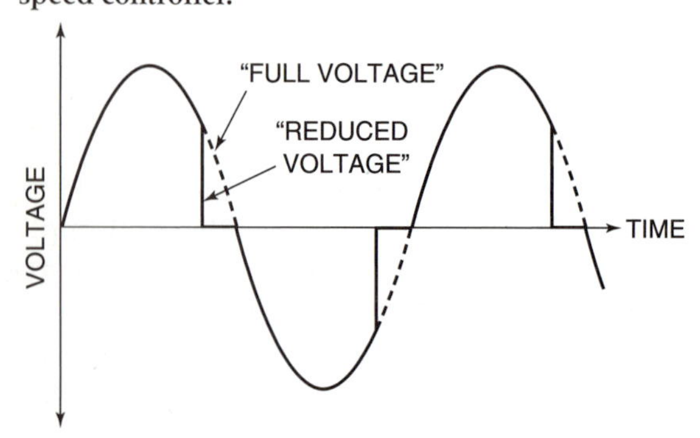

RELAY/CONTACTOR CIRCUIT NO. 1

Figures 2–34 and 2–35 show different types of relays and contactors. They work on the same principle of operation, but the contactor is used in circuits for larger loads. A schematic is shown in figure 2–36. The relay in figure 2–34(c) is called a **plug-in relay.** It plugs into a base, and the wiring is connected to the base instead of directly to the relay. You can determine which terminals are which by looking inside the clear case, and following the wires from each switch and the coil until they emerge at the base of the relay. The terminal arrangement shown is popular, but not universal. The principal of operation of a relay or contactor is as follows:

1. A voltage is applied to a coil of wire (the coil), creating a magnetic field.

2. The magnetic field draws a metal bar into the magnetic field, causing a switch (or multiple switches) to operate. The number of switches that operate are called **poles.** Thus, you can have a single-pole, two-pole, or three-pole control relay or contactor.

3. When the metal bar is drawn into the magnetic field, it may cause the switch to close (normally open) or it may cause the switch to open (normally closed). Contactor switches are always normally open.

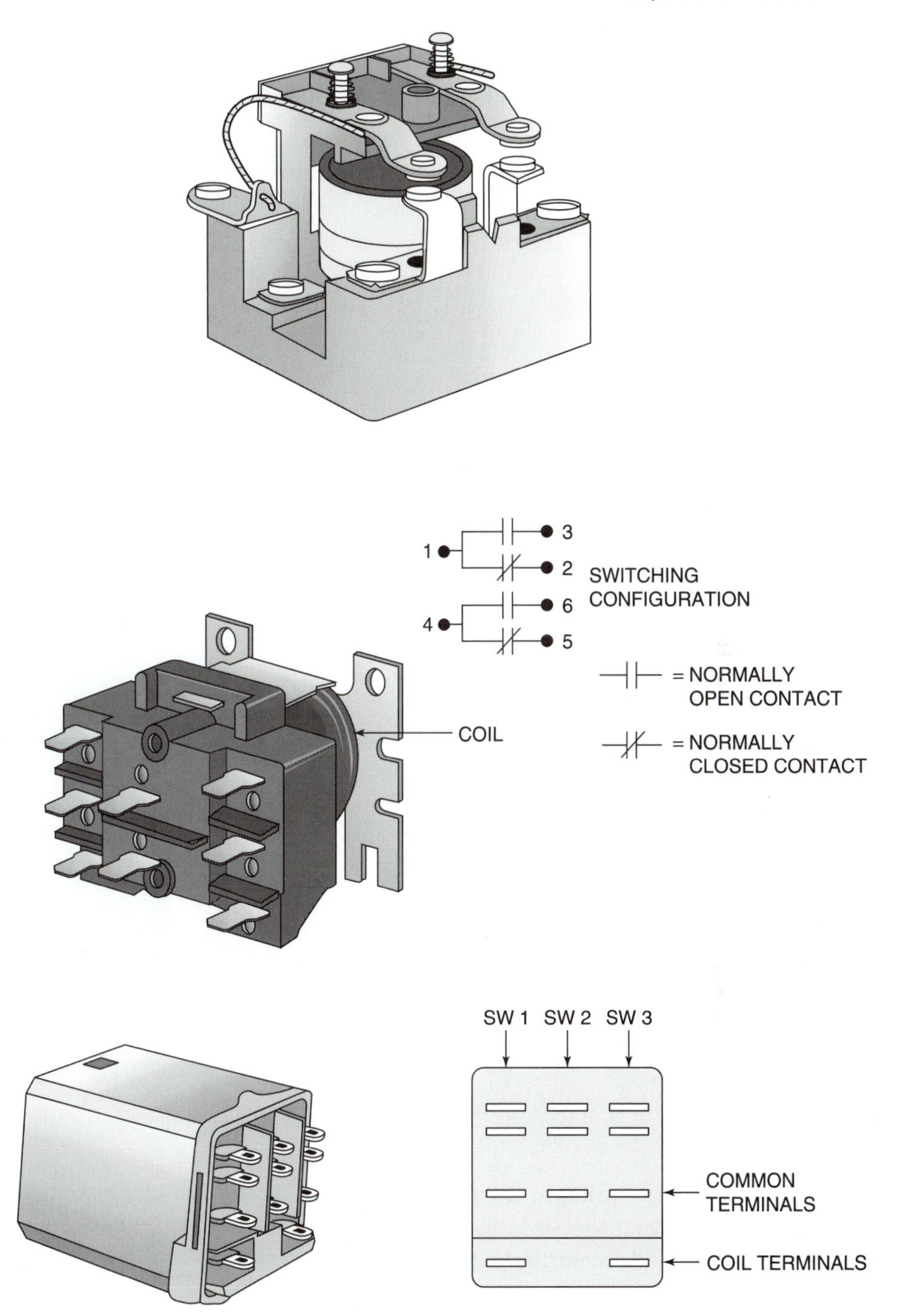

Figure 2–34 Control relays.
(a) Relay-open contacts.
(b) Relay-enclosed contacts.

A circuit using a single-pole contactor to start a motor is shown in figure 2–37. During the "off" cycle, as the box temperature rises, the corresponding pressure being sensed by the LPC also rises. When the cut-in setting of the LPC is reached, the switch closes, energizing the coil in the compressor contactor. The magnetic field produced by the coil closes the switch in the compressor circuit. There are two aspects of this device that sometimes cause confusion to the student:

1. The circuit through the LPC and the coil is *completely separate* from the circuit through the switch contacts and the compressor. The current flow through the LPC and coil is very low, while the current flow through the switch contacts and compressor may be as high as the amp rating of the switch (20, 30, 40 amps or more). There is *no electrical connection* between the coil of a contactor and the switch that is operated from its magnetic field. Later in the text, you will see contactors where the coil

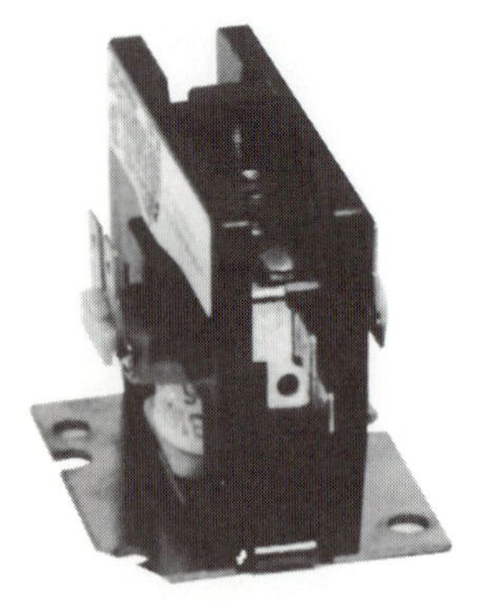

Single Pole
(25 & 30 amps)

Three Pole
(25, 30, 40 amps)

Single Pole with Bus Bar
(25 & 30 amps)

Three Pole
(50, 50, 75 amps)

Two Pole
(20, 25, 30 amps)

Figure 2–35 Contactors. *(Courtesy Joslyn Clark Controls, Inc.)*

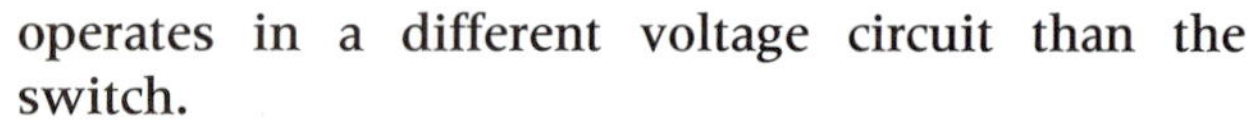

operates in a different voltage circuit than the switch.

2. Even though the ladder wiring diagram shows the switch contacts in a different place than the coil, both are actually located in the same physical device. Note that both the coil and the switch are labeled CC. In circuits with several control relays or contactors, you will be able to determine which switches are operated by which coils, because each switch will be identified the same as the coil that operates it.

Note that the LPC controls the operation of the compressor, but only indirectly. The contactor is being used as an intermediary device, so that the LPC does not have to carry the full current draw of the compressor. When a relatively light-duty switch is used to control the operation of a much larger device in the manner shown, the switch is called a **pilot-duty** device. Note that for a smaller compressor that draws far less current, the very same LPC could be used to control the compressor directly. Then, the very same LPC would be called a **line-duty** device.

EXERCISE:
Using correct symbols, draw two wiring diagrams for a 230V compressor, one with a line-voltage thermostat, and the other with a pilot-duty thermostat.

QUESTION:
If you were to take a contactor off the shelf and apply 230V to the coil, the switch would close. Don't try it, but explain why you would or would not get a shock if you touched the *switch* terminals.

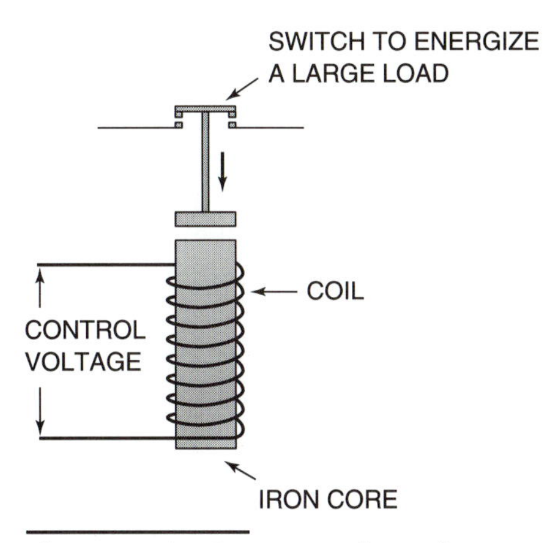

Figure 2–36 Contactor schematic.

> **EXERCISE:**
> Take a contactor off the shelf and wire the coil to a proper voltage source (read the proper voltage for the coil off the contactor itself). After the switch closes, use a volt meter to determine if there is any electrical pressure in the switches (measure between the switch terminals and the ground).

CONTACTOR CIRCUIT NO. 2

Contactor circuit no. 1 used a coil to operate a single switch. In 230V single phase applications, even though breaking one of the legs is sufficient to stop the operation of the load, many safety codes require breaking both legs (L1 and L2). Figure 2–38 shows how an unsuspecting technician can get shocked from compressor wiring, even if the contactor has been de-energized.

A compressor motor being controlled by a two-pole contactor is shown in figure 2–39(a). Figure 2–39(b) shows the exact same system, but in a pictorial wiring diagram. It sometimes confuses students because the two switches, which are actually right next to each other, are shown in different places on the ladder diagram. But remember, the

Figure 2–37 Single-pole contactor circuit.

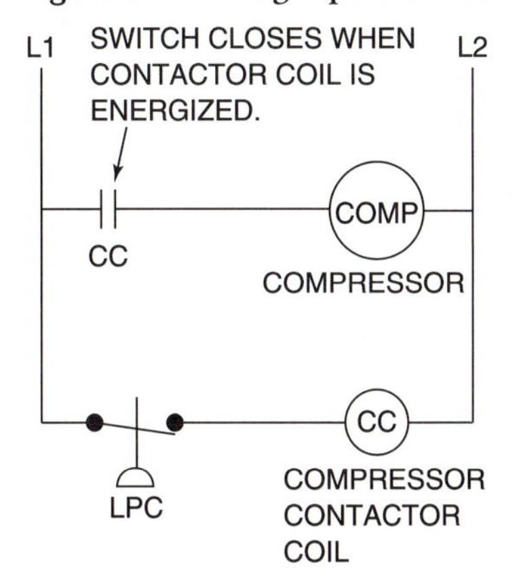

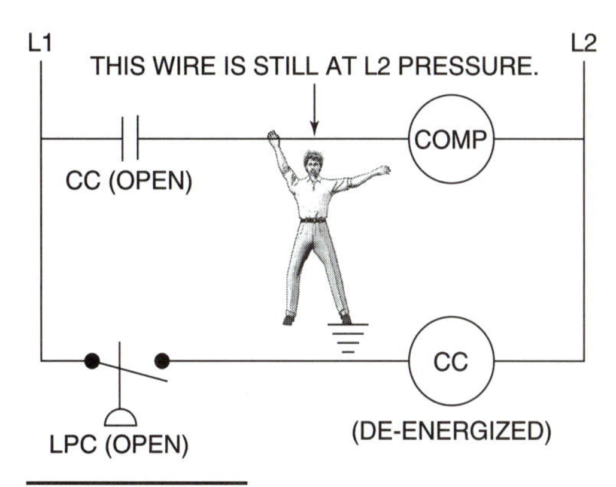

Figure 2–38 Unsafe condition caused by a single-pole contactor.

Figure 2–39 Two-pole contactor circuit.

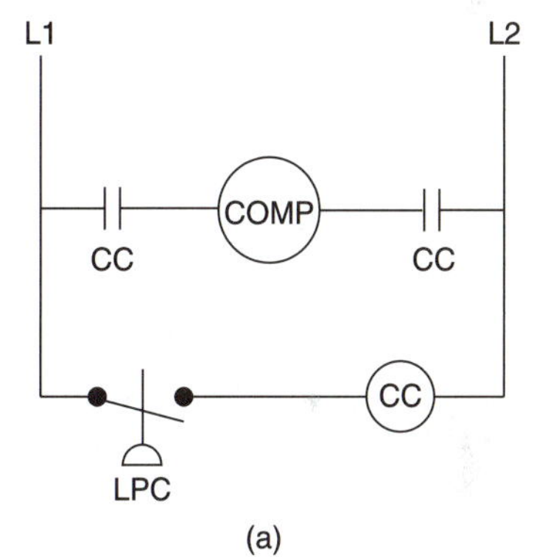

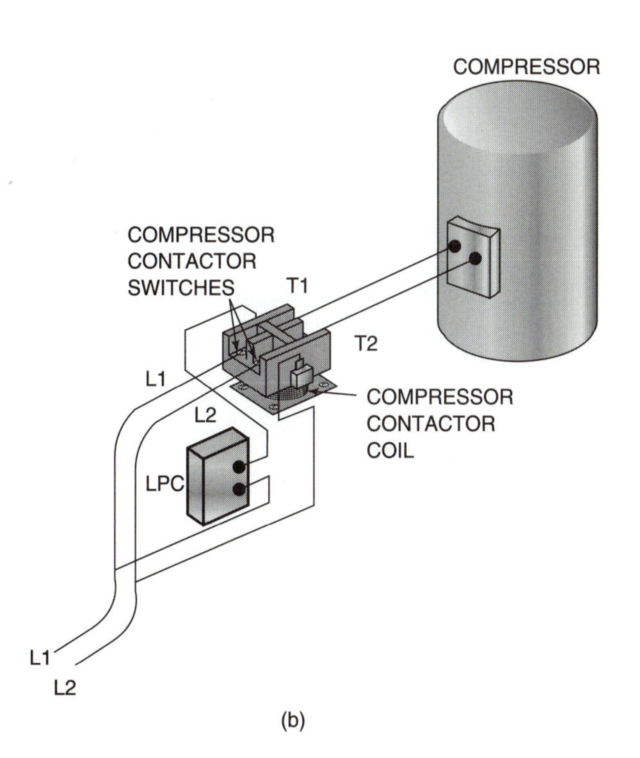

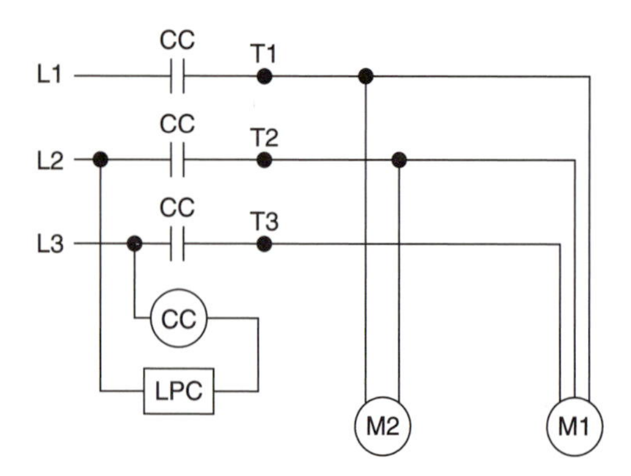

M1 = COMPRESSOR MOTOR
M2 = CONDENSER FAN MOTOR
CC = COMPRESSOR CONTACTOR
LPC = LOW PRESSURE CUT-OUT

Figure 2–40 Three-pole contactor circuit.

ladder diagram makes no attempt to show physical locations of devices. The common designation for terminals on a contactor are L1, L2, and L3 for the power coming into the switches, and T1, T2, and T3 for the power going out to the motor.

CONTACTOR CIRCUIT NO. 3

Figure 2–40 shows a wiring schematic for a large three-phase cooling system. Note that the contactor has three poles. Some electrical codes permit the use of a two-pole contactor to stop a three-phase load, but using a three-pole contactor is safer for the same reasons that the two-pole contactor is safer on a single-phase 230V system.

Figure 2–41 Multiple loads on a three-phase contactor.

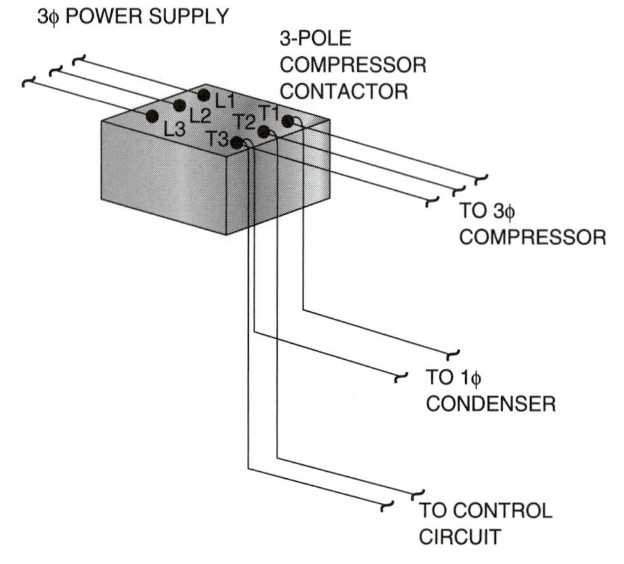

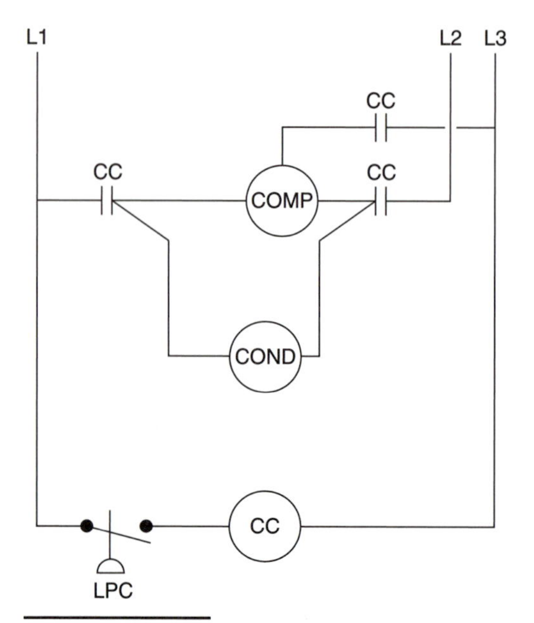

Figure 2–42 Three-phase ladder diagram.

The compressor has a 230V three-phase motor, but the condenser fan motors are 230V single-phase. The condenser fan motor and the control circuit each take their power from two of the three legs of the three-phase power supply. The contactor would be wired as in figure 2–41. Most contactors have multiple connectors on the T1, T2, and T3 terminals to allow for the connection of multiple loads.

A three-phase system can also be shown in a ladder diagram, as in figure 2–42).

CONTACTOR CIRCUIT NO. 4

Figure 2–43 shows a circuit for an even larger refrigeration system. Instead of operating the compressor and the condenser fan(s) from a single contactor, each of the condenser fans has its own contactor.

Each condenser fan motor is tied into a different pair of legs of the three-phase. This is done to attempt to balance (as much as possible) the amp draws on the three legs. If there were a severe load imbalance, it could cause a voltage imbalance on the three wires going to the compressor motor, resulting in motor overheating. Note also that each condenser fan contactor is controlled from a pressure switch called a **fan cycling switch.** Although the fan cycling switch senses pressure on the high-pressure side of the refrigeration system, it is different from an HPC. The HPC *opens* when the sensed pressure rises. The fan cycling switch *closes* on a rise in pressure, thereby bringing on the fans one at a time, as needed. The ladder diagram for this circuit is shown in figure 2–44.

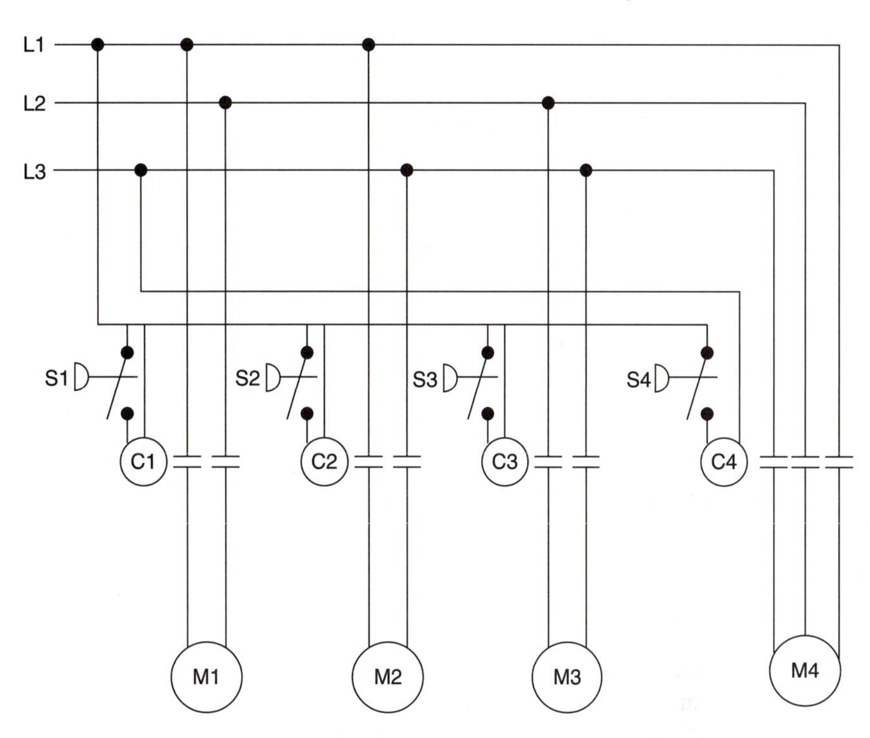

L1
L2
L3
S1
S2
S3
S4
C1
C2
C3
C4
M1
M2
M3
M4
M4 = COMPRESSOR MOTOR
M1, M2, M3 = CONDENSER FAN MOTORS
C1,C2, C3, C4 = CONTACTORS
S1, S2, S3 = FAN CYCLING PRESSURE SWITCH
S4 = LOW PRESSURE CUT-OUT

Figure 2–43 Circuit with multiple contactors.

Figure 2–44 Multiple contactor circuit ladder diagram.

L1
C4
L2
L3
C4
M4
C4
C1
M1
C1
C2
M2
C2
C3
M3
C3
C1
S1
C2
S2
C3
S3
C4
LPC

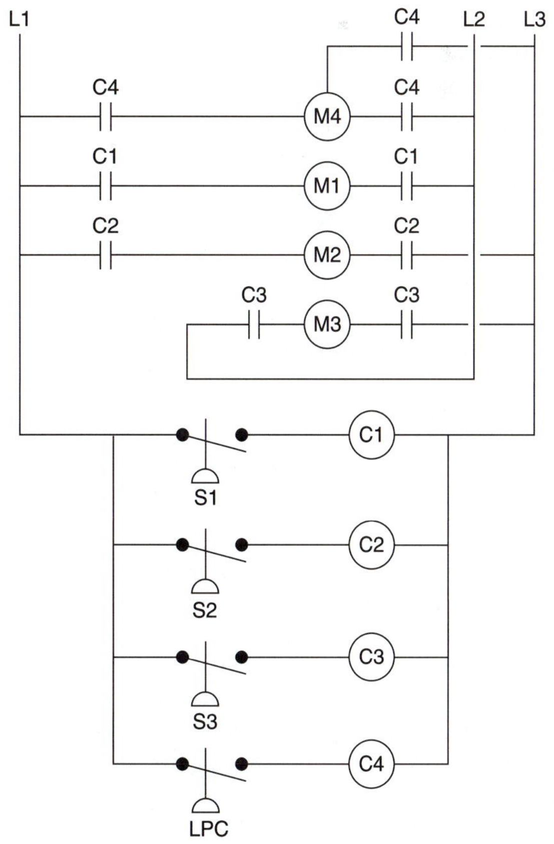

QUESTION:
Why is the symbol for the fan cycling switch the same as for the LPC? What is different about these switches?

TROUBLESHOOTING—BOX WARM, COMPRESSOR/CONDENSER NOT RUNNING

Assume that the circuit is a two-pole contactor, as in figure 2–39.

1. Check for voltage between T1 and T2. It is probably zero, as neither the condenser fan motor nor the compressor are running, and it is unlikely that both have failed.

2. Assuming a zero voltage reading between T1 and T2, there are only two possible reasons:

a. Power is available between L1 and L2 but the contactor switches are open.

b. No power is available to L1–L2.

Measure the voltage between L1 and L2. If it is zero, go to the fuses or the circuit breaker box to find out if a circuit breaker has tripped. If it is 230V, then we know that the contactor switches are not closing. There are only two possible reasons:

a. Voltage is not being applied to the contactor coil.

b. Voltage is being applied to the contactor coil, but the contactor coil has failed.

Measure voltage across the terminals of the contactor coil. If you read 230V, replace the contactor. If you read zero, you must determine why voltage is not getting to the coil. In this simple example, the only obvious reason is that the LPC is open. If you go to the LPC and measure 230V across the terminals, your suspicion will be confirmed. However, you are not ready to replace the LPC quite yet. There are two possible reasons that the LPC switch might be open:

1. The LPC has failed.

2. There is insufficient refrigerant in the system. The pressure actually is below the setting of the LPC, and the LPC is merely doing its job.

The service technician can measure the pressure in the system, but this is inconvenient because s/he probably does not have refrigerant gauges without making an extra trip to the truck. More likely, the service technician will place a jumper wire across the LPC switch. If the system then runs and cools normally, then replacement of the switch is all that is required. If the system runs, but still doesn't appear to be cooling normally, further mechanical troubleshooting is necessary.

In order to safely jumper across the switch, the power to the unit should be turned off before the jumper is placed across the switch terminals. Whenever a jumper wire is used, the wire used must be sufficiently heavy to carry the load. In this case, the jumper will only be carrying the amps to the compressor contactor coil. A light-duty wire will be sufficient.

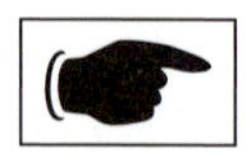

An alternative to placing a jumper across the switch is to mechanically close it. Simply find the part of the moving mechanism that is getting pushed on by the pressure-sensing element, and duplicate this force with your screwdriver. Most switches can be closed in this fashion.

Make sure you complete this last step of troubleshooting to ensure that the fix you implement is actually going to solve the problem. It is embarrassing and bad business to quote a price for repair to the customer based on a new pressure switch, and then have to tell the customer that you were only kidding, and that you need to repair a refrigerant leak and recharge the system too!

PUMPDOWN CIRCUIT

On some refrigeration systems, especially where the evaporator is located at a higher elevation than the compressor, an **automatic pumpdown** circuit is used to empty all of the refrigerant out of the evaporator before shutting down the compressor. This prevents liquid refrigerant from migrating to the compressor during the "off" cycle, which would have the undesirable effect of diluting the compressor oil. Figure 2–45 shows how an automatic pumpdown is accomplished. During normal "freeze" operation, the solenoid valve in the refrig-

Figure 2–45 Automatic pumpdown no. 1.

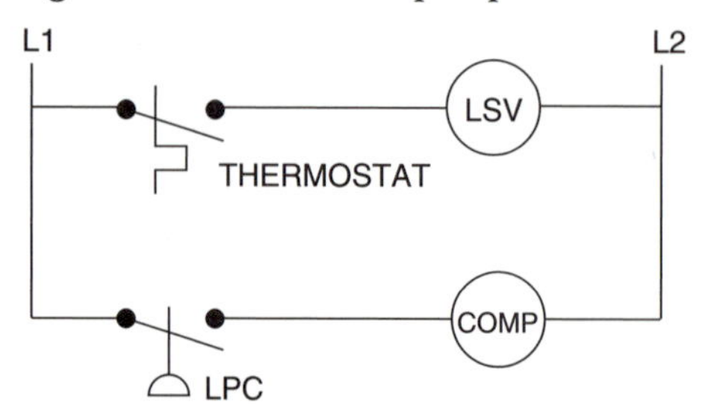

erant liquid line (**LSV**) and the compressor are both energized. When the thermostat becomes satisfied, it opens, de-energizing the LSV coil, causing the LSV to close. The compressor, however, continues to run. The liquid in the liquid line downstream of the LSV and the liquid in the evaporator are exposed to a decrease in pressure, and all the liquid vaporizes. When all of the liquid has vaporized, the suction pressure sensed by the LPC decreases rapidly, and the LPC opens. This de-energizes the compressor.

When the thermostat once again calls for cooling, it energizes the LSV causing it to open. The low side pressure rises, closing the LPC, and energizing the compressor.

Figure 2–46 shows the same sequence of operation, but for a larger compressor system. Instead of the LPC operating the compressor, it operates a compressor contactor coil that, in turn, energizes a large three-phase compressor.

QUESTION:
During the "off" cycle on an automatic pumpdown circuit, the compressor starts, operates for a few seconds, and then turns off. This may happen every five minutes, until finally the compressor turns on and stays on for a reasonable period of time. What is the problem?

SOLUTION: The liquid solenoid valve is leaking. During the "off" cycle, the refrigerant leaking through the valve causes the low side pressure to rise slowly, until it finally reaches the cut-in setting on the LPC. This causes the compressor to start, but because the LSV is still closed, the low-side pressure drops to the cut-out setting within a few seconds, and the compressor turns off. When the thermostat finally closes, then the LSV opens and the compressor stays on-line. ■

Figure 2–46 Automatic pumpdown no. 2.

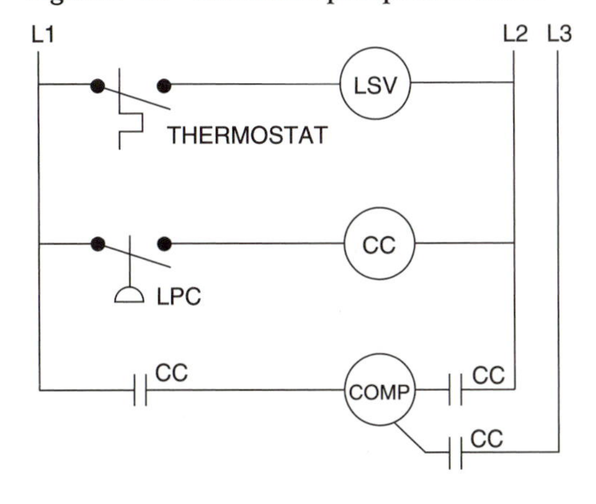

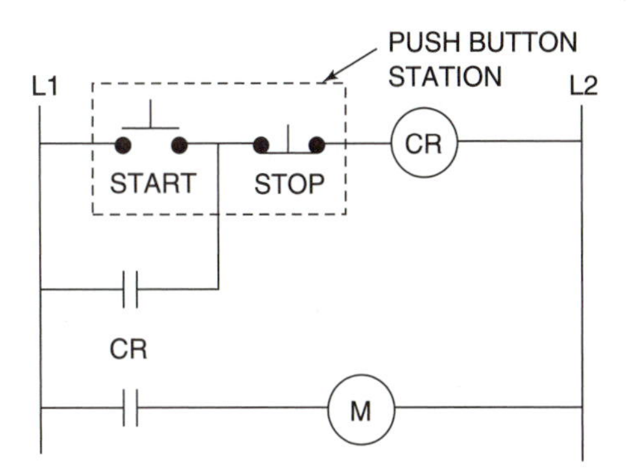

Figure 2–47 Seal-in circuit.

SEAL-IN CIRCUIT

Figure 2–47 shows an interesting circuit that has many variations in HVAC/R applications. The pushbutton station consists of two momentary switches. The normally open switch is labeled "Start" and the normally closed switch is labeled "Stop" (figure 2–48). It is sometimes referred to as a "Start-Stop Push Button Station."

When the "Start" switch is depressed, it energizes the control relay coil. When energized, the coil operates (closes) *two* switches. One switch starts a motor. The other switch is wired in parallel with the "Start" switch. When the "Start" switch is then released, it opens (remember, a normally open momentary switch only remains closed for as long as it is being pressed). However, even though the "Start" switch opens, the control relay coil remains energized through the contacts that have *sealed-in* the circuit around the "Start" switch. In order to shut this system off, the "Stop" button needs to be pressed to de-energize the control relay coil. This will drop out the control relay contacts, and the system will not restart when the "Stop" button is released.

With this system, if there is a momentary power outage, the motor will not restart automatically when power is restored. It must go through its normal start-up sequence (pushing the "Start" button).

Figure 2–48 Push-button station.

> QUESTION:
> The motor is running. The "Start" switch is open. What voltage would you measure across each of the following:
> 1. The "Start" switch.
> 2. The "Stop" switch.
> 3. The control relay coil.
> 4. The control relay contacts.
> 5. The motor.

Another way that a Start-Stop Push Button Station may be wired is with a contactor with an **auxiliary contact** (figure 2–49). To visualize the auxiliary contact, touch the three middle fingers on your left hand to the corresponding three fingers on your right hand. These are like the three poles on a contactor that carry current to the motor. Now, when you touch your middle three fingers together, also touch your pinky fingers together. Your pinky fingers are like the auxiliary switch on the contactor. It closes at the same time as the three contactor poles, but it is a **dry switch.** That is, the contactor coil makes the switch close, but the auxiliary switch is not connected electrically to any other part of the contactor.

When the "Start" button is pushed, the auxiliary switch on the contactor also closes. This switch is then used to seal-in the circuit around the "Start" switch.

LOCK-OUT RELAY

A **lock-out relay (impedance relay, reset relay)** is used to provide lock-out and remote reset in refrigeration, air conditioning, and other systems (figure 2–50). During normal operation, the normally closed contacts of the pressure controls and motor overloads short out the relay coil so that the contactor pulls in. If one of the pressure controls or overloads opens, the impedance relay coil is energized

Figure 2–49 Push-button station with contactor and auxiliary contact.

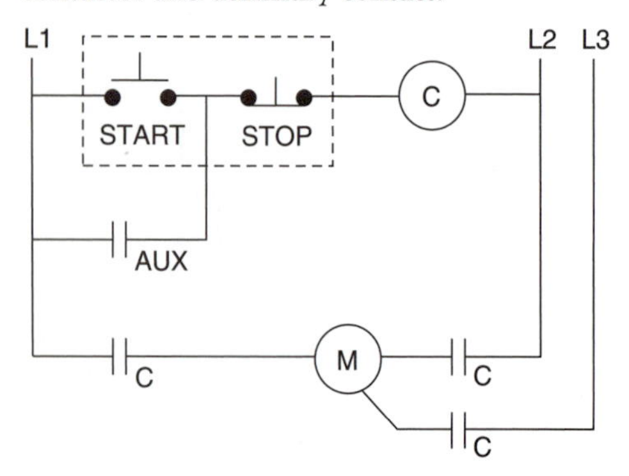

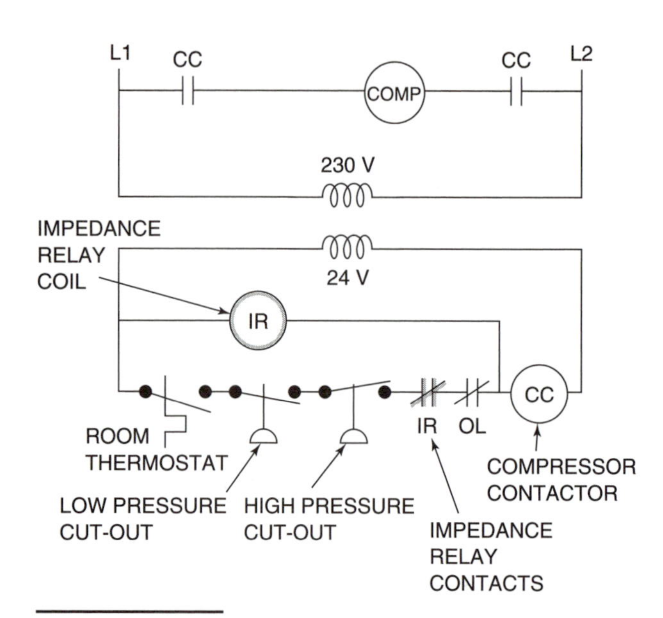

Figure 2–50 Lock-out relay.

in series with the contactor coil. When this occurs, the high impedance of the relay coil uses most of the voltage available, leaving insufficient voltage for the contactor coil and causing the contactor to drop out. As the impedance relay pulls in, its normally closed contacts open to keep the contactor out, even though the pressure control or overload (automatic reset) remakes.

On some models, the impedance relay has an additional normally open contact that can be wired to operate a check light to signal the homeowner that the system has shut down.

LEVEL CONTROL

Figure 2–51 shows a liquid level switch. It has three wires, so we know that this is an SPDT switch. Depending on which two wires are selected, it can be used to open on a rise in level, or close on a rise in level (figure 2–52). Most level switches are a simple mechanical float type device, similar to the float that controls a toilet valve. But

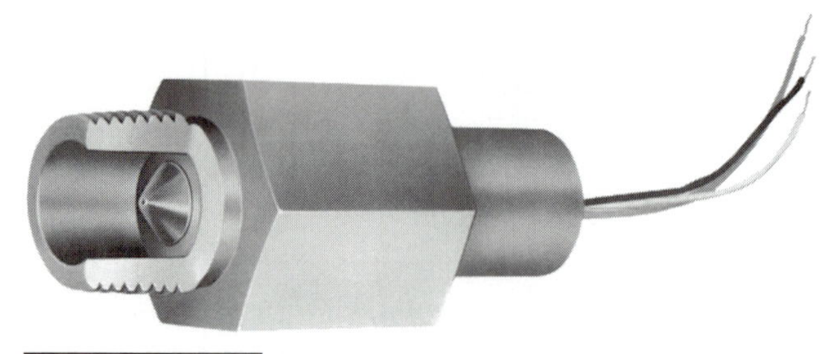

Figure 2–51 S-9400 series liquid level switch. *(Courtesy AC & R Components, Inc.)*

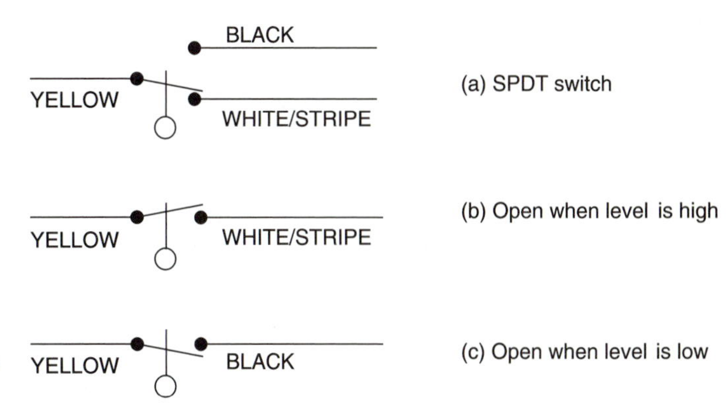

Figure 2–52 Level switches. (a) SPDT switch. (b) Open when level is high. (c) Open when level is low.

the level switch shown in figure 2–51 works on a principle of sending out a light signal to a mirror to detect whether liquid is present or not.

A level switch can be used to act as a high level alarm, low level alarm, or differential control using two switches. It can be installed in oil separators, oil reservoirs, liquid refrigerant receivers, suction accumulators, and compressor crankcase applications. Figure 2–53 shows an interconnection diagram for controlling the level between a certain range. When the tank is empty, both level switches are closed. This completes a circuit from L1, through the bottom switch, through the top switch, energizing the solenoid valve, back to L2. With the solenoid valve energized, it opens, allowing liquid to flow into the tank. As the liquid level rises, the bottom switch opens. However, the liquid solenoid remains energized because of the relay coil that was energized at the same time as the solenoid valve (this is a seal-in circuit). When the level reaches the top switch, it opens and the solenoid and relay coil become de-energized, closing the solenoid valve. When the liquid then drops below the bottom switch, the bottom switch closes again, opening the solenoid valve and filling the tank.

EXERCISE:
Redraw the level control diagram shown in figure 2–53 as a ladder diagram.

INTERLOCKS

Often, it is necessary to prevent one device from starting until after another device has started. Consider, for example, the circuit shown in figure 2–54(a). The water pump circulates chilled water through a building and a water chiller. When the room thermostat closes, it energizes both the contactor for the water chiller and the pump. This will work fine, so long as nothing goes wrong. But it is desirable to not start the water chiller until after the pump has started. In figure 2–54(b), the thermostat energizes the pump and a control relay. The chiller

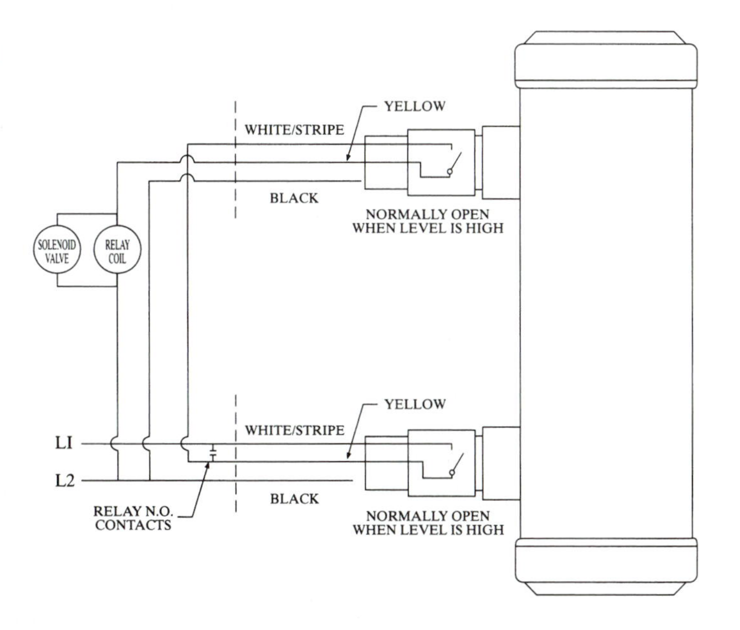

Figure 2–53 Interconnection diagram to control a liquid level within a range. *(Courtesy AC & R Components, Inc.)*

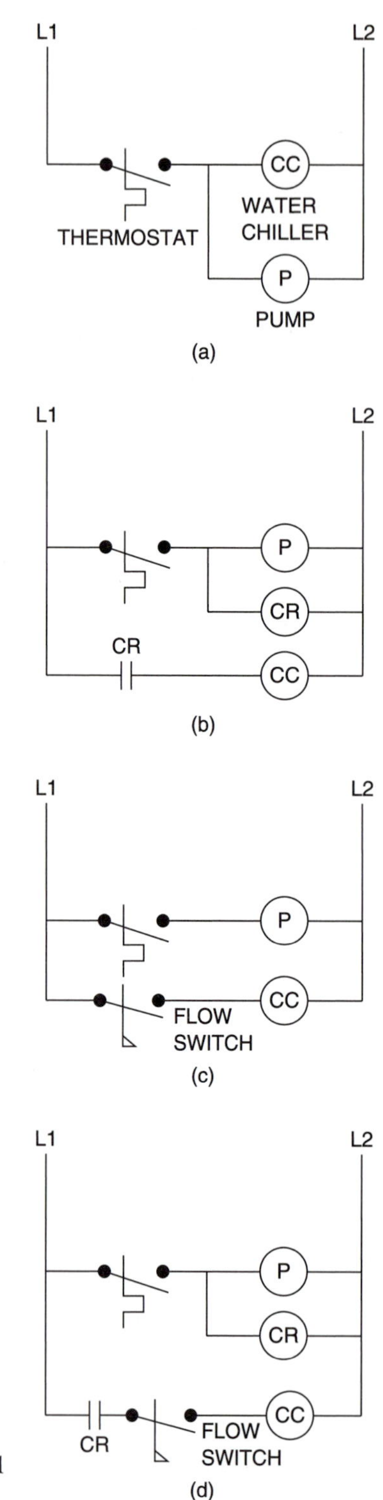

Figure 2–54 Electrical and mechanical interlocks.

contactor is energized through the control relay. This is called an **electrical interlock**.

Suppose that the pump fails. The control scheme in figure 2–54(b) would still allow the chiller to run, and the results would be catastrophic. The chiller would cause the still water in evaporator tubes to freeze, causing the evaporator tubes to rupture. There must be additional protection provided to prevent this damage. In figure 2–54(c), instead of using the control relay contacts to energize the water chiller, we have used a flow switch. The flow switch is located in the chilled water line, and closes only if the water is actually flowing in the pipe. This is called a **mechanical interlock**. And, if the designer of the system was really serious, both the mechanical and the electrical interlock may be used [figure 2–54(d)].

CHAPTER THREE

Standing Pilot Furnaces

A standing pilot furnace is one in which there is a small pilot flame burning continuously in a pilot burner. It is located physically close to the main burner. When main gas is allowed to enter the main burner, it is ignited by the pilot flame. This heats a heat exchanger. When the room-air side of the heat exchanger (sometimes referred to as the **bonnet**) is sufficiently warmed (usually around 130°F), a fan turns on to circulate room air over the heat exchanger. The fan is not turned on until the heat exchanger is warmed so that there will not be an initial blast of cool air that could be uncomfortable for the occupants.

The variations in standing pilot furnace control schemes are as follows:

1. Different methods are used to prove that the pilot flame is actually burning before main gas is allowed to be introduced into the main burner.
2. Different methods are used to determine when to turn on the fan that will circulate the room air over the heat exchanger.

FURNACE CIRCUIT NO. 1

Figure 3–1 shows the wiring for a central standing pilot gas-fired furnace for a residence. The modern trend is away from the use of standing pilot systems, in favor of electronic ignition furnaces (see Chapter fourteen). But it will be many years until all of the standing pilot furnaces have been replaced.

The devices in this diagram that have not been previously described are as follows:

1. A **transformer** is used to supply 24V.
2. A **gas valve** that has a 24V coil. When the coil is energized, the gas valve opens.
3. The thermostat is designed to operate in the 24V circuit. It is called a **low-voltage thermostat.** 24V is a standard low voltage in heating and air conditioning systems.
4. A **pilot safety switch** is located at the pilot flame. If it does not sense heat (indicating that the pilot flame is present), the contacts will open.
5. A **limit switch**, located in the **bonnet** of the furnace. The bonnet is the side of the heat exchanger through which the room air passes. If the temperature of the air reaches approximately 180°F, something is wrong and the limit switch will open.
6. A **bonnet switch** located in the bonnet of the furnace.

Transformer

The transformer (figure 3–2) works on the principle of electric induction. You will recall that when current passes through an electric wire, it creates a magnetic field. When the wire is wound into a coil, the magnetic field increases with each additional turn. You will also recall that when a moving mag-

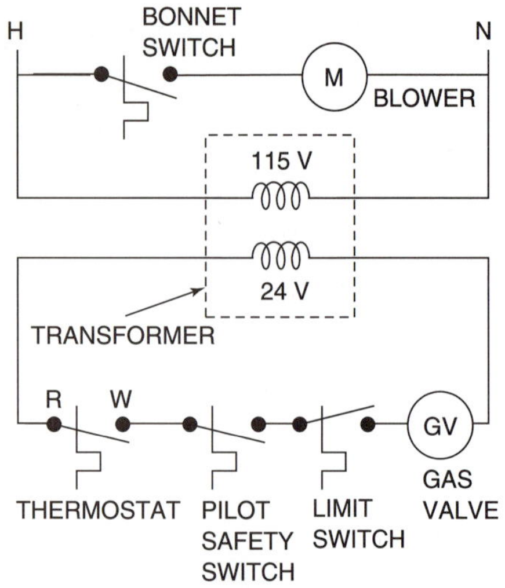

Figure 3–1 Furnace circuit no. 1.

netic field passes through a coil of wire, a voltage is induced in that coil of wire. In the transformer, there are two coils of wire. The first coil is called the **primary winding.** It is in the 115V circuit. Because the 115V circuit is alternating current, the magnetic field around the primary winding builds and collapses 60 times each second. A second coil (the **secondary winding**) is placed physically close to the primary coil. As the magnetic field from the primary coil builds and collapses, it induces a voltage in the secondary coil. Depending on the number of turns in the primary and the secondary, the voltage in the secondary may be higher or lower than the voltage in the primary.

When the secondary voltage is higher than the primary, it is called a **step-up transformer.** When the secondary voltage is lower than the primary, it is a **step-down transformer.**

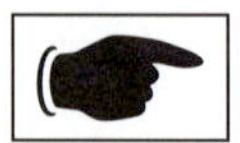

As more loads are added in parallel on the secondary side, the current increases in both the secondary winding and the primary winding. If there is a short circuit in the circuit being powered by the secondary winding, the most likely result is that the primary winding will burn out. This is because the primary winding is designed for lower current than the secondary winding, and therefore is constructed from thinner wire than the secondary winding.

Gas Valve

The gas valve in this circuit is called a **single-function** gas valve (figure 3–3). Other gas valves are **combination gas valves** (figure 3–4). When the coil of a single-function gas valve is energized, the valve will open, allowing main gas to flow, without checking to see if a pilot flame is present. The single-function gas valve must be used with some other device (such as a pilot safety switch) to check

ADVANCED CONCEPTS

The ratio of the primary voltage to the secondary voltage is equal to the ratio of the turns. For example, if the primary has 100 turns, and the secondary has 20 turns, the voltage ratio will be 100/20 = 5. If the voltage on the primary is 120V, then the output voltage from the secondary will by 120V/5 = 24V. The formula for this relationship is:

$$\frac{Turns_{pri}}{Turns_{sec}} = \frac{Volts_{pri}}{Volts_{sec}}$$

where

$Turns_{pri}$ = number of turns in the primary coil
$Turns_{sec}$ = number of turns in the secondary coil
$Volts_{pri}$ = voltage applied to the primary coil
$Volts_{sec}$ = voltage output from the secondary coil

The current in the primary and the secondary follow the same ratio, but inversely. In the above example, the 24V secondary coil would produce five times more current than the primary coil. The power output from the secondary will theoretically equal the power input to the primary, as follows:

$$Volts_{sec} \times Amps_{sec} = Volts_{pri} \times Amps_{pri}$$

In actual practice, the power output from the secondary will be slightly lower because some power is wasted because of the heat being produced by the wiring.

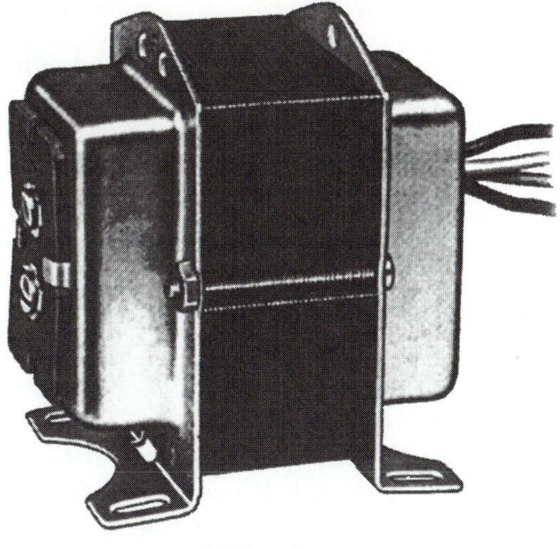

(a) Transformer

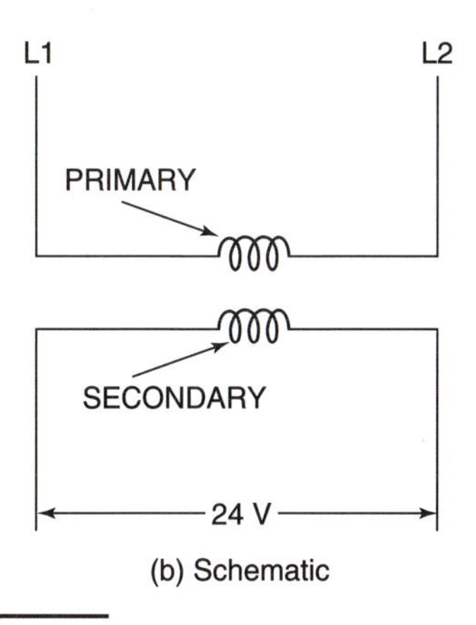

(b) Schematic

Figure 3–2 Control transformer. (a) Transformer. (b) Schematic. *(a. Courtesy of Honeywell, Inc.)*

whether or not a pilot flame is present before the main gas valve is allowed to open.

The combination gas valve actually consists of two valves in series (figure 3–5). It uses a **thermocouple** (figure 3–6) to sense heat from the pilot flame. The thermocouple converts this high temperature into a small electrical voltage (12–18mV, where 1000mV = 1V). The thermocouple output is used to energize a small coil that holds the pilot valve open. In a system using a combination gas valve, the pilot safety switch is not used. The pilot valve is pushed open manually to allow pilot gas to flow. While the pilot valve is being manually held open, the pilot flame must be lit manually. After approximately one minute, the thermocouple generates sufficient voltage to hold the pilot valve open. The button that was manually pressed (or pulled, or pushed or twisted) may then be released, and the pilot valve will stay open for as long as the thermocouple stays hot from the pilot flame. Neither the thermocouple nor the coil on the pilot safety valve shows up on the wiring diagram.

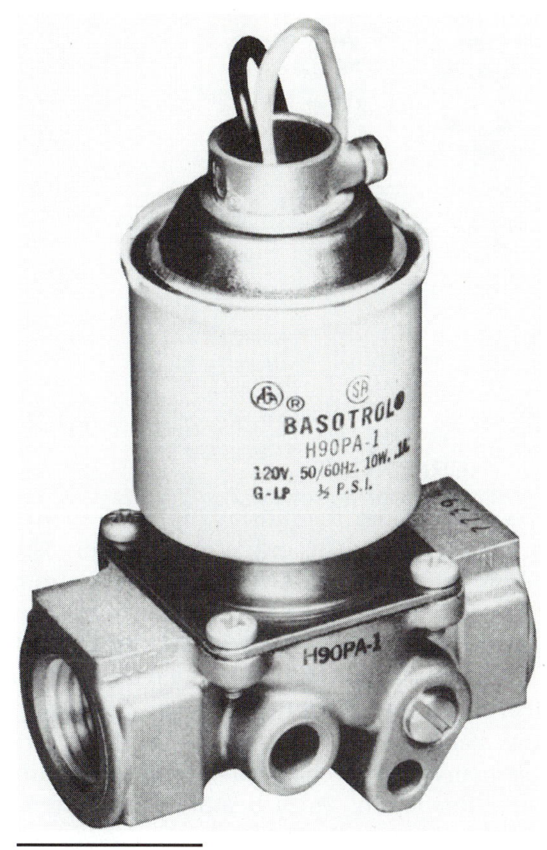

Figure 3–3 Single-function gas valve. *Courtesy of Johnson Controls, Inc.)*

If the main valve in a combination gas valve gets energized while there is no pilot flame, the main valve will open but no gas will flow. Main gas cannot flow unless *both* the pilot valve and main valve are open.

Figure 3–4 Combination gas valve. *(Courtesy of White-Rodgers Div., Emerson Electric Co.)*

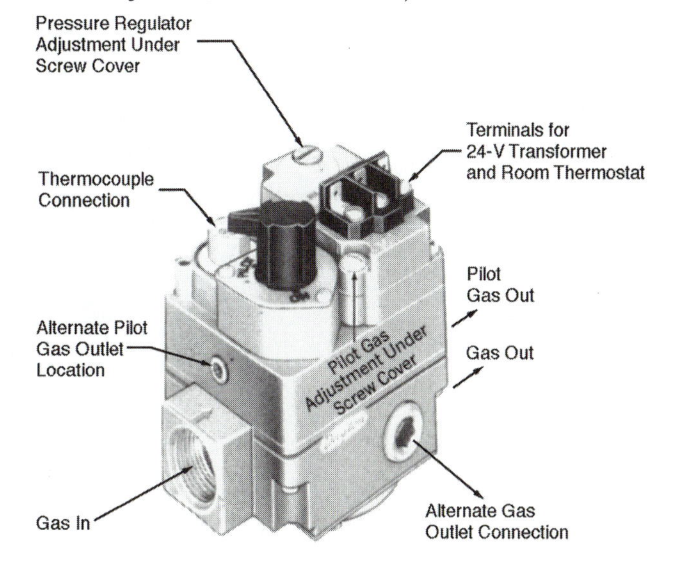

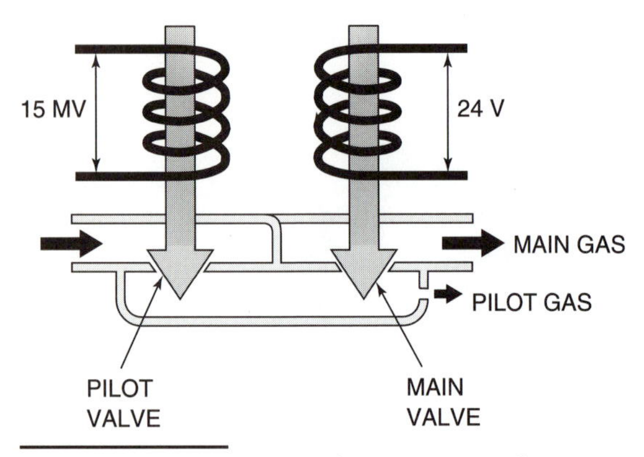

Figure 3–5 Cutaway-combination gas valve.

Room Thermostat

The low-voltage room thermostat (figure 3–7) has two terminals, R (for red) and W (for white). The room thermostat is located in the room, remote from the furnace. It is connected to the furnace by a pair of thermostat wires inside a common sheath (figure 3–8). Thermostat wire is commonly referred to as **two-wire, three-wire, four-wire,** etc., depending upon how many wires are inside the sheath. There may be as many as ten conductors, each with a different color insulation. For a heating-only system, two-wire would be used.

Pilot Safety Switch

The pilot safety switch is usually combined with the pilot burner (figure 3–9). The heat from the pilot flame impinges on a bimetal element that then bends, closing a switch (the pilot safety switch). It is popular on Carrier, Payne, Day Nite, and Bryant furnaces. Most other standing pilot furnaces use the combination gas valve system with a thermocouple.

Another type of pilot safety switch that is sometimes used is the **thermopilot relay** (figure 3–10). It uses a thermocouple, like the combination gas valve that uses a thermocouple to hold open a valve. But with the thermopilot relay, the thermocouple holds a switch closed. When you manually press the button, you close the pilot safety switch and, at the same time, allow pilot gas to flow. After the pilot flame has been lit and allowed time to heat the thermocouple, the button is released. If the pilot flame then goes out, the switch will be pushed open by an internal spring.

Limit Switch

The **limit switch** is a safety control that should remain closed during the entire operating life of the furnace, unless there is some malfunction that causes

Figure 3–6 Thermocouple. *(b. Courtesy of Honeywell, Inc.)*

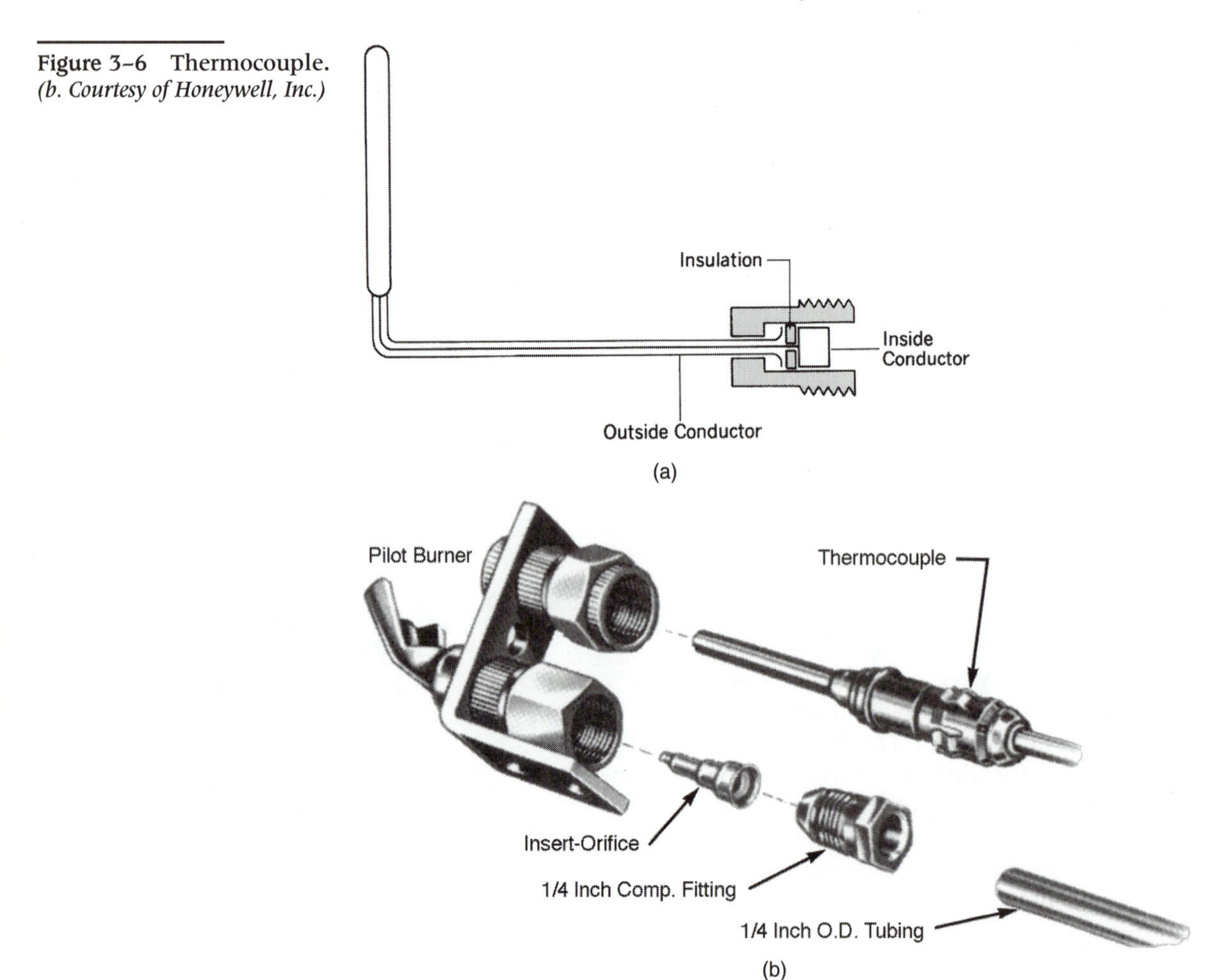

Figure 3–7 Low-voltage room thermostat. *(Courtesy of Honeywell, Inc.)*

it to open. Several types of limit switches are shown in figure 3–11.

The limit switch is a thermostat that senses the temperature on the room-air side of the heat exchanger. Normally, this temperature should not exceed 130 to 150°F. However, if the flow of room air is restricted (or stopped as in the case of a failed fan motor), or if the furnace fires at higher than its normal Btu rating, the room air temperature side of the heat exchanger will rise. If the limit switch reaches its set temperature (usually 180 to 200°F), it will open, de-energizing the gas valve, and extinguishing the main flame. Most limit switches are **automatic reset.** That is, when the sensing portion of the limit switch cools, the contacts will re-close automatically. Some limit switches are **manual reset.** Once tripped, they will not re-close until someone pushes on the reset button.

Bimetal Fan Switch

The bimetal fan switch senses the same bonnet temperature as the limit switch. Its appearance can also be the same as a limit switch. However, its function is the opposite of a limit switch. A bimetal fan switch *closes* when the bonnet temperature rises. It is usually set for an "On" temperature of approximately 130°F, and an "Off" temperature of approximately 100°F.

Figure 3–8 Thermostat wire.

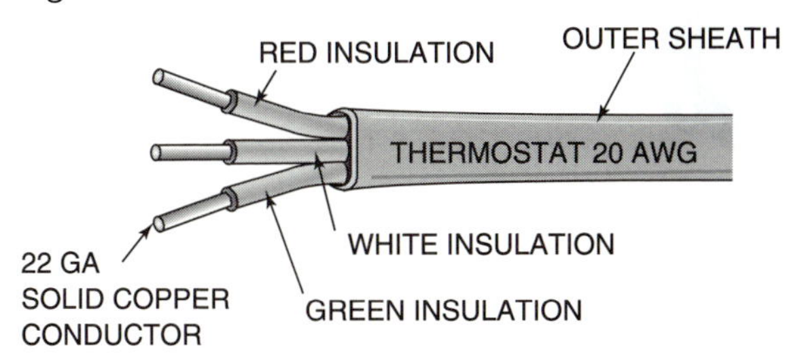

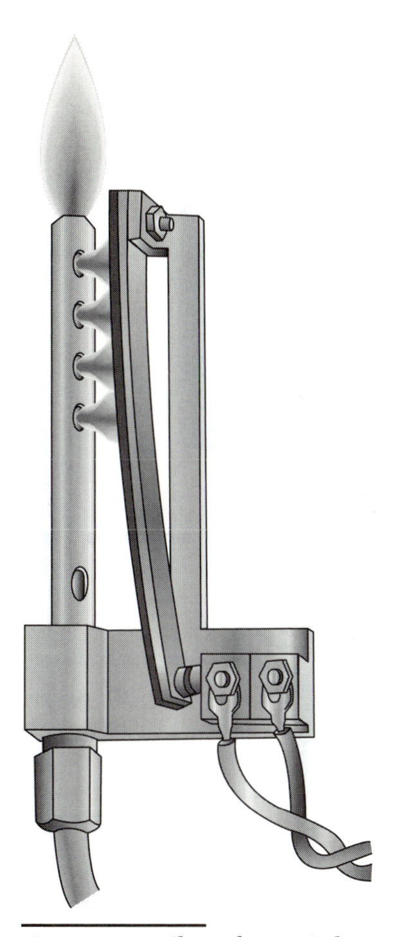

Figure 3–9 Pilot safety switch.

Sometimes, as a manufacturing cost savings, the limit switch and the fan switch are combined together into a single unit, sharing a single sensing element. This is sometimes referred to as a **fan-limit switch** (figure 3–12). Even though the two switches share the same sensing element (figure 3–13), the switches are completely separate electrically, and each has its own set point.

Note that the bimetal element movement is what causes the round face plate to rotate. Do not attempt to rotate the face plate by hand. You may ruin the calibration of the device, so that it will not actually operate at temperatures that match the set temperatures.

Sequence of Operation

When the room gets cold, the room thermostat closes, completing a circuit through the gas valve. If the pilot flame is lit, the pilot safety switch will be closed, and the solenoid coil of the gas valve will be energized. The gas valve opens, admitting main gas into the combustion chamber where it ignites from the standing pilot flame. After approximately one

(a)

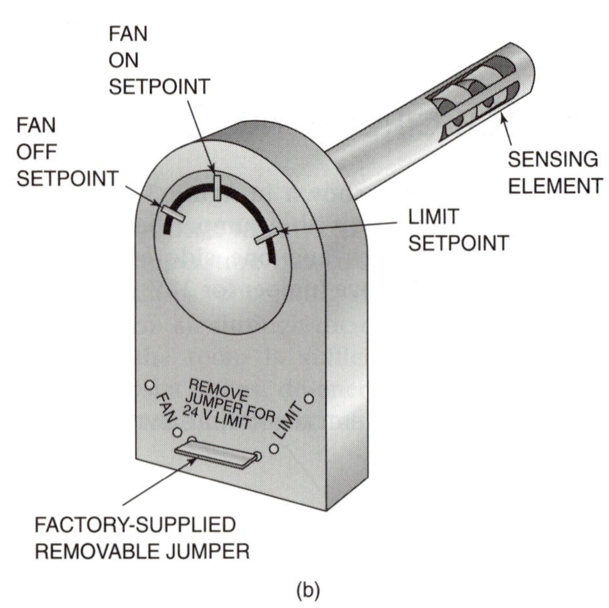

(b)

Figure 3–16 Fan-limit switch with jumper. *(a. Courtesy of Honeywell, Inc.)*

ure 3–17 shows how this limit switch would be used when the jumper is not removed. If the limit switch opens, it will de-energize the transformer. The 24V on the secondary of the transformer will be lost, and the gas valve will be de-energized. In this way, the limit switch in the 115V circuit shuts down the gas valve just as surely as it would if the limit switch were wired in series with the gas valve.

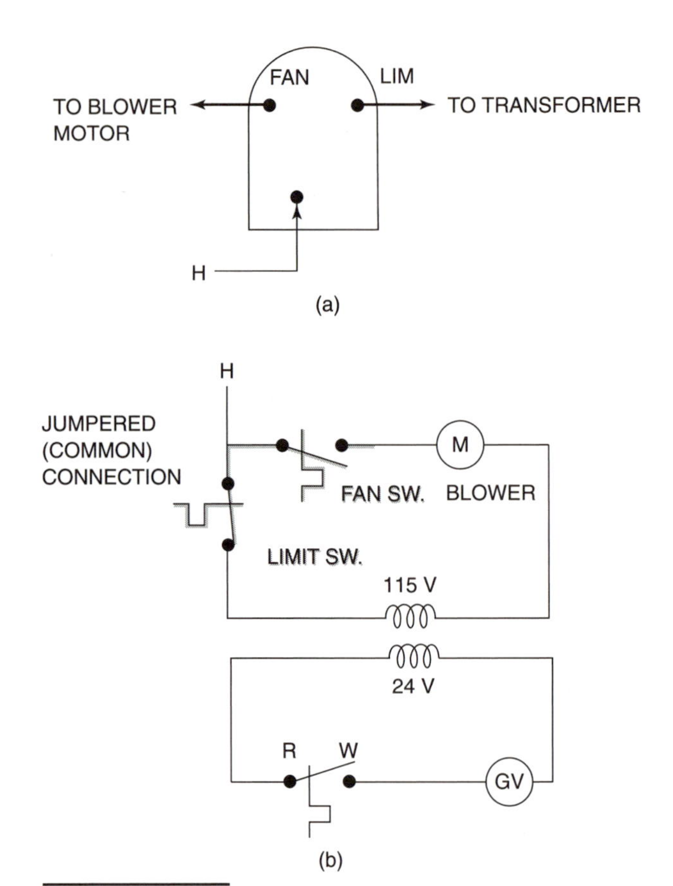

Figure 3–17 Jumper in place.

If the jumper on the fan-limit switch is removed, then the fan switch and the limit switch may be wired independently into the 115V and 24V circuits, as in figure 3–18.

EXERCISE:
Redraw the furnace components shown in figure 3–19. The fan/limit switch is the type shown previously in figure 3–16. Add the necessary wiring, with the limit switch operating in the line voltage circuit. Repeat, with the jumper on the fan/limit switch removed, and the limit switch operating in the 24V circuit.

FURNACE CIRCUIT NO. 3

The circuit shown in figure 3–20 shows yet another way that the operation of the blower motor is controlled. It uses a **time-delay fan switch** (figure 3–21). This device contains a bimetal switch and a 24V heater. The bimetal is wrapped with **Nichrome** wire. Nichrome is a material that has a significant resistance, and when a current is passed through it, it heats up because of the I × R drop. There is no electrical connection between the heater wire and

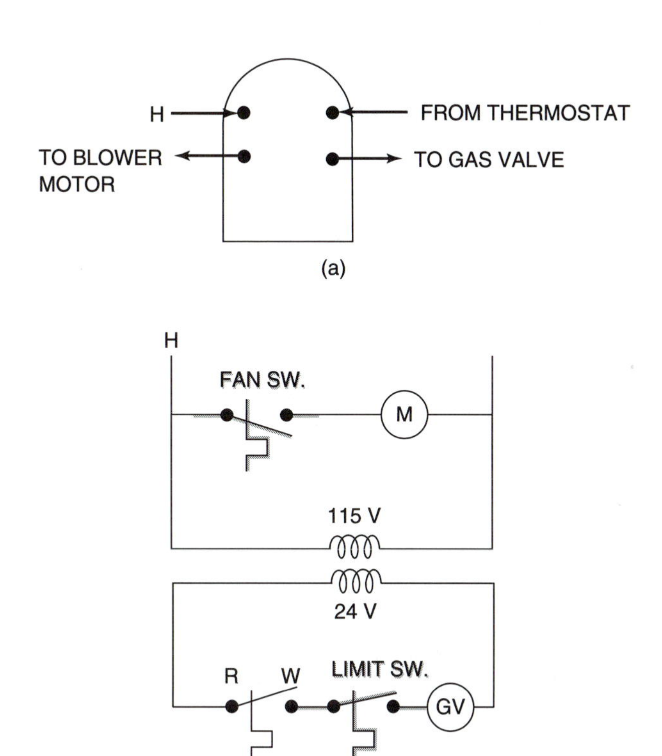

Figure 3–18 Jumper removed.

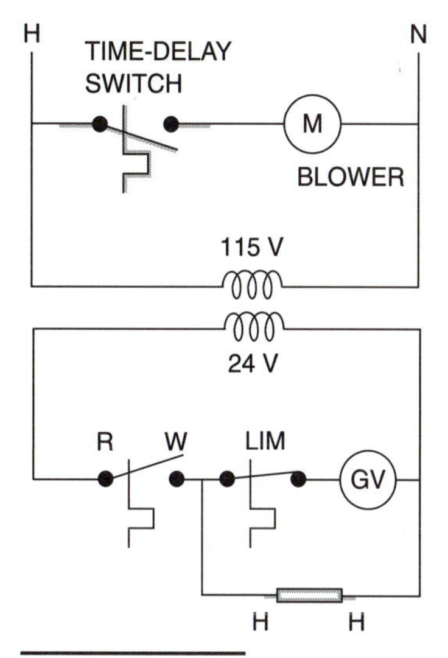

Figure 3–20 Time-delay fan switch circuit.

the switch. Note that the ladder diagram in figure 3–20 shows the heater and the switch in completely different parts of the diagram, even though they are contained within the same device (figure 3–21). Remember, the ladder diagram does not attempt to show you any of the physical relationships of the locations of electrical devices.

The heater is wired in parallel with the gas valve. Whenever the gas valve becomes energized, the heater will also become energized. Approximately one minute after the heater wire is energized, it will have generated enough heat to cause the bimetal switch to close.

Other variations of the time-delay fan switch are shown in figure 3–22. In figure 3–22(a), a bonnet-sensing temperature element has been added. This switch will close after a fixed time delay *or* whenever the bonnet temperature reaches the set point of the switch, whichever occurs first. In figure 3–22(b), a limit switch has also been added, sharing the same bimetal sensing element. There is no electrical connection between the limit switch, the fan switch, or the heater element.

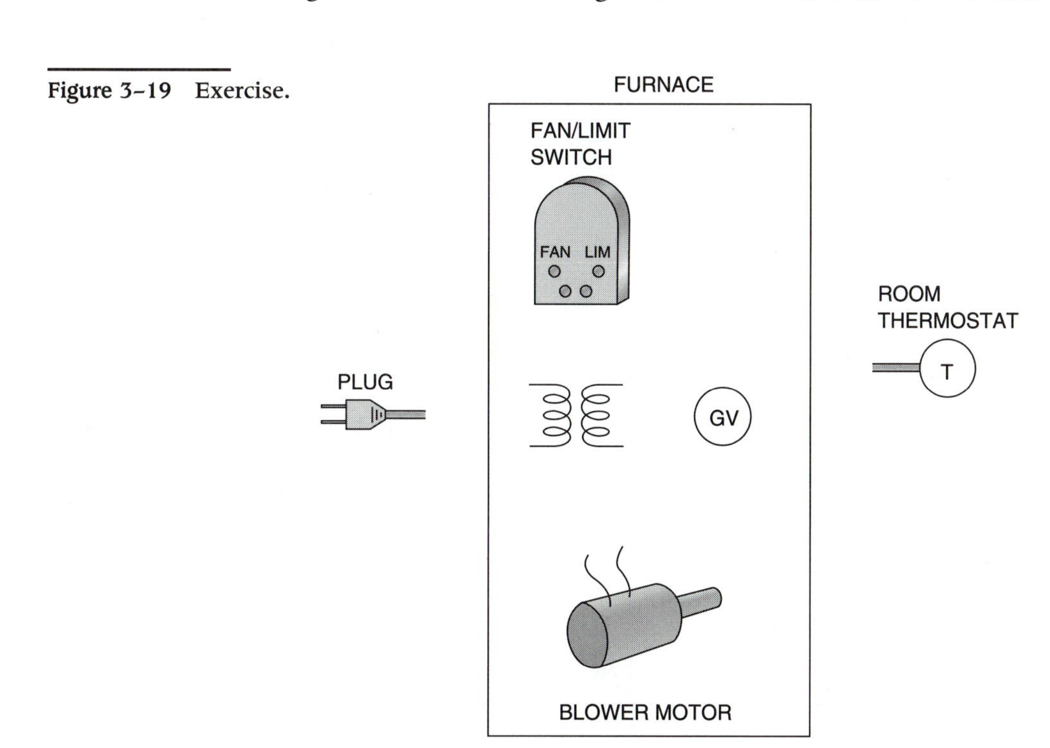

Figure 3–19 Exercise.

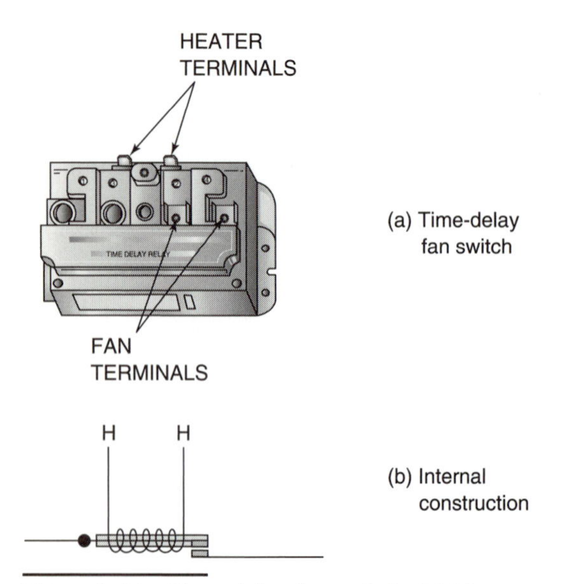

Figure 3–21 Time-delay fan switch. (a) Time-delay fan switch. (b) Internal construction.

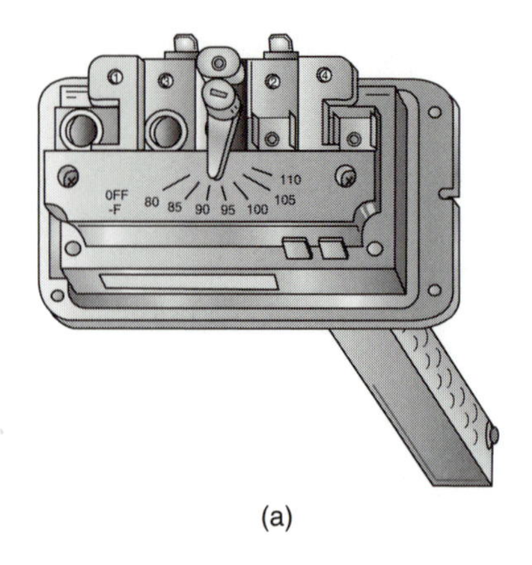

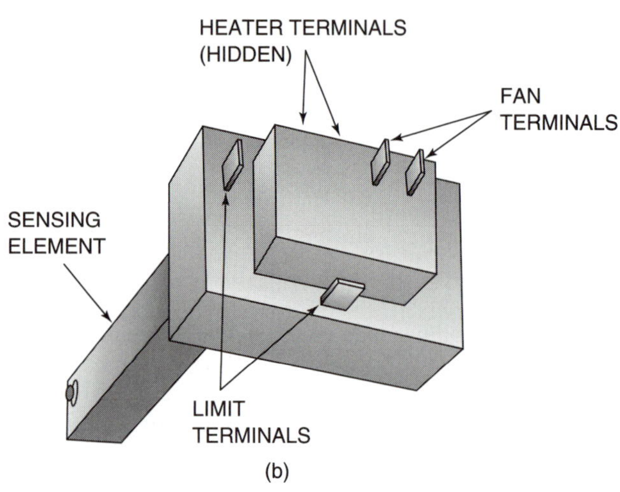

Figure 3–22 (a) Time-delay fan switch with temperature sensing element. (b) Time-delay fan/limit switch.

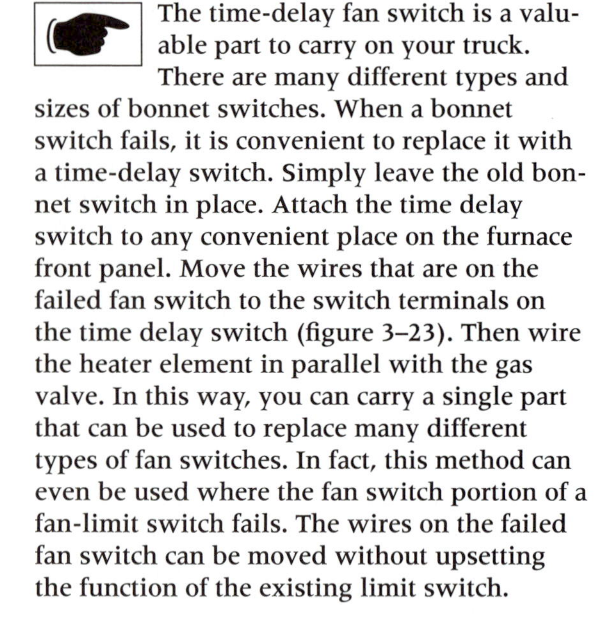

The time-delay fan switch is a valuable part to carry on your truck. There are many different types and sizes of bonnet switches. When a bonnet switch fails, it is convenient to replace it with a time-delay switch. Simply leave the old bonnet switch in place. Attach the time delay switch to any convenient place on the furnace front panel. Move the wires that are on the failed fan switch to the switch terminals on the time delay switch (figure 3–23). Then wire the heater element in parallel with the gas valve. In this way, you can carry a single part that can be used to replace many different types of fan switches. In fact, this method can even be used where the fan switch portion of a fan-limit switch fails. The wires on the failed fan switch can be moved without upsetting the function of the existing limit switch.

EXERCISE:
Redraw the furnace components shown in figure 3–24. Add the necessary wiring for a fully operational furnace with a time-delay fan switch.

TIME DELAY WITH THREE-WIRE LIMIT SWITCH

Figure 3–25 shows a wiring scheme used on unit heaters that are mounted near the ceiling. In a limit condition caused by a gas valve stuck in the open position, normally the room thermostat would open and the fan would turn off. The flame would continue to burn, but the occupants would be unaware of the problem because the heat would stratify near the ceiling. However, with the three-wire limit switch shown in figure 3–25, the fan will remain on

Figure 3–23 Replacing a bonnet temperature sensing fan switch with a time-delay switch.

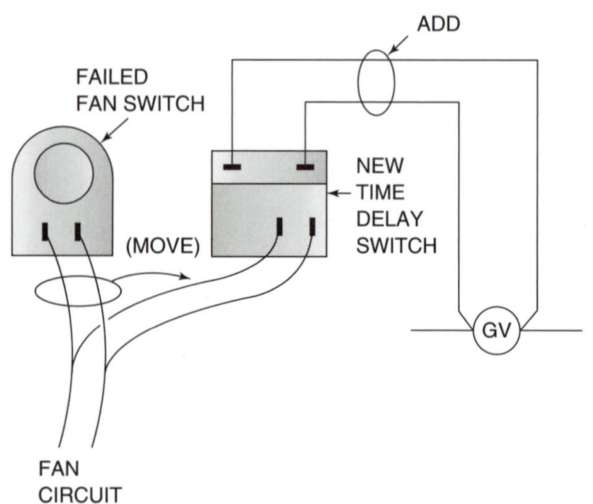

Figure 3–24 Exercise.

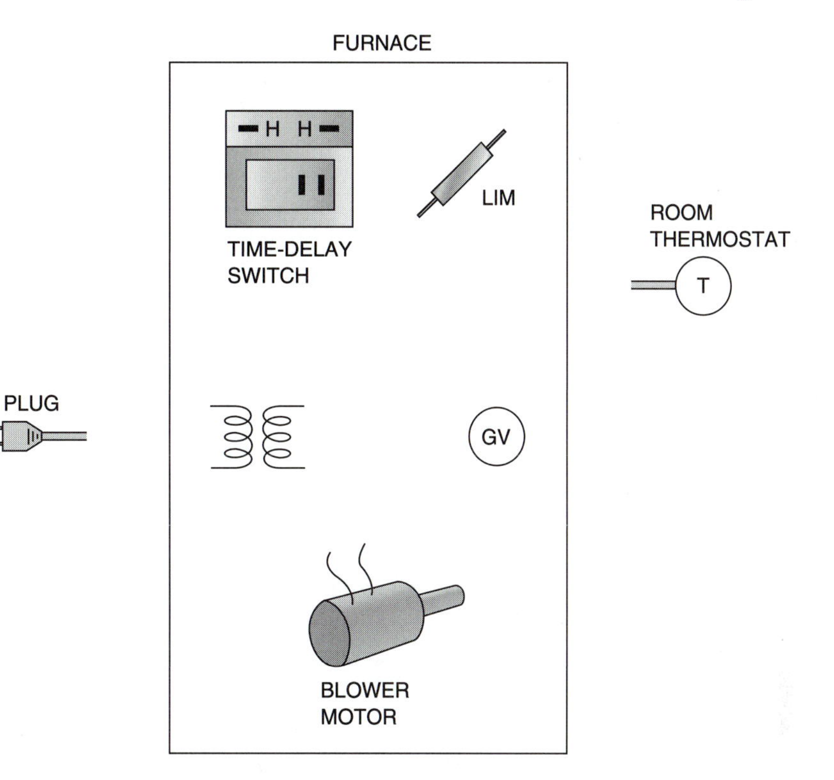

during the limit condition, even if the room thermostat turns off. This will alert the occupants (who will become too warm) that there is a problem with the heater.

INDUCER FAN

In some furnace applications (figure 3–26), the manufacturers have tried to gain better efficiency by increasing the number of passes that the flue gas makes through the heat exchanger. This allows the room air to absorb more heat from the products of combustion, but it also increases the pressure drop through the heat exchanger. At the same time, the cooler flue gas has less buoyancy, and the result is a furnace that needs an **inducer fan** to help pull the products of combustion through the furnace. When an inducer fan is used, generally the following sequence of operation is used:

Figure 3–25 Three-wire limit switch.

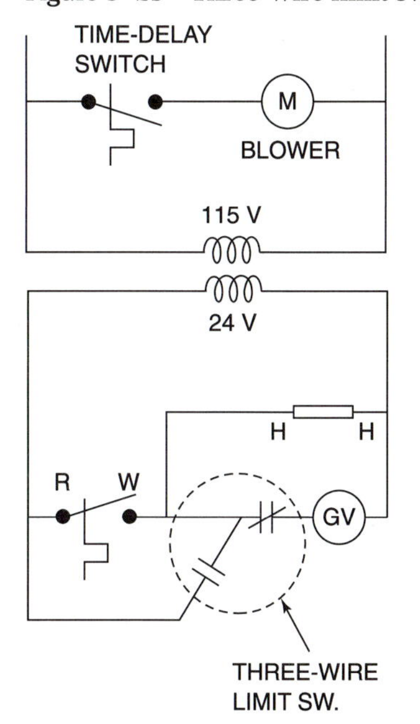

1. On a call for heat, the room thermostat energizes an inducer fan relay, bringing on the inducer fan.

2. The operation of the inducer fan is proved by a pressure switch, a sail switch, centrifugal switch, or other means, which, in turn, supplies power to the ignition sequence. A simplified schematic is given in figure 3–27.

QUESTIONS:

1. If the inducer fan is running but there is no voltage to the gas valve, which two components would you check?
2. If the limit switch is open when the room thermostat calls for heat, what loads (if any) will run?
3. During normal heating operation, what voltage would you measure across each device in figure 3–27?

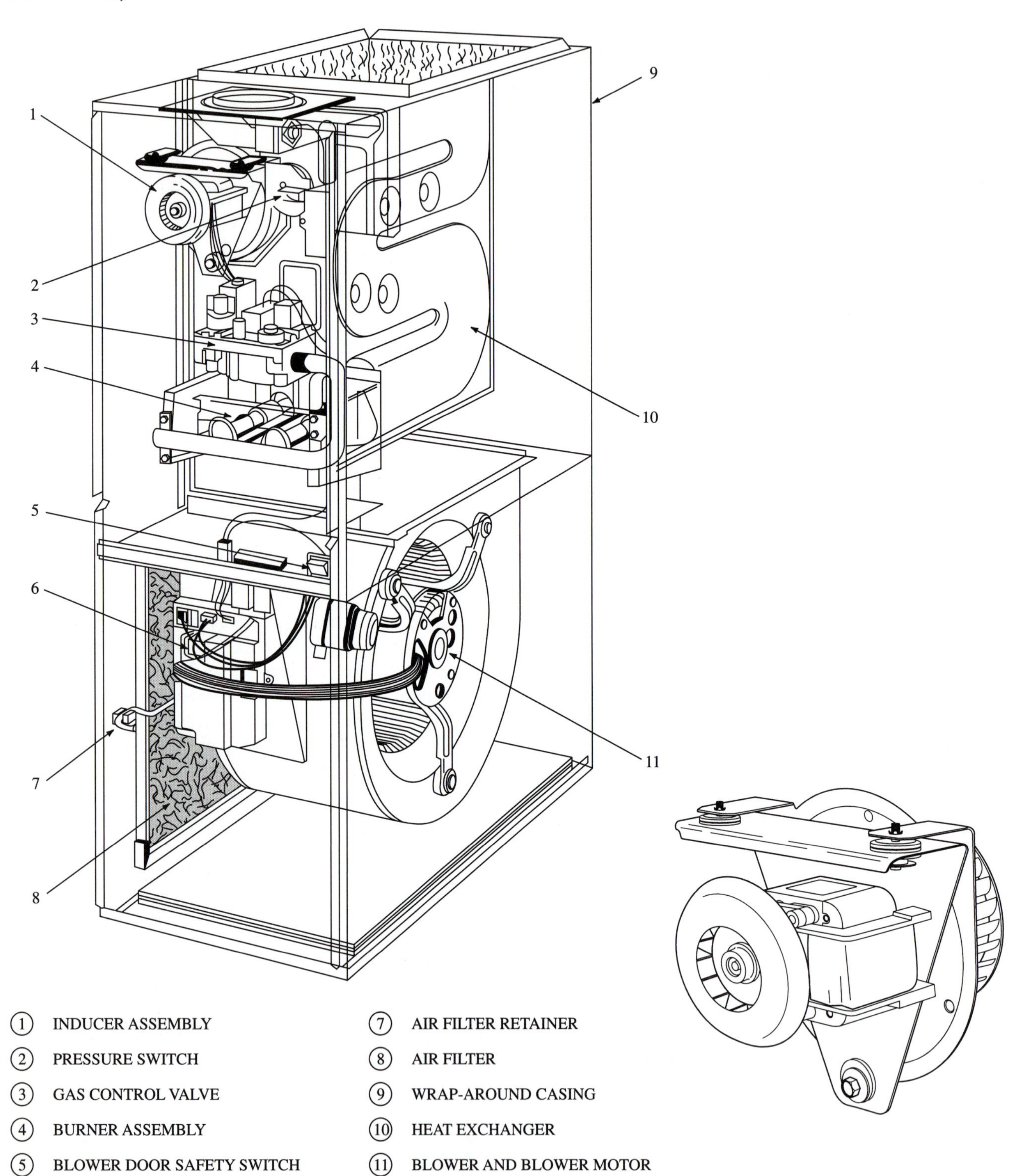

Figure 3–26 Furnace with inducer fan. *(Courtesy of Carrier Corp.)*

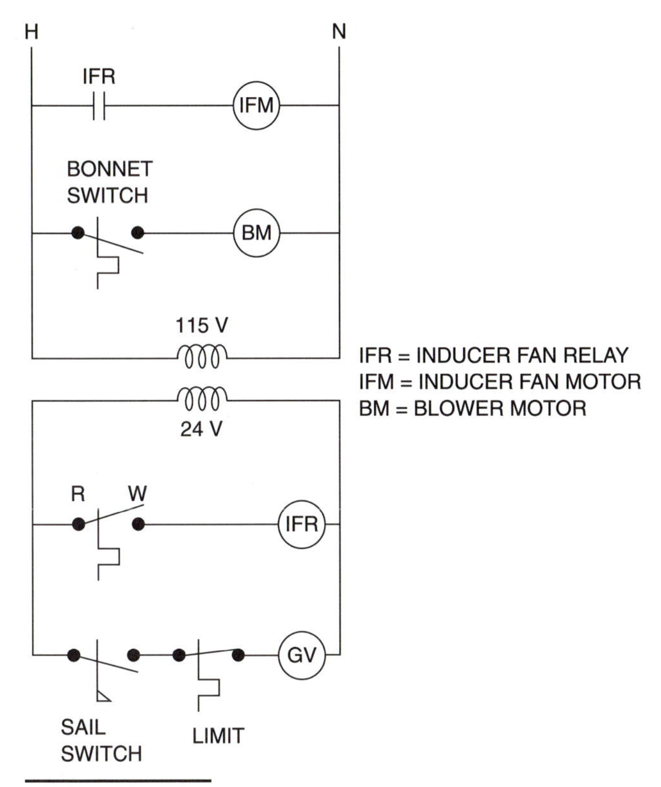

Figure 3–27 Inducer fan circuit.

MILLIVOLT FURNACES

A millivolt furnace is one that uses a **pilot generator** (also referred to as a **powerpile**) to sense the pilot flame and provide a source of power to the millivolt gas valve (figure 3–28). The pilot generator works on the principal that two dissimilar metals, when joined together, will generate a small voltage that is dependent upon the temperature of the junction. The hotter the junction, the higher the voltage output. In a pilot generator, a number of these junctions are wired together in series to produce an output as high as 600 mV (less than one volt). The millivolt furnace does not require any external source of power, and is therefore well suited for use on wall furnaces and floor furnaces.

The inexperienced technician may mistake the pilot generator for a thermocouple because they are each positioned next to the pilot flame. The end of the pilot generator is physically larger than the end of the thermocouple. The voltage is transmitted over two wires, protected by metal shielding. The end of the pilot generator usually has normal wire terminations rather than the screw-in arrangement familiar to the thermocouple.

The gas valve used on a millivolt system is also very different from the 24V gas valve. Although similar in appearance, the millivolt valve is a **diaphragm-type valve**, shown schematically in figure 3–29. When the valve is closed, there is inlet gas pressure available on both the top and the bottom of the diaphragm. In fact, the area of the top of the diaphragm that "sees" the gas pressure is larger than the area on

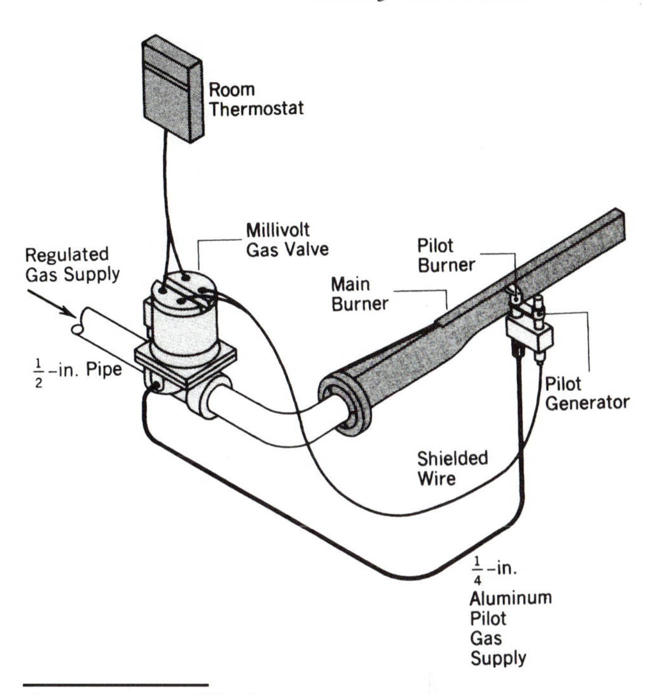

Figure 3–28 Millivolt system.

Figure 3–29 Diaphragm-type millivolt gas valve.

Terminals for
Pilot Generator
Millivolt Coil
Iron Core
Pilot
Operated
Valve
Pilot Gas Outlet
2
Spring
Metal Disc
Rubber
Diaphragm
Main
Gas
Inlet
1
3
Main
Gas
Outlet

(a)

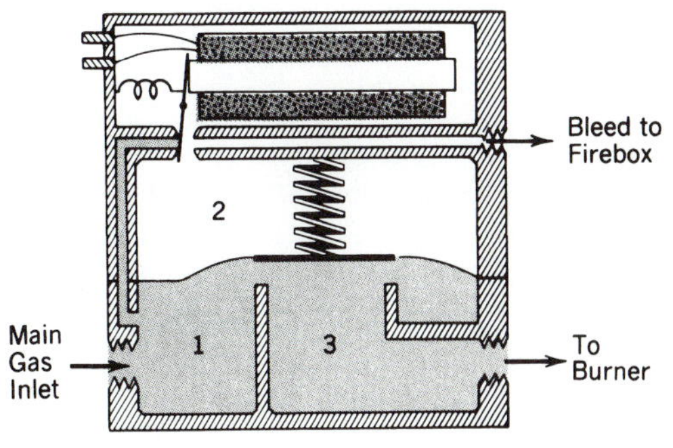

(b)

the bottom. Therefore, the gas pressure actually tends to help the small spring hold the valve closed. When the room thermostat closes, the millivoltage from the pilot generator is supplied to the coil, creating a magnetic field. The top of the small pilot valve is pulled towards the coil. The passage from the gas inlet is shut off, while the trapped gas pressure above the diaphragm is allowed to bleed off through the vent tube. The only remaining pressure on the diaphragm is from chamber #1, lifting the diaphragm, and opening the main gas valve. In this way, the small voltage available from the pilot generator can be used to operate a valve simply by directing the forces already available from the gas pressure.

CHAPTER FOUR

HEATING/AIR CONDITIONING CIRCUITS

Comfort cooling circuits control the same types of devices as refrigeration and freezer circuits: compressor, condenser fan, and evaporator. The differences between comfort cooling systems and refrigeration systems are:

1. Comfort cooling systems (except for window air-conditioners) almost always use low voltage (24V) controls, while refrigeration systems use line voltage controls.

2. Comfort cooling systems are commonly combined with furnace systems.

3. Comfort cooling systems operate with saturated suction temperature (evaporator coil temperature) above 35°F, so there is no need for defrost systems, except for heat pumps (see Chapter eighteen).

WINDOW AIR CONDITIONER

The window air conditioner shown in figure 4–1 is typical of the type that plugs into a standard 115V or 230V outlet. The major components are the compressor and a double-shaft motor (figure 4–2) that turns both the evaporator blower and the condenser fan. Power comes into a selector switch. Depending on which selection is made by the occupant, the selector switch will send the power out to one of the fan motor speed windings and the compressor. For example, if the selector switch is set for Fan only, it will complete the circuit from terminals 7–4. If it is set for High Cool, it will make 7–4 and 7–2, bringing on the compressor. Low Cool will cause the selector switch to make 7–3 and 7–2. When the selector switch is set for cooling, the fan will run continuously, and the compressor will cycle off and on as the room thermostat opens and closes.

The compressor motor and the fan motor each has its own run capacitor. Sometimes the two capacitors will be combined into a single casing as shown in figure 4–3. The fan is connected between terminals C (common) and F (Fan). The compressor is connected between terminals C and H (Hermetic).

In figure 4–1, the side of the power supply that goes to the run capacitor is labeled "marked." Often, one side of the power supply will have a ridge on the insulation or some other marking that makes it easier for you to identify which side of the power supply you are looking at.

EXERCISE:
When purchasing a replacement double-shaft motor, it is not necessary to worry about the direction of rotation. Explain why.

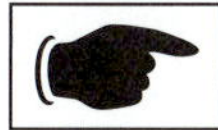

The fan motor that uses multiple taps on the run winding to produce different speeds will only produce different speeds when the motor is under a load. If you remove the fan from the motor

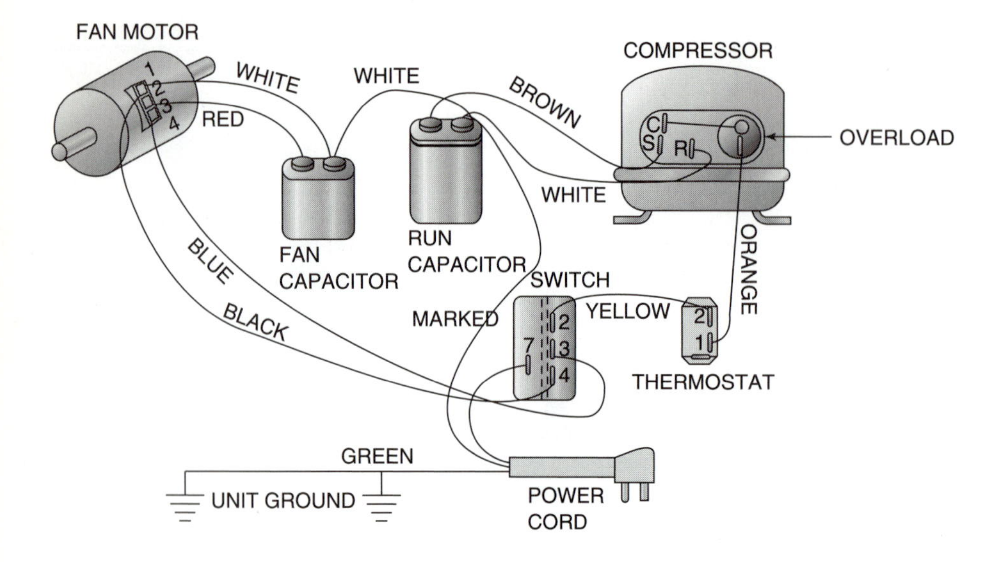

SWITCH POSITION	CONTACTS 2	3	4
OFF	O	O	O
NORMAL FAN	O	C	O
SUPER FAN	O	O	C
NORMAL COOL	C	C	O
SUPER COOL	C	O	C
C-CLOSED O-OPEN			

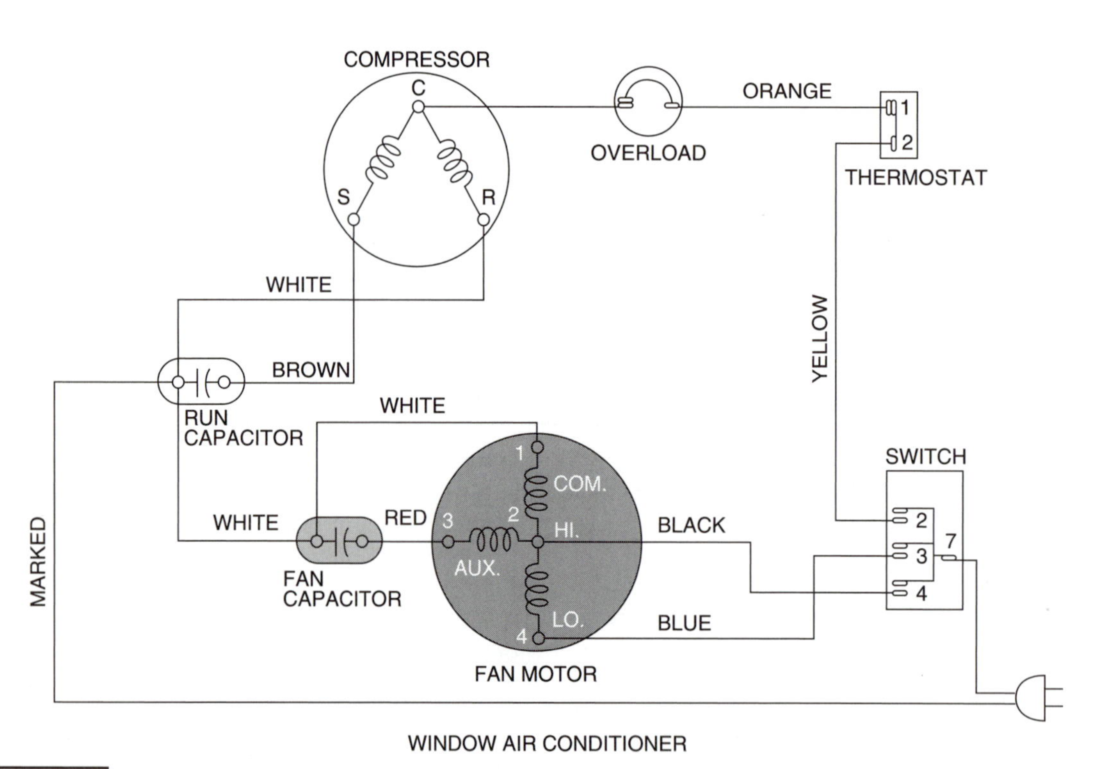

Figure 4–1 Window air conditioner.

Figure 4–2 Double-shaft motor.

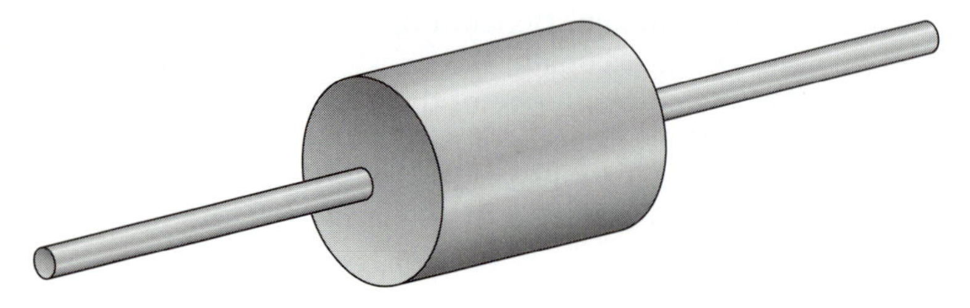

ADVANCED CONCEPTS

The fan motor shown has two speeds. Different speeds can be achieved by tapping the run winding in different places. When the selector switch makes 7–4 (High speed), it uses only the portion of the run winding in the motor between motor terminals 2 and 1. However, when the selector switch makes 7–3 instead, another portion of run winding (between motor terminals 4–2) is inserted in series with the run winding between motor terminals 2–1. Adding this coil to the circuit reduces the current flow through the run winding, and therefore reduces the strength of the magnetic field. The motor produces less torque, and the motor speed will decrease. A three-speed motor is shown in figure 4–4. It allows yet another portion of run winding to be inserted, reducing the current flow still further and producing a still slower speed.

shaft and operate the motor alone, it will run at full speed even if the medium or low speed taps are used.

EXERCISE:
Redraw the interconnection diagram in figure 4–5 as a ladder diagram. It is not necessary to show the internal parts of the fan motor or compressor. Show all wire colors.

Plug and Receptacle Configurations

The standard house outlet shown in Chapter one is not always suitable for a window air conditioner. Some window air conditioners require more current than can be supplied from a standard outlet. Higher-amperage and higher-voltage outlets are available for use on larger circuits. Each different rating uses a different plug and receptacle configuration to prevent a higher-voltage or -amperage device from being plugged into an outlet that cannot handle the load. Figure 4–6 shows some of the standard plug and receptacle configurations.

Window Air Conditioner with Strip Heat

The window air conditioner in figure 4–7 has, in addition to its air conditioning components, an electric heater (commonly called a **strip heater**). The differences between this unit and a cooling-only unit are:

1. There are additional switch positions in the selector switch to accommodate different Heat settings.

2. The thermostat is an SPDT switch, using one side to control the heating and the other side to control the cooling.

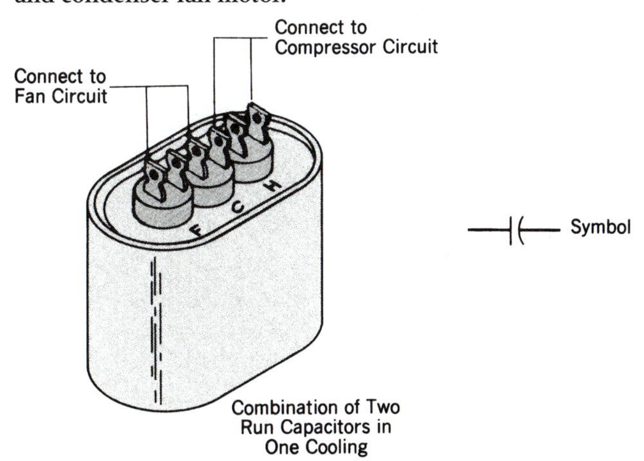

Figure 4–3 Dual capacitor for compressor and condenser fan motor.

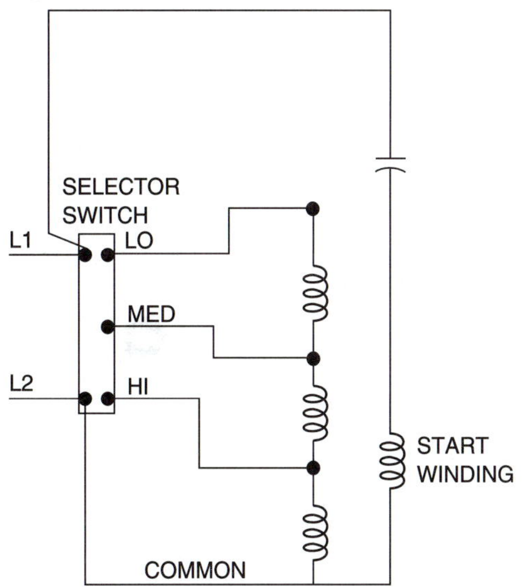

Figure 4–4 Three-speed motor circuit.

SWITCH POSITION	TERMINALS	READING
OFF	L1 TO 1-2-3-4	OPEN
LO-COOL	L1 TO 1-4	CLOSED
MED-COOL	L1 TO 2-4	CLOSED
HI-COOL	L1 TO 3-4	CLOSED
FAN	L1 TO 2	CLOSED

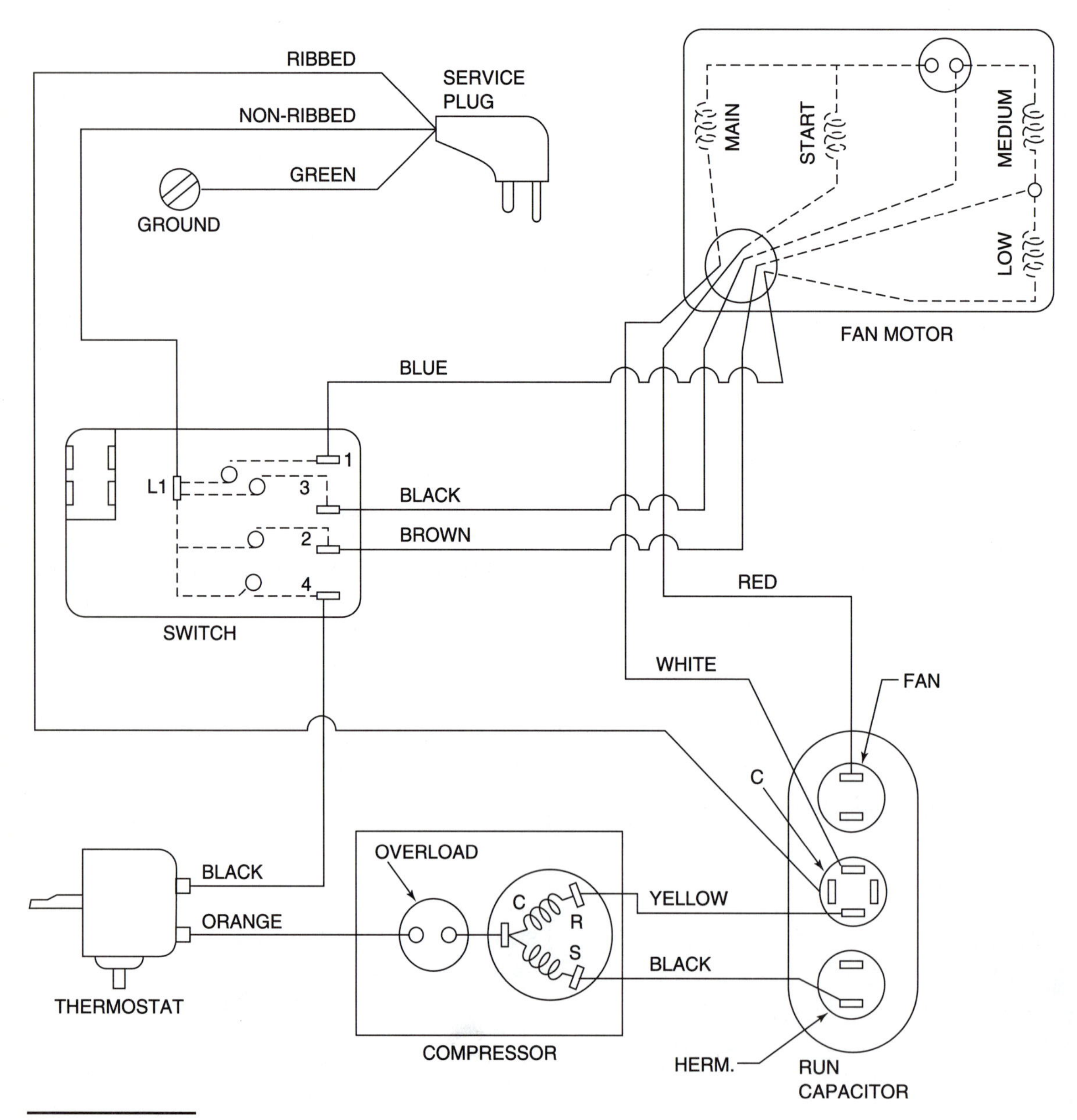

Figure 4–5 Exercise.

RECEPTACLE CONFIGURATION	RATING
	15 A 125 V
	20 A 125 V
	30 A 125 V
	50 A 125 V
	15 A 250 V
	20 A 250 V
	30 A 250 V
	50 A 250 V
	15 A 277 V
	20 A 277 V
	30 A 277 V
	50 A 277 V

Figure 4–6 Standard receptacle configurations.

EXERCISE:
State what voltage reading you would get between each of the terminals on the selector switch and terminal 2 on the thermostat for normal operation on Cool-Med. Repeat for Heat-Med. Assume 115V supply voltage.

ROOFTOP AIR CONDITIONER

One of the simplest control schemes using a control voltage that is different from the line voltage is for a packaged rooftop air conditioner. All components except for the room thermostat are located in a single unit, mounted on the roof. Electrical installation couldn't be simpler. The L1 and L2 power supply are connected to the L1 and L2 terminals in the unit and the R, Y, and G terminals on the room thermostat are connected to the R, Y, and G terminals in the air conditioning unit, using thermostat wire.

SEQUENCE OF OPERATION

Figure 4–8 shows the ladder diagram of a simple packaged air conditioner. The transformer is always energized. When the room thermostat senses that the room is too warm, it closes a switch (inside the thermostat) between R-Y and another switch between R-G. The R-Y switch energizes a compressor contactor, whose contacts energize the compressor and condenser fan. The R-G contacts energize a blower relay, whose contacts energize the blower fan motor. In air conditioning systems, the evaporator fan is commonly called a blower, and the blower relay is nothing more than an SPNO (single pole normally open) control relay. All of the cooling devices come on at the same time, and when the thermostat is satisfied, they all turn off at the same time.

The blower motor can also be operated continuously, even when the compressor turns off. This is sometimes desirable for continuous circulation, a non-changing noise level, and equalizing temperatures between rooms that tend to have non-uniform temperatures. There is a switch on the thermostat labelled FAN-AUTO and FAN-ON. When it is set to FAN-AUTO, R-G is made whenever R-Y is made. When it is set to FAN-ON, R-G is made regardless of the position of R-Y.

QUESTION:
You come upon the packaged rooftop air conditioner in figure 4–8. The room is much warmer than the thermostat set point, but the compressor, condenser fan, and blower are all off. Would you suspect that the problem is the compressor contactor? Explain.

QUESTION:
If both fans are running but the compressor is not, could the problem be the compressor contactor? Explain.

EXERCISE:
Redraw the rooftop air conditioner components shown in figure 4–9. Add the required wiring.

PACKAGED HEATING AND AIR CONDITIONING SYSTEM

Figure 4–10 shows the same cooling system, but the packaged rooftop also now includes a furnace. The room thermostat is also different, as it now includes

SWITCH POSITION	TERMINALS	READING
OFF	L1 TO 1-2-3-4-5	OPEN
COOL-LO	L1 TO 1-4	CLOSED
COOL-MED	L1 TO 1-3	CLOSED
COOL-HI	L1 TO 1-2	CLOSED
HEAT-LO	L1 TO 5-4	CLOSED
HEAT-MED	L1 TO 5-3	CLOSED
HEAT-HI	L1 TO 5-2	CLOSED
FAN	L1 TO 3	CLOSED

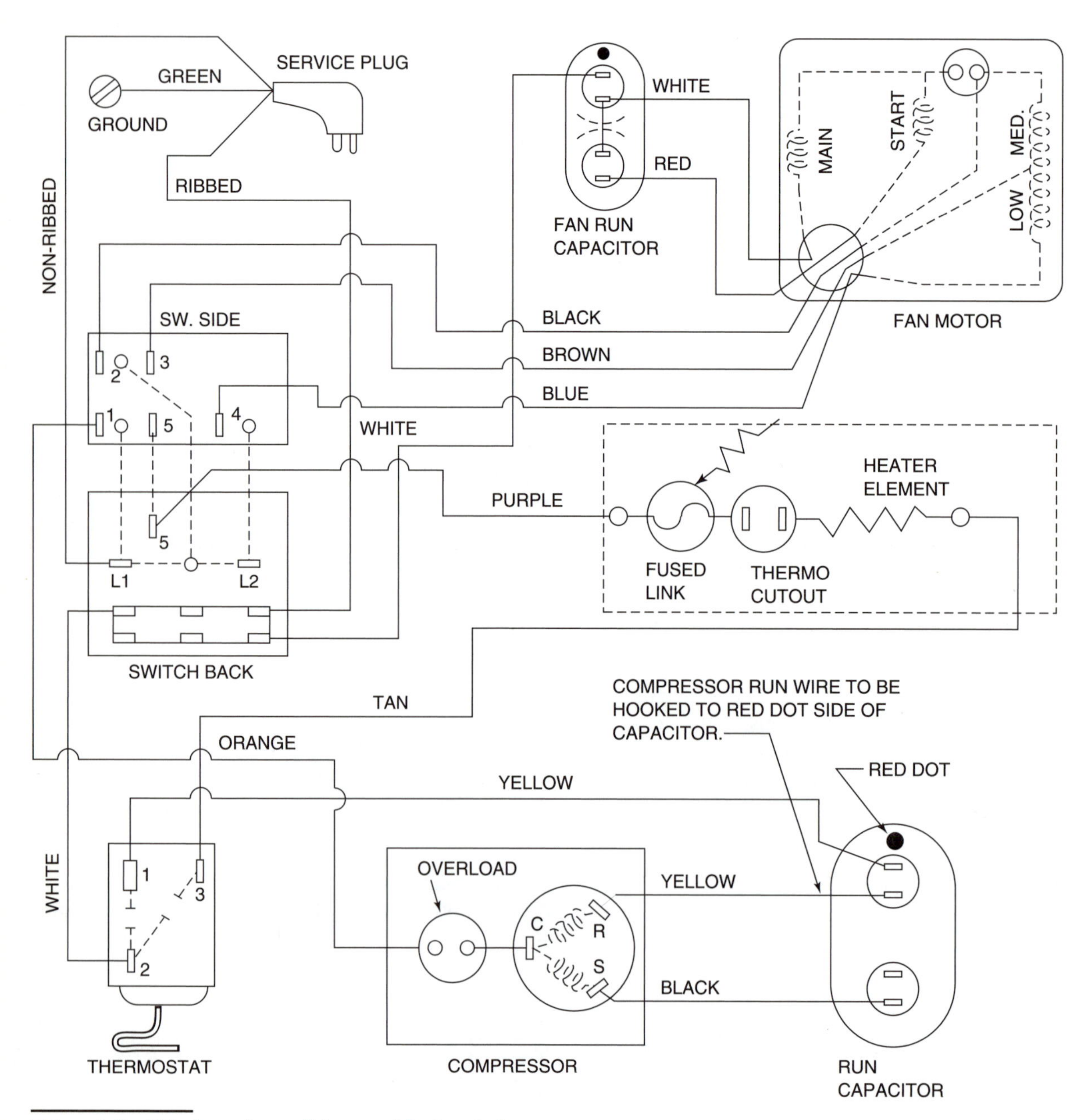

Figure 4–7 Window air conditioner with electric heat.

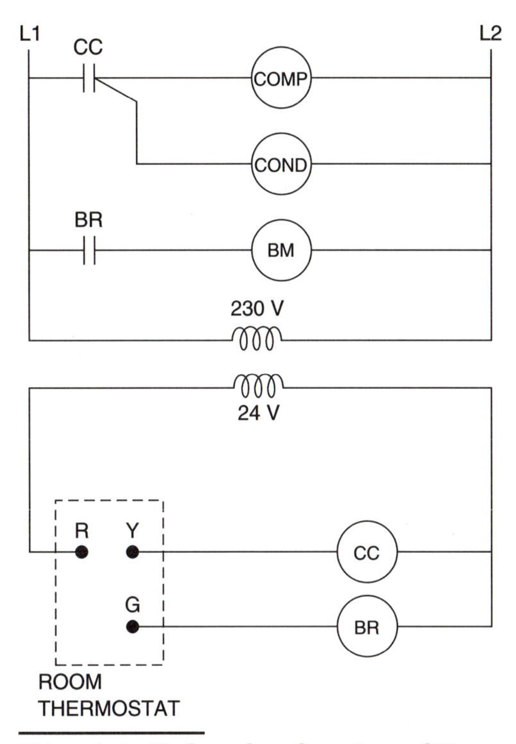

Figure 4–8 Packaged rooftop air conditioner.

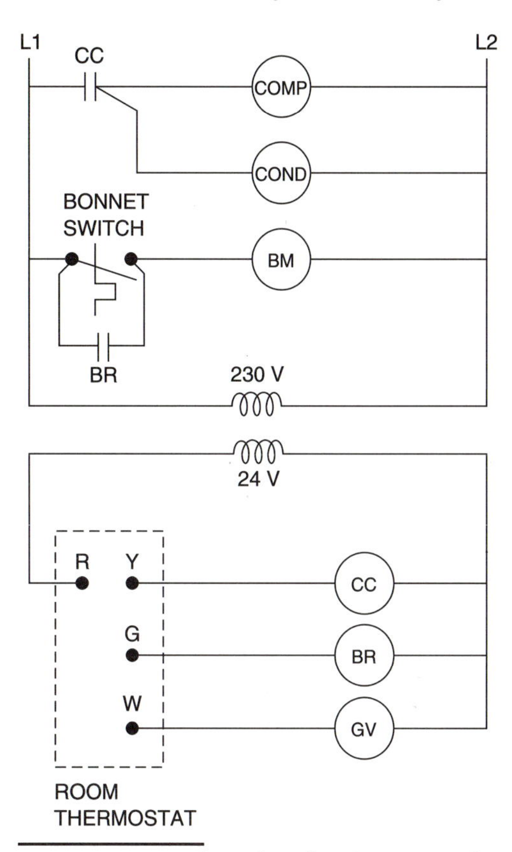

Figure 4–10 Packaged rooftop heating and air conditioning unit.

the functions of both the heating thermostat and the cooling thermostat(four-wire thermostat). When the thermostat needs heating, it makes R-W only. The gas valve comes on right away, but the blower doesn't come on until a minute later when the bonnet switch closes. When the thermostat needs cooling, it makes R-Y and R-G, and all components come on right away as described above.

QUESTION:
The unit is operating. The compressor, condenser, and blower motors are all running. What voltage would you measure across the bonnet switch? (Be careful—this is a tricky question.)

Figure 4–9 Exercise.

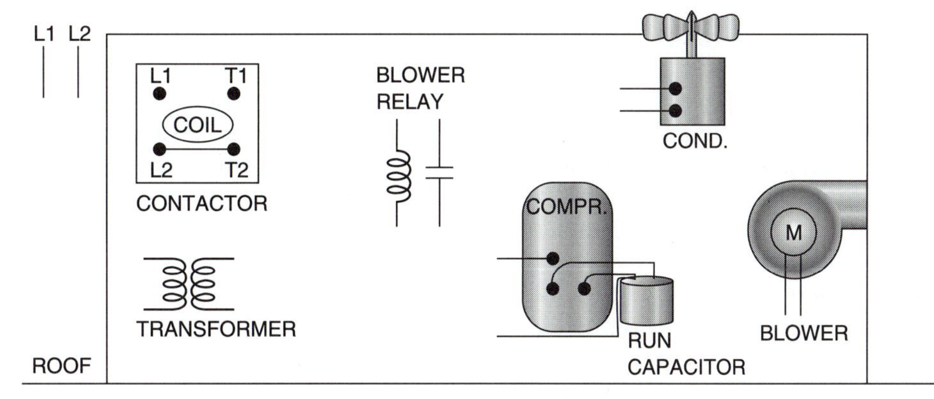

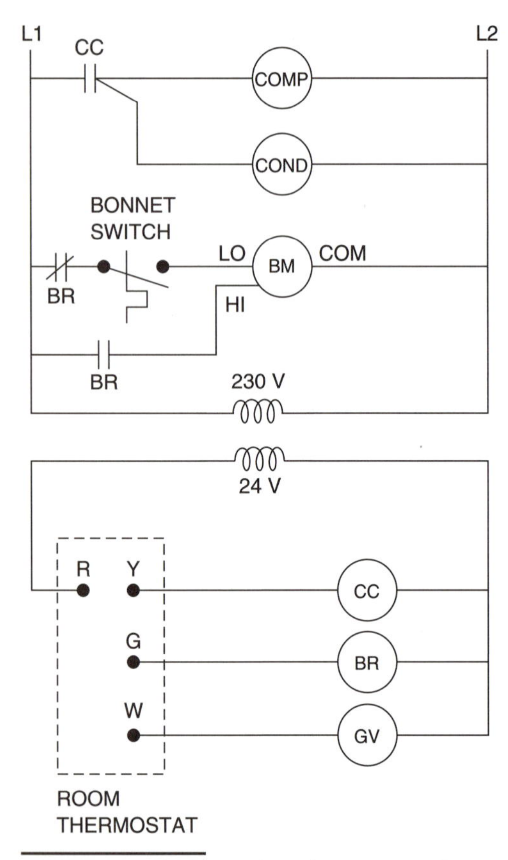

Figure 4–11 Packaged rooftop unit with two-speed blower.

Sometimes, the rooftop unit is furnished with a two-speed blower motor. When that is the case, the high speed is used for cooling while the lower speed is used for heating. This is done because on heating, air is supplied to the room at 50°F higher than the room temperature. But on air conditioning, the air is supplied to the room at only 20°F lower than the room temperature. Therefore, more air is needed for air conditioning than for heating. Figure 4–11 shows how the low-speed winding of the blower motor is energized by the bonnet switch, while the high-speed winding is energized by the blower relay contacts. Note that the blower relay has changed. It is now an SPDT switch, with the normally closed contacts being wired in series with the bonnet switch. If this normally-closed switch were not added, there would be the following potential problem. Suppose that during heating (bonnet switch closed), an occupant switched the room thermostat from FAN-AUTO to FAN-ON. The motor would have both its High and Low speed windings energized at the same time. This would cause the motor to burn out. With the addition of the normally closed switch, when the thermostat was switched to FAN-ON, the Low speed winding would be de-energized.

RESIDENTIAL CIRCUIT NO.1—THREE-VOLTAGE SYSTEM

Residential central heating/air conditioning systems usually use a 115V furnace in which the furnace fan also serves as the blower for the air conditioning. Within the furnace is a 24V control system. Outside is the condensing unit, which is supplied from a 230V power source. The physical location of the components for this three-voltage system is shown in figure 4–12, and the ladder diagram is shown in figure 4–13.

EXERCISE:
Redraw the component location diagram in figure 4–14. Using the ladder diagram of figure 4–13 as a guide, add the interconnection wiring to the physical components, and answer the following questions:

1. If the furnace is unplugged, will the compressor contactor still work?
2. If the 230V circuit breaker has tripped, will the thermostat be able to operate the compressor contactor?

RESIDENTIAL CIRCUIT NO. 2

Figure 4–15 shows an interconnection diagram and a ladder diagram for a system that provides heating, air conditioning, humidification, and electronic air cleaning. Some of the unique features of this system are as follows:

1. Instead of the room thermostat energizing the compressor contactor when R-Y is made, the thermostat energizes a thermal delay heater. This is simply a time-delay switch similar to the one that is commonly used to provide a time delay on furnace blower motors. It prevents the compressor from short-cycling. **Short-cycling** is a term used to describe a compressor that starts, runs for a few seconds, turns off, and then repeats the cycle. For example, if the refrigeration unit were low on charge, as soon as the compressor started, it would cut out on the low pressure cut-out. After the compressor stops, the low-side pressure increases, the LPC recloses, and the compressor starts again. In this circuit, use of the time delay heater is probably redundant to prevent short cycling because the impedance relay already provides that protection. In this circuit, the time delay switch would prevent the compressor from restarting immediately after shutting down if somebody happened to adjust the thermostat downward. This will allow pressures to

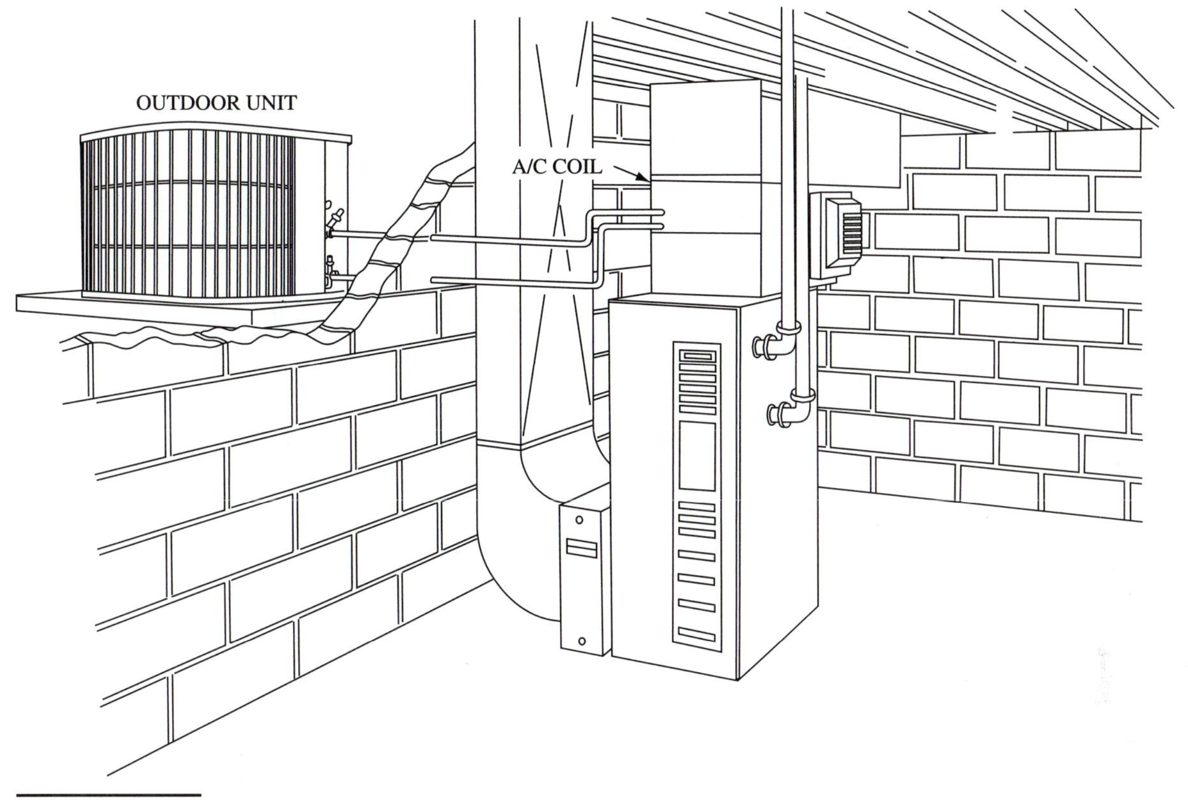

Figure 4–12 Three-voltage system. *(Courtesy of Carrier Corp.)*

equalize within the refrigeration system before the compressor attempts to restart.

2. The impedance relay (lock-out relay) has two sets of contacts instead of one. The normally closed set of contacts provides the manual-reset feature if any of the safety controls open. The normally open set of contacts completes a circuit through a trouble-indicator light in the room thermostat. This alerts the occupants that something is wrong with the system.

Figure 4–13 Ladder diagram three-voltage residential heating/air conditioning.

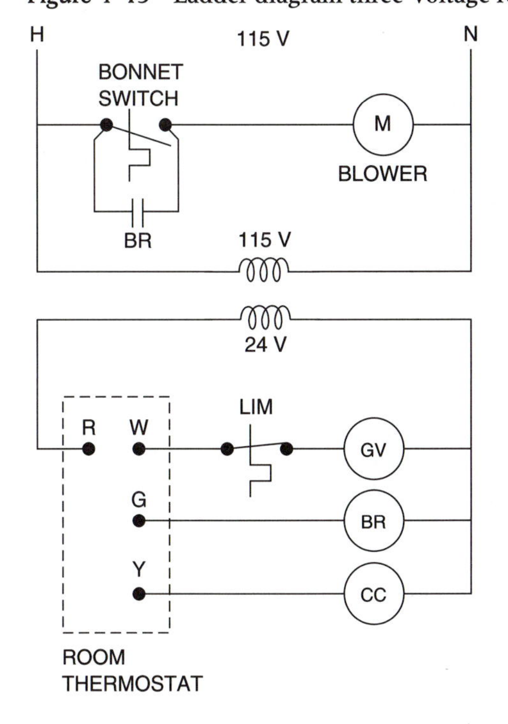

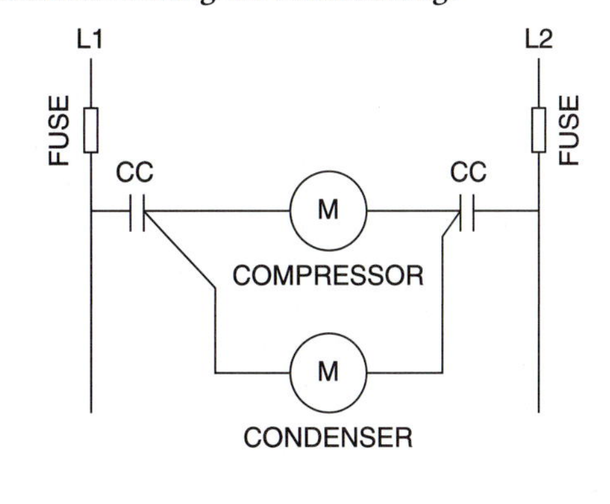

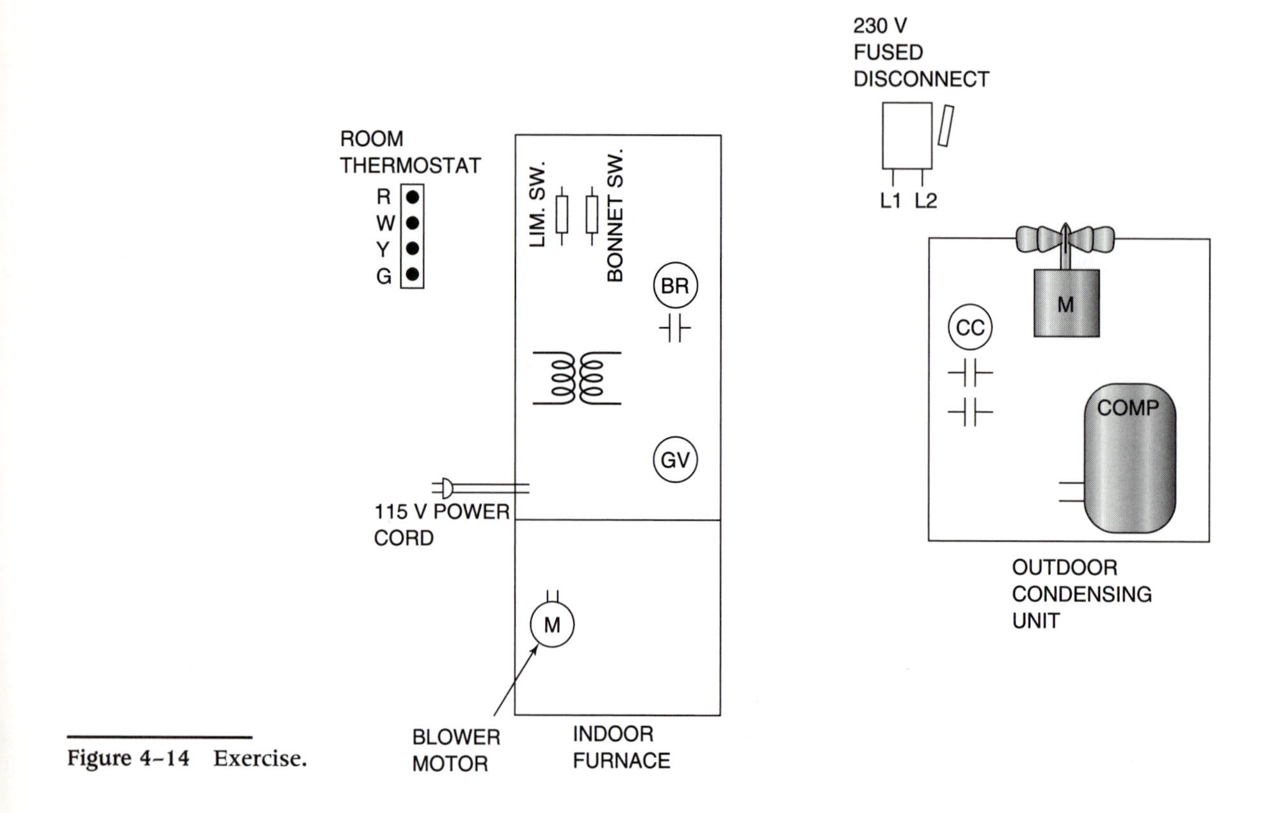

Figure 4–14 Exercise.

Figure 4–15 Residential circuit no. 2.

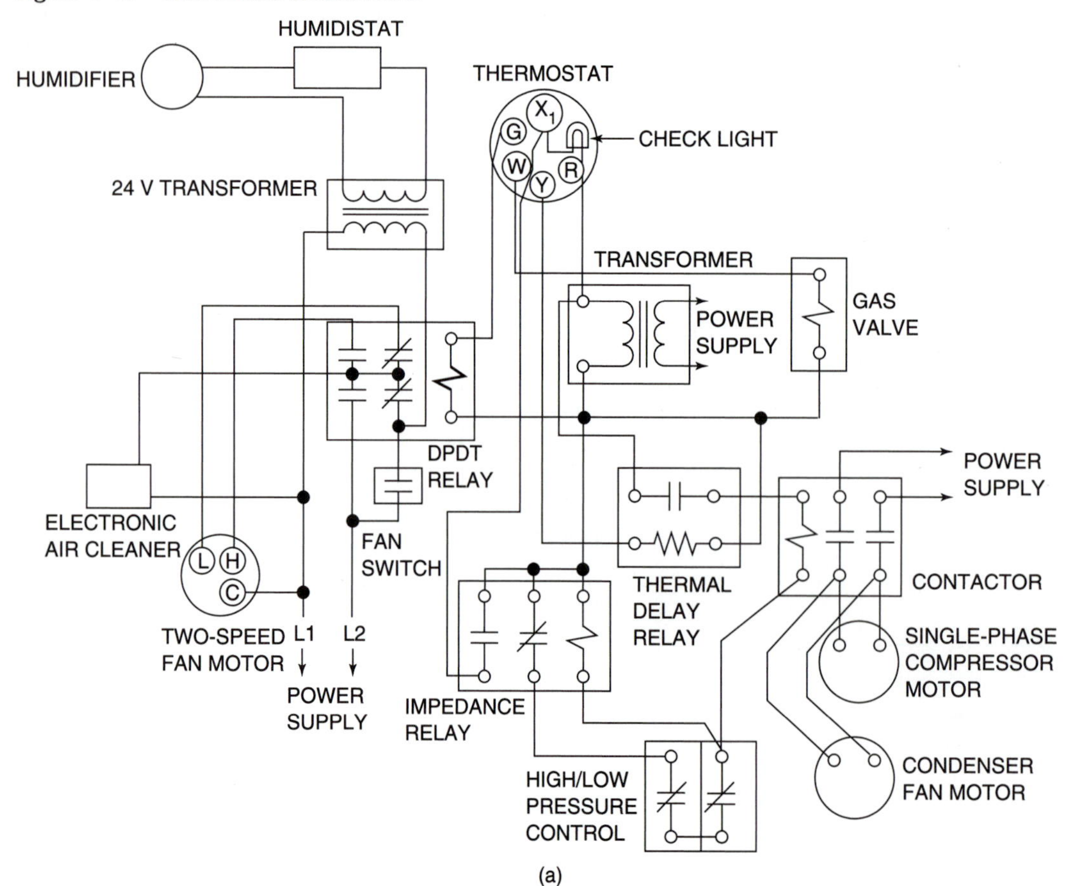

3. The humidifier operates by a low-voltage motor driving a pad through a water pan so that the room air can pass through a continuously wetted pad. A separate transformer is provided to supply the correct voltage for the humidifier motor.

4. The DPDT fan relay operates the blower motor on high speed for cooling and low speed for heating.

RESIDENTIAL CIRCUIT NO. 3

Figure 4–16 shows a circuit for controlling a total electric packaged rooftop heating/cooling system.

HEAT SEQUENCER

The heat is provided by an electric resistance heater located in the duct or in an electric furnace. The elec-

Figure 4–15 Continued.

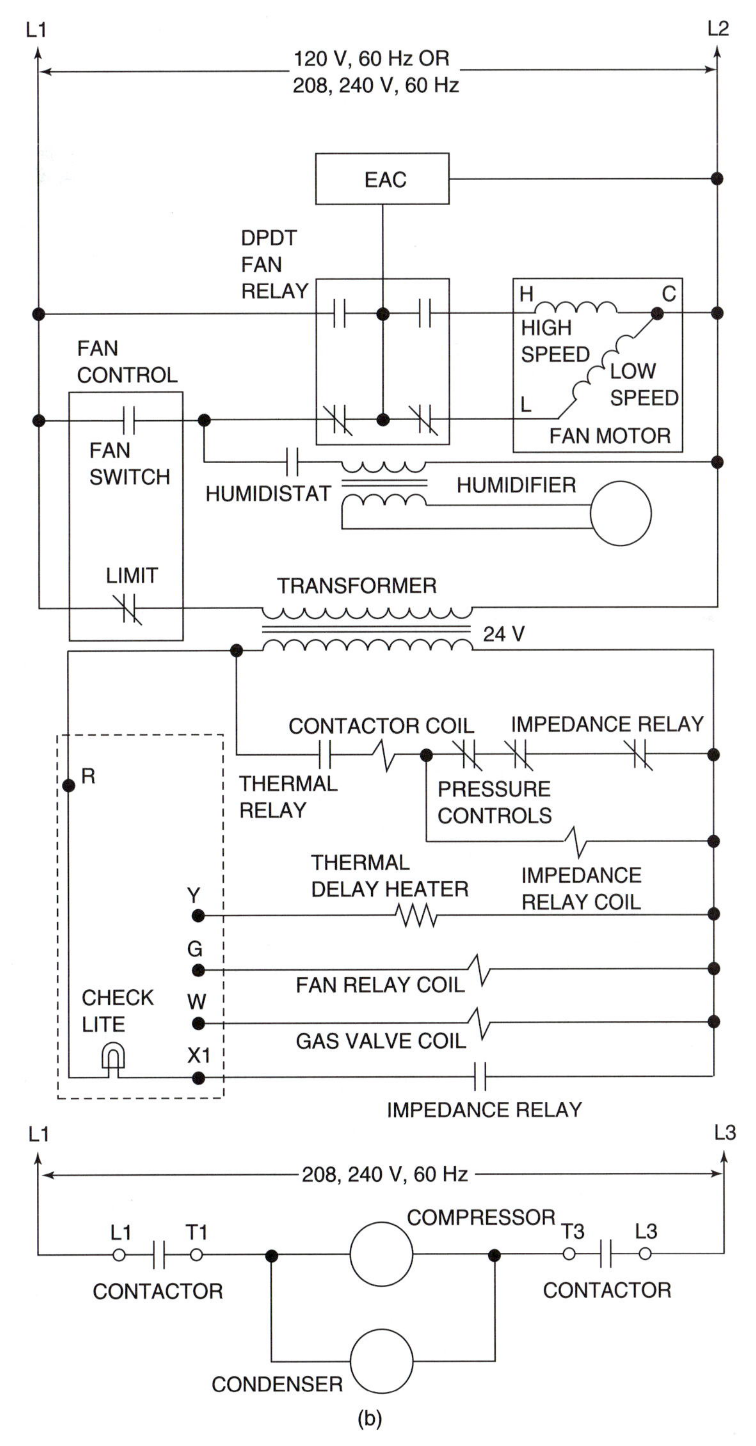

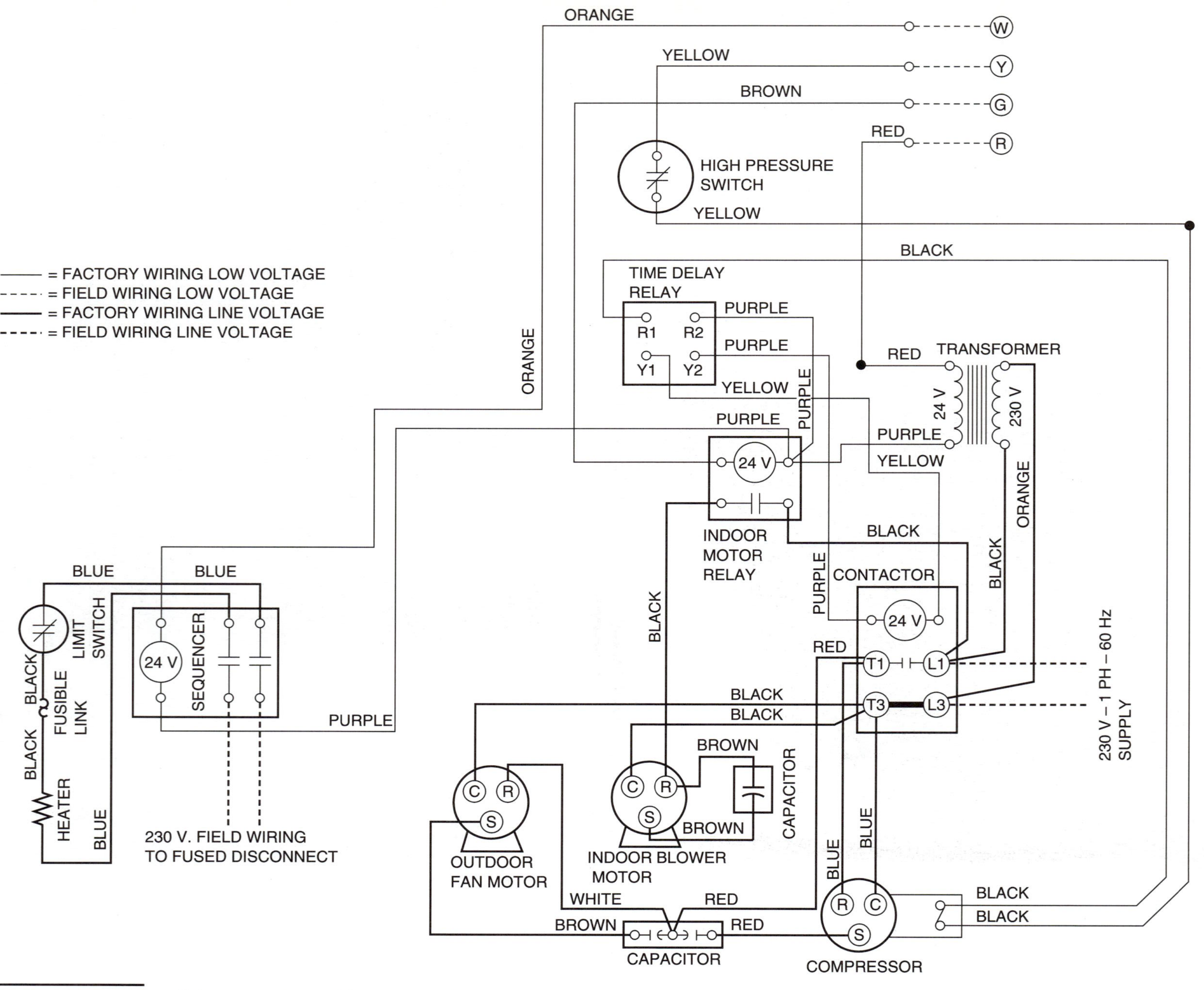

Figure 4–16 Air conditioning system with resistance heaters.

tric heating element is switched on and off through a **heat sequencer.** A heat sequencer (figure 4–17) acts like a time-delay contactor. Instead of a control signal energizing a magnetic coil which immediately closes the line-voltage switches, the control signal energizes a small heater inside the sequencer. After a time delay, a bimetal switch inside the heat sequencer operates the switch(es) that complete the circuit through the electric heating element.

Time Delay Relay

On simpler systems, when the room thermostat makes R-Y, it completes a circuit through the compressor contactor coil. On this system, instead of energizing the contactor coil, R-Y completes a circuit through the time-delay relay (TDR), terminals R1 and R2. Some time after 24V are applied to R1-R2 (usually about one minute), 24V come out of the relay at terminals Y1 and Y2. The output voltage from the time-delay relay energizes the compressor contactor. This prevents short cycling of the compressor. For example, suppose an occupant adjusts the set point of the room thermostat downward. Suppose that by chance, this causes the thermostat to make R-Y just seconds after the compressor turned off because the thermostat was satisfied. The compressor would have a hard time starting because the pressures in the high side and low side of the refrigeration system will not have had time to equalize. The result could be either an overload opening, or a fuse in the supply power wiring blowing. The time delay will avoid this situation by keeping the compressor from trying to start for at least a minute after the thermostat calls for cooling. Some of these time delays are adjustable, and may be set by the service technician for any period from one to ten minutes.

Compressor Overload

The compressor overload is the switch that is shown between the two black wires emerging from the compressor. It is not known from this diagram whether the overload is internal or external. However, if it opens, it will interrupt the circuit that includes the room thermostat (R-Y), the high pressure switch, and the time-delay input.

Figure 4–17 Heat sequencer.

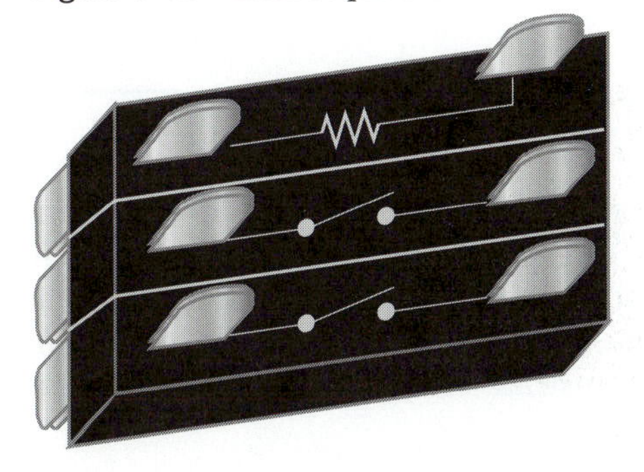

Room Thermostat

Although the room thermostat is not shown on the wiring diagram, we can tell that an electric heat type of thermostat is required. We can figure this out, because if you look at the black wire on the red terminal of the indoor blower motor, it goes through the switch of the indoor motor relay before reaching L1. Therefore, we know that the indoor motor relay *must* be energized in both heating and cooling. The indoor relay coil gets energized from the G terminal of the room thermostat. A thermostat designed for electric heat will make R-G on both heating and cooling, whereas a thermostat not designed for electric heat will only make R-G on a call for cooling.

QUESTION:
The room thermostat is set for 65°F, but the room is warm, and the compressor and condenser are not running. The indoor blower is running. How would you test the time-delay relay to see if it has failed? If you do not determine that the time delay has failed, what other devices would you check, and how would you test them?

EXERCISE:
Redraw the wiring diagram in figure 4–16 as a ladder diagram. Include all wire colors and terminal identification.

CHAPTER FIVE

TROUBLESHOOTING STRATEGIES

The trick to effective troubleshooting is to zero in as quickly as possible by testing the components that are most likely to be the cause of the problem. This chapter presents several of the techniques that are second nature to experienced technicians, and some that may even be helpful to experienced troubleshooters.

TROUBLESHOOTING RULES

There are only two rules for troubleshooting. They are simple to state, and they are always true:

1. *If you measure a voltage across a switch, the switch is open.*
2. *If you measure a correct voltage across a load and the load doesn't work, the load has failed.*

Some comments are in order regarding the application of these two rules.

1. With digital meters, voltage readings that we have always considered to be zero when using analog meters will indicate some very small voltage. A very small voltage reading across a switch could indicate either a meter inaccuracy or a very slight resistance across the contacts of a closed switch. If you read a voltage of less than one or two volts, assume that it is zero volts.

2. Rule #1 does *not* say that if you read zero volts across a switch, the switch is closed. There are many situations in which you might read zero volts across an open switch.

3. Rule #2 indicates when the load has failed. This only means that the problem is with the load, and not with any other part of the circuit. The "fix" might be as simple as turning a knob or resetting an overload switch on the load. It does not necessarily mean that the load must be replaced.

CHECK THE EASY ITEMS FIRST

Always look for the easy fix first. Whatever electrical components there are that might explain the symptom that you observe, go first to the devices that are most accessible, and easiest to check. If a 230V 1ϕ unit is dead, check for any fuses or disconnect switches at the unit. Check to see that the unit is supplied with the proper incoming voltage. However, if the condenser fans are running and the compressor is not, don't bother checking the fuses no matter how easy it is. It is a waste of time, because the running condenser fans tell you that the fuses are good.

USE VOLTS

Theoretically, you can troubleshoot a problem using either volts or ohms. However, it is far more practical to choose volts. When you troubleshoot using an ohm meter, you must first disconnect the load being tested. To demonstrate why that is necessary, refer to figure 5–1. Suppose the blower motor does not operate. You think that the coil in the control relay may have failed, and you want to use your ohm meter to see if there is continuity across the coil. If you simply place your ohm meter probes across the coil contacts, you may get fooled. The coil you are trying to measure may actually be open, but if you don't disconnect it from the circuit before measuring it, you will measure continuity through the transformer secondary, and you might mistakenly conclude that the control relay coil has continuity.

So it is necessary to disconnect a device from the circuit before attempting to measure its resistance. But when you remove the wiring from the device, you will be moving, removing, pushing, and jostling wires, with several potential undesirable results:

1. There may be an intermittent problem that resolves when you move the wires. Then you will not be able to find it. The problem will most likely reoccur sometime later, and you will have a call-back.
2. Some wiring is old, and its insulation is brittle. If you start moving wires around, the insulation may become damaged. You may wind up needing to replace some or all of the wiring.
3. Whenever you remove wires, you need to identify them with tape so that you won't forget on which terminals they get reconnected. This is extra work, and is not required if you troubleshoot with volts.

When you use your volt meter to troubleshoot, you will either find a switch that is open (you measured voltage across it) or a load that has failed (you measured proper voltage across it, and it doesn't work). You will be able to do this without moving any wires, and without changing the circuit in any way. You may then remove the device, and double-check it with ohms if you like.

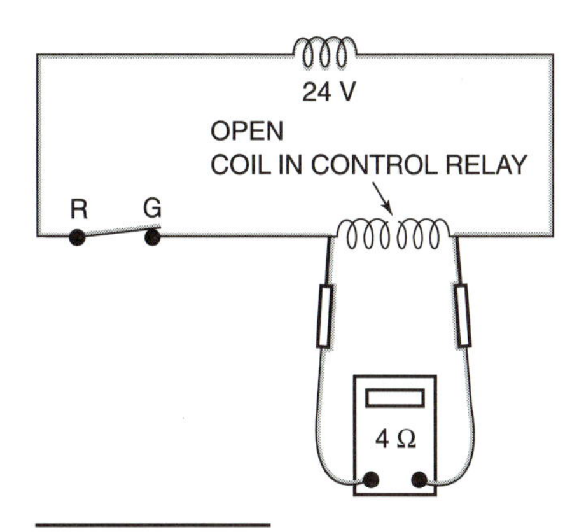

Figure 5–1 An erroneous ohm measurement.

CHECK THE SWITCHES AND PUSH THE BUTTONS

Sometimes the wiring for the unit you are troubleshooting is either very complicated, or the wiring diagram is missing (or both). It is not always necessary to figure out everything about how a system is supposed to work before you can find the problem.

Sometimes there is a variety of pressure switches, temperature switches, or other operating or safety controls that are easily visible and accessible. You may have no idea which switches may or may not be wired in series with the non-working load that you are trying to troubleshoot. But you can spend a few seconds measuring voltage across each of the switches to see if it is open. When you find an open switch, you can either force it closed mechanically, or press the manual reset if there is one, or place a jumper wire across it to see if that makes the load operate. If it does, you are well on your way towards isolating the problem. Some technicians will, as a matter of course, press on all the reset buttons to see if anything happens. If the switch is not already open, pressing the reset button will have no effect. However, if the switch is already tripped (open), then pressing on the reset button will produce an audible "click," and releasing it will cause some load to operate.

OH DEAR, WHAT CAN THE MATTER BE?

In most cases, you have been called to troubleshoot a system because something is not running. Maybe everything is inoperable. Maybe only the compressor or only the condenser fan isn't running. When you first arrive at the non-functioning unit, your observations will determine what is the logical sequence to follow. You will note that this text does not use troubleshooting charts that present a wooden-like step-by-step rigid approach. There is no single best way to approach the troubleshooting, but there are certainly some incorrect ways that you should avoid. To be a good troubleshooter, you must be flexible, and constantly alter your approach depending upon what you observe at each job.

The concept of "What can the matter be" means that you should first look for a single failed electrical component that could be causing all of the symptoms that you observe, and not bother with guesses that would only partially explain your observations. For example, suppose you had an air

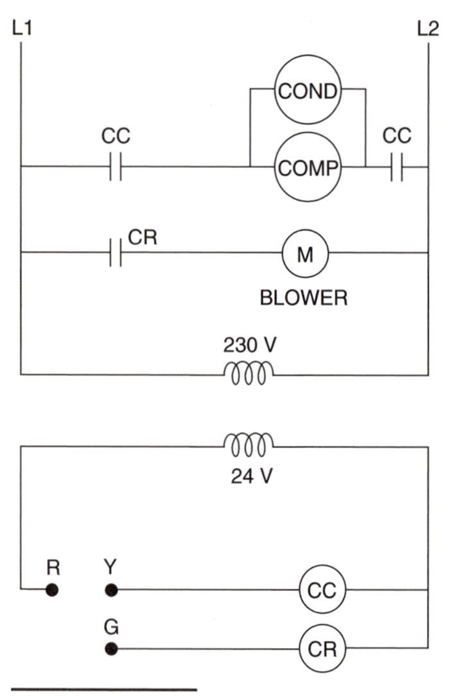

Figure 5–2 Rooftop air conditioner.

conditioner as in figure 5–2. The low-voltage room thermostat controls a compressor and condenser from a single contactor CC and a blower motor from a control relay CR.

Suppose that the condenser fan was running, but the compressor was not. The logical first step would be to go directly to the compressor, because that is the only device not working. It would not make sense to check the contactor, because we know that its switches are closing because we can observe that the condenser fan is running. It would not make sense to check the transformer, or the thermostat, or the fuses in the power supply for the same reason. It would not make sense to check the blower relay, or the blower motor, or the condenser fan motor, because none of these components could fail in a way that would cause the compressor to not run.

If neither the compressor nor condenser were running, but the blower motor was, your troubleshooting approach would be different. While it is certainly possible that both of these loads could have failed, it is extremely unlikely, and your initial approach should be to proceed as if there is only a single device that has failed. What devices in this circuit, if they failed, could cause all of the problems you observe? It could be a failed compressor contactor, or a failed room thermostat (not making R-Y). But the problem could not be a failed transformer or a blown fuse in the power supply (the blower motor is running).

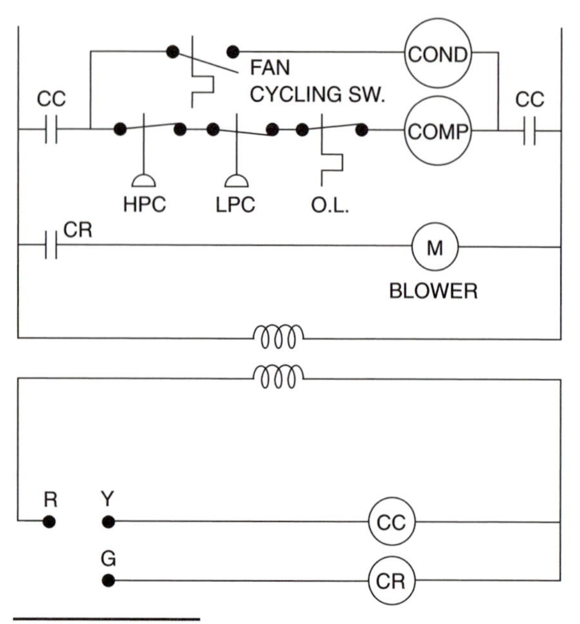

Figure 5–3 Example.

EXAMPLE:

Suppose the wiring for the above problem were slightly more complicated, as shown in figure 5–3. There is a high pressure cut-out (HPC) and a low-pressure cut-out(LPC) and a thermal overload (OL) in the compressor circuit. There is a fan-cycling switch in the condenser fan circuit. If the condenser fan and blower are both running but the compressor is not, what devices would be logical to check first?

SOLUTION: Compressor, HPC, LPC, and OL. No other single device could have failed that would explain the symptom that has been observed. ■

JUMPER WIRES

When used properly, a jumper wire (a wire with an alligator clip at each end) is a valuable troubleshooting tool. Used improperly, a jumper wire can cause damage to an electrical circuit, and just as importantly, can cause embarrassment to you and your company when you create a short circuit and make all the lights go out in your customer's place of business. The most important rule for use of a jumper wire in the field is this:

Never place a jumper wire across a load. Only place a jumper wire across a switch.

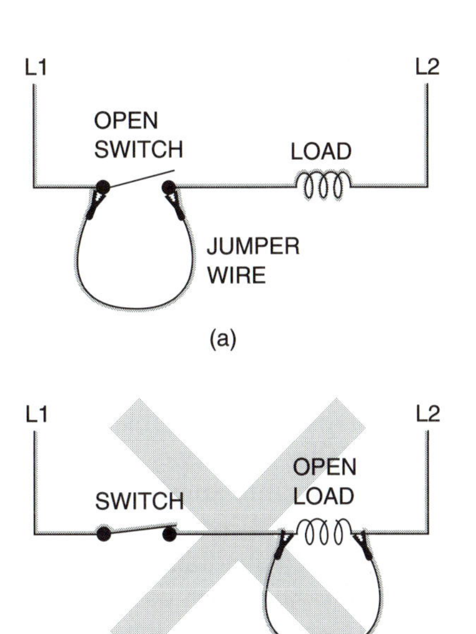

Figure 5–4 Never place a jumper across a load.

The reason for this rule is illustrated in figure 5–4. Suppose that the switch has failed. If you place a temporary jumper wire across the switch (5–4a), it will complete a circuit through the load, and the load should run. However, suppose the switch is still good, and the load has failed. If you place your jumper wire across the load (5–4b), it will create a short circuit. Aside from blowing a fuse or a circuit breaker, the high current flow through the switch could easily ruin the switch because it was never designed to carry so much current. When you place a jumper wire across a switch, you will not get into that type of trouble. You are only making contact between two electrical points that normally get connected together when the switch closes normally.

There are two other precautions to observe when using a jumper wire:

1. Make sure that your jumper wire is sufficiently heavy gauge to carry the current. If it is as heavy as the wiring that is attached to the switch you're jumping, that is sufficient.

2. If the alligator clips on the ends of your jumper wire are not insulated, make sure the power is off when you attach them, and make sure that they are securely fastened and will not move and create unwanted contact when you turn the power back on.

3. If you jumper across a switch in a live circuit, there will be a spark created when you attach the second alligator clip. That occurs because you are completing a circuit. Don't get startled and drop the alligator clip.

4. Do not place a jumper wire across a switch that carries a high current. For example, do not jumper across the switches of a contactor that is supplying current to a compressor.

Figure 5–5 shows an alternative way of jumping across a switch when you don't have or can't use a jumper wire. The wire(s) on one terminal can be moved onto the other terminal. This creates the same effect as if the switch had closed.

Why Jumper Wires?

A jumper wire is not to be used to find the problem. You use your volt meter for that. But after you have found an open switch with your volt meter, you should jumper across the switch to make sure that everything will work properly after you replace the switch. If you don't do that extra step, the following sad story can happen to you.

You find that the compressor in figure 5–6 won't operate because the LPC is open. You tell the customer that it will cost $80 to replace it, and the customer approves. After replacing the LPC, you find that the compressor runs, but it doesn't cool. Further investigation reveals that there is no charge in the unit. You find the leak, and advise the customer that it will cost a lot more than $80 to fix the

Figure 5–5 Jumpering a switch without a jumper wire.

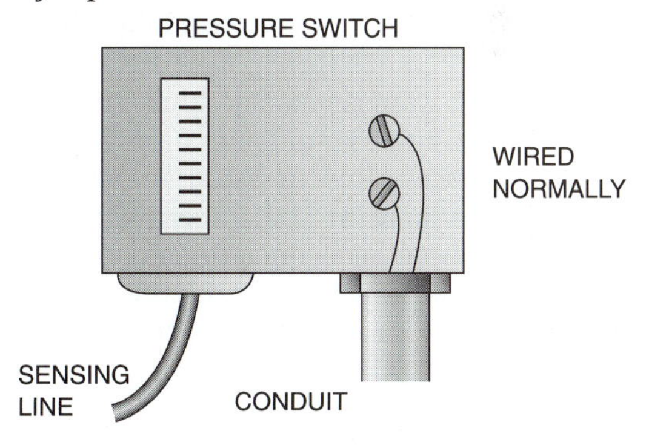

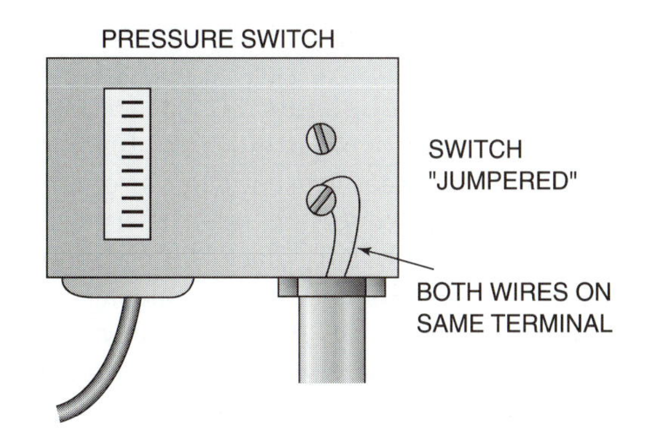

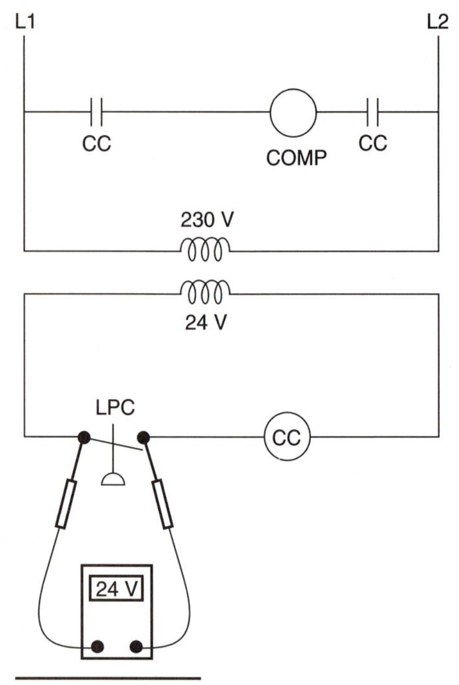

Figure 5–6 The LPC is open, but is that the problem?

leak and evacuate and recharge the unit. Your customer will be upset, because if had you given him or her all of that information before you replaced the pressure switch, s/he might have decided to replace the unit instead of repairing it. At the very least, you will appear to be incompetent if you quote a price to repair a unit, and then go back to the customer to say that you were only kidding.

The same story can happen in a number of different ways. You can replace a high pressure switch only to find that your condenser fan has failed. You can replace a low pressure switch only to find that there is a restricted metering device. You can replace a thermostat only to find that the load that it is controlling has shorted, and the new thermostat fails for the same reason as soon as it is installed.

Jumper Wire Caution

The last situation described (a shorted load) can cause a problem when you use a jumper wire. Assume that the circuit in figure 5–2 has a shorted coil in the compressor contactor. The high current through the thermostat caused it to fail open. When you arrive, you discover the open thermostat, but not the shorted coil. You place your jumper wire across the R-Y terminals of the thermostat, and you recreate the short circuit. If you do not remove the jumper wire immediately, you can burn out the transformer. Therefore, as a general rule, if you place the jumper wire and it doesn't produce the anticipated result immediately, turn the power off (or remove the jumper) and look for another problem.

KEEP CHANGING THE QUESTION

Figure 5–7 shows a compressor that is energized by a contactor, which, in turn, is energized by other controls. The compressor does not run. This troubleshooting strategy involves starting with the disabled load, and asking "Why isn't the load running?" There are only two possible answers—either the load is not getting energized, or if the load is getting energized, then it must have failed. Measure the voltage across the load. If it's 230V, you're done. If it's zero, the new question becomes, "Why isn't the compressor getting energized?"

The only possible answers are that either the compressor contactor switch(es) is (are) not closing, or there is no power available to the circuit. Measure voltage across L1-T1. If it's zero, you need to check the incoming voltage between L1 and L2. If you read voltage across the switch, it's open, and the new question becomes, "Why is the contactor switch open?"

Again, there are only two possible answers—either the contactor coil is not getting energized, or if the contactor coil is getting energized, then the contactor must have failed. If the contactor coil is not getting energized, the new question becomes, "Why isn't the contactor coil getting energized?"

You may find that the contactor coil is not getting energized because the normally open contacts from CR are open. Then you ask, "Why are the contacts open?" You may find that the contacts are open

Figure 5–7 Keep changing the question.

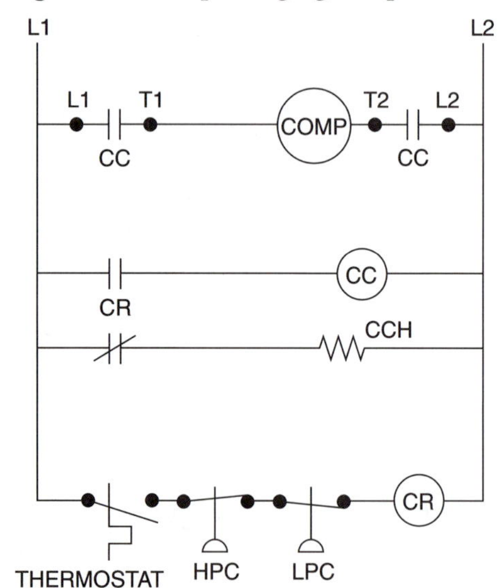

because the CR coil is not getting energized, and then you ask, "Why is the CR coil not getting energized?" Eventually, you will get to a point where there are no more questions. For example, you find that the LPC is open because the unit has leaked out all of its refrigerant. At that point, your troubleshooting ends, and the repair can go forward.

CUTTING THE PROBLEM IN HALF

If you had to guess a secret number between 1 and 16, random guessing would usually get the right answer in about 8 guesses. However, if you could guess in the middle of the possible numbers each time, and you were told whether the actual number was higher or lower, you would guess the right answer in no more than 4 guesses. This is the principle involved with cutting the electrical troubleshooting problem in half.

In a complex wiring diagram, there can be dozens of potential culprits. Figure 5–8 shows a contactor in series with eight switches. These may be pressure switches, temperature switches, safety switches, contacts from other relays. You have read zero volts across the compressor contactor coil, and you suspect that one of the eight switches is open. If you take hopscotch readings as shown, you may wind up having to follow the wiring and identifying eight different devices and taking 16 different readings before you get to the actual culprit, switch #8. Why not take your first reading at switch #4 or switch #5? If you get 230V, then you know all of the switches ahead of it are closed, and the problem is to the right of the measured switch. If you get 0V, the open switch is one of the ones to the left.

Another way of cutting the problem in half is to go directly to the contactor for the motor that is not running. Manually depress the contactor switches using a *well insulated* screwdriver. If the motor still does not run, the problem is in the power wiring part of the circuit. If it does run, then the problem is in the control voltage part of the circuit.

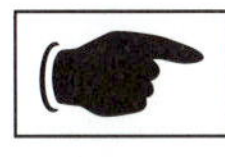

When the customer or the maintenance manager is watching you work, this is a particularly good strategy. It is very impressive to the customer when the service technician can make the system run within just a few seconds after removing the access panels!

Don't hold the contactor switches closed for more than a few seconds. The problem might be that the contactor coil is not getting energized because of an open safety control. By holding the contactor switches closed, you are bypassing the safety control, and subjecting the system to whatever hazard the safety control was protecting it against.

IS THE COMPRESSOR WARM?

You have come upon a system whose compressor is not running. You test the compressor motor wind-

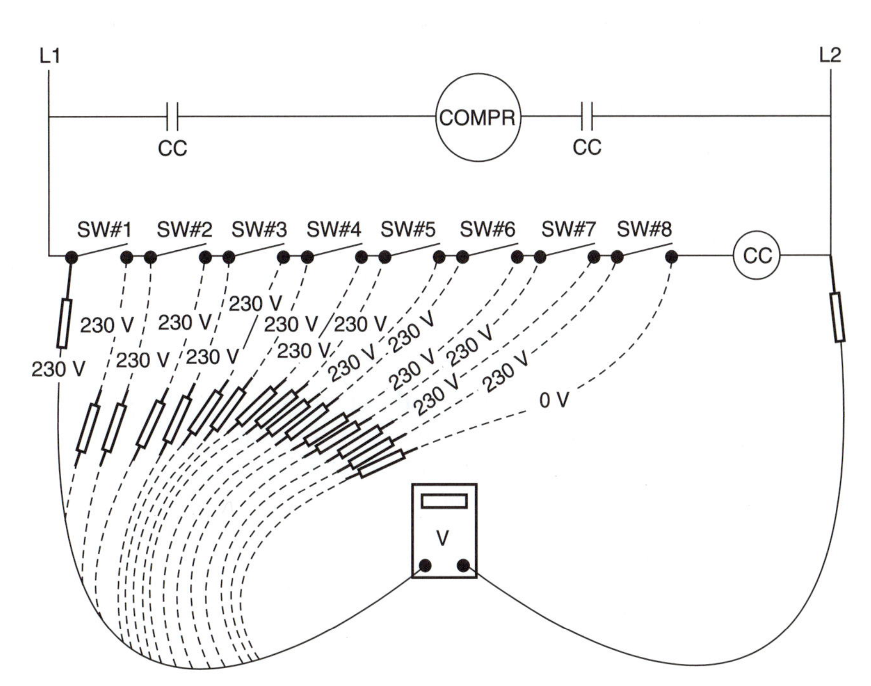

Figure 5–8 Cutting the problem in half.

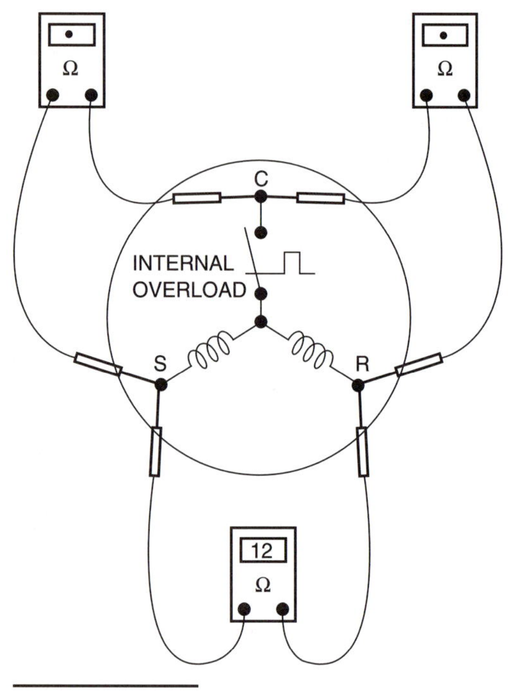

Figure 5–9 Resistance readings indicate that the windings are OK but the internal overload is open.

ings. The readings between the pins are infinite, infinite, and 12Ω. This indicates that the windings are continuous, but the internal overload is open (figure 5–9). On a hermetic compressor, the overload cannot be replaced or bypassed. If it does not re-close, the compressor must be replaced.

Feel the compressor. If it is not warm, replace it. However, if it is at all warm, that indicates that the compressor tried to start not too long ago, and it went out on the thermal overload. It may take an hour or more for the motor windings to cool sufficiently to make the overload re-close. Lock out the unit, and go to another job. Come back when the compressor is at ambient temperature. If the overload still has not closed when the compressor is cold, then the compressor must be replaced.

TURN THE UNIT OFF, AND THEN ON

If you come upon a unit that has nothing running, but voltage is available coming into the unit, the unit may be in a lock-out condition. Before even trying to determine if there is a lock-out relay, simply turn the unit off at the disconnect, and then back on. If the unit starts up, but then stops again, there is a good possibility that there is a lock-out relay. You may be able to troubleshoot during the short period when the unit runs, before going into lock-out. Otherwise, you may need to bypass the lock-out system by disconnecting a wire from the lock-out relay.

MAKE DECISIONS FROM NON-ZERO VOLTAGE READINGS

You always need to be aware of the possibility that your volt meter is either not functioning properly, or set incorrectly. For example, test leads do not last forever. Sometimes, a test lead on your meter can appear to be perfectly normal, but inside the insulation, the wire has broken. If that happens, every voltage reading you take will be zero. Make sure that your meter actually reads a non-zero voltage before you get involved in each troubleshooting project. Then, using the first two rules of troubleshooting, you will eventually get to a switch or a non-functioning load that has a voltage across it, and you will have found the problem.

CYCLING—TIMING

Sometimes, you come upon a unit that has a component cycling. That is, it is turning on, then turning off, and repeating the cycle continuously. For example, you might have a compressor that cycles on for five seconds, off for 30 seconds, on for five seconds, and so on.

Whenever a load is cycling, the problem is most likely not at the load. There is a switch that is opening and closing to cause the load to be energized intermittently. Sometimes you can get a clue where to look by simply observing the length of the cycles.

If a compressor shuts down within just a few seconds of starting, and then restarts within a minute, check the low pressure cut-out. There is a good chance that the system is low on charge, and the low-side pressure reaches the cut-out setting very quickly. When the compressor turns off, the low-side pressure rises to the cut-in pressure, and the cycle repeats.

Sometimes a motor cycles, but over a much longer time period. For example, it might stay off for two or three minutes or more before restarting. This sounds like a thermal overload is opening and closing. When an overload must cool down before re-closing, it takes much longer to react than a pressure switch.

Cycling on fan motors can be fast or slow. For example, a condensing fan motor that cycles quickly might have a pressure switch with the differential set too close. But a condensing fan motor that cycles over a longer period might have a faulty run capacitor. A run capacitor makes a motor operate more efficiently. When it has failed, the motor draws too many amps, and the internal thermal overload opens. When it cools, the cycle repeats.

When a load cycles, make sure that you take your voltage readings across switches while the motor is de-energized. Otherwise, you might go right past the offending switch if you happen to measure across it while it is closed.

TWO FAULTS

Occasionally you will encounter a situation where two devices in the same circuit are open. For example, in figure 5–10, suppose that both of the limit switches are open. If you are using an analog meter, testing voltage across either of the limit switches will give you a zero reading, not because there is not a voltage drop from one side of the switch to the other, but because of limitations on how the meter works. The wire between the two switches is an "island." That is, it is not connected to either side of the transformer, or any other part of the circuit. When one probe of the volt meter is touched to the island, it is almost as if that probe is not touching the circuit at all.

If you are using a digital meter, the result will be somewhat different. Digital meters are more sensitive than analog meters, and reading across either switch will give you a "funky" reading. Usually, when you measure voltages, you expect a reading equal to the voltage available (230V, 115V, 24V, etc.) or a zero reading. But in this case, the digital meter will give you a reading that is more than zero, but less than the available 24V.

EXERCISE:
For each type of volt meter available to you, set it on AC volts, and place one probe in the Hot connection of an outlet. Do not connect the other probe to anything else. What voltage do you read? Then, hold the other probe between your fingers. What voltage do you read? Why did the voltage reading increase? (Hint: The volt meter relies upon a flow of current through the meter itself in order to measure voltage. The higher the current flowing through the meter, the higher the voltage reading.)

So what do you do if you have 24V available at the transformer secondary, but you cannot find any device in the circuit that has a 24V drop? The answer is to take hopscotch voltage readings. Park one probe on the right side of the wiring diagram as shown in figure 5–11. Starting with reading #1, there are 24V available. Keep moving one probe until you lose voltage. In this illustration, you will lose voltage between the third and the fourth readings, indicating that the first limit switch is open. Then, you can approach the problem again from the other direction as in figure 5–12. With one probe parked on the left side of the wiring diagram, move

Figure 5–10 Two faults.

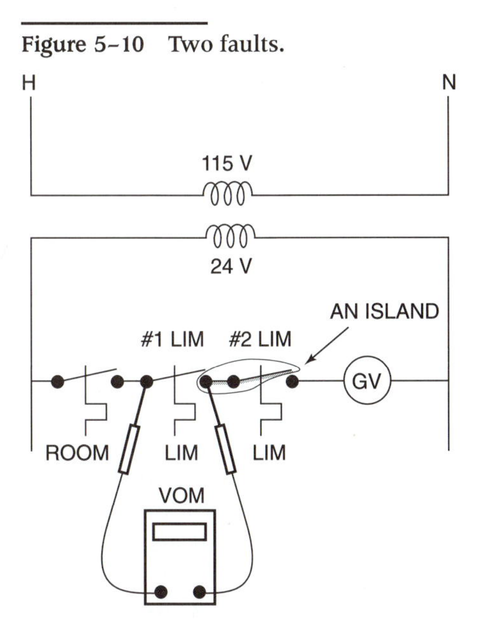

Figure 5–11 Parking one probe.

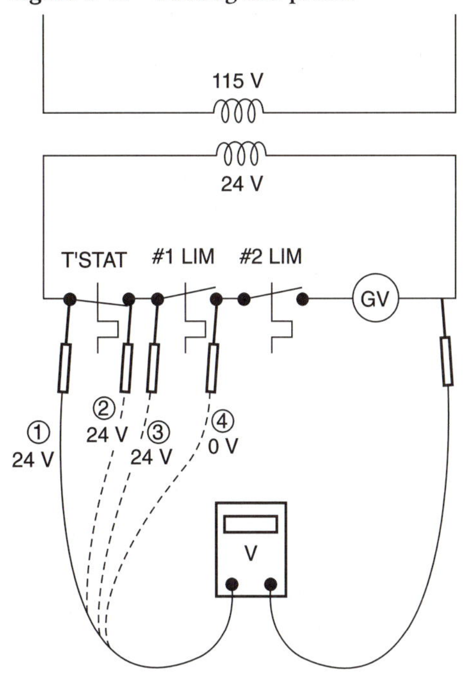

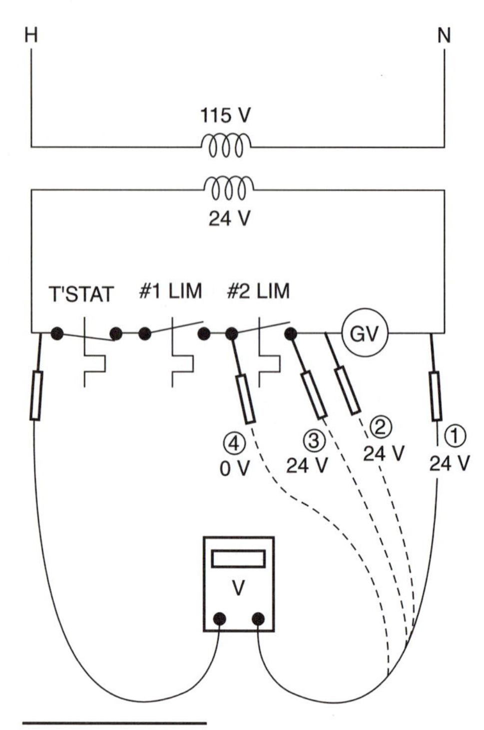

Figure 5–12 Parking the other probe.

the right probe. You will lose voltage between the third and fourth readings, indicating that the second limit switch is also open.

BROKEN WIRE

A broken wire presents a situation similar to the one above. You have voltage available, but there is not voltage at the load, and the voltage readings across each of the switches is zero (figure 5–13a). Start hopscotching (figure 5–13b). Reading #1 tells you that there is voltage available. Reading #2 tells you that the thermostat is closed, because it is reading the same electrical pressure as the wire coming into the thermostat. Reading #3 tells you that between the connection leaving the thermostat and the connection entering the HPC, you have lost electrical pressure. This reveals that there is a broken wire (or a poor connection) between the thermostat and the HPC.

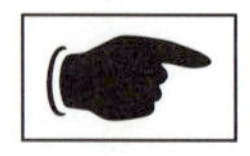

Some technicians will park a probe on a ground, instead of the right side of the diagram which happens to be Neutral in this case. This is OK for 115V circuits only. Ground is at the same zero electrical pressure as Neutral. However, this will not work on a 230V circuit. See figure 5–14. Suppose the thermostat is open. Readings on either side of the thermostat will give 115V.

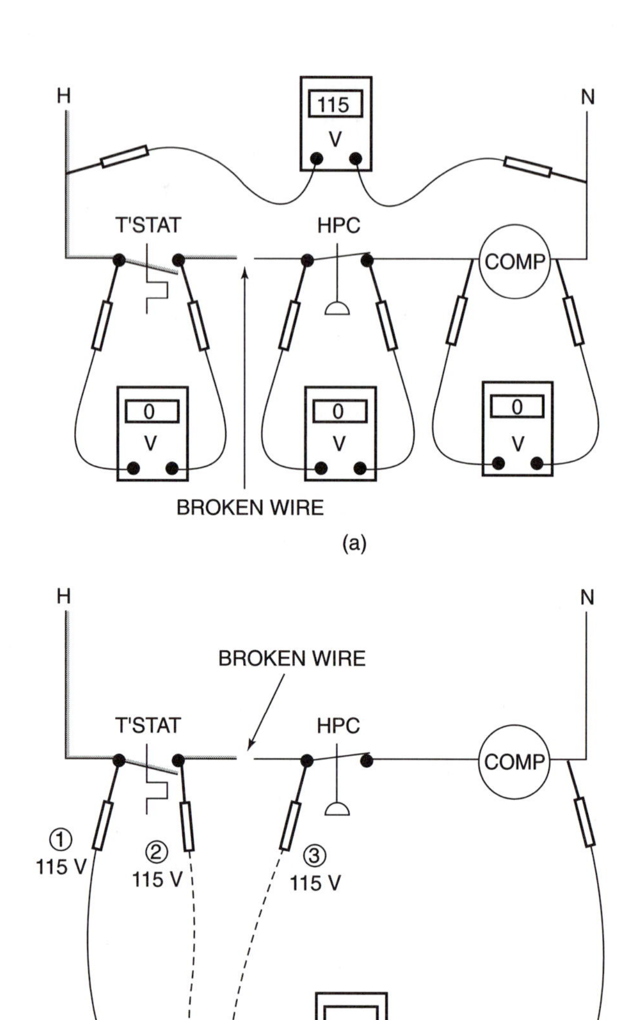

Figure 5–13 Diagnosing a broken wire.

The technician would not realize that reading #1 is actually reading L1-Ground, and reading #2 is actually reading L2-Ground.

FINDING A GROUND

Finding a ground is one of the trickiest troubleshooting problems. A ground exists when an electrical conductor touches a metal casing. What makes it tricky to troubleshoot is that if there is a ground fault in one device, every point that is electrically connected to that device will also appear to be grounded. See figure 5–15. If the control relay coil is grounded, then measuring from M1 to ground will indicate continuity, even though there is nothing wrong with M1.

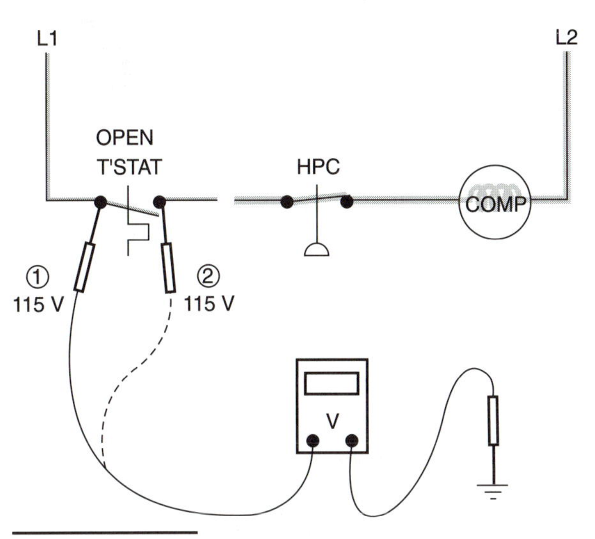

Figure 5-14 You can't use ground to troubleshoot a 230V circuit.

The ground condition could be caused by a switch or a load that has failed, or it could be a wire whose insulation has worn or been cut, and the conductor inside is touching the casing. The way to find it is to select one place where you are measuring continuity to ground. Then, start disconnecting devices, one by one, rechecking for the ground condition each time until the ground condition disappears. Then you will have isolated the ground to the part of the circuit that you just disconnected, and you can start looking for a ground condition there.

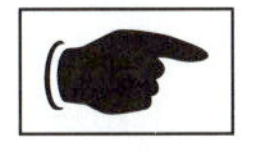

When disconnecting more than one or two wires, use masking tape on each wire, and identify where each goes. It's not hard to lose track of where everything goes, and figuring it out is a potentially time-consuming chore that can easily be avoided by proper wire-marking. You can simply assign a number to each wire, and make a sketch for yourself showing which numbered wires attach to which terminals.

Figure 5-15 Finding a ground fault.

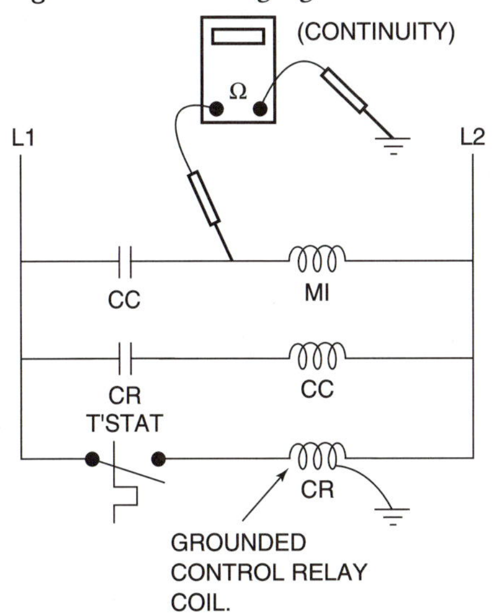

WHICH ELECTRICAL MEASUREMENTS CAN YOU TAKE TO THE BANK?

You may be familiar with the terms false-negative or false-positive when referring to medical testing. These terms refer to medical tests where, for example, if the test shows positive, you can be sure that you have the condition tested for, but if the test shows negative, it is likely (but not sure) that you don't have the condition.

A similar situation arises when taking electrical measurements for troubleshooting. Sometimes, your reading is one that gives you a sure answer. Other times, your answer is just a maybe. Following are some examples.

Infinite Resistance or Zero Volts

Any reading of infinite resistance or zero volts, by itself, cannot be relied upon with certainty. Either one may be an incorrect reading if your meter is not set on the correct scale, an internal fuse in your meter has opened or if your meter has a test lead that is broken inside the insulation. Check all of these factors by touching your probes together when using the ohm scale, or measuring a place where you know there is voltage present when using the volt scale. Note that none of these factors could be present if you had a non-infinite resistance reading or a non-zero voltage reading.

Some meters have very limited ohm ranges that are unsuitable for HVAC/R troubleshooting. One high-tech-looking digital meter only has an ohm scale that ranges from 0-2000 ohms. Resistances over 2000 ohms will display as infinite resistance. This is unsuitable because a clock motor (for example) might have a resistance of 3000 ohms. If you checked the motor windings with this meter, you might mistakenly conclude that the motor winding is open.

Other inexpensive analog meters might have several ohm scales, but the lowest is the 1K scale. On these meters, you would not be able to distinguish between zero ohms and just a few ohms. Therefore, you would not be able to tell if a motor winding was good or if it was shorted.

Compressor Windings

If you measure an open winding (infinite resistance), you can be sure that the compressor has failed. But beware of internal overloads that may be open. If you are not aware, an open internal overload can fool you into thinking that you have measured an open winding.

If you measure a ground condition (less than infinite resistance) between any compressor pin and the compressor casing (with all external wiring disconnected) the compressor has failed. But beware—the new, very accurate digital ohm meters can read out millions of ohms whereas older meters would display millions of ohms as infinite ohms. If your ohm meter shows millions of ohms when checking for a ground condition, you can take this as an infinite resistance reading.

Mechanical Devices

Many devices such as motors, solenoid valves, contactors, and relays use electricity to produce some mechanical motion. Electrical testing of these devices can tell you for certain that the device has failed. However, even if all of the electrical readings show that the device is good, that's no guarantee. Even though these devices may check out OK electrically, they can still have failed mechanically. Compressors can seize, bearings can fail, and any moving mechanical device can be stuck.

Shortcut Readings

You have learned that when taking an ohm reading across a device, you must remove the wiring from at least one side of the device to obtain a reliable reading. But moving wiring is time consuming, and on old wires with crumbling insulation or brittle conductors, you may be getting yourself into an unwanted rewiring job. Some technicians will measure ohms *without* removing the wiring. If it reads infinite ohms, you can be sure that the device is open. If it reads continuity, then *that* reading cannot be believed until one side of the device is disconnected.

EXERCISE:
Explain, using wiring diagrams as examples, why, when the wiring has not been removed from a device, an infinite ohm reading can be believed, but a continuity reading cannot.

FAILED TRANSFORMERS

If, while troubleshooting, you discover a zero voltage reading on the secondary of a transformer, and a correct voltage reading on the primary, you can be sure that the transformer has failed, and must be replaced. But every once in a while, when you replace the transformer, it will immediately fail also. If this happens, you must locate a short within the rest of the control wiring that is causing the transformer to fail.

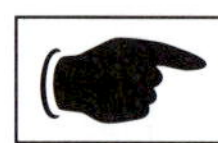

Some technicians may ohm out the transformer windings, and find that it is the primary winding that has opened. They will then mistakenly assume that there is a problem on the primary side of the circuit. However, there is nothing that can go wrong on the primary side of the circuit that would cause the transformer to burn out. The problem *must* be on the secondary. A short on the secondary side will draw more amps from the transformer secondary than the transformer can supply. The secondary winding will, in turn, draw more current from the primary winding. Even though both transformer windings are carrying higher than design amps, it is the primary that is more likely to fail first because the primary wire is thinner than the secondary wire.

In order to locate the short, disconnect the secondary portion of the circuit from the transformer. Using an ohm meter, measure what the transformer secondary is "seeing" (figure 5–16). If there is a

Figure 5–16 Testing the control circuit for a short.

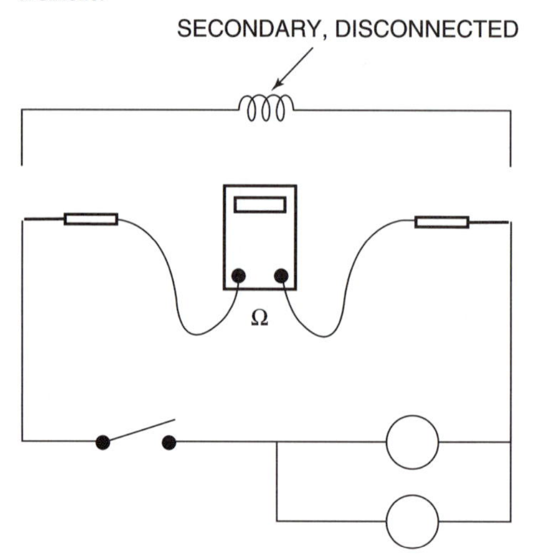

short, the ohm meter will read zero ohms. Disconnect devices from the circuit until the short disappears. When the shorted device or wiring has been repaired, then the transformer can be replaced.

If the transformer you are replacing is particularly expensive, you should perform this check before replacing it. But with most small systems, the transformer costs less than $10. Many technicians will save time by not performing this check. If once each 20 times the replacement transformer also fails, the cost of the extra transformer is probably worth having avoided spending the extra time on the other 19 times.

Technicians have been known to replace a part such as a transformer, only to find that the new one is also failed. They say to themselves, "How do you like that! This new replacement part was defective, right out of the box." They then get another replacement part from their truck, and after installing it, find that it also has failed. Some will say to themselves, "Isn't that something! What are the chances that I would have two new replacement parts that were defective right out of the box?" Don't fall into this trap. The chances of a new part being defective are very small, and certainly much smaller than some technicians would have you believe. The chances of it happening twice on the same unit are small enough to say it's almost impossible. If you find a new replacement part is failed in the same way as the part you are replacing, you need to look for something in the system that is causing the part to fail.

EXERCISE:
Suppose you find a voltage reading of zero volts on the secondary, and 120V on the primary. You know that the transformer has failed, but do you know which winding has failed? Explain your answer.

CHAPTER SIX

TESTING AND REPLACING COMMON DEVICES

After you have completed your troubleshooting using voltage, you will want to retest the individual device that you have identified as the problem. This serves two purposes:

1. It confirms your troubleshooting conclusion, eliminating the occasional "brain failure" experienced by all technicians at one time or another.

2. It identifies the specific mode of failure for the component. This is sometimes required for warranty replacement of failed parts.

MANUAL SWITCHES

Usually, you have determined that a manual switch has failed because you have read voltage across it, indicating that it is open. Remove the manual switch from the circuit. Using your ohm meter across the terminal contacts, you should measure almost zero ohms when the switch is supposed to be closed, and infinite ohms when the switch is supposed to be open. If you measure infinite ohms across the switch when it is in the closed position, the switch has failed.

When you replace a manual switch, check the amp rating. This is the maximum current that the switch contacts can safely carry. Also, check your actual circuit volts. The replacement switch *amp rating* must be the same or higher than the amp rating of the failed switch. The replacement switch *volt rating* must be at least as high as the actual circuit volts. There is no electrical problem if the replacement switch has a much higher amp or volt rating.

AUTOMATIC SWITCHES

Automatic switches are those that sense something (pressure, temperature, humidity, flow, level, etc.) and open or close a set of contacts when a certain "set point" value of the sensed variable is reached. A voltage reading across the switch terminals indicated that the switch is open. After the switch has been disconnected from the circuit, the switch contacts can be tested again with an ohm meter to determine if they are open or closed. However, with an automatic switch, you also need to determine if the switch should be open or closed while you are testing it.

HIGH PRESSURE CUT-OUT

The HPC will usually fail in the open mode. If it failed in the closed mode, and it was the only device that had failed, there would be no service call. Test ohms across the switch while the compressor is not operating. The switch should be closed. If you measure infinite ohms across the switch, it has failed. Only after you determine that the HPC has failed do you remove the pressure sensing line from the system.

Condenser Fan Cycling Switch

The condenser fan cycling switch closes when the high pressure rises high enough that the condenser fan is needed. When the compressor is not running, it should be open because the high-side pressure is too low to close it. If the condenser fan cycling switch fails, it will usually fail in the open position, causing the condenser fan to not run, and maybe causing the compressor to cut out on the HPC. If the condenser fan switch fails in the closed position, there will probably be no service call.

As with manual switches, when replacing an automatic switch, the switch contacts must be rated for amps at least as high as the failed switch, and for volts at least as high as the actual circuit voltage.

Transformer

When troubleshooting why a circuit doesn't work, you may eventually be led to the transformer, and you find that you are not getting 24V from the transformer secondary. You must then measure voltage at the transformer primary. If the correct voltage (usually 115V or 230V) is present, then the transformer has failed. From these readings, many novices would mistakenly conclude that the secondary has failed. But, in fact, either winding could have failed. As shown in figure 6–1, with either winding open, you will get the same set of voltage readings.

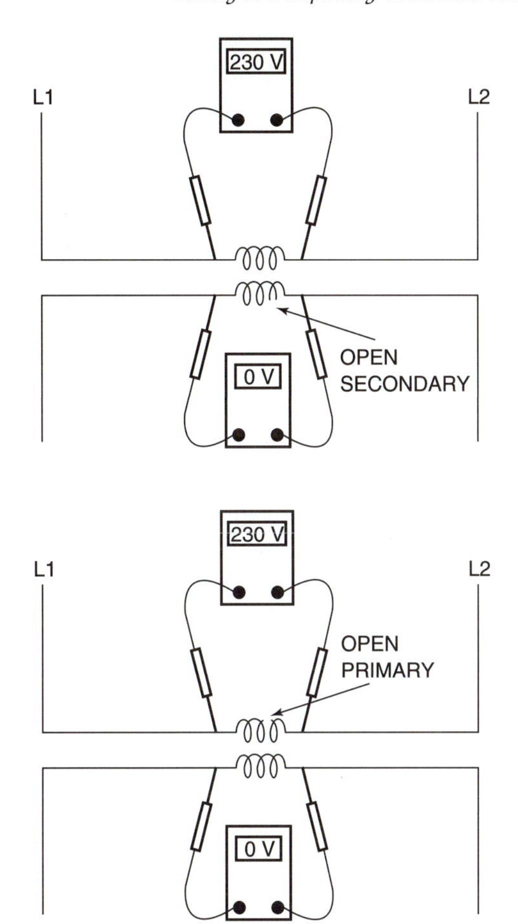

Figure 6–1 Open primary gives same readings as an open secondary.

Sometimes you need to know *which* winding has failed. On warranty replacements, you normally need to fill out a card describing the failure. The manufacturer who is receiving return of the failed part would much prefer a description that says "primary coil open" or "secondary coil open" than a simple "broken" or "no 24V."

In order to determine which coil has failed, you need to remove the transformer from the circuit and measure the ohms across the secondary and across the primary. It is possible for either winding to be either shorted or open, but you will usually find that the failure is an open coil, and more often than not it is the primary rather than the secondary that has opened.

When replacing the transformer, you must check three numbers: the primary voltage, the secondary voltage, and the VA rating. The primary voltage and secondary voltage must match the failed transformer. The VA rating of the replacement transformer must equal or exceed the failed transformer.

Transformer VA Rating

The **VA (volt-amps) rating** is a measure of how much power the transformer can supply at the rated voltages. In other words, a transformer with a 115V primary and a 24V secondary and rated for 20 VA might be sufficient to supply enough power to supply two contactors. However, if the circuit required enough 24V power to supply four contactors, a 40 VA transformer might be used.

ADVANCED CONCEPTS

Manufacturers request return of failed parts so that the failure mode can be analyzed by their engineers. If they detect a pattern of failures, they will take steps to modify the design to improve the reliability of the product line.

ADVANCED CONCEPTS

The VA rating is the product of the secondary voltage times the maximum number of amps that can be supplied at the secondary voltage. It can be expressed as:

$VA = Volts_{sec} \times Amps_{sec}$

The above 20V transformer would be capable of supplying slightly less than one amp at 24V, as follows:

$20VA = 24V \times Amps$
$Amps = 20/24 = .83$

EXERCISE:
You have diagnosed a transformer with an open primary. You replace it with another transformer of the proper rating that you know to be good. Several seconds after the circuit is energized, the new transformer fails with an open primary winding. You suspect a short in the system. Will you search for it in the primary voltage portion of the circuit or in the secondary portion of the circuit? Explain your answer.

OVERLOADS

An overload may be of the **line duty** or the **pilot duty** type. A line duty overload (figure 6–3)is used on compressors in small systems. It is wired in series with the common terminal of the compressor, and therefore it carries the total current draw of the compressor. If the current draw of the compressor exceeds the rating of the overload, the switch opens. On larger systems, a pilot duty overload is used (figure 6–4). The sensing portion of the pilot duty overload carries the compressor current, but the switch portion of the overload does not. The switch on the pilot duty overload is wired in series with the compressor contactor coil. On some air conditioning systems, a thermal pilot duty overload is used (figure 6–5). With this system, the overload sensing element is a thermostat that is imbedded into the stator windings of the compressor. If the windings draw too much current, the winding temperature will rise above the set point of the overload switch. When the overload switch opens, the compressor contactor is de-energized, and the compressor stops.

When you find an open overload switch (indicated by a voltage measured across it), there are two possible failures:

1. The overload has failed.
2. The compressor was drawing too many amps, and that caused the overload to trip. When it reset, it tripped again, and the cycle continued until the overload gave up.

If you don't check, you could wind up replacing the overload, only to find that the replacement overload also trips when the compressor runs. Before you replace the overload, you must check to see if the problem is simply a defective overload, or if there is also a defective compressor. You can do this by temporarily bypassing the overload with a jumper wire. Start the compressor and measure the

MORE TRANSFORMER ADVANCED CONCEPTS

Even though a transformer secondary is rated at 24V, its actual output may be a few volts higher or lower. As the load on a transformer increases (more 24V loads are added in parallel with existing loads), its output voltage decreases. The larger the load, the more a transformer's voltage output will drop below its open (no load) circuit rating. Most systems will work over a voltage range that varies some from the rated input voltage. However, an electronic control can burn out if the circuit voltage exceeds the maximum for that control. The standard for acceptable output voltage variation is set by the National Electrical Manufacturers Association (NEMA). It is shown in figure 6–2. At rated load, the output voltage should be 23 to 25V. At minimum load, it may be as high as 27V.

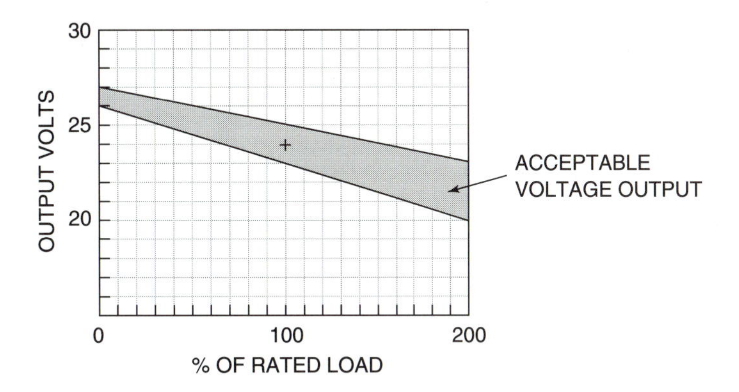

Figure 6–2 Voltage variation for a 24V transformer.

amp draw. *Do not run the compressor for more than a few seconds with the overload bypassed.* Compare the amp draw of the compressor with the rated full load amps. If the compressor amp draw is below the FLA, you can simply replace the overload. If the compressor draws more than the rated FLA, further investigation is required to determine why. The compressor may be low on oil, the condenser may be dirty, there may be a refrigerant restriction, etc. If the compressor does not start at all, there is another problem besides the open overload.

Universal Replacement Overload

While it is always best to replace overloads with an exact replacement, sometimes you are unable to determine the model of compressor or overload that you are working on. In this case, you can use a universal replacement based on the locked rotor amperage of the compressor. To find the locked rotor amps, do the following (the compressor motor windings must be at ambient temperature to use this method):

1. Disconnect the power supply and temporarily bypass the defective overload with a jumper.

2. Temporarily remove the wire from the start terminal of the compressor.

3. Place a clamp-on amp meter around the wire to the run terminal or the common terminal.

4. Momentarily reconnect the electrical power and read the amp meter.

5. If the unit does not have a start capacitor, multiply the amp reading by 1.33 to obtain the locked rotor amps.

6. If the unit does have a start capacitor, multiply the amp reading by 1.10 to obtain the locked rotor amps.

7. Select a new universal replacement overload from the locked rotor amps determined above.

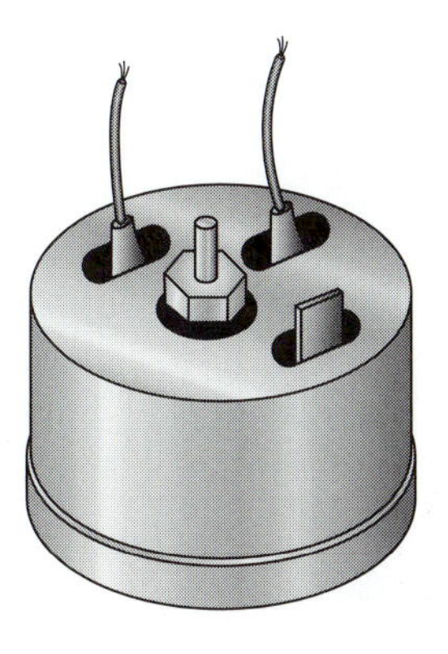

Figure 6–3 Line-duty overload.

CIRCUIT BREAKER

The circuit breaker (figure 6–6) protects the wiring that supplies power to the air conditioning or refrigeration unit you are troubleshooting. If you measure zero voltage at the unit, go to the circuit breaker panel (you might need some help from the customer to find it).

You may see individual circuit breakers, pairs of circuit breakers with their trip levers connected, or groups of three circuit breakers with their trip levers connected (figure 6–7). These are for single-phase 115V circuits, single-phase 230V circuits, and three-

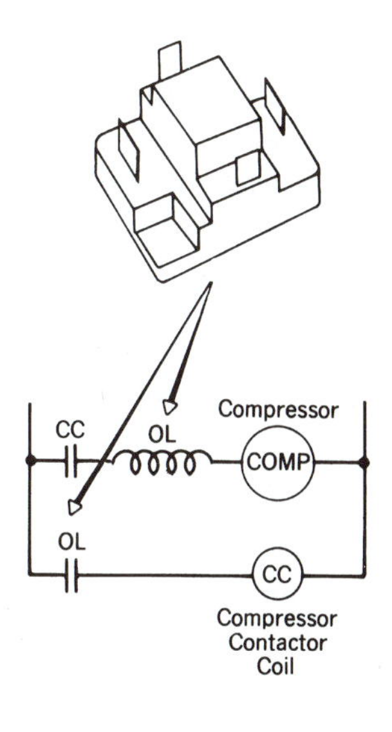

Figure 6–4 Pilot duty overload (current sensing).

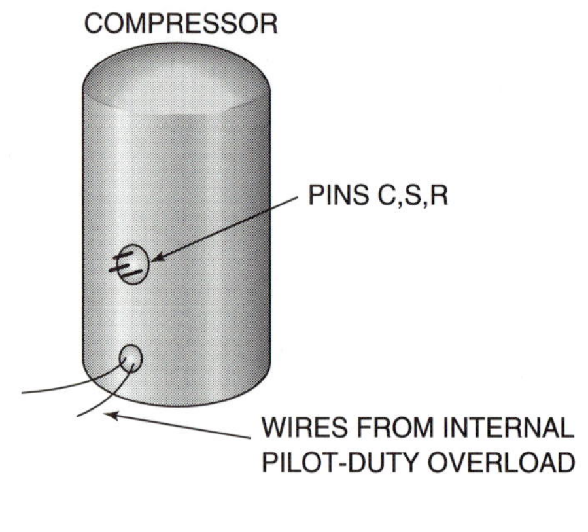

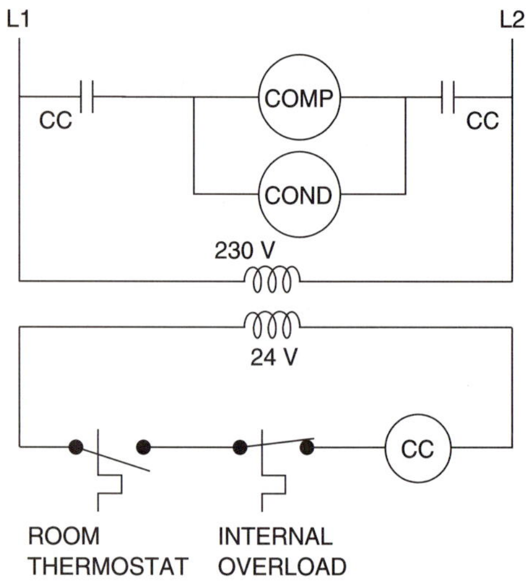

Figure 6–5 Pilot duty overload (temperature sensing).

phase circuits, respectively. You must determine which circuit breaker is the one that supplies power to your unit. If, for example, you're working on a 115V unit, you don't need to consider the circuit breakers that are ganged together in pairs or in threes.

Look at the circuit breakers. You will note that the trip levers are all aligned. However, if one of the breakers is tripped, its trip lever will be *slightly* out of alignment. Inspect the alignment carefully to determine which circuit breaker has tripped. If you cannot identify the tripped breaker, push each trip lever towards the ON position. Each breaker that is not tripped will feel solid. However, the tripped breaker will be "soft," and will allow you to push it a slight bit towards the ON position. That is your clue that the circuit breaker has tripped.

Sometimes a technician will turn each circuit breaker off and then back on, resetting each one. In commercial applications, this can be a disaster. Many stores, for example, will have cash registers that are also computers, and they

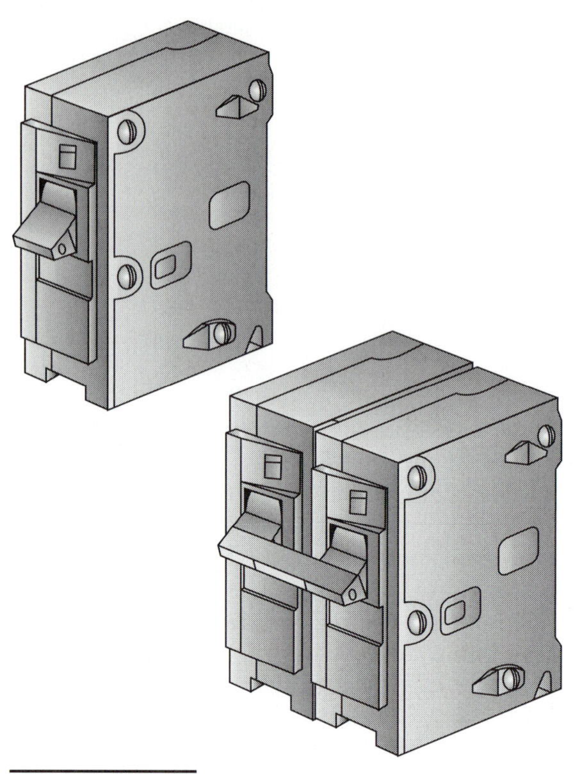

Figure 6–6 Single-pole and two-pole circuit breakers.

keep records of the day's transactions. By indiscriminately turning off circuit breakers, all the data could be destroyed. This is only one example of the problems you can cause by momentarily interrupting power to circuits when you don't know what's connected. You

Figure 6–7 Circuit breaker arrangements.

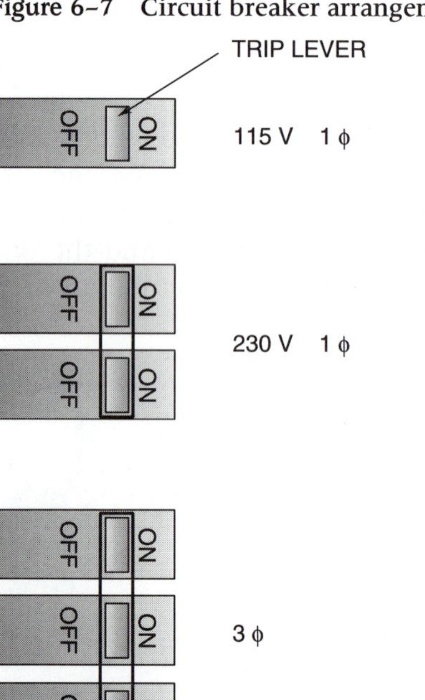

could cause significant grief to the customer and embarrassment to yourself. Don't do it.

To reset the circuit breaker, push the trip all the way to the OFF position, and then push it back to the ON position. If there is still no power to the unit, there are two potential problems:

1. There is a short in the unit that caused the circuit breaker to trip as soon as you reset it.
2. The circuit breaker is defective, and it will not reset.

Disconnect the power wiring from the unit (or trip the disconnect switch to Off if there is one). Then reset the breaker again. If it resets this time, then there is a short in the unit. If it does not reset, the circuit breaker is defective.

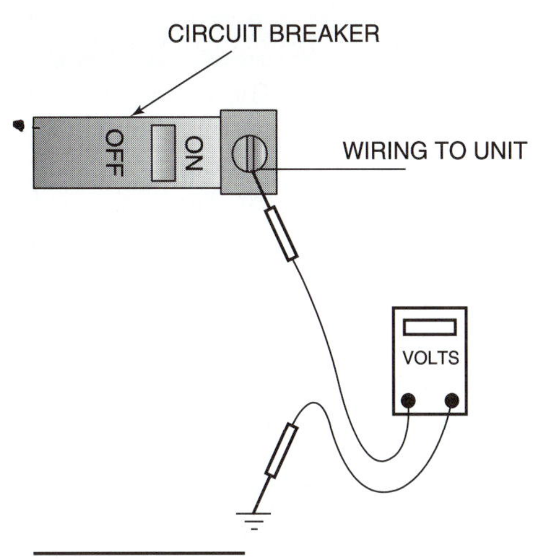

Figure 6–8 A voltage reading indicates that the circuit breaker is making contact.

Testing the Circuit Breaker Electrically

All of the above assumes that if the circuit breaker stays in the ON position, that it is supplying power to the unit. This is not necessarily so. If you remove the panel that covers the wiring connected to the circuit breakers, you will see that there is one wire connected to each circuit breaker (figure 6–8). Using your volt meter, place one probe on a ground connection, and move the other probe to each wire attached to a circuit breaker. If there is voltage present, the circuit breaker switch is closed. A zero reading between any circuit breaker and ground indicates that the switch inside the circuit breaker is not making contact, even if the trip lever indicates that it is ON.

Sometimes the circuit breaker is functioning normally, but you must turn it off so that you can do some service work on a unit. How do you identify which circuit breaker to turn off? Usually, you will find that each circuit breaker is labelled on the door of the circuit breaker box, identifying what loads are served by each circuit breaker. But on some systems, the identifications have been lost, or not kept up to date as the system was modified. There is a tool now available that consists of two electronic parts. One part is a transmitter. It is fastened to the power wiring of the unit that you wish to deenergize. The other part of the tool is a receiver. It is taken to the circuit breaker box, where it will provide an audible or visible signal when it is attached to the circuit breaker that matches the circuit where the transmitter is attached.

Contactors

Contactors contain both a load (the contactor coil) and a switch or switches (contacts). Sometimes the contacts are visible, and you can tell from simply looking at the contactor if the switches are closed ("pulled in"). Other contactors have the switches concealed beneath a removable cover. You can either remove the cover to observe the contacts, or you can check for voltage across the contacts by measuring across L1–T1, L2–T2, and L3–T3. If you read voltage across the switch, it is open. If you measure proper voltage across the coil, and you also determine that the switches have remained open, the contactor has failed. It may be a failed coil, or the coil could be normal and the moving switch mechanism is mechanically stuck. Or it could be that the switches moved to the closed position, but because of badly pitted contacts, one or more of the switches has not actually made electrical contact.

Ohm out the coil. If it has infinite resistance, it has failed. Otherwise, the failure is probably mechanical. On small contactors, you will simply replace the entire contactor. On larger contactors, the coil and the switches may be individually replaceable. However, in some cases, the extra time required to obtain and change the coil or contacts is more costly than it would be to simply replace the entire contactor with one that you stock on the truck.

When you replace a contactor, the coil voltage (24V, 120V, 230V) must match. The number of poles for the replacement must equal or exceed the failed contactor. You can use a three-pole contactor to replace a two-pole contactor by simply leaving one of the poles unused. The amp carrying rating on the switches (20a, 30a, 40a, etc.) must equal or exceed the amp rating on the failed contactor.

ADVANCED CONCEPTS ABOUT CONTACTOR CONTACTS

The current carrying capability of contacts is rated several ways, depending on the load being switched. *Inductive full load amps* are the maximum amps a contact can switch when it is controlling an inductive load such as a motor. This rating is for continuous duty. The inductive contact rating may also be expressed in horsepower, being the approximate maximum size of electric motor that may by switched by the contact. *Inductive locked rotor amps* are the rating of maximum current for short periods of time when a motor is first started. The initial current drawn by a motor is 4 to 6 times higher at start up than the full load.

The *resistive rating* is the maximum amps a contact is capable of switching when it is controlling a resistive load such as an electric heater. The resistive load rating is normally higher than the inductive rating since the resistive load causes less arcing at the contacts during switching and inrush is very low. Sometimes the resistive load rating is also given in equivalent kilowatts of electric heating for various voltages.

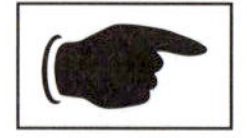

Suppose you have discovered a failed single-pole 40a contactor. You don't have a 40a contactor on the truck, but you do have a two-pole 20a contactor. You can use the two individual switches in parallel to replace a single 40a switch. Each of the two poles will carry half of the load (see figure 6–9). This arrangement will also be found on some factory-wired units.

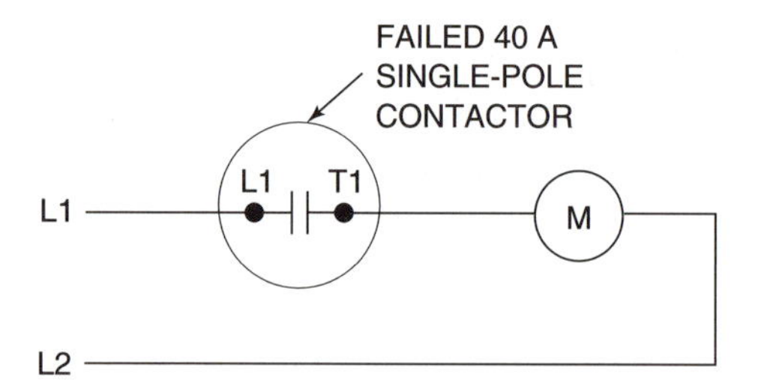

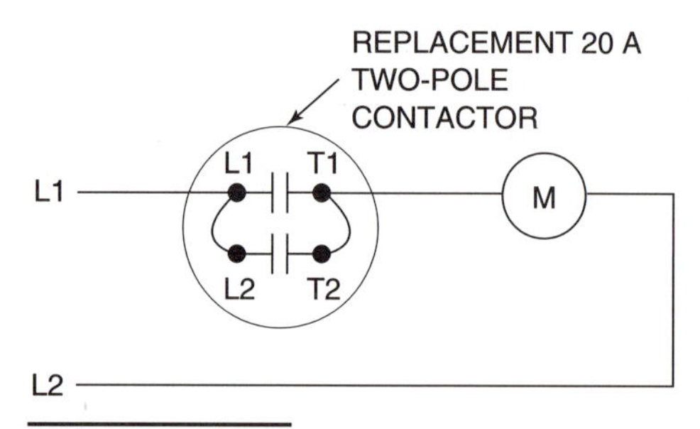

Figure 6–9 Using a two-pole contactor to carry twice its rated amps.

CONTROL RELAYS

Control relays are, in theory, the same as contactors, with the following differences:

1. Contactors have switches usually rated for 20a or more, while control relay contacts are usually rated for 18a or less.
2. The switches on contactors are always normally open, but the switches on control relays come in many variations, including NO, NC, SPDT, DPDT, 3PDT.

The troubleshooting and replacement guidelines for control relays is the same as for contactors.

CAPACITORS

Capacitors can fail in one of four ways:

1. The plates are shorted together (shorted capacitor).
2. There is a discontinuity between the two capacitor terminals (open capacitor).
3. There is a path from one of the capacitor plates to the capacitor casing (grounded capacitor).
4. The capacitor plates do not hold the correct amount of charge.

Of course, it is not necessary to diagnose or test the capacitor if you find that it has blown itself apart, as is sometimes the case.

Many technicians use an ohm meter (set to the highest resistance scale) to check capacitors. If the ohm meter shows that either terminal is grounded to the casing, or that there is zero or infinite resistance between the pins, you can take that reading to the bank—the capacitor has failed. If your ohm

ADVANCED CONCEPTS

The principal of operation of the ohm meter is that a battery inside the meter sends electrons out through one of the probes, through the device whose resistance is being measured, and then back through the other probe to the meter. The volt meter converts the rate of electron flow to a resistance reading. When the ohm meter is first attached to the terminals of the capacitor, the battery inside the ohm meter is able to push electrons onto the capacitor plates relatively quickly. However, as the capacitor plate approaches its limit in its ability to store electrons, the electron flow out of the ohm meter diminishes. This shows up as an increasing resistance.

reading between pins shows a resistance that initially falls towards zero and then switches so that the resistance is increasing, the capacitor is *probably* good. However, some capacitors may check out OK with an ohm meter, but then fail when they are under load in a circuit. This resistance test on capacitors is best done with an analog ohm meter. You can watch the needle swing to the right (towards zero ohms), and then turn around and swing to the left as the resistance increases. Early models of digital ohm meters (and some modern models as well) are not suitable for using to test capacitor resistance. Because the actual resistance is continuously changing, some digital ohm meters cannot "zero in" on one correct reading.

Capacitors are best checked with a capacitor tester, and they will almost always give you a correct determination of whether a capacitor is good or failed. It not only measures whether or not the capacitor stores electrons, but it also determines the microfarad capability of the capacitor. Some of the modern digital multi-meters include a capacitor checking option. When set on this option, the multimeter will read out the microfarad capability of the capacitor being tested. You must still check to see if the capacitor has grounded to the casing.

When replacing a capacitor, you can replace it with one of an equal or higher voltage rating. However, the capacitance rating of the new capacitor should match (within 10%) the capacitance of the failed capacitor. Figure 6–10 shows what failures you can expect if you don't use the correct rated replacement capacitor.

Figure 6–10 Effect of not matching the ratings of a failed capacitor.

Replacement Capacitor Compared to Original	Possible Results
mfd rating too low	Capacitor fails
mfd rating too high	Stalled motor, lower running amps, low motor rpm
Voltage rating too low	Capacitor fails
Voltage rating too high	No problem

ELECTRONIC "BLACK BOXES"

Each year, more and more control functions are being accomplished with electronic "black boxes." The term "black box" refers to an electronic box. We don't need to know how it works; we only need to know what it is supposed to do. For example, figure 6–11 shows a circuit that uses an electronic time delay. The time delay is a "black box." All we know is that when voltage is applied to terminals 1–2, one minute later, the same voltage should appear at terminals 3–4. If you measure 24V at terminals 1–2,

Figure 6–11 4-wire electronic time delay.

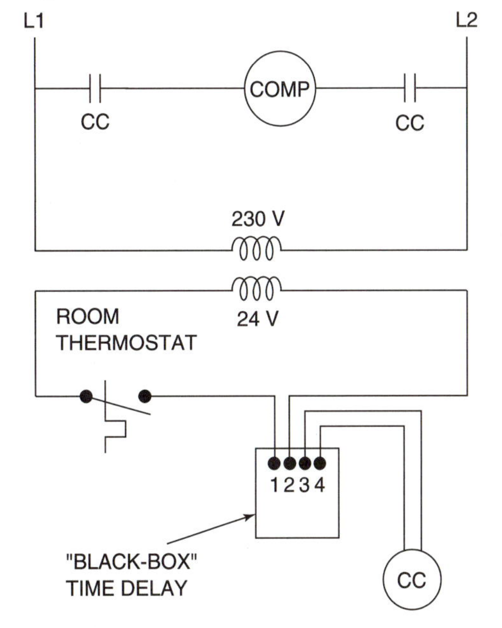

and after one minute you measure zero volts at terminals 3–4, the "black box" time delay has failed, and must be replaced. Do not attempt any repairs on a "black box" device.

Other "black box" devices may have more than one input signal. The general strategy is to know what all of the input signals must be, and what the output signal should be. If you check all of the inputs, and they are all proper, and if you are not getting the correct output, replace the box. There is usually no testing (i.e., no resistance checks) done on the box itself to determine if it has failed.

CHAPTER SEVEN

Repair Strategies

REPAIR OR REPLACE?

Whenever customers are faced with a major repair such as a compressor replacement, they will have to consider whether to go ahead with the repair, or simply discard the entire unit and replace it.

What's It Worth?

Before undertaking any repair, the cost of the repair will have to be compared with what the unit or appliance will be worth after it is repaired. For example, if a 10-year-old residential refrigerator needs a new compressor, the cost of the repair will probably exceed the value of the refrigerator when you are done. It would not make sense to repair it. For less money, the customer could simply replace it with an equivalent model that is operable, or they have the option to upgrade to a new unit. For this reason, many technicians that work on relatively inexpensive units such as domestic refrigerator/freezers or window air conditioners will restrict their repairs to replacing only components that do not require opening the refrigeration system (switches, heaters, timers, thermostats, etc.). When opening the refrigeration system is required, the cost of repair is very likely to approach or even exceed the value of the appliance.

DIRTY MOTOR BEARINGS

It is common to find a motor that has failed mechanically, but is OK electrically. Often, motors in furnaces, evaporator blower-coil units, and condenser fan motors will operate for years without receiving any maintenance. Some of these motors have provisions for oiling the bearings (figure 7–1), while others do not. The "permanently lubricated" motors are ones that use brass bushings instead of bearings. Brass is a rather "slippery" material when used as a bushing, and requires no further lubrication. However, all bearings can become dirty after time, and present resistance to the free rotation of the motor.

The rotating part of the motor (rotor or armature) is supported by either bearings or bushings (figure 7–2). The symptoms of dirty motor bearings (or bushings) are as follows:

1. With the motor de-energized, the fan does not rotate freely when you give it a spin with your finger.
2. When you energize the motor, it gets warm, but it will not rotate.

When a motor has failed due to dirty bushings or bearings, you have two choices:

1. Replace the motor.
2. Attempt to repair the motor.

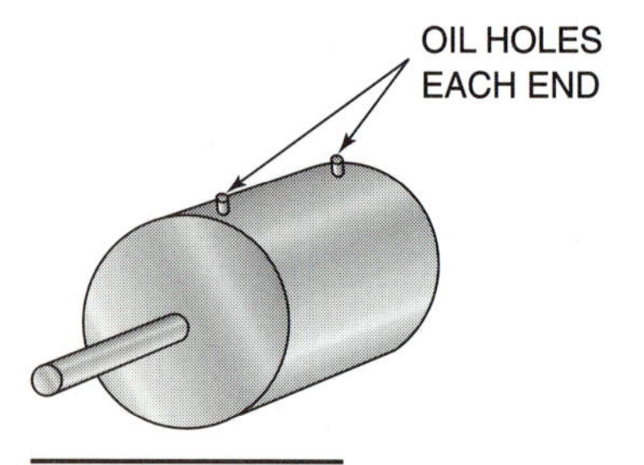

Figure 7–1 Oil ports on a motor.

Replacing the Motor

Motor replacement is the safest choice for the service technician. You should explain to the customer that while the motor might be repairable, you don't recommend it because you might waste your time (and the customer's money) in an attempt that is sometimes futile. Further, even if you are successful, there is no way to know if the motor bearings (or bushings) have been damaged, or how long they will last. If you don't explain this to the customer and you simply say that the motor failed and needs to be replaced, you run the risk of incurring some bad customer relations. Consider the following scenario. You install a new motor, and you leave the failed motor with the customer (which you should, unless you specifically ask if they want you to discard it). They have a "mechanic" friend who later inspects the motor, and repairs it by simply lubricating it (see below). The customer thinks you "ripped them off" because you replaced a "perfectly good motor" when all it needed was a few drops of oil! Obviously, it's better to discuss the options with the customer at the beginning to avoid this kind of misunderstanding.

Repairing the Motor

Most times (but certainly not always) a motor that has dirty bushings or bearings can be restored. This is done by lubricating it liberally with a light-weight oil such as WD-40, until it is clean and rotates freely. Remove the motor if required to get free access to it. Remove the fan blade, or whatever other device it is driving.

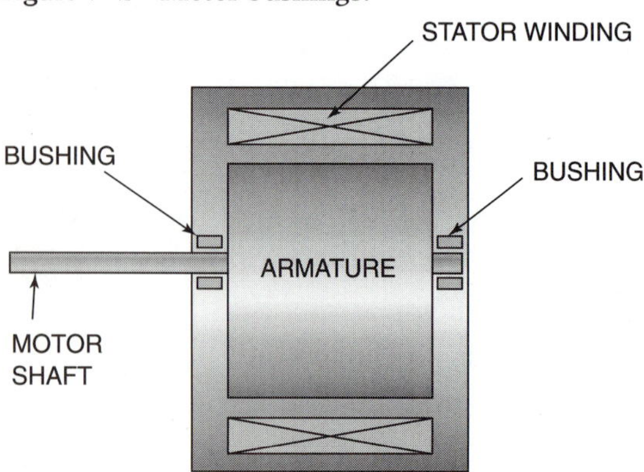

Figure 7–2 Motor bushings.

Try to "wiggle" the shaft. There should be very little "play" or movement. A lot of play indicates that the bushing or bearing is worn, and repair should not be attempted. Note that for all except very large motors, it is usually more cost effective to replace an entire motor rather than to replace the bushings or bearings.

With the motor de-energized, rotate the shaft by driving it with a variable-speed drill motor (figure 7–3). While the rotor is turning, spray the WD-40 into the oil ports provided, or if no ports are provided, spray the lubricant directly into the shaft-bushing area (figure 7–4). Some technicians will do this while the motor is energized and rotating (having started it rotating with a manual spin). This can be dangerous. While most times this is a successful approach, there have been cases where a spark from the motor ignited the WD-40, and caused the entire can to explode.

It may take five minutes or more of rotating the motor and adding lubricant until it finally rotates freely. When (and if) the motor loosens to where it rotates freely, lubricate the motor with oil of the recommended viscosity. WD-40 is a great cleaner, but is too light to provide adequate lubrication for most motors.

Before you undertake this repair, you should explain to the customer what is wrong. Recommend

Figure 7–3 Restoring a stiff motor with external lubrication ports.

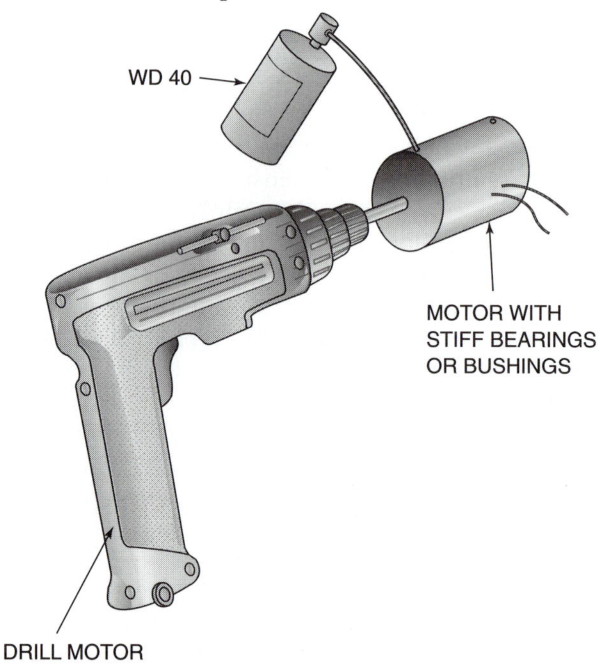

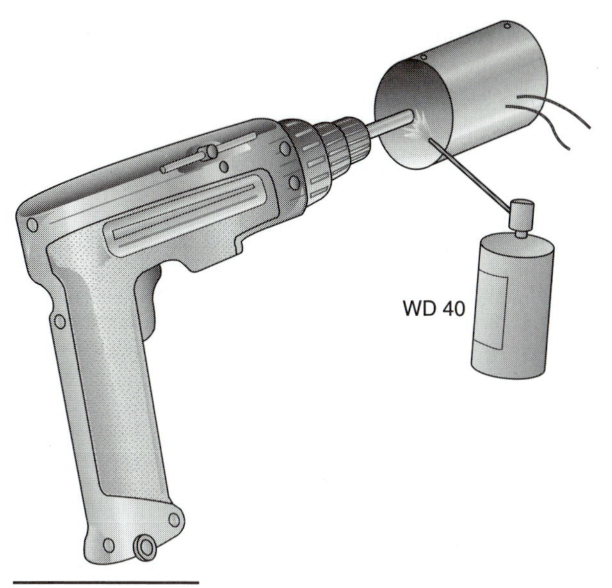

Figure 7–4 Restoring a stiff "permanently lubricated" motor.

that they replace the motor, but give them the option of attempting a repair. Do not place yourself in a position of recommending a repair, because if it doesn't work, the customer will only remember that it was your idea. Explain that even if the repair does work, you cannot say how long it will last. And lastly, when you make out your service ticket, write, "Recommended motor replacement. Customer declined." This way, if the motor does fail one month later, the customer will call and say, "You were right, the motor should have been replaced." If you don't write that you recommended replacement, when the motor fails, the customer will call you and say, "You just fixed this motor a month ago and now it's broken again." No amount of verbal explanation will serve your purpose as well as having the customer sign a ticket that states "Recommended motor replacement."

Even though occasionally a bushing/bearing lubrication repair will not work, sometimes taking the gamble is a good choice. This is particularly true on some specialty motors on blower coil units inside walk-in coolers or freezers. These are sometimes fairly small motors, but can be very expensive to replace. Another situation where repair makes especially good sense is when a replacement motor is not readily available and you must get the system running to avoid a costly loss of product.

MULTI-SPEED MOTORS

On residential furnaces, packaged rooftop units, and many other applications, the evaporator fan motor (sometimes referred to as the blower motor) is sometimes supplied with several speeds, some of which are unused (figure 7–5). This is done so that the manufacturer can use one standard motor for several different size models and simply select the appropriate speed to match the heating or cooling capacity of the model on which it is installed. Suppose you have determined that the low-speed winding has failed. Certainly, you can replace the motor to effect a repair. But you may want to try removing the wire that is energized on heating from the low-speed connection and connect it to the medium low speed, which may still be operational.

What are the drawbacks to using an incorrect speed? Whenever you go to a higher speed, there will be more noise. Run the unit on that speed, and ask the customer if the increased noise level is acceptable. Higher fan speed will also consume more electricity, but it also improves heating or cooling efficiency, so the overall increase is usually not significant. Also, at higher speed, a cooling system will not produce as much dehumidification in the room, but this is usually acceptable in air conditioning applications. When a heating system is operated at higher speed, the discharge air temperature will be slightly lower, and that, in combination with the higher airflow, may be perceived by the occupants to be a draft. When going to a lower-than-design motor speed, all of the effects are the opposite of those described above. The efficiency and the capacity of the heating or cooling system will be reduced. On cool-

Figure 7–5 Multi-speed motor.

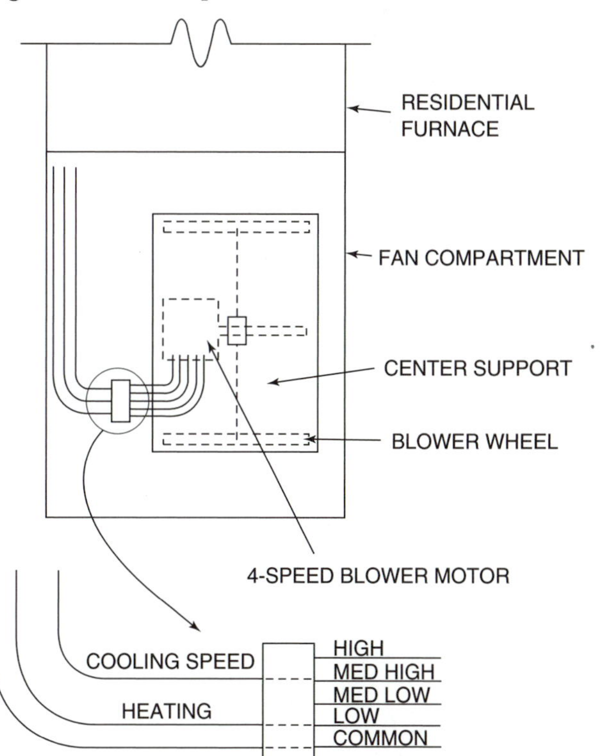

ing, the lower airflow may not be sufficient, and it could cause the evaporator coil to freeze up. But this can be checked by measuring the operating pressures at the lower speed. On heating, the higher discharge air temperature and lower airflow could create problems with temperature stratification, where the difference in temperature between the air near the ceiling and the air near the floor becomes exaggerated.

Having described all the potential drawbacks, most times the effects of using a different fan speed are imperceptible to the customer, and the savings realized in not having to replace the motor will be appreciated.

REPLACE COMPRESSOR OR CONDENSING UNIT?

A condensing unit consists of a compressor, condenser, condenser fan, controls, and sometimes a receiver. When the compressor fails, it can be replaced. But you should also get pricing from the wholesaler on a new condensing unit. While it will be more costly to the customer to replace the entire condensing unit, it may be attractive for the following reasons:

1. Replacement of the entire condensing unit will take you less time than replacement of the compressor alone. The reduction in labor cost will partially offset the increased material cost.

2. There is a risk that the compressor failed due to a restriction elsewhere in the condensing unit. You cannot know this before you replace the compressor, and you avoid this risk by replacing the entire condensing unit.

3. The newer model of condensing unit may be higher thermal efficiency than the existing unit, and will offset the higher equipment cost through lower operating cost.

4. The newer model of condensing unit may be designed to operate with one of the newer "ozone friendly" refrigerants.

5. The risk of future failure of the condenser fan, start relay, contactor, capacitors, etc. is significantly reduced when they are all changed out together with a new condensing unit.

REPLACE OR BYPASS THE SWITCH?

When a switch fails in the open position, you can make the system run by placing a jumper wire across the switch. If the jumper wire makes the system run normally, the switch should be replaced. However, there is another option. The switch may sometimes be bypassed. This suggestion will alarm some people. Some technicians believe (correctly so) that the safest strategy is to leave the unit in the same condition as the manufacturer designed it. Some will be afraid to modify the system for fear of incurring legal liability. This also is a valid reason to reject bypassing a switch. But in spite of these concerns, with the willing and informed consent of the customer, you may wish to sometimes bypass a switch, depending on what it does. Following are some examples of where it might make sense.

CONDENSER FAN CYCLING SWITCH

Suppose that you have determined that an ice machine is not making ice because the condenser fan cycling switch (figure 7–6) has failed in the open position. You know that the fan cycling switch has been provided by the manufacturer for low-ambient operation. The manufacturer does not know whether the unit will be installed indoors or outdoors, so the fan cycling switch is installed just in case the unit is installed outdoors. If yours is an indoor installation, the condenser fan cycling switch is really not crucial, and it may be bypassed. Indeed, many ice machines are supplied from the factory without a fan cycling switch.

MOTOR SPEED SELECTOR SWITCH

Some older furnaces and other applications have a manual switch that the owner can use to choose a low-speed or a high-speed operation for a fan motor (figure 7–7). Suppose you find that the motor will not run because the speed selector switch is not making contact. Inquire if the owner actually uses

Figure 7–6 Fan cycling for a condenser fan motor.

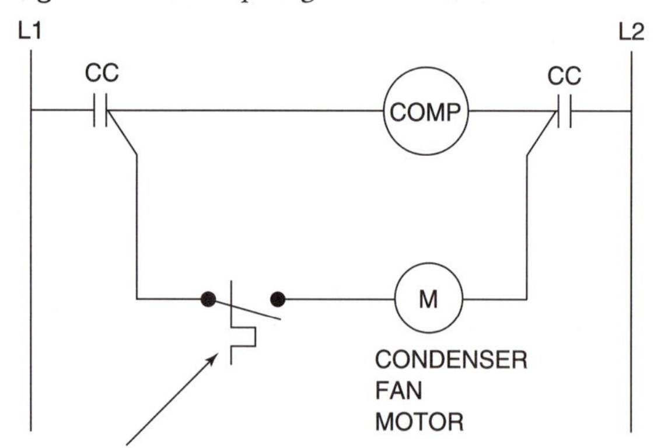

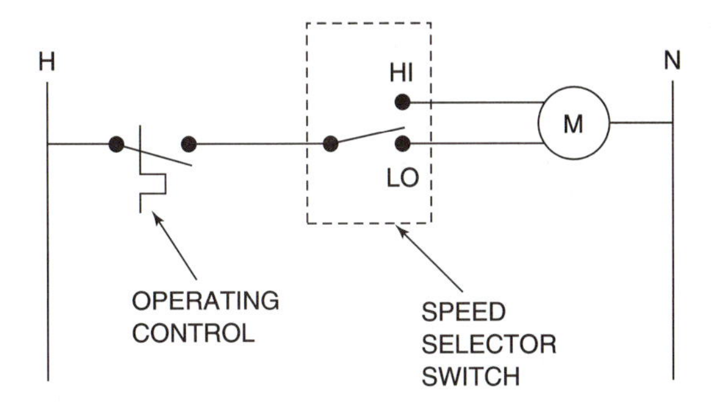

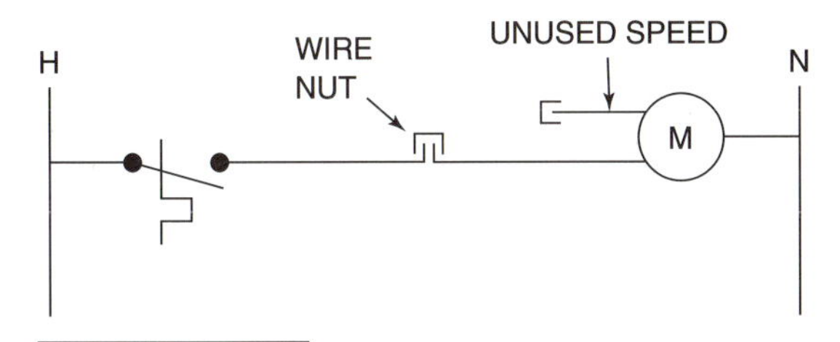

Figure 7–7 Removal speed selector switch.

the speed selector switch (most owners will be unaware that it even exists). If the speed selector switch is never used by the owner, simply remove it, choosing one motor speed to use, and capping off the unused speed.

Internal Overload

You have arrived at the job site, and determined that the condenser fan run capacitor has failed, and it caused the condenser fan to fail. That, in turn, caused the compressor to go out on internal overload, and it will not reset.

The internal overload has externally accessible wiring (figure 7–8). The internal overload switch cannot be replaced. You can replace the entire compressor, but that will be very expensive. What is the risk if you bypass the internal overload by cutting the wires leading to it, and placing them together in a wire nut (figure 7–9)? Obviously, you will lose the protection that is provided by the overload. If the condenser fan fails again for any reason, it may cause the compressor to fail also. But is that so bad? The cost of replacing the compressor at that time will be the same as replacing it today, so there is nothing lost in delaying that repair. But the advantage in bypassing the overload is that maybe the condenser fan will never fail, and the unit will be replaced entirely after it has reached the end of its useful life. In that case, the customer will have avoided the expense of replacing the compressor. The risk that the customer assumes in bypassing the overload is small in comparison to the potential benefit.

Retrofit/Upgrade?

Sometimes a needed repair presents an opportunity for upgrading equipment that would otherwise not be justified. For example, suppose that a customer with a standing pilot furnace asks you if it should be upgraded (modified) to an electronic ignition system (see Chapter fourteen on electronic ignition controls for gas-fired furnaces). In truth, the savings that will be realized by turning off the pilot flame when room thermostat is satisfied will probably not be sufficient to justify the expense of retrofitting the furnace. However, suppose you have diagnosed that this same standing pilot furnace has a failed gas valve. If you have to replace the gas valve anyway, it may now be worth the *incremental* cost to upgrade. Only the difference in cost between the gas valve replacement and the retrofit must be justified by the savings.

UNIVERSAL REPLACEMENT PARTS

Once the service technician has completed troubleshooting, there is usually an electrical part that needs to be replaced. If you are working for a company that provides service exclusively on one brand of equipment, chances are that you will have most of the exact replacement parts that you need. However, many technicians will need to be prepared to

ADVANCED CONCEPTS

The internal overload may be a thermostat that opens on a temperature rise, or it may be a **thermistor.** A thermistor is an electronic device whose resistance changes dramatically with a change in temperature. Some will increase resistance with a rise in temperature (**positive temperature coefficient**), while others will decrease resistance with a rise in temperature (**negative temperature coefficient**). When used as an overload, the thermistor is imbedded in the motor stator windings. If they become too warm, the thermistor resistance gets very high, acting similar to an open switch.

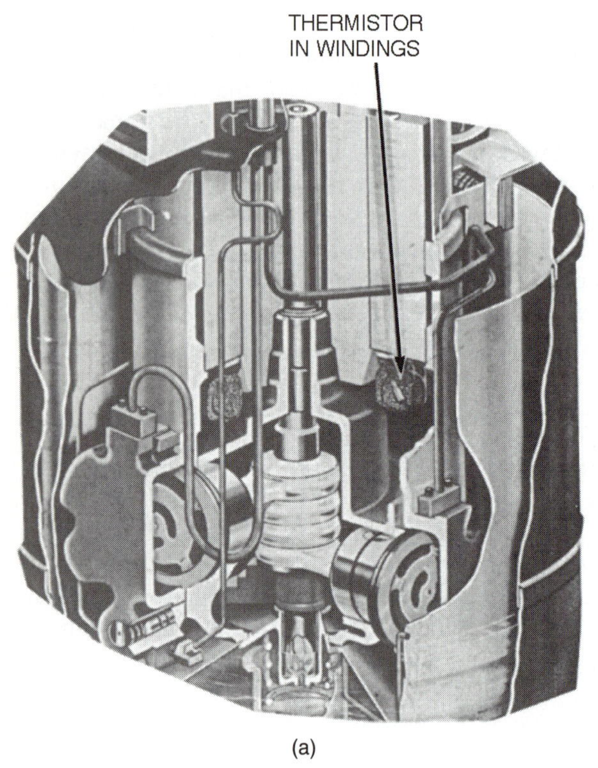

(a)

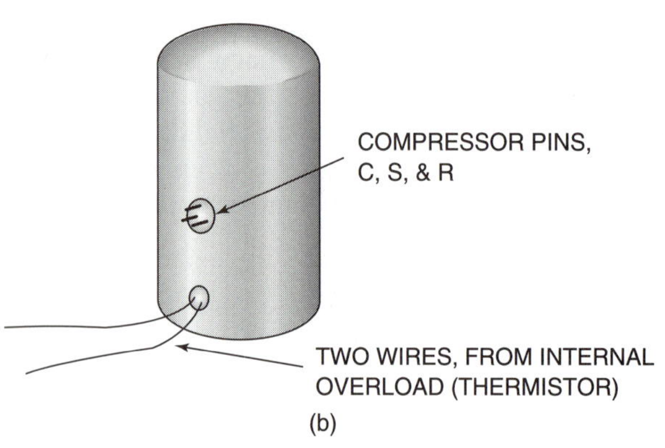

(b)

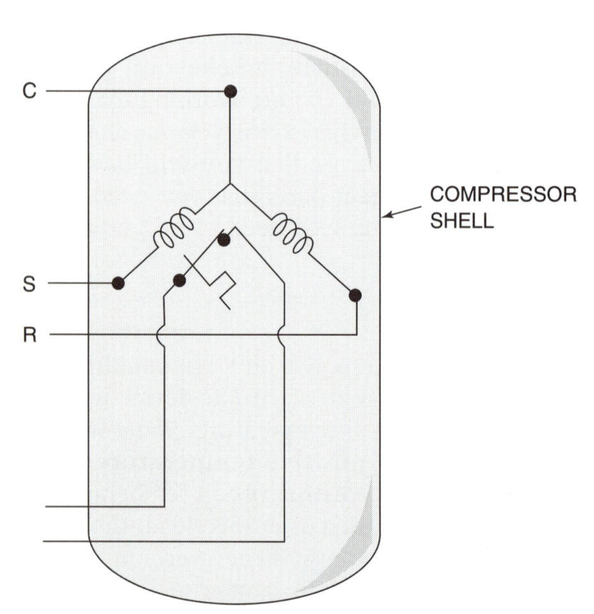

Figure 7–8 Internal pilot-duty overload. *(Reprinted from Refrigeration and Air Conditioning, 3rd Ed. by Air Conditioning and Refrigeration Institute, Copyright © 1998, Prentice Hall, Inc. Reprinted by permission.)*

Figure 7–9 Compressor with bypassed overload.

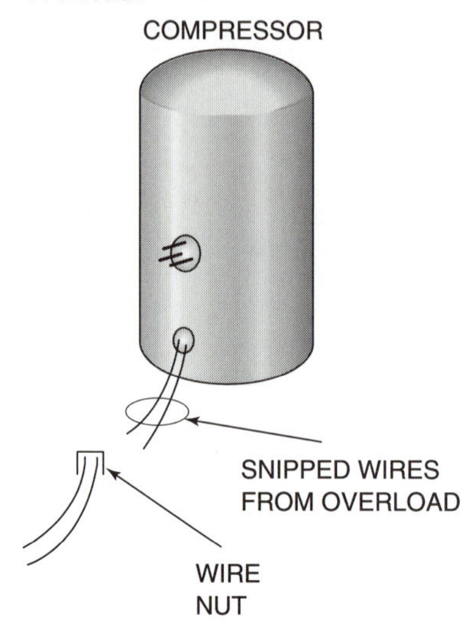

perform repairs on any brand of equipment. There are three strategies for having repair parts on hand:

1. Carry as many different parts as possible on your truck. You will need a very large truck, you will likely spend a lot of money on gas, you will have a lot of money invested in your inventory, and you will probably need an inventory system to help you find the exact replacement part that you need.

2. Whenever you need a part, go to the wholesale parts house to purchase the exact replacement. You will spend a lot of time on the road.

3. Carry parts on your truck that can be used for many different applications. You will avoid many (not all) trips to the wholesaler while minimizing the number of parts that you need to stock.

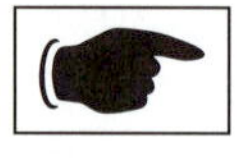

Following are some suggestions for parts to carry:

1. A 40 VA transformer with a 120/208/230V primary and a 24V secondary. This will replace 20VA and 40VA transformers on residential and small commercial air conditioners.

2. A 115/230V reversible motor in 1/4 and 1/3 HP sizes. These will replace many evaporator blower and condenser fan motors.

3. A cold control (box thermostat) set at 35°F and another set at 0°F, with a differential of less than 5°F. Many cold controls sense evaporator coil temperature, and the cut-in and cut-out temperatures are specific to the design of that particular box. However, the 35°F and 0°F cold controls can be installed to sense box temperature instead of coil temperature, and therefore can be used on all coolers and freezers.

Commercial Systems

Commercial systems are distinguished from residential systems because they are generally larger and more expensive, and they sometimes must deal with more rigorous requirements and provide more flexibility than residential systems. Commercial systems dealt with in this chapter include walk-in coolers and freezers, commercial packaged refrigeration systems, and air conditioning systems used on commercial type occupancies.

COMMERCIAL DEFROST TIMER CIRCUIT NO. 1

Figure 8–1 shows a defrost timer that is used in commercial freezer applications, rather than the type shown in a previous chapter which is limited to smaller units. The commercial defrost time clock looks far more complicated, especially when you look at the wiring diagram that is pasted on the timer enclosure door (figure 8–2). However, a little investigation will reveal to you that the circuit for the commercial defrost timer shown in figure 8–3 is exactly the same as for the smaller defrost timers (figure 2–21). The commercial defrost timer has an outer ring that rotates once each 24 hours, and an inner ring that rotates once each two hours. There are multiple trippers located on the outside ring, and a single tripper located on the inner ring. The trippers on the outer ring determine the start times for "defrost." The tripper on the inner ring determines the duration of the "defrost."

While the circuit of this commercial defrost time clock is similar to the smaller defrost timers, there are several differences:

1. By the placement of tripper pins on the outer ring (they unscrew), the technician may choose how many times the "defrost" will be initiated in a day, and at what times they will occur. The intervals do not need to be the same. If the time of "defrost" initiation is important, then the timer must be set to the correct time of day. Find the present time on the outer scale. Rotate the inner wheel clockwise, until the outer wheel has moved to a position where the present time is aligned with the time pointer.

2. The duration of the "defrost" may be adjusted by moving the pointer on the inner wheel to the duration desired (push in and rotate the pointer).

3. The switches are accessible. The whole "works" of the timer may be pulled out by removing one screw at the top, and two posts at the bottom (they unscrew).

☛ If you find that a switch is not making contact, and you want to get the freezer running while you run for a replacement timer, you can clean the contacts. With the power off, hold a non-glossy business card or matchbook cover between the

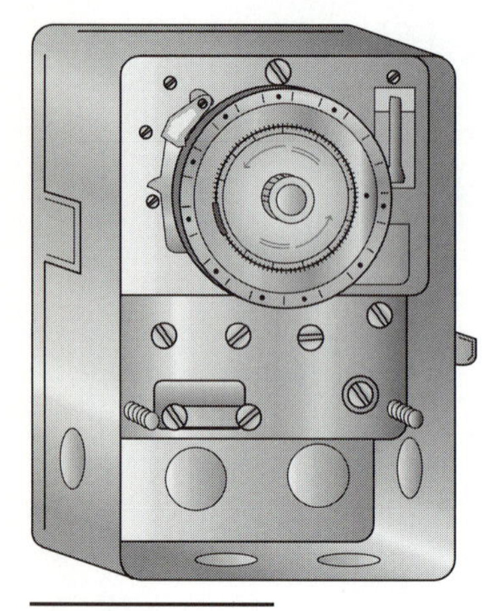

Figure 8–1 Commercial defrost timer.

contacts. Press the contacts together, and move the business card back and forth. It will provide a gentle cleaning for the contacts. The contacts should now work until you are able to return with a replacement time clock. [Note: Contacts that have been cleaned have not been repaired. They may last for days or even weeks, but the protective silver coating on the original contacts has been worn away. The cleaned contacts will deteriorate rapidly.]

COMMERCIAL DEFROST TIMER CIRCUIT NO. 2

All of the defrost timer circuits that have been presented to this point have been time-initiated, time-terminated. That is, it was only the passage of time

Figure 8–2 Factory-supplied wiring diagram.

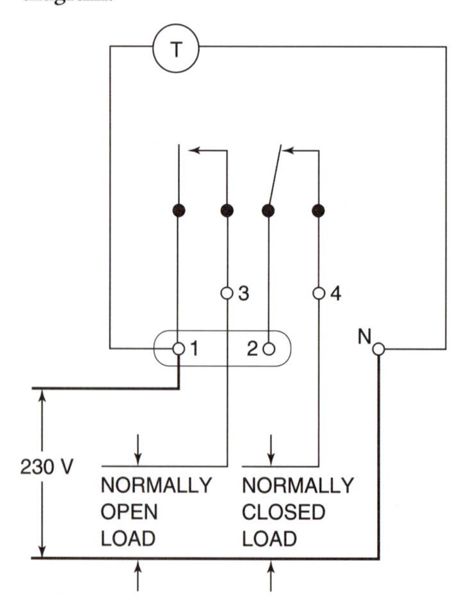

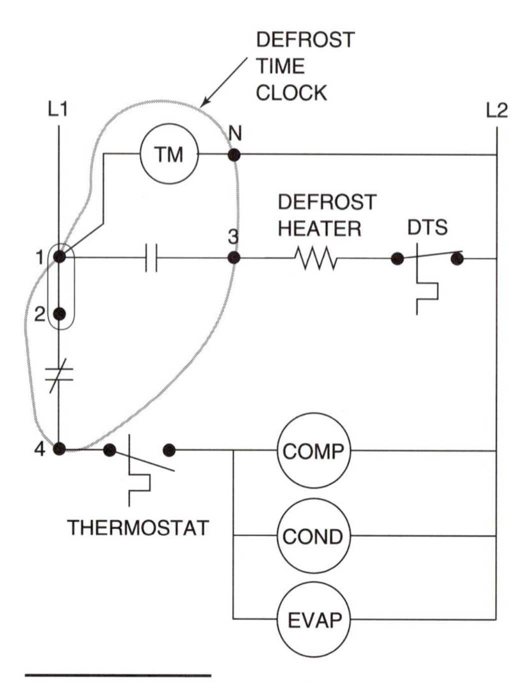

Figure 8–3 Commercial defrost timer circuit no. 1.

that caused the timer to place the SPDT switch into the "defrost" mode, and it was only the passage of time that caused the timer to return the SPDT switch to the "freeze" mode. The circuit shown in figure 8–4 is typical of the manufacturer's diagram for a time-initiated, *temperature*-terminated system. When the evaporator is defrosted, the system will go back to the "freeze" mode, without waiting for the rest of the "defrost" mode to time-out. There are two differences from the prior circuit:

1. The defrost timer includes a solenoid coil. When it is energized, it pulls on a trip-lever that returns the SPDT switch to the "freeze" mode.

2. The defrost termination thermostat is a three-wire, SPDT switch, usually with a remote sensing bulb (see figure 8–5).

The manufacturer's diagram may be confusing to you, but except for the solenoid, it is the same as the previously discussed defrost timers. In figure 8–6, it has been redrawn as a ladder diagram that should be easier to follow. During the "freeze" cycle, the normally closed switch is closed between terminals 2–4, and the normally open switch is open between 1–3. When the "freeze" cycle has timed out, the gears in the defrost timer operate the switches into "defrost" mode, opening 2–4 and closing 1–3. The freezing components are de-energized, and the electric heater is energized. After some time, when the evaporator temperature rises to 45°F, the defrost termination switch operates, completing a circuit through the solenoid coil of the de-

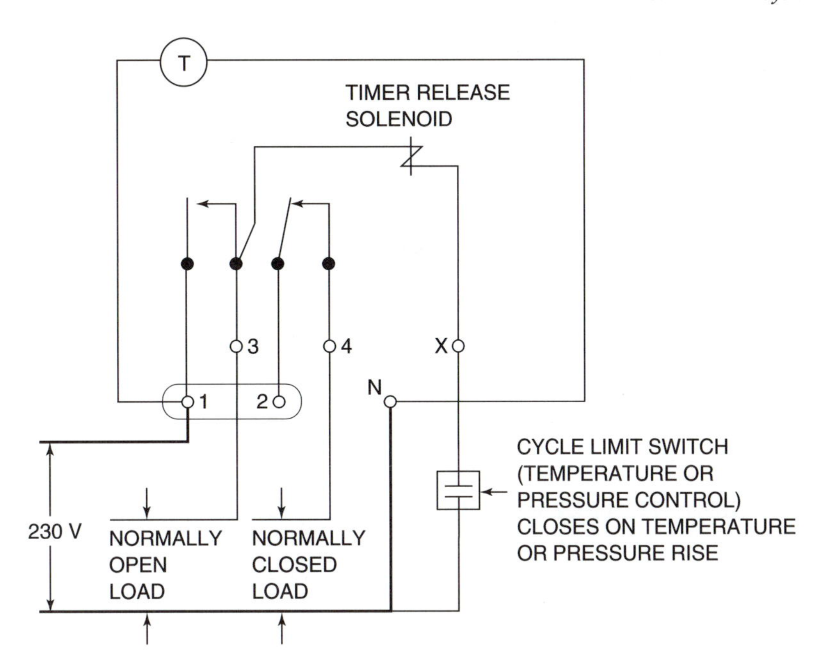

Figure 8–4 Defrost time clock with timer release solenoid.

frost timer through terminal X. However, the solenoid coil only stays energized for a second. As soon as the solenoid is energized, it returns the normally open and normally closed switches to their normal positions ("freeze" mode), once again de-energizing the solenoid coil. The compressor and condenser will run, but the evaporator fan will remain de-energized until the defrost termination switch senses that the evaporator coil temperature has returned to normal operating conditions.

Note that the power supply in figure 8–6 is labeled L1 and L2 instead of H and N. This just indicates that it is a 230V system rather than a 115V system. Commercial defrost timers are also available for use on 115V systems. When replacing a commercial defrost timer, make sure that you purchase

Figure 8–5 Defrost termination switch with remote sensing bulb.

Figure 8–6 Defrost timer circuit no. 2.

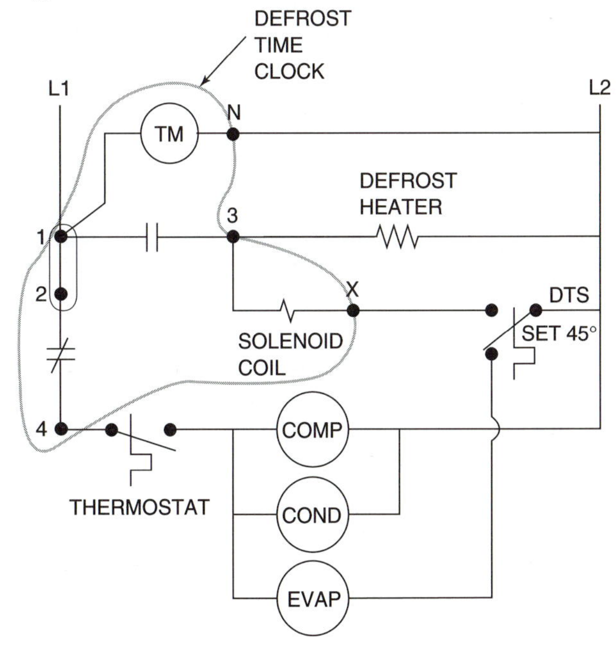

ADVANCED CONCEPTS

Note that the defrost timer can return to the "freeze" mode on *either* time or temperature. If the duration time setting times out before the evaporator coil reaches 45°F, the "freeze" cycle will resume. Eventually, the coil will become covered in ice. But the time-termination is necessary as a safety feature in case the solenoid coil or the defrost termination switch fails. Without this time-termination feature, a failed solenoid coil would prevent the compressor from operating. A failed defrost termination switch would be devastating. If its contacts were welded in the "cold" position, the heaters would stay on, and all the product in the box would become warm and spoil quickly.

the correct replacement. The clock motors are different to correspond to the available voltage.

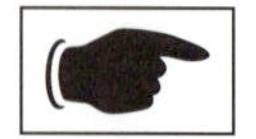

Assume you have a walk-in freezer with the compressor and condenser fan running, evaporator fans not running. There is a three-wire defrost timer mounted on the side of the casing. As there are multiple evaporator fans not running, you suspect that the defrost termination switch has failed. However, there are three wires emerging from the switch. You would like to take a voltage reading across the "cold" switch of the defrost termination switch to see if it has failed, but unfortunately, the actual wires on the DTS are not identified. You don't know which wire is which.

It's 20°F in the freezer, and you'd like to get out of there sooner rather than later. With the power turned off, take the three wires of the DTS, and put them all together in the same wire nut (or jumper them together). When you restore the power, this should energize the evaporator fans. If it does, replace the DTS.

QUESTION:

You have removed the non-functioning defrost termination switch above. Your new DTS comes with the terminals A, B, and C identified as follows:

A-C makes on a rise in temperature.

B-C makes on a fall in temperature.

There are three unidentified wires hanging out of the evaporator unit, which were connected to the failed DTS. Which wires will you connect to which terminals on the replacement DTS?

SOLUTION: Place the defrost timer in the "freeze" mode. Label the wires that were connected to the failed DTS as #1, #2, and #3. Connect them together, two at a time, until you find the pair that makes the evaporator fans run. The one wire not involved in making the evaporator fans run must be the one that comes from terminal X in the defrost timer. For ease of discussion, let's say that you found that to be wire #2. Next, go to the defrost timer, and turn the knob until it goes into the "defrost" mode. Return to the evaporator and measure voltage between wire #2 and each of the other two wires. One of the pairs will give you a voltage reading equal to the line voltage. Let's say that a line-voltage reading was obtained between wires #2 and #3. That means that wire #1 must be the one that comes from the evaporator fan, and wire #3 must be the one that connects to L2. Connect wire #1 to terminal B on the replacement DTS, connect wire #2 to terminal A, and connect wire #3 to terminal C. ■

EXERCISE:

Redraw the wiring diagram in figure 8–7 as a ladder diagram. Show all of the terminals on the terminal strip. Answer the following questions:

1. Where does the power come into this system? (It is not shown on the diagram.)
2. What would you guess is the purpose of the heat limiter (HL)?

ELECTRONIC REFRIGERATION CONTROL

The newest trend in controlling walk-in boxes and other commercial refrigeration systems is to use a "black box" electronic control similar to the Beacon control system shown in figure 8–8. This "black box" is factory mounted on the blower coil unit inside the box. It has temperature inputs that tell the controller what is happening. Temperature is sensed by thermistors, whose resistance changes

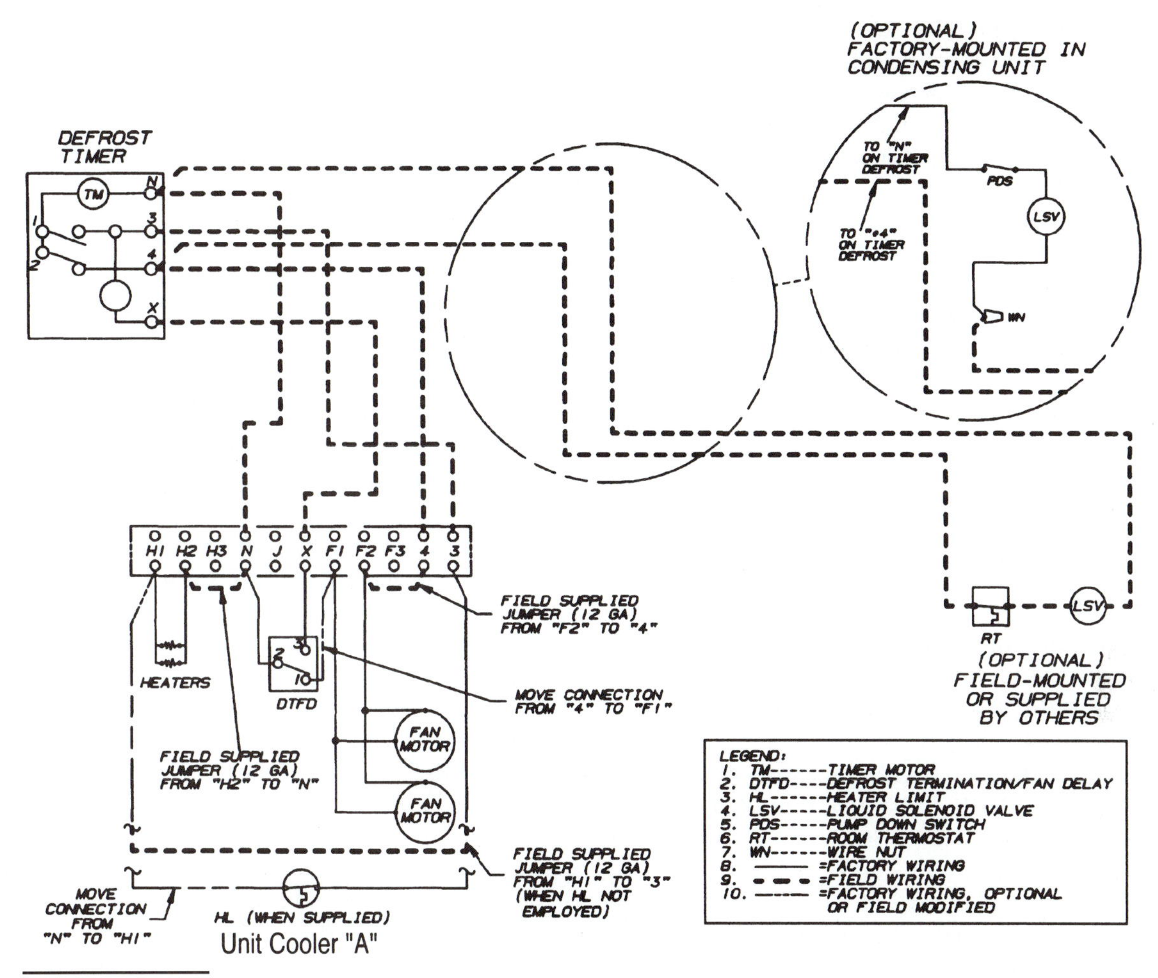

Figure 8-7 Exercise. *(Courtesy of Heatcraft.)*

Figure 8-8 Beacon system diagram featuring Smart Controller. *(Courtesy of Heatcraft.)*

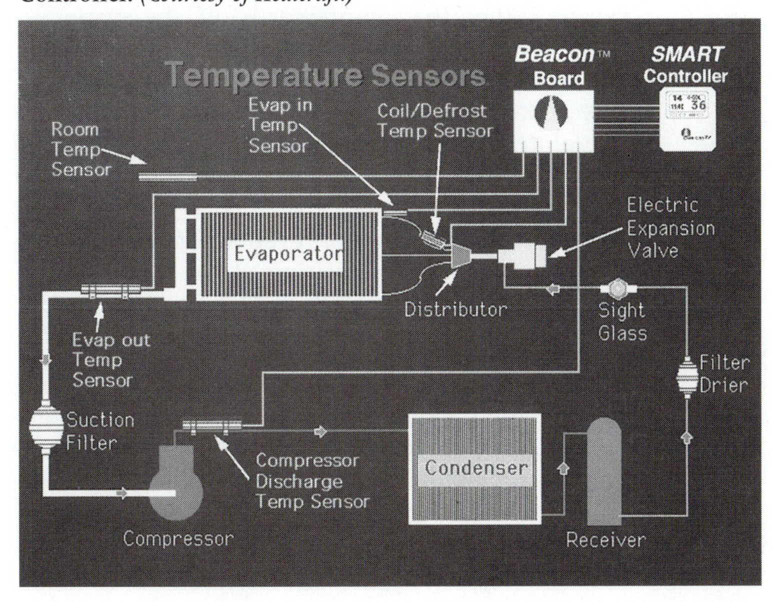

with temperature. The controller then has outputs to the electrical components (including an electronic expansion valve) which accomplish all of the conventional functions of refrigeration, and more. Following are some of the functions of this controller:

1. The controller can cause the electronic expansion valve to close, without regard to superheat, allowing the expansion valve to operate like a liquid solenoid valve. There is no need for a separate LSV to accomplish a pumpdown.

2. During pull-down from a warm start, the electronic expansion valve will be modulated so that the suction pressure (as sensed by saturated suction temperature) is limited to a preset maximum. This eliminates the need for a crankcase pressure regulator which might otherwise be required to prevent overloading the compressor on start-up.

3. If the discharge line reaches 225°F for four minutes, the superheat setting will be temporarily lowered to keep the discharge temperature from going any higher. If the discharge line reaches 275°F for four minutes, the compressor will be shut down and locked out. These are safety functions that were rarely (if ever) done prior to the advent of electronic controls.

4. There is a minimum run time of four minutes for the compressor to eliminate the possibility of short cycling.

5. Pump down can be accomplished by closing the electronic expansion valve, and then shutting down the refrigeration system one minute later. This eliminates the need for a low-pressure switch (although one can be used if desired).

6. The controller can be used to cycle the condenser fan off if needed to get sufficient flow through the evaporator. This is more accurate than conventional head pressure control which is only concerned with maintaining a minimum high-side pressure.

7. Status indicator lights on the controller board provide information as to what the system is doing. It also performs an error check to determine if any of the temperature sensors are open or shorted.

8. The controller has a set of **dry contacts** that are normally closed. The term "dry contacts" refers to a switch that is not connected electrically to any other device within the controller. It is simply a set of contacts that can be used in a different circuit. In this case, during normal operation, the controller will open the dry contacts. However, if an alarm condition exists, the dry contacts will return to their normally closed position to activate a light, buzzer, or bell to indicate trouble.

Optional Smart Controller

The Smart Controller connected to the controller in figure 8–8 provides a means for remote programming of the controller. The following functions can be set by this controller:

1. Box temperature.
2. Defrost frequency, start times, fail-safe times, and termination temperature.
3. Evaporator superheat setting.
4. Alarm set points.

Additionally, with the push of a button, the Smart Controller allows the operator to monitor box temperature, actual superheat, the expansion valve position, the evaporator coil temperature, compressor discharge temperature, outdoor temperature, compressor run time, compressor cycling rate, and accumulated defrost time. These are powerful diagnostic tools that can indicate when a problem is developing long before an emergency situation arises. For example, if maintenance personnel note that the total defrost time each day is gradually increasing, it might indicate that the door seals or the door closer needs to be checked.

USING AN LPC INSTEAD OF A THERMOSTAT

In walk-in boxes, it is very common to use a low pressure cut-out (figure 8–9) instead of a thermostat to control the operation of the compressor. Note the symbol used to represent an LPC. The switch closes on a rise in sensed pressure. The LPC senses the pressure on the low side of the refrigeration system. When the compressor is off, the evaporator is at the same temperature as the box. The refrigerant inside the evaporator is at the saturation pressure that corresponds to the box temperature. Therefore, an LPC can be used to sense pressure instead of using a thermostat to sense temperature. When the box temperature rises to its maximum allowable temperature, the evaporator pressure rises to the cut-in pressure setting of the LPC, starting the compressor. This method of control is popular for two reasons:

1. The LPC is located at the condensing unit, sensing low-side pressure at the compressor suction. It is not subject to the mechanical damage that might knock a thermostat off the wall as the refrigerated box is being loaded or unloaded.

2. The wiring cost is lower with the LPC because it is unnecessary to run wire between the condens-

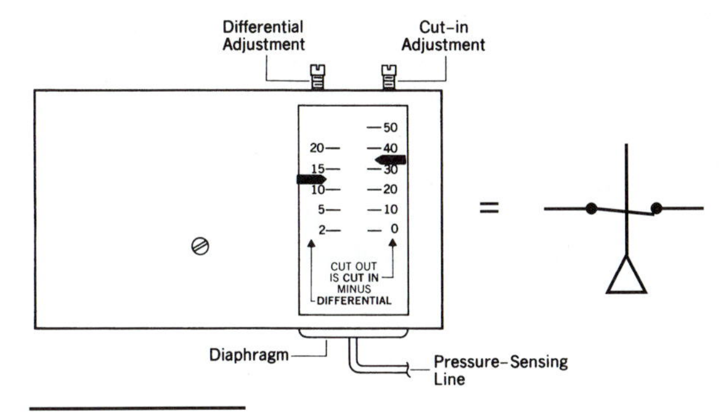

Figure 8–9 Low-pressure cut-out.

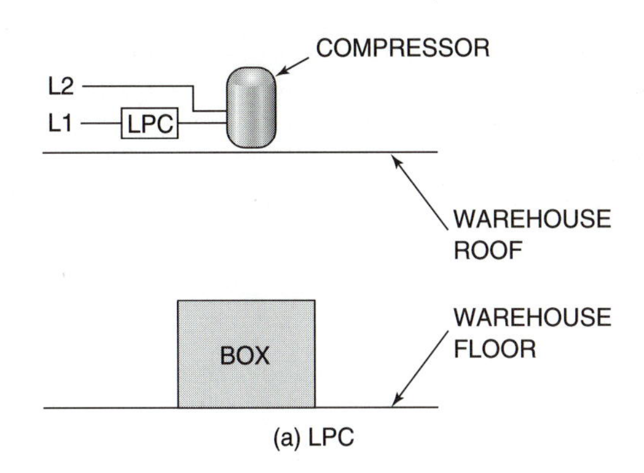

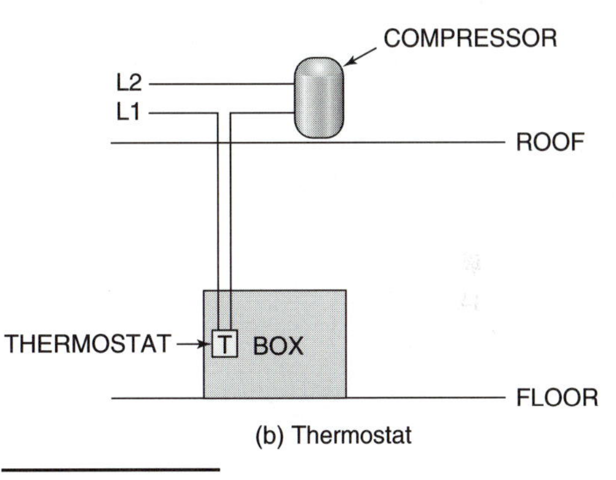

Figure 8–10 Comparison of LPC and thermostat control. (a) LPC. (b) Thermostat.

ing unit and the box. See figure 8–10 for a comparison of the two methods of wiring.

When an LPC is used to sense box temperature, the evaporator fan is usually on a separate circuit, and the evaporator fan runs continuously. This assures good heat transfer between the box and the refrigerant inside the evaporator coil. The pressure of that refrigerant will then accurately reflect the actual box temperature.

FREEZER WITH HOT-GAS DEFROST

Figure 8–11 shows a wiring diagram for a walk-in freezer that uses hot gas from the compressor to accomplish the defrost. This is different from the more popular electric defrost as follows:

1. During the defrost, the compressor continues to run.

2. Instead of energizing an electric heating element to provide heat for defrost, we energize a hot-gas defrost valve (figure 8–12) to allow hot gas from the compressor to bypass the condenser and metering device and reenter the system between the metering device and the evaporator.

During the freeze portion of the timer operation, the timer switch 2–4 is closed, bringing L1 pressure to terminal 4 in the defrost timer. This powers terminal 4 on the terminal block of the evaporator coil unit, and the temperature control. Terminal 4 energizes the fan motors, and the temperature control energizes the liquid solenoid valve, which will cause the compressor contactor C to energize through the LPC pressure control (automatic pumpdown system). Note that the defrost termination switch makes R-B when it senses a cold evaporator, and R-W when the evaporator is warm. Also, note the symbol for the pressure control. It shows two sensing elements operating a single switch. It senses pressure on both the high side and the low side of the refrigeration circuit. If the high-side pressure gets too high, or if the low-side pressure gets too low, the switch will open, shutting down the compressor.

When the timer motor advances the switches to the "defrost" mode, switch 2–4 opens, and switch 1–3 closes.

When the hot gas from the compressor completely defrosts the evaporator, the defrost termination switch opens R-B, and closes R-W. The R-W switch completes a circuit through the terminating solenoid in the defrost time clock (through terminal X). The solenoid mechanically opens switch 1-3,

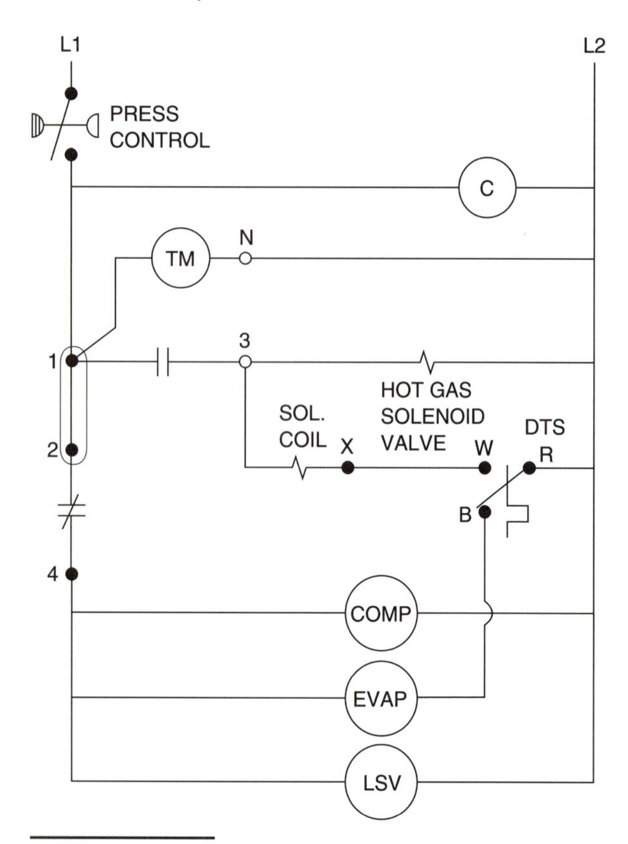

Figure 8–11 Hot gas defrost.

and closes switch 2-4. This restarts the compressor, but not the evaporator fan. The evaporator fan will not restart until the defrost termination switch makes R-B.

You have been called to repair a walk-in freezer. It has been down for many hours, and the box is at 75°F. When you start the system after you have completed your repair, you note that the evaporator fans do not run. Before you start troubleshooting the system again to find out why, give the compressor time to bring down the evaporator temperature. If you give the system a few minutes to run, the defrost termination switch will close, and the evaporator fans should start.

Figure 8–12 Location of hot gas defrost valve.

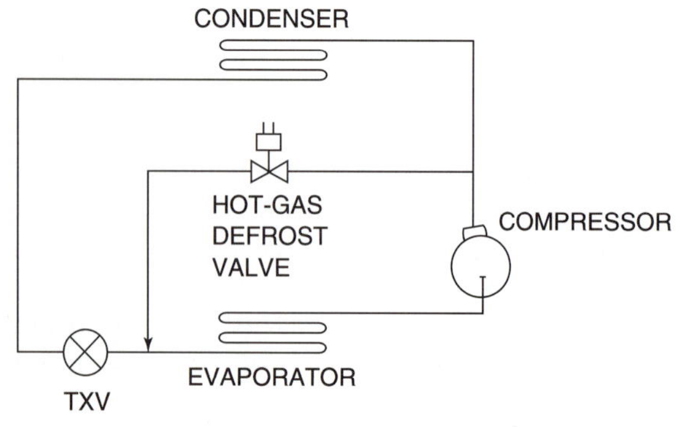

EXERCISE:
Draw a line encircling all of the portions of figure 8–11 that would be found inside the defrost time clock.

OTHER DEFROST SCHEMES

In figure 8–13, the compressor is operated through a 115V liquid solenoid valve and a box thermostat, while the electric heaters are operated on a separate 230V circuit. The time clock and the fan motor both operate on 115V.

Figure 8–14 shows the same defrost time clock, but with the electrical bridge moved from terminals 2-N to terminals N-1. With this scheme, the time clock motor, the compressor contactor, and the electric heaters all operate on the same voltage.

Figure 8–15 shows a slightly different defrost timer, in which one of the internal switches is SPDT while the other is SPST. This allows the defrost timer to open both legs of the electric heater circuit when the defrost system is not being used.

EXERCISE:
For each of the manufacturer's time clock wiring diagrams in figures 8–13, 8–14, and 8–15, draw an equivalent ladder diagram.

OVERLOADS

3-Phase Line Duty

Many walk-in refrigeration systems use a semi-hermetic three-phase compressor with a line-duty overload (figure 8–16). This arrangement can be confusing to those technicians unfamiliar with it. The fusite connector looks exactly like the C, S, R or L1, L2, L3 power connections on a welded hermetic compressor. However, if you remove the wires from the fusite connector and use your ohm meter on those pins to check the compressor windings, you will read infinite ohms, and you are likely to condemn a perfectly good compressor.

The compressor windings are connected in a y-shape, but the connection point where each winding joins together is brought external to the compressor through the fusite connector. The windings are then joined together in the overload, but not before first going through an overload switch in

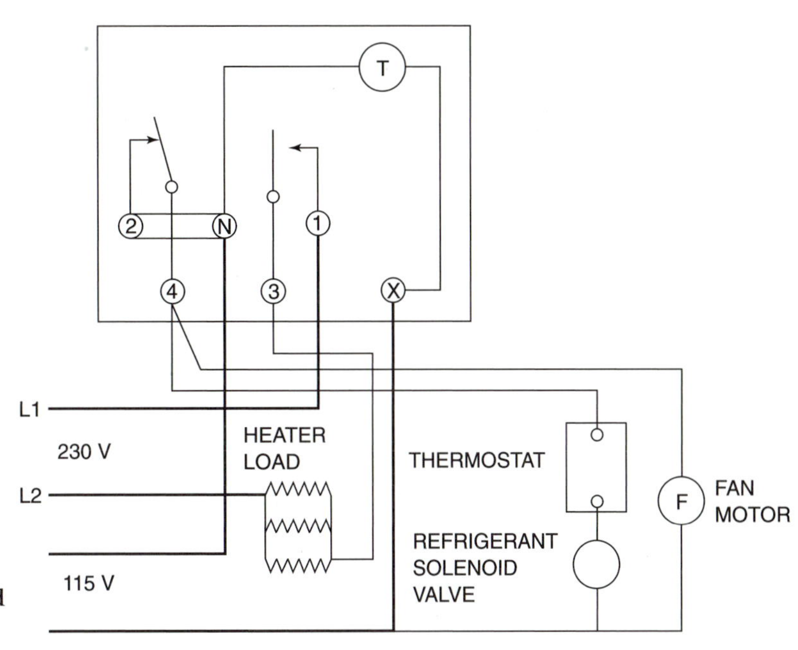

Figure 8–13 230V system with 115V liquid solenoid valve.

each leg. The three-phase overload operated just like the single-phase overload, operating on current or temperature. The only difference is that when the three-phase overload senses an overload condition, it opens three switches instead of one switch.

EXERCISE:
You have been called to diagnose a three-phase compressor with a line-duty overload. Describe how you would check to determine if the overload has failed.

Figure 8–14 Time clock with jumper moved.

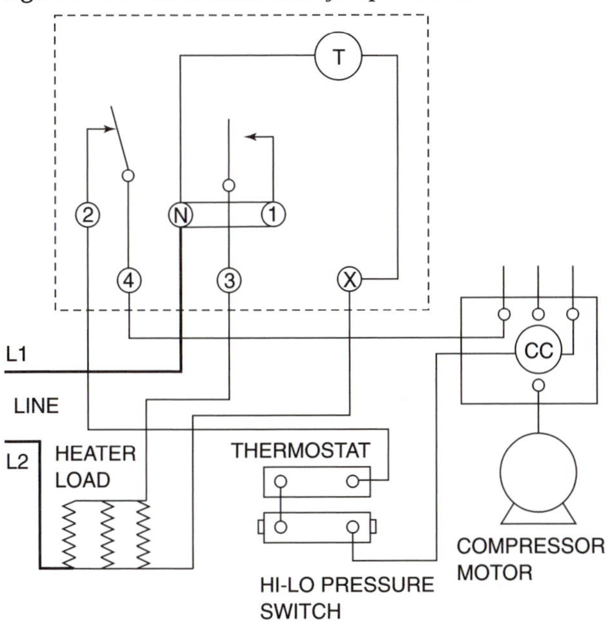

Three-Phase Pilot Duty

A pilot duty overload can also be used on three-phase systems (figure 8–17). Three separate overloads are used. Each overload senses the current on a different leg of the power supply to the compressor. But the three different overload switches are wired in series with the compressor contactor. If any of the three legs draws higher than the design current, it will be sensed by the overload on that leg and will open the corresponding switch. If any one of the overload switches opens, the compressor contactor will be de-energized.

STARTERS

Some larger applications will use a motor starter. Sometimes called a magnetic starter, it is really nothing more than a combination of a contactor and a pilot-duty overload combined into a single unit (figure 8–18). A schematic for a motor starter is given in figure 8–19. This type of starter has overload switches that are spring loaded to open. When a reset button is pressed, the overload switches are closed, and they are held closed against the spring pressure by a ratchet mechanism. The ratchet mechanism is held secure by an enclosed solder pot (figure 8–20).

When the operating control closes, it energizes the starter coil, closing the starter contacts and energizing the motor. The current in each power leg passes through a heater, which is wrapped around the solder pot in the starter. As long as the current draw of the motor is within limits, the heater will not generate sufficient heat to have any effect. How-

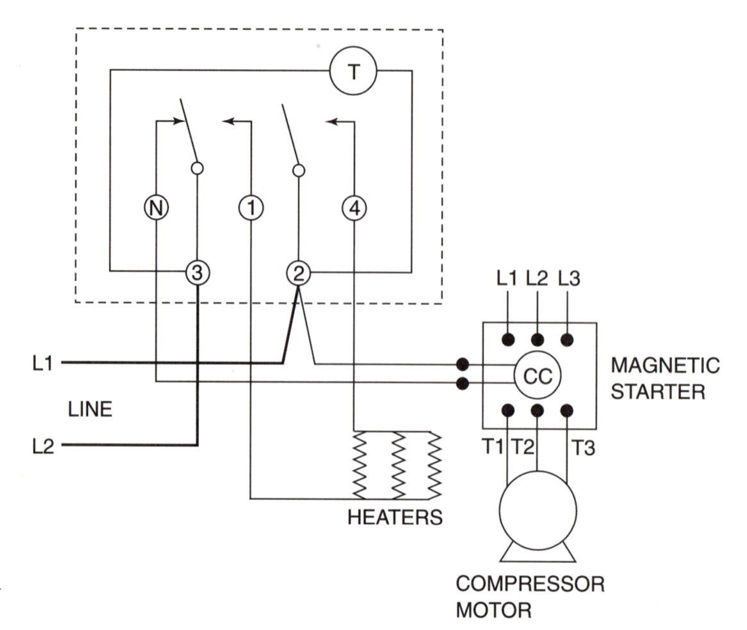

Figure 8–15 Time clock breaks both L1 and L2 in heater circuit.

ever, if the motor current in any leg exceeds the motor rating, that current will be sufficient to cause the heater to generate enough heat to make the solder in the solder pot melt, thus causing the spring-loaded overload switch to open. The starter coil will be de-energized, and the motor will stop until the overload switch is manually reset.

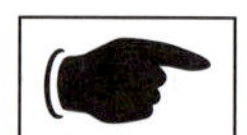

One size starter is suitable for use with a range of motor sizes. The starter is matched to a particular motor by the selection of the proper heater. Sometimes, on new installations, the wrong heater size has been selected by the electrical contractor. This can cause the starter

Figure 8–16 Three-phase compressor with line-duty overload.

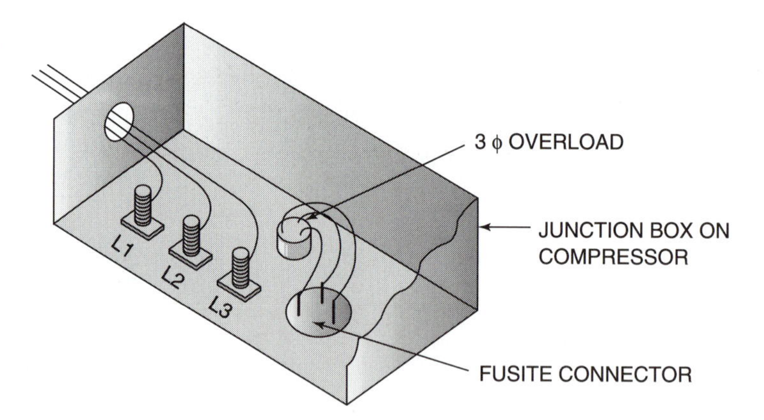

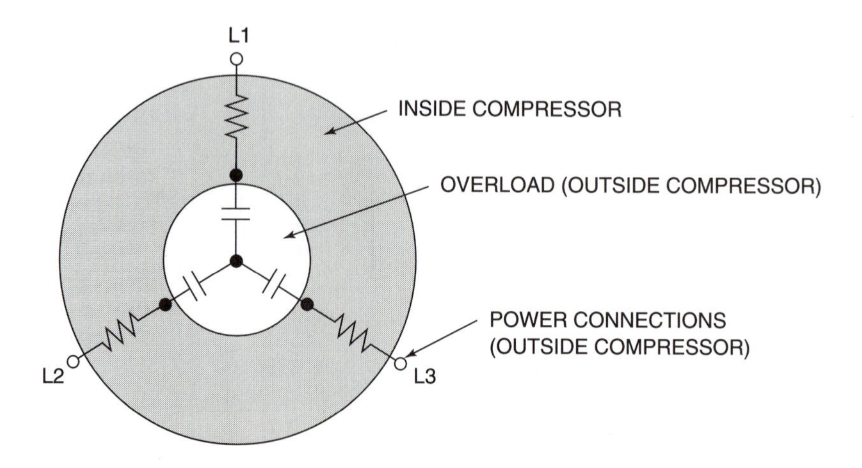

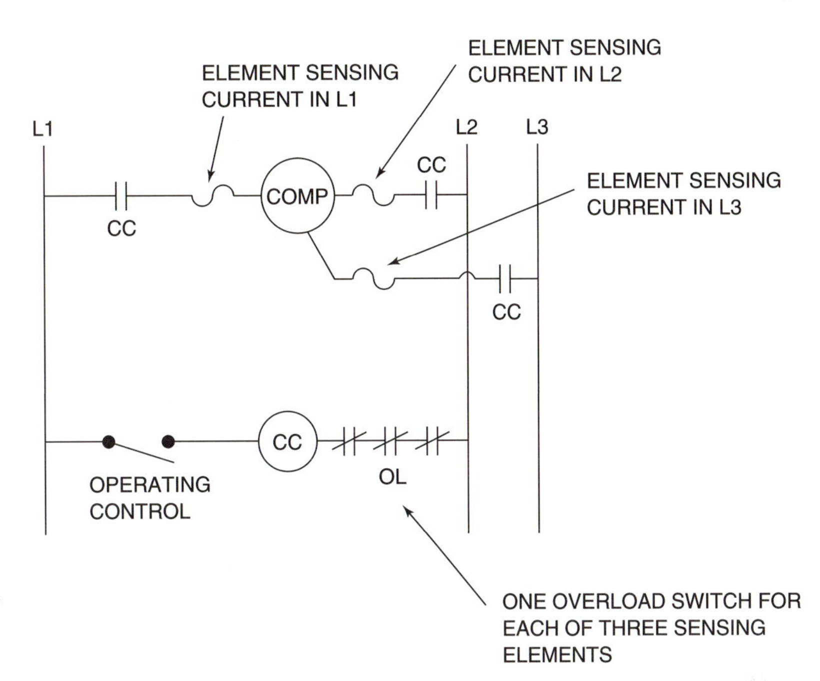

Figure 8–17 Three pilot-duty overloads used on a three-phase compressor.

overload to trip, even though the motor is operating within its acceptable range of amp draw. In this case, you need to advise the electrician on the job to check the heater for proper sizing.

Figure 8–18 Magnetic starter. *(Courtesy of Siemens & Furnace Controls Business Unit, a division of Siemens Energy & Automation.)*

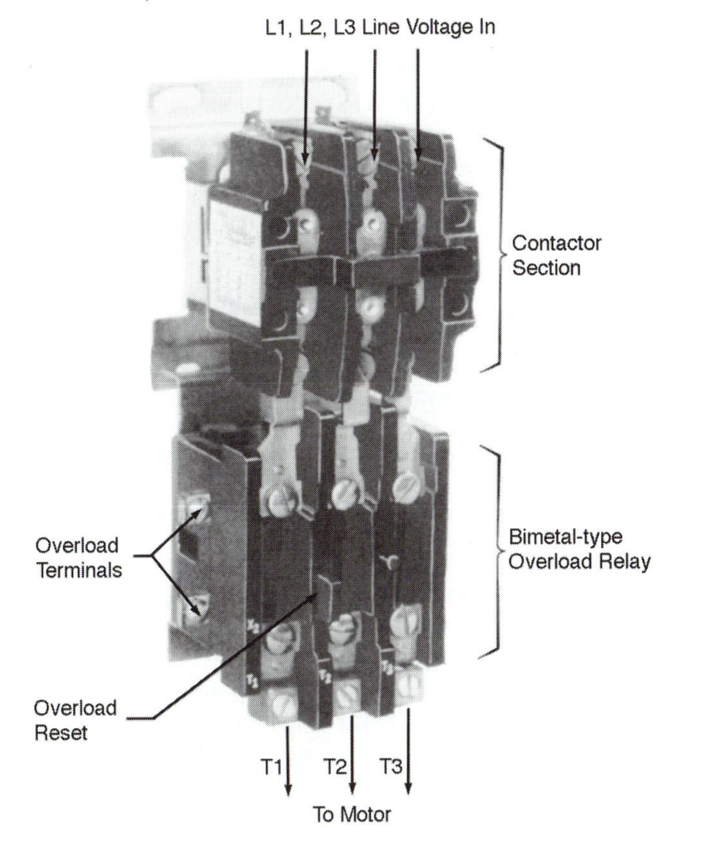

SMALL CONDENSING UNIT

Figure 8–21 shows the ladder diagram for a condensing unit used on air conditioning systems in the 7.5-ton range. When the room thermostat makes R-Y1, the compressor contactor coil CC is energized. This starts the three-phase compressor motor, as well as the condenser fan (OFM) when it is required. The **fan cycling control** (FCC) senses high-side pressure. During periods when it is cold outside, the FCC will turn off the condenser fan if the head pressure drops too low. When the head pressure rises, the FCC will turn the fan back on.

QUESTION:

1. When the condensing unit is running, what voltages would you measure:
   a. Across the HPC?
   b. Across CC coil?
2. What type of motor is the OFM?

LARGE CONDENSING UNIT

Figure 8–22 shows the ladder diagram for a condensing unit used on air conditioning systems in the 15-ton range. The numbers on the wires correspond to labels that are actually attached to the wires in the unit. This makes it especially easy for the service technician to identify wires on the unit that correspond to the wiring diagram.

There is a three-phase compressor motor, a crankcase heater that is energized all the time wired

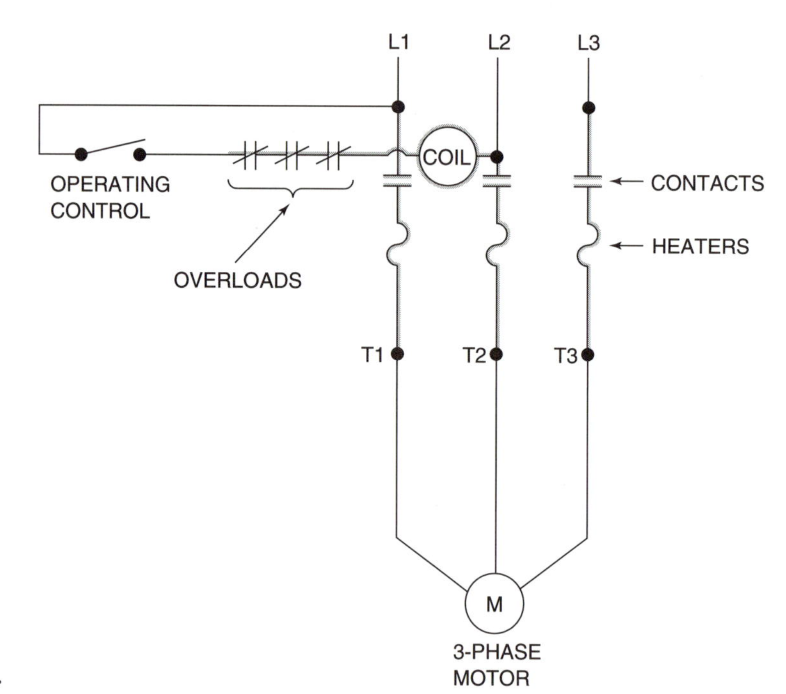

Figure 8–19 Magnetic starter schematic.

between L1 and L3, two single-phase condenser fan motors wired between L2 and L3, and a control power is wired between L1 and L2. The single-phase loads are all wired to different pairs of legs of the supply power in an attempt to equalize the loads between legs as much as possible. The following information about some of the components will assist you in understanding the sequence of operation:

1. The **unloader solenoid** (US) is located on the compressor. When it is energized, it mechanically holds open the suction valve on one or more of the compressor cylinders. This renders those cylinders ineffective, as no compression can take place if the suction valve doesn't close.

Figure 8–20 Overload protection (only one leg shown).

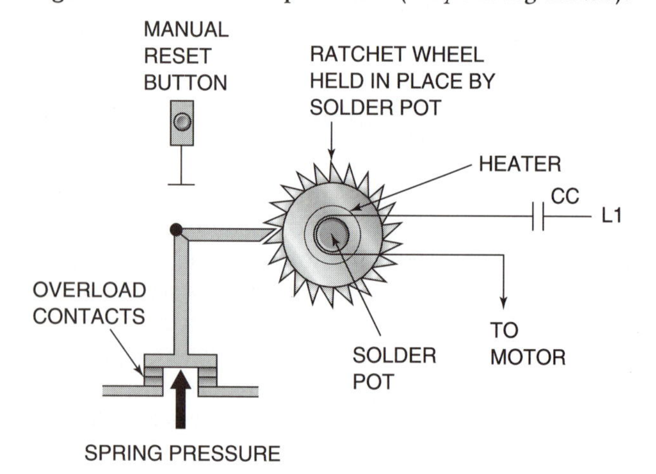

2. The **oil safety time control** (OSTC) is shown in two places. Between wires no. 16 and 17, there is a pressure switch sensing two pressures. The crankcase pressure acts on the top of the switch, and the oil pump discharge pressure acts on the bottom of the switch. The unlabeled rectangle in series with the pressure switch is a small resistance heater located in the OSTC. If this heater remains energized for more than about one minute (45 to 90 seconds, depending on the model), it will open the bimetal switch of the OSTC located between wires no. 4 and 5, de-energizing the compressor contactor. However, if sufficient oil pressure is established soon enough, it will open the pressure switch, de-energizing the heater, and the compressor contactor will remain energized.

3. The solenoid between terminals BB and CC (terminals are identified by hexagons) is a **liquid line solenoid valve.** When it is de-energized, it shuts off the flow of refrigerant in the liquid line. This is used to provide **automatic pumpdown** each time the room thermostat is energized.

4. The time delay control (TDC) is a factory option. You can tell by looking at the key for wiring information on the diagram. When it is used, instead of the thermostat energizing the cooling relay (CLR) directly, the thermostat energizes the TDC. The TDC may be an on-delay or an off-delay. We can't tell from simply looking at the wiring diagram. The output from the TDC, in turn, will energize the CLR. When the TDC is used, the wire between the Y1 terminal and the CLR coil, and the wire between the CC terminal and the CLR coil do not exist.

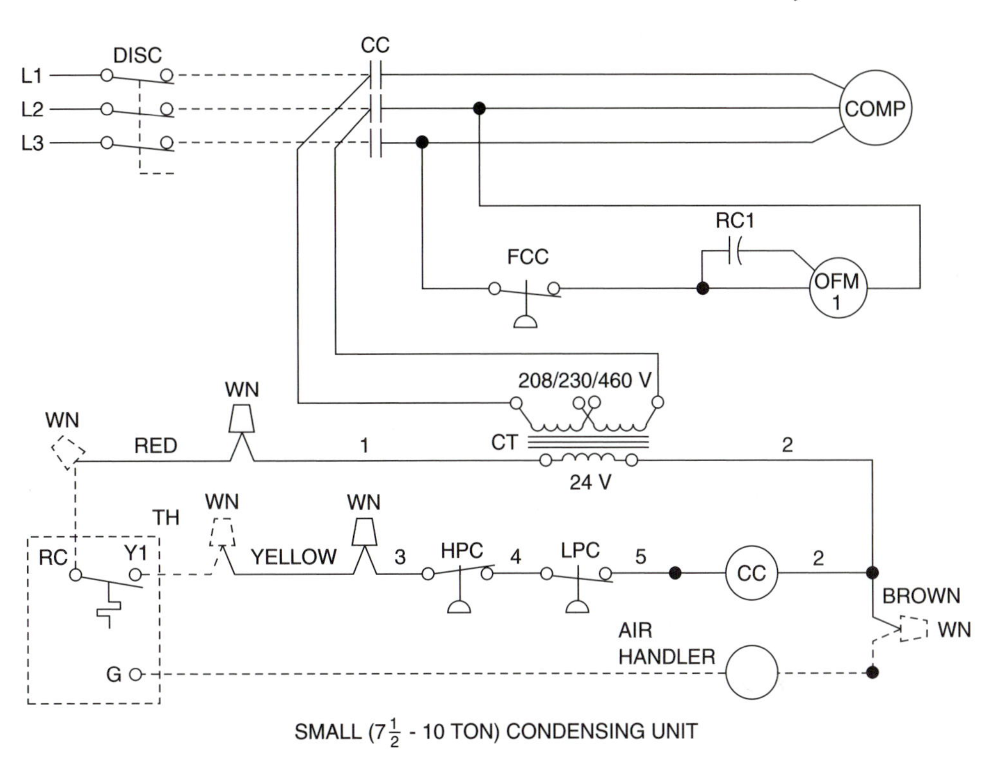

COMPONENT CODE

CC	COMPRESSOR CONTACTOR
COMP	COMPRESSOR
CT	CONTROL TRANSFER
FCC	FAN CYCLE CONTROL
HPC	HIGH PRESSURE CONTROL
LPC	LOW PRESSURE CONTROL
OFM	OUTDOOR FAN MOTOR
RC	RUN CAPACITOR
TH	THERMOSTAT
WN	WIRE NUT

WIRING INFORMATION

1. LINE VOLTAGE
   - FACTORY STANDARD ————————
   - FACTORY OPTION —— — ——
   - FIELD INSTALLED - - - - - - - - -
2. LOW VOLTAGE
   - FACTORY STANDARD ————————
   - FACTORY OPTION — — —— — —
   - FIELD INSTALLED - - - - - - - - -

Figure 8–21 Small (7 1/2–10 ton) condensing unit.

Sequence of Operation

On a call for cooling, the room thermostat (TH) makes contact from RC to Y1 (first stage cooling). This energizes the CLR coil, closing CLR contacts between wires 27 and 28, thus energizing the liquid line solenoid valve. When the liquid solenoid valve opens, refrigerant enters the low-side of the system from the receiver, and the low-side pressure rises. This causes the low-pressure control (LPC) between wires 8 and 9 to close, energizing the compressor, the unloader solenoid, and the fan motor contactor. The compressor starts (in an unloaded condition). Some time later, after the head pressure reaches the cut-in setting of the FCC, the condenser fans will start.

If the first stage of cooling is insufficient to keep the room temperature from rising, the room thermostat will make RC to Y2 (second stage cooling). This will energize the unloader relay, de-energizing the unloader solenoid, thus allowing the compressor to operate at full capacity.

Field Wiring

The dashed lines indicate field wiring, that is, wiring that is installed by you, the installing technician. The manufacturer of the condensing unit anticipates that you will be installing a room thermostat that has an RC and RH terminal, so that a separate control power transformer can be used to power the heating controls.

The condensing unit manufacturer has also made it possible to operate this system without the automatic pumpdown. If automatic pumpdown is not used, there would be no liquid solenoid valve, and the jumper wire (J1) between terminals E and F would be removed. This allows the CC coil to be operated directly from the CLR contacts between terminals E and F.

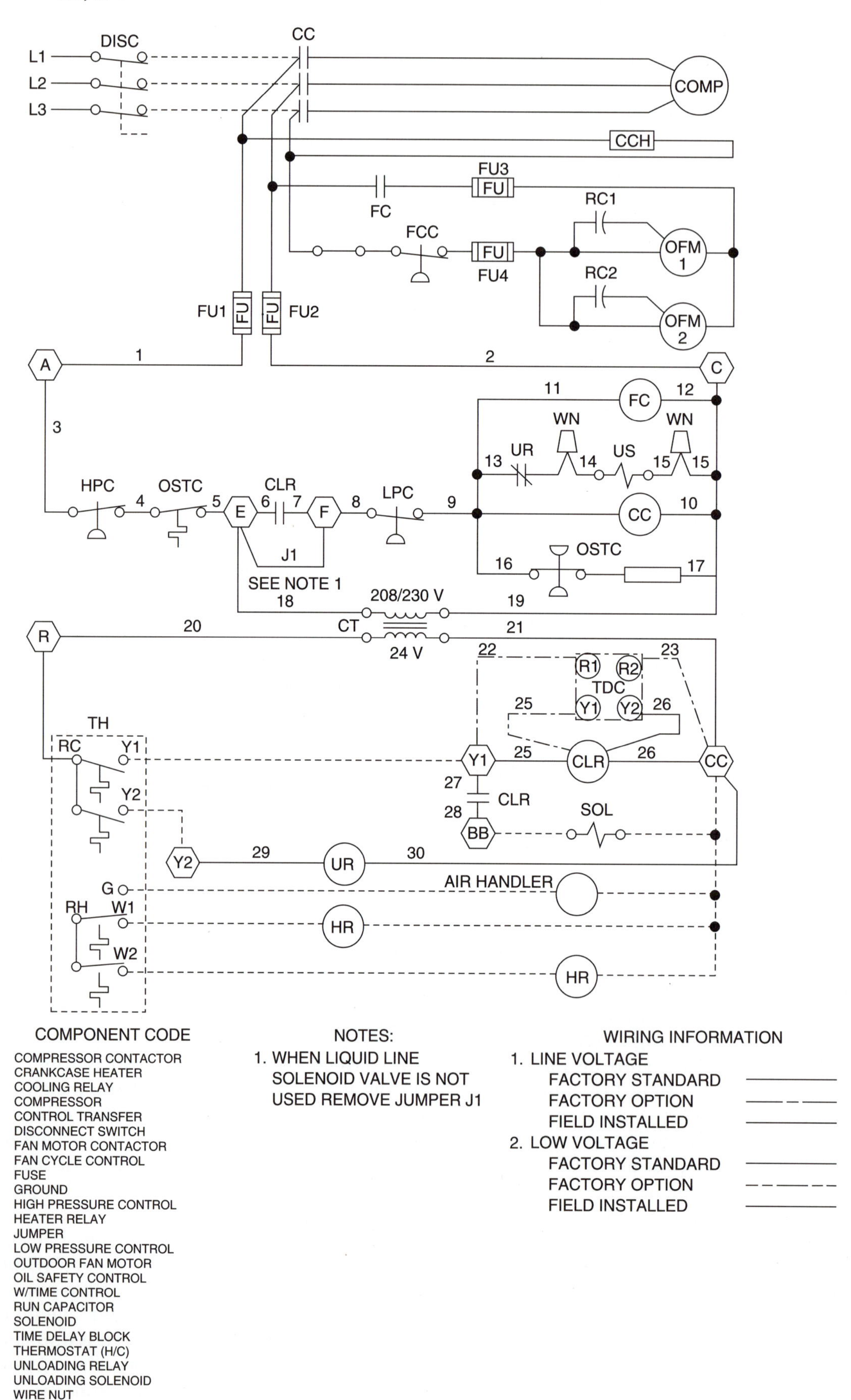

Figure 8–22 Large (15-ton) condensing unit.

ADVANCED CONCEPTS

The oil pressure safety control is more commonly called an oil pressure control or oil pressure cut-out (OPC). A schematic of the OPC is shown in figure 8–23. A diagram typical of that usually found on the OPC itself is shown in figure 8–24. There is a "dropping resistor" wired in series with the heater. If the controller is to be used on a 230V system, the entire length of the dropping resistor is used. However, if used on a 110-120V control system, only half of the dropping resistor is used in the circuit. In this way, the same amount of current is allowed to flow through the heater, and the same time delay is produced for either control voltage.

EXERCISE:
Redraw the OPC diagram in figure 8–24 similar to the format used in figure 8–23. It will be slightly different. Show all terminal identification. [Hint: The switch between terminals 1–2 in the OPC is the switch that opens when sufficient oil pressure is established.]

QUESTION:
When the condensing unit is running at maximum capacity, what voltages would you measure:

1. Between terminals Y2 and CC?
2. Between terminals A and E?
3. Across the unloader solenoid?
4. Between terminals R and CC?

Figure 8–23 Oil pressure cut-out.

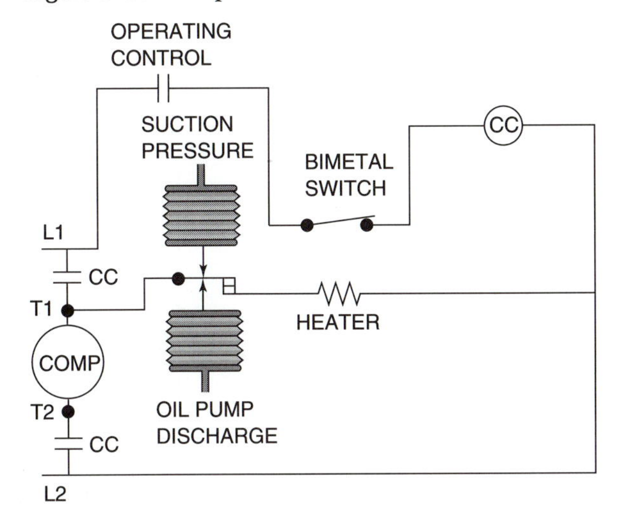

CONDENSING UNIT WITH MECHANICAL TIMER AND PUMP-OUT RELAY

Figure 8–25 shows an older model condensing unit. It has the following unique features:

1. Mechanical **anti-recycle** circuit.
2. **Automatic pumpdown.**

An anti-recycle circuit is one that prevents the compressor from restarting for a period of time after it has shut down. This is important on air conditioning systems that use a PSC compressor. The PSC compressor does not have a lot of starting torque, and it may not be able to start against the pressure of the high-side. The system requires some "off-time" to allow the refrigerant pressures to equalize between the high-side and the low-side. Without the anti-recycle timer, if an occupant moves the thermostat to a lower temperature just after the compressor has shut off, it could cause the compressor to trip out on overload, or worse, blow a fuse.

The anti-recycle timer is a unique device, designed specifically for this type of circuit. The timer starts with the switches made between A-A2 and B-B2. The operation of the timer is that if the timer motor is continuously energized, the switches remain in that position for 15 seconds, then switch down to A-A1 and B-B1 for 10 minutes. After that, the switches return to the original position. But in the circuit, the timer motor does not stay energized all the time.

Sequence of Operation

When the room thermostat closes, it energizes the CR coil. The liquid line solenoid valve gets energized through the CR normally open contacts, the internal thermostat on the compressor, the two overload switches, the high-pressure switch, a second set of CR normally open contacts, and then through the solenoid valve coil. When the solenoid

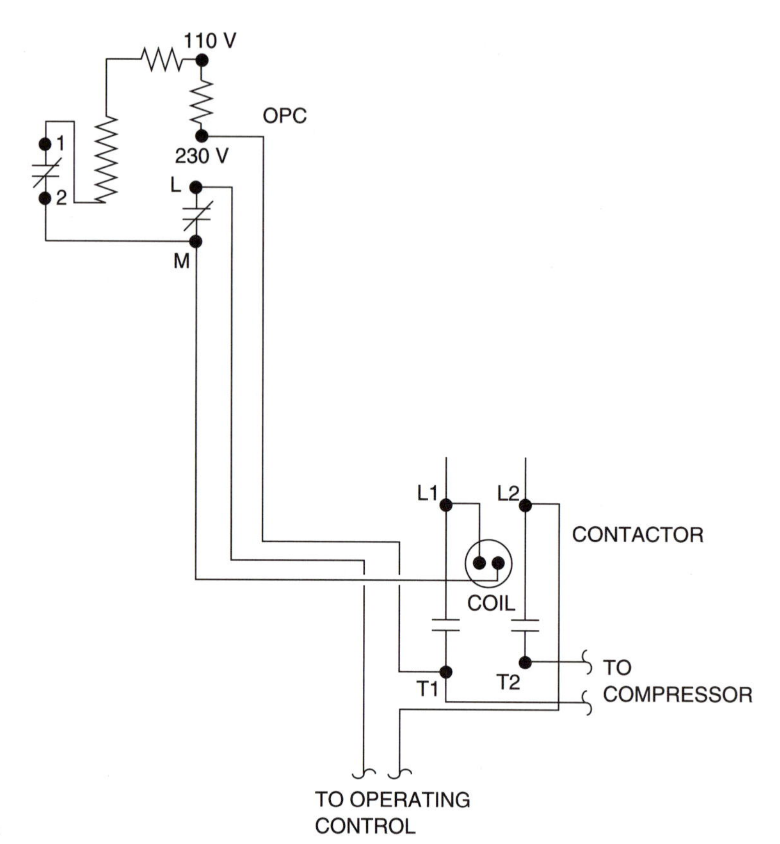

Figure 8–24 Oil pressure cut-out.

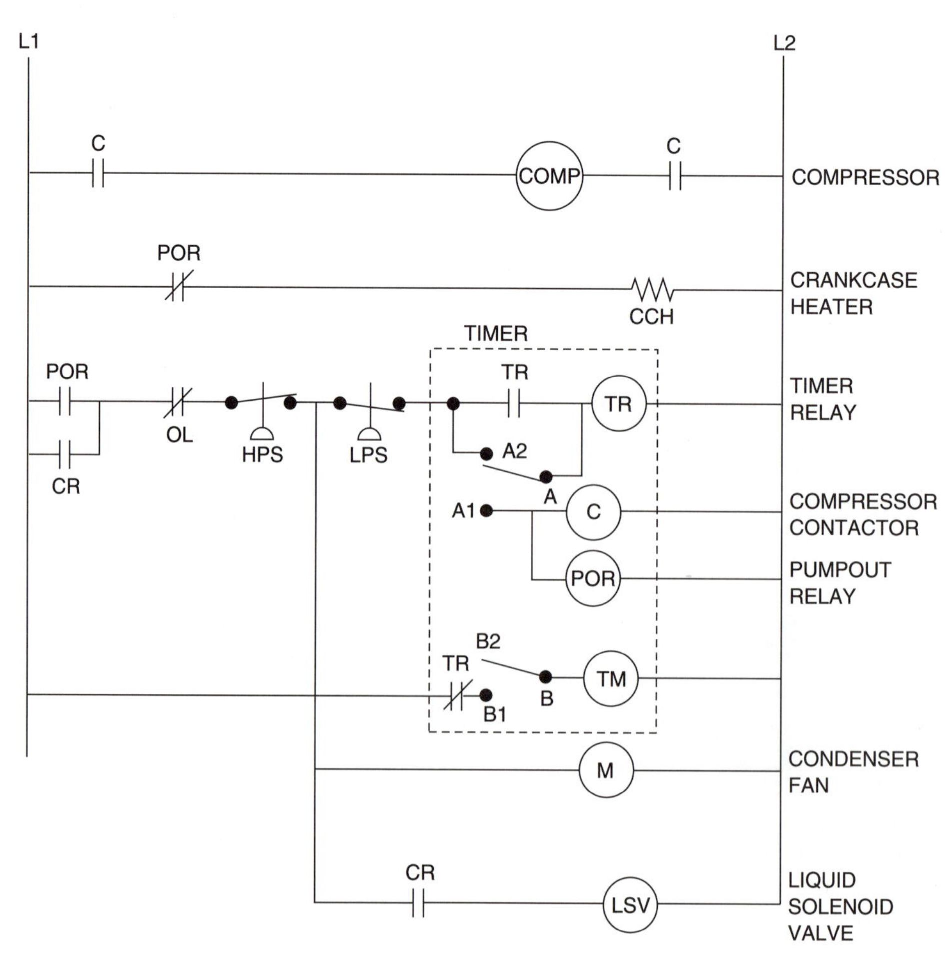

Figure 8–25 Condensing unit with mechanical anti-recycle timer.

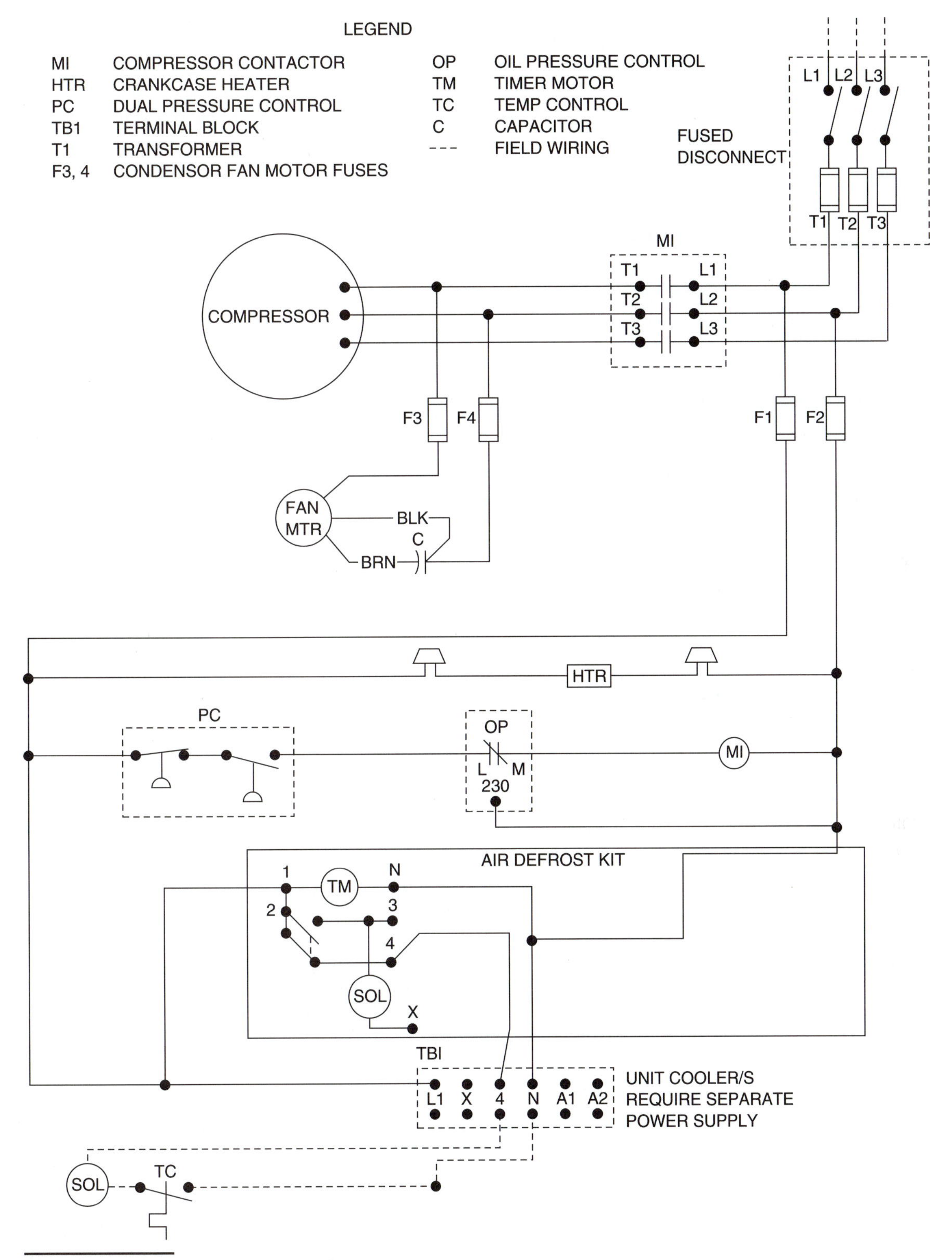

Figure 8–26 Walk-in cooler-air defrost.

valve opens, refrigerant is allowed to pass into the low-side, closing the low-pressure switch. This energizes the timer relay (TR) coil through A2-A. The timer motor TM is also energized, through switch B2-B. After the timer motor runs for 15 seconds, the switches change position to make A1-A and B1-B. This energizes the compressor contactor through the normally open TR contacts. The pump-out relay (POR) is energized in parallel with the contactor coil, and its contacts seal-in around the first set of CR contacts. The timer motor becomes de-energized, because the normally closed TR contacts are

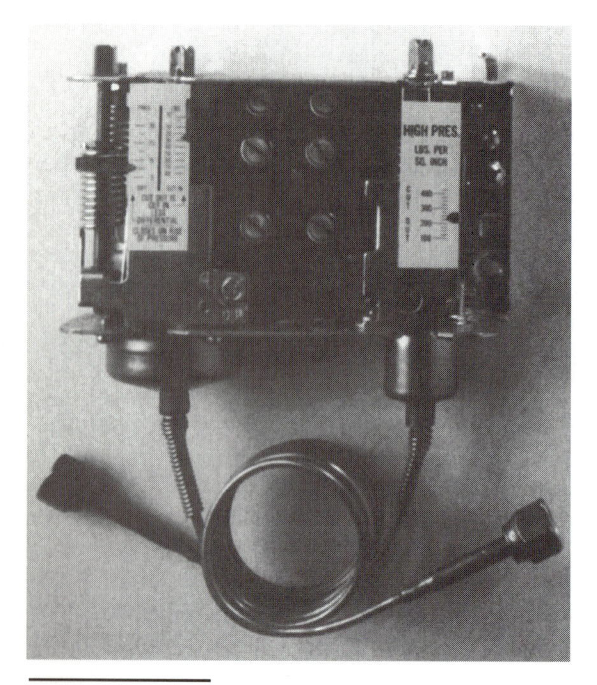

Figure 8–27 High-low pressure cut-out.
(Courtesy of Johnson Controls/PENN.®)

open. The compressor continues to run (and the timer motor continues to not run) as long as the thermostat continues to call for cooling.

When the thermostat opens, the compressor continues to run (through the normally open POR contacts), but the solenoid valve becomes immediately de-energized as the second set of CR contacts opens. With the compressor running and the liquid solenoid valve closed, the low-side refrigerant gets pumped into the receiver, and the low-side pressure drops. When the low-pressure switch opens, the compressor stops, the TR coil is de-energized, and the timer motor starts. After the timer motor runs for 10 minutes, the switches will return to the A2-A and B2-B positions, ready to start again. The compressor cannot restart during that 10-minute period, even if the thermostat re-closes.

QUESTIONS:
1. If the POR coil failed, how would the operation of this system change?
2. If the solenoid coil failed, what would you see operating when you arrived on the job?

WALK-IN COOLER— AIR DEFROST

Figure 8–26 shows a split system that might be used in a walk-in cooler (a cooler is a box that operates above freezing temperatures).

Sequence of Operation

When the box temperature rises above the set point of the temperature control (TC), TC closes, energizing the liquid line solenoid valve (SOL). This allows refrigerant to pass from the receiver into the low-side, allowing the low pressure switch in the combination high-low pressure control (PC) to close (see figure 8–27). This will energize the compressor contactor M1, which in turn, energizes the three-phase compressor motor and the single-phase condenser fan motor.

Air Defrost

The air-defrost kit is an optional accessory. Most coolers depend upon having sufficient off-cycle time for the evaporator coil to defrost. The defrost kit assures that a defrost period will occur at the frequency dictated by the defrost time clock. The timer motor TM is always energized between T1 and T2. When the cams and switches inside the defrost time clock determine that it is time for a defrost, the switch between terminals 2 and 4 open, de-energizing the liquid line solenoid valve, and throwing the condensing unit into pumpdown. Note that terminals 3 and X in the defrost time clock are unused with an air-defrost.

QUESTIONS:
1. What type of motor is used to drive the condenser fan?
2. What is the purpose of the wire between the OP and L2?
3. What is the SOL located inside the air-defrost kit?

EXERCISE:
Redraw this interconnection diagram as a ladder diagram.

WALK-IN FREEZER WITH ELECTRIC DEFROST

Figure 8–28 shows a wiring diagram similar to the previous system, but it is used for a split-system on a freezer. Therefore, it requires an active defrost system. Following are the differences between this diagram and the previous diagram:

1. When the defrost time clock initiates the defrost, making the switch between terminals 2–3, it applies power to terminal A1 on terminal board TB1. There is an auxiliary contact on the compressor

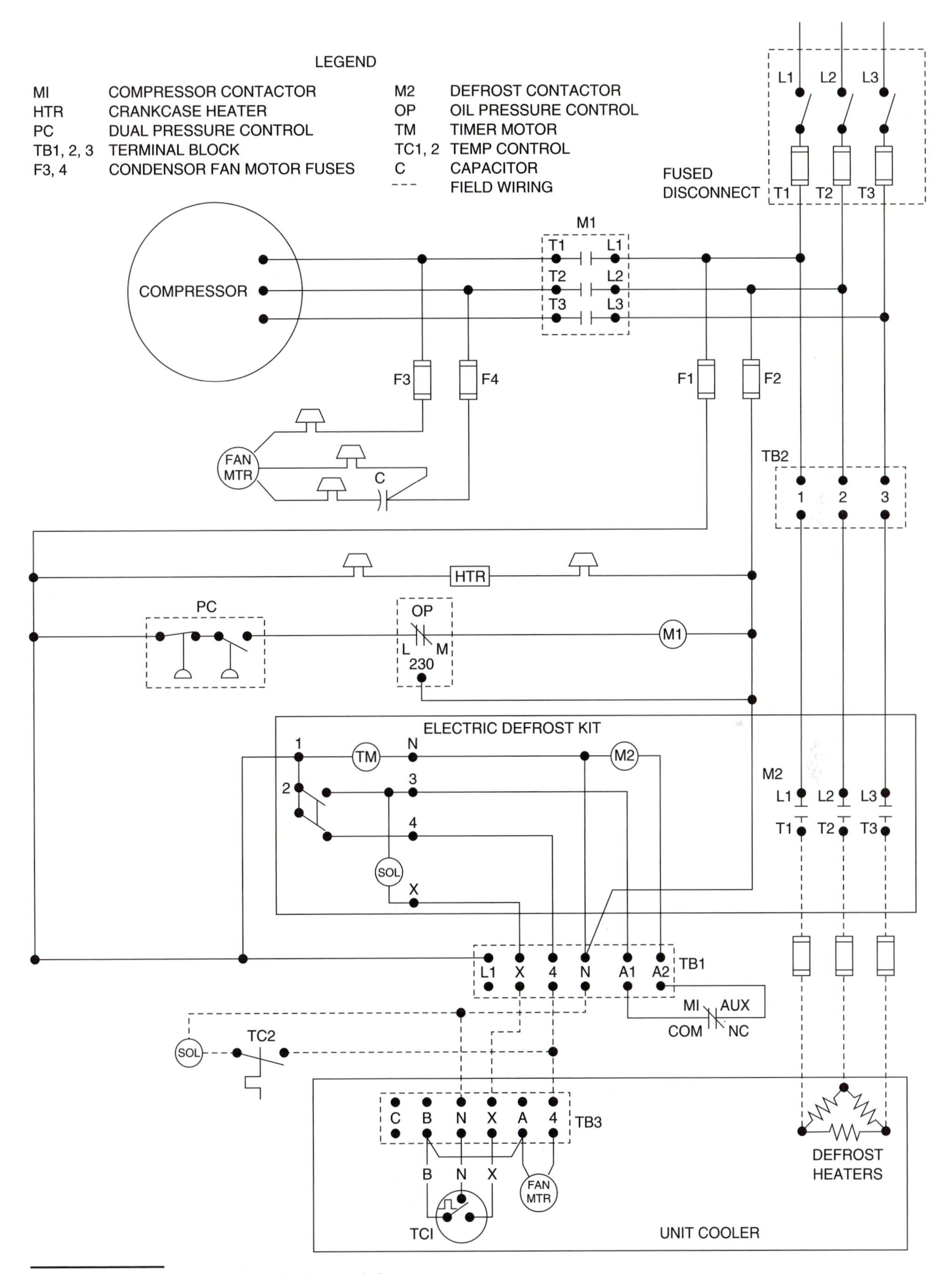

Figure 8–28 Walk-in cooler with electric defrost.

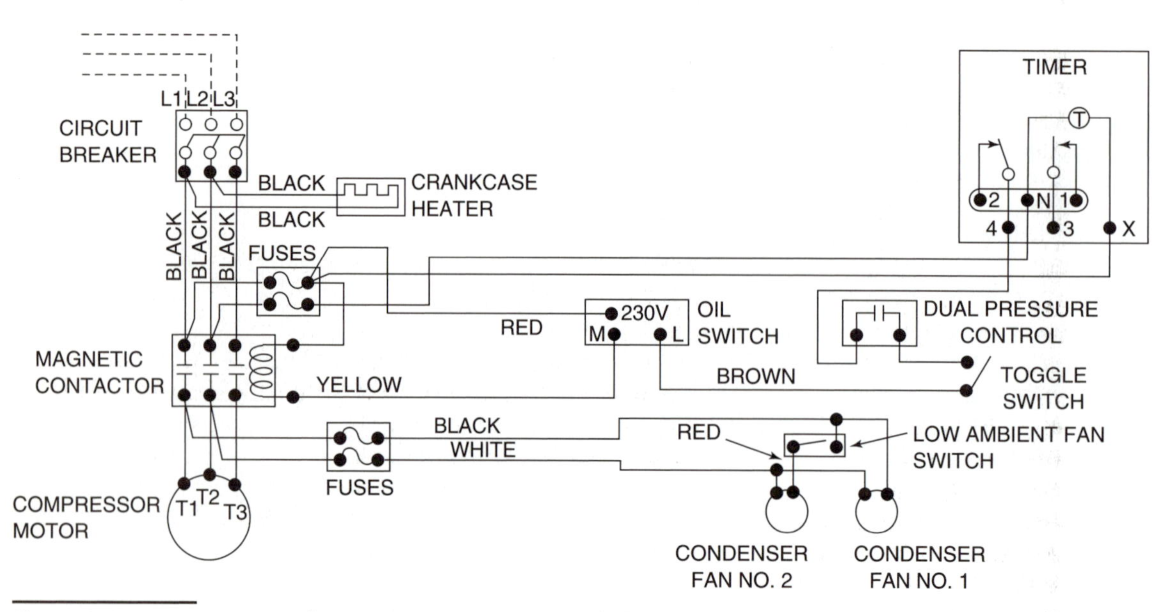

Figure 8–29 5–7.5-ton condensing unit.

contactor M1 that takes the voltage on terminal A1 and applies it to terminal A2, where it energizes contactor M2. When the M2 contacts close, the electric defrost heaters will be energized.

2. When the defrost has continued long enough for the indoor coil to reach 45°F (approximately), the defrost termination switch (TC1) opens the switch between terminals B and N, and closes the switch between terminals X and N. This energizes the solenoid valve inside the defrost time clock, mechanically returning the switches to the "freeze" position (2–3 open, 2–4 closed).

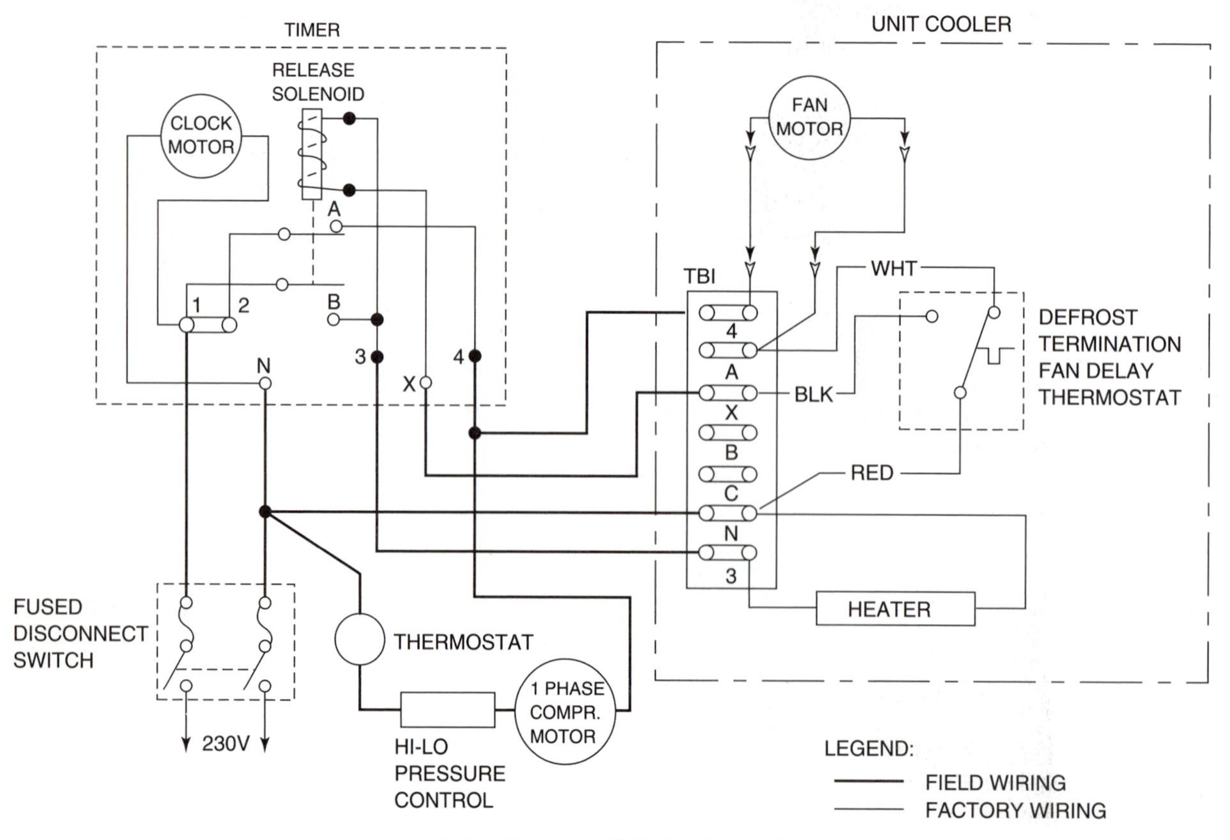

Figure 8–30 Refrigeration condensing unit.

QUESTIONS:

1. When does the fan motor in the unit cooler re-close following the defrost?
2. What temperature is sensed by TC1? TC2?
3. You have found the compressor and fans running, but the box is too warm because the evaporator coil is covered with ice. You move the defrost time clock into the defrost position. The compressor and fans stop, but the ice on the coil does not begin to melt. You measure the following voltages at TB1:

   L1–X = 0V, L1–3 = 0V, L1–4 = 230V, L1–N = 230V, L1–A1=0V, L1–A2 = 230V. What is the problem?

EXERCISE:

Redraw this interconnection diagram as a ladder diagram. Show all of the terminals with the proper identification.

QUESTIONS:

See the wiring diagram in figure 8–29 for a refrigeration condensing unit in the 5–7.5-hp range:

1. Is this condensing unit for a walk-in cooler or a walk-in freezer? How do you know?
2. What is the purpose of the low-ambient fan switch?
3. What is the purpose of the toggle switch?
4. How many switches (other than the disconnect) must be closed for the magnetic contactor to become energized?

QUESTIONS:

See the wiring diagram in figure 8–30 for a refrigeration condensing unit:

1. The fan motor in the unit cooler (inside the walk-in) does not run. If you measured 230V between terminals A and 4 on TB1, what voltage would you measure between terminals 4 and N? Explain your answer.
2. Is the problem with the fan motor or with the defrost timer? How do you know?
3. What switch must close on the defrost termination fan delay thermostat in order to energize the release solenoid (identify the wire colors that become connected)?

INDUSTRIAL CONTROL TRANSFORMERS

The control transformers used in many air conditioning systems for commercial or industrial applications are designed to operate on either 230V or 460V on the primary, depending upon how they are wired. There are two primary windings, and they may be wired in series or in parallel. When wired in series, a 460V input will provide the required control voltage. When wired in parallel, a 230V input will provide the required control voltage. The control voltage can be either 24V or 120V. You will recognize this type of transformer by the terminal markings (figure 8–31):

1. One of the primary windings is located between terminals H1-H2.
2. The other primary winding is located between terminals H3-H4.
3. The secondary winding is located between terminals X1-X2.

Figure 8–31 Industrial control transformer.

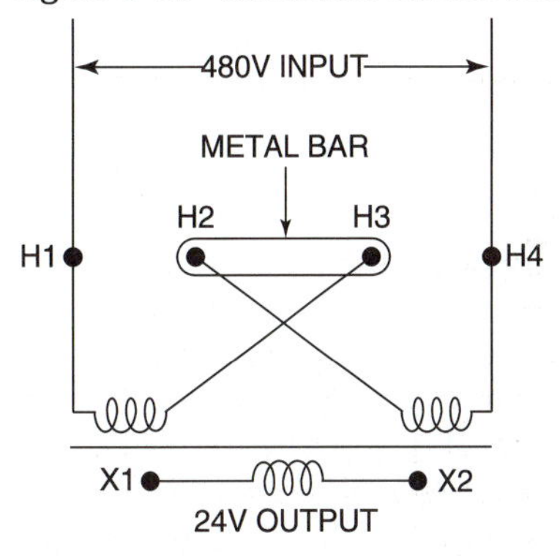

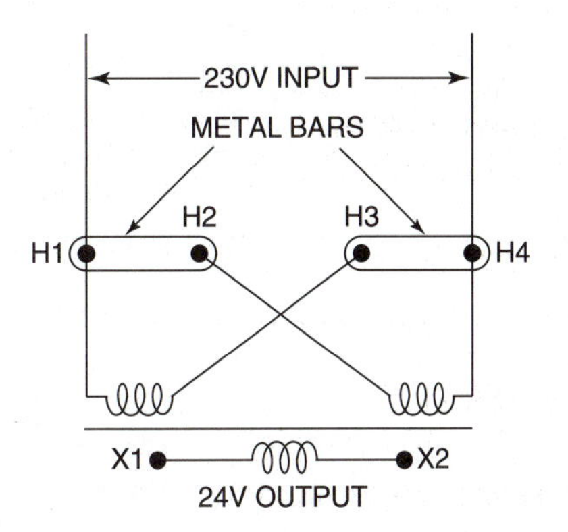

CHAPTER NINE

Motor Applications

MOTOR TYPES

There are three general types of motors used in heating, air conditioning, and refrigeration applications. They are:

1. **Shaded-pole**
2. **Split-phase**
3. **Three-phase**

The theory of operation of these different motor types is discussed in Chapter twelve. This chapter is concerned only with the wiring of these different motors.

Shaded-Pole Motors

Figure 9–1 shows an open type of shaded-pole motor. This is typical of the small evaporator fan motor that will be found in domestic refrigerators and other small appliances. Figure 9–2 shows a shaded-pole motor commonly used as a condenser fan motor or a blower motor in the 1/5-hp to 1/3-hp range that would be seen on a furnace. Shaded-pole motors are limited in their application to the smallest horsepower requirements up to approximately 1/3hp. The wiring of the shaded-pole motor is quite simple. There is a single winding (coil of wire) inside the shaded-pole motor (figure 9–3). There are two motor leads that must be connected to a single-phase source of power (two wires). It doesn't matter which of the power source wires is connected to which of the motor wires.

Split-Phase Motors

Split-phase motors include many different types of motors, but with one common characteristic. The split-phase motor, unlike the shaded-pole motor, has two windings inside. One is called the **run winding,** and the other is called the **start winding.** They are connected together at one end, and three wires are connected to the two windings, as shown in figure 9–4. The wire that is connected to the point where the two windings connect to each other is called the **common** wire. The wire connected to the other side of the run winding is called the **run** wire, and the wire connected to the other side of the start winding is called the **start** wire. Split-phase motors are used in all single-phase compressors. The three wires penetrate the shell by means of pins as shown in figure 9–5. They are called the **start pin,** the **run pin,** and the **common pin.** The ceramic insert that the pins pass through in the compressor shell is called a **fusite** connector.

The wiring connectors that attach to the compressor pins are specially made for this purpose. They are of two types (figure 9–6). Part (a) 9–6a shows the connector used for a simple straight pin on the compressor. Part (b) 9–6b shows the connector used for a pin attached to a flat plate. This connector may look like a simple spade connector, but it is actually of different dimensions. A simple spade connector will not fit onto the compressor pin.

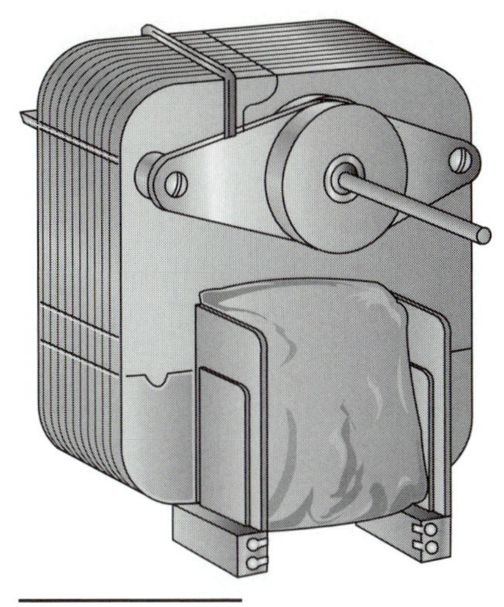

Figure 9–1 Open-type shaded-pole motor (evaporator fan motor).

The addition of a start-winding gives the split-phase motor more **starting torque** (twisting force produced by the shaft) than a shaded-pole motor. A shaded-pole motor driving a fan generally doesn't need very much starting torque, because there is very little inertia to overcome in starting a fan blade moving. But a compressor requires far more starting torque, especially if the compressor is to start up against a pressure difference that already exists between the high-pressure side and the low-pressure side of the refrigeration system. The start winding is comprised of more turns of smaller diameter wire than the run winding. Therefore, the resistance of the start winding is significantly higher than the resistance of the run winding. In most split-phase motor applications, the start winding is only in the circuit for a few seconds on start-up. Thereafter, a switch opens to take the start winding out of the circuit, and the motor operates with current flow through the run winding only. The wiring of the split-phase motor, and how the start winding is taken out of the circuit is described later in this chapter.

Figure 9–2 Open shaded-pole condenser fan or evaporator fan.

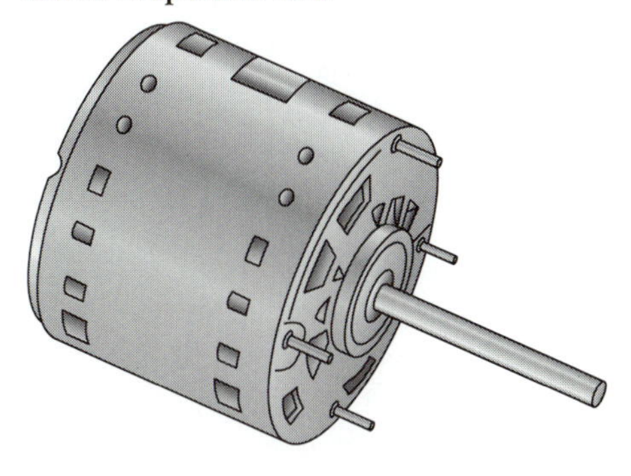

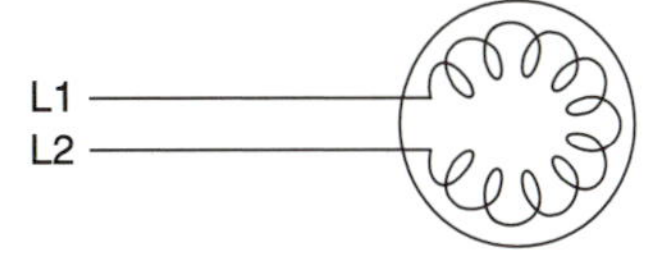

Figure 9–3 Wiring of a shaded-pole motor.

The direction in which the start winding is wound determines the direction of rotation in which the motor will run. Reversing the polarity of the incoming power will have no effect on the rotation direction. Split-phase motors are generally used in the range of 1/4hp to 5hp.

Three-Phase Motors

Three-phase motors consist of three identical windings. They may be wired internally in either a **delta** configuration [figure 9–7(a)] or a **wye** configuration [figure 9–7(b)]. The names come from the letters that these wiring schemes resemble (the Greek letter delta looks like a triangle). They are used in larger size applications than the single-phase motors, usually 5hp and larger (sometimes they are used in smaller sizes, but not often). The wiring of the three-phase motor is quite simple. Each of the motor terminals is connected to one leg of a three-phase power supply, usually 230V or 460V. A three-phase motor has the highest starting-torque of any type of motor. When you connect the wiring, you usually have no idea which way the motor is going to turn. However, if it rotates in the wrong direction, all you need to do is switch any two of the power leads. Then the motor will run in the opposite direction. When the three-phase motor is used in a compressor motor, it usually doesn't matter which direction it rotates.

There are two different voltage supply systems that the utility company uses to supply three-phase power. The difference stems from the type of transformer that the utility company uses to step down the voltage from the high voltage transmission lines to the building supply voltage. They are called **delta connected** or **wye connected,** similar to the two different motor winding types.

Some very large three-phase motors actually have two separate sets of motor windings inside the

Figure 9–4 Wiring for a split-phase motor.

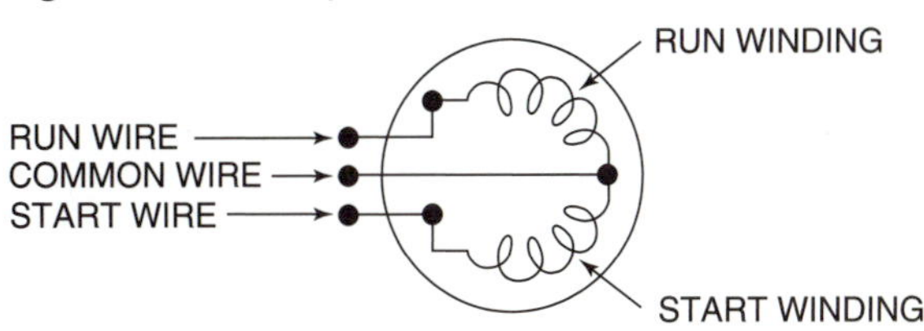

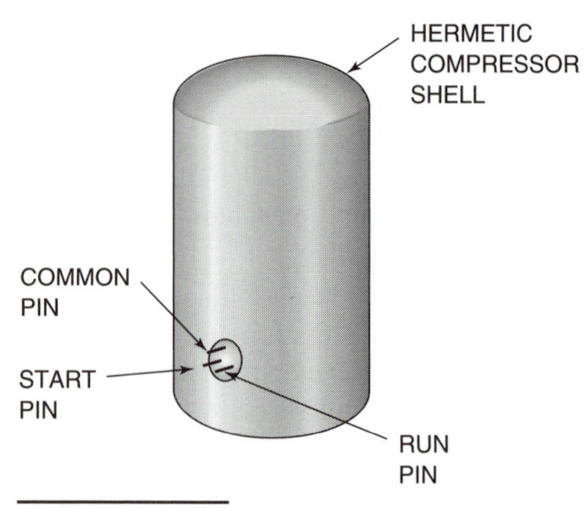

Figure 9–5 Electrical connections for a hermetic compressor.

housing to permit starting the motor in two steps (figure 9–8). Initially, only one set of windings is energized. At the same time that the contactor (C1) for the first set of windings is energized, a time-delay relay is also energized. Within a few seconds, the time-delay relay contacts close, energizing the second contactor (C2), which brings on the second set of motor windings.

MOTOR SUPPLY VOLTAGE

All motors have a nameplate voltage rating, such as 115V, 230V, or 460V. In most cases, the motor will operate satisfactorily so long as the voltage supply is within plus or minus 10% of the rated voltage. This means that the acceptable supply voltage to a 115V motor is 103.5 to 126.5V and the acceptable supply voltage to a 230V motor is 207V to 253V. If the supply voltage is above or below these limits, excessive heat may be generated in the motor windings, resulting in premature motor failure.

For three-phase motors, there is an additional consideration. The voltage imbalance between the three phases should never exceed 2%. Voltage imbalance is calculated as follows:

1. Take voltage readings between each pair of pins (three readings).

2. Calculate the average of the three readings.

3. Find the maximum difference between one of the voltage readings and the average of the three readings.

Figure 9–6 Compressor terminal connectors. (a) Compressor pin. (b) Compressor pin with plate attached.

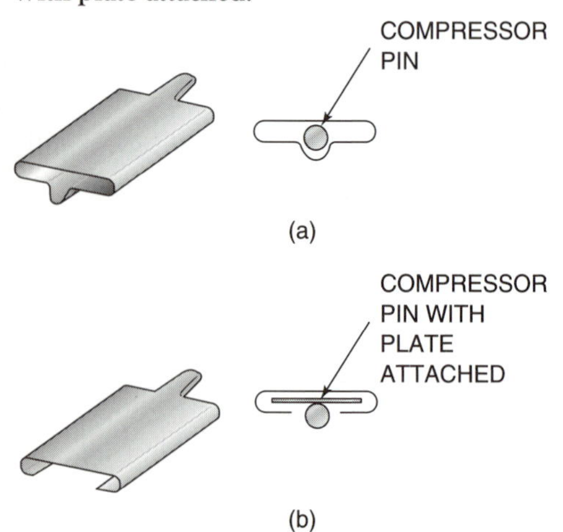

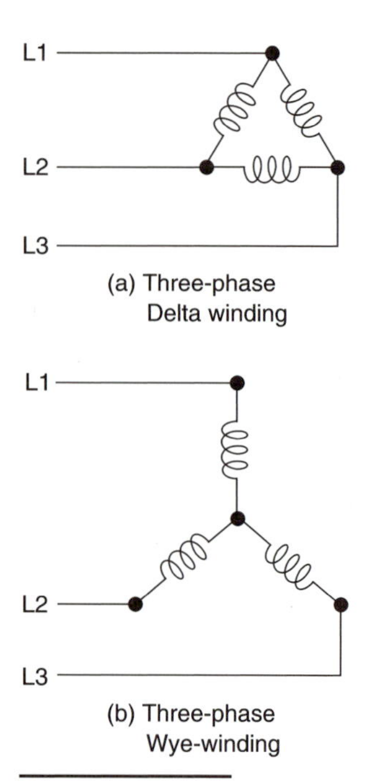

Figure 9–7 Three-phase motor wiring configurations. (a) Three-phase Delta winding. (b) Three-phase Wye-winding.

Figure 9–8 Part-winding motor.

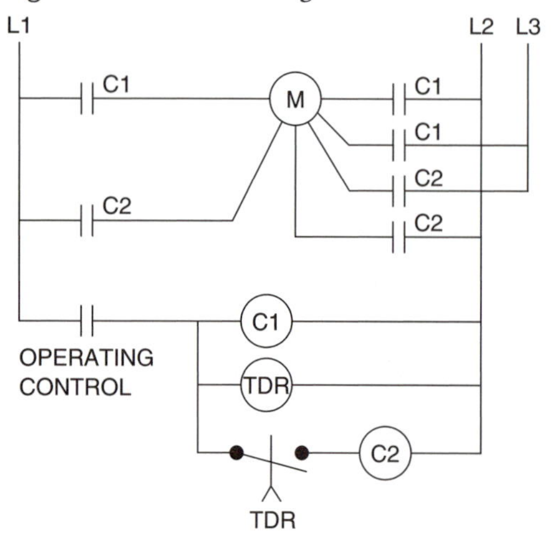

4. Divide this difference by the average voltage. If it exceeds 0.02, you will need an electrician to determine whether there is a problem with the utility company supply voltage or the building's power transformer.

EXAMPLE:

You have taken power supply readings of 235V, 231V, and 222V. The average of the three readings is:

$$V_{avg} = \frac{235 + 231 + 221}{3} = 229V$$

The individual deviations from the average for each phase is:

$235 - 229 = 6V$
$231 - 229 = 2V$
$229 - 221 = 8V$

Dividing the highest difference by the average, the voltage imbalance is:

$$\frac{8V}{229V} = 0.0349, \text{ or } 3.49\% \text{ imbalance (unacceptable).}$$

MOTOR SHAFT
WEIGHT
CLOSED SWITCH

(a) Centrifugal switch at rest

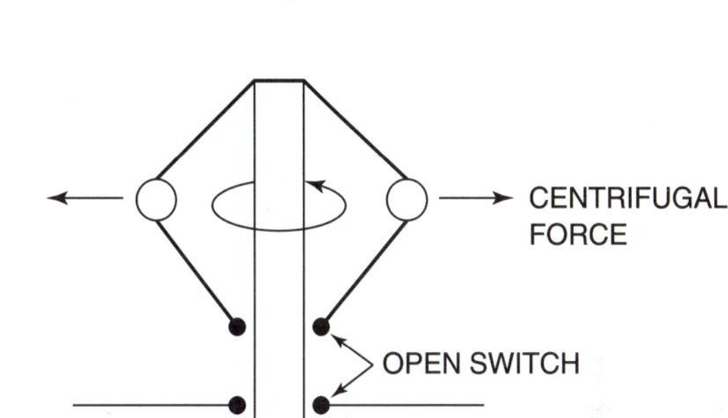

(b) Centrifugal switch spinning

Figure 9–9 Centrifugal switch. (a) Centrifugal switch at rest. (b) Centrifugal switch spinning.

START RELAYS

The split-phase motors (except for the PSC motors discussed below) require a means to automatically disconnect the start-winding from the circuit after the motor gets up to speed. One way that start windings can be taken out of the circuit is by means of a **centrifugal switch** (figure 9–9). It is a switch that senses the centrifugal force being produced by the rotation of a motor. When the motor gets up to almost full speed, the centrifugal switch will open, taking out the start winding. When the motor is de-energized, it will slow, and the centrifugal switch will close. The centrifugal switch can be used in applications where the switch is not exposed to refrigerant. It cannot be used for a hermetic compressor motor which is located inside the refrigeration system, and it is surrounded with the same refrigerant that circulates through the system. If a centrifugal switch were mounted on the end of a hermetic compressor motor, each time the switch opened it would produce a spark, breaking down a small quantity of refrigerant and producing an acid. Over the long term, the acid would attack the motor winding insulation, causing the motor to fail.

In order to avoid the sparking inside the refrigerant system, split-phase compressors use a **start relay** to take the start winding out of the circuit. Three types of start relays will be described:

1. **Current relay**
2. **Potential relay**
3. **Solid state relay**

ADVANCED CONCEPTS

The reason that there is a spark each time the switch opens is because when the switch contacts separate, there is an instant where the contact is broken, but the contacts are sufficiently close together so that the electrons can jump from the contact on the higher potential side to the contact on the lower potential side.

CURRENT RELAY

Two types of current relays are shown in figure 9–10. The characteristics that are common to all the current relays are:

1. There is a coil of large-diameter wire that has few turns and a very low resistance.
2. There is a normally open switch that closes due to the magnetic field produced by the coil, but only if the coil is carrying sufficient current.

Most current relays have three terminals, labelled L, M, and S. This type of current relay is wired into a circuit as shown in figure 9–11. When power is first applied, there is a circuit from H, through the current relay coil, then through the run winding, and then back to N. At the moment that power is applied, the inrush current flow through the run winding is very high. This is called the **locked rotor amps** (LRA). Locked rotor amps are four to six times higher than the **running load amps** (RLA), sometimes called **full load amps** (FLA). Running load amps or full load amps are the maximum currents that the motor will draw when it is running normally at its design load and speed.

Some current relays (push-on type) simply push on to the start and run pins of the compressor (figure 9–12). Others are mounted remotely from the compressor and are connected to the start and run pins by wiring.

Figure 9–10 Current relays. (a) Push-on current relay. (b) Remote mounted current relay.

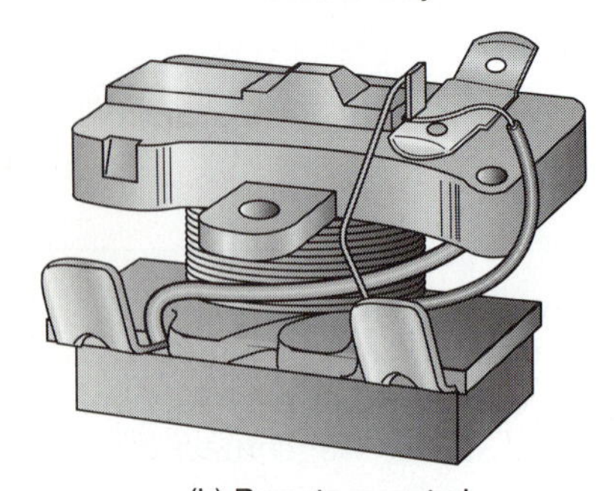

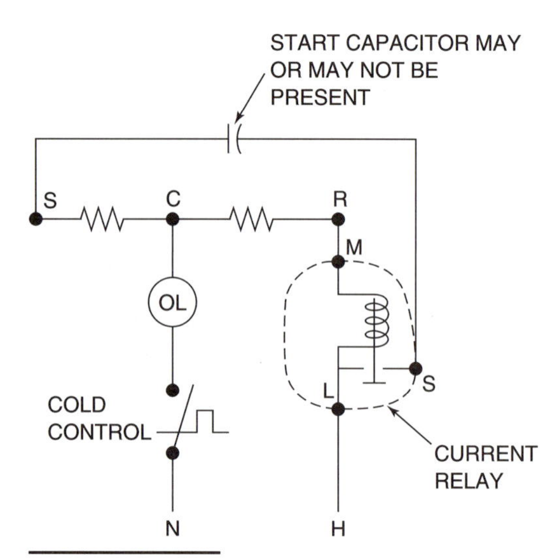

Figure 9–11 Current-relay schematic.

The locked rotor amps through the run winding are high enough to cause enough magnetic field in the current relay coil to cause the switch between L and S to close. This completes a parallel circuit through the start winding a fraction of a second after the run winding is energized. With both windings energized, the motor starts. As it comes up to speed, the current draw diminishes. When the motor reaches approximately 75% of its rated rpm, the current flow through the current relay coil diminishes to a point where the magnetic field produced by the coil is no longer sufficient to hold the switch closed, and it drops open due to the force of gravity. However, at this point, the motor is rotating fast enough that the run winding alone will be able to bring it up to rated speed. The closing and then opening of the current relay switch all happens within a second or two after voltage is applied to the motor circuit.

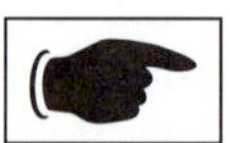

The current relay switch is opened by the force of gravity. Therefore, it is important that the current relay be mounted with the shaft that moves this switch perfectly vertical. If it is mounted out-of-vertical, the switch may not open after the motor achieves sufficient speed.

Some current relays have four terminals. Figure 9–13 shows how the four-wire current relay is wired, providing terminals for convenient attachment to the start capacitor. Note that in this wiring diagram, the relay switch is located between the start capacitor and the start pin on the compressor. This is different from the prior diagram, in which the capacitor is located between the switch and the

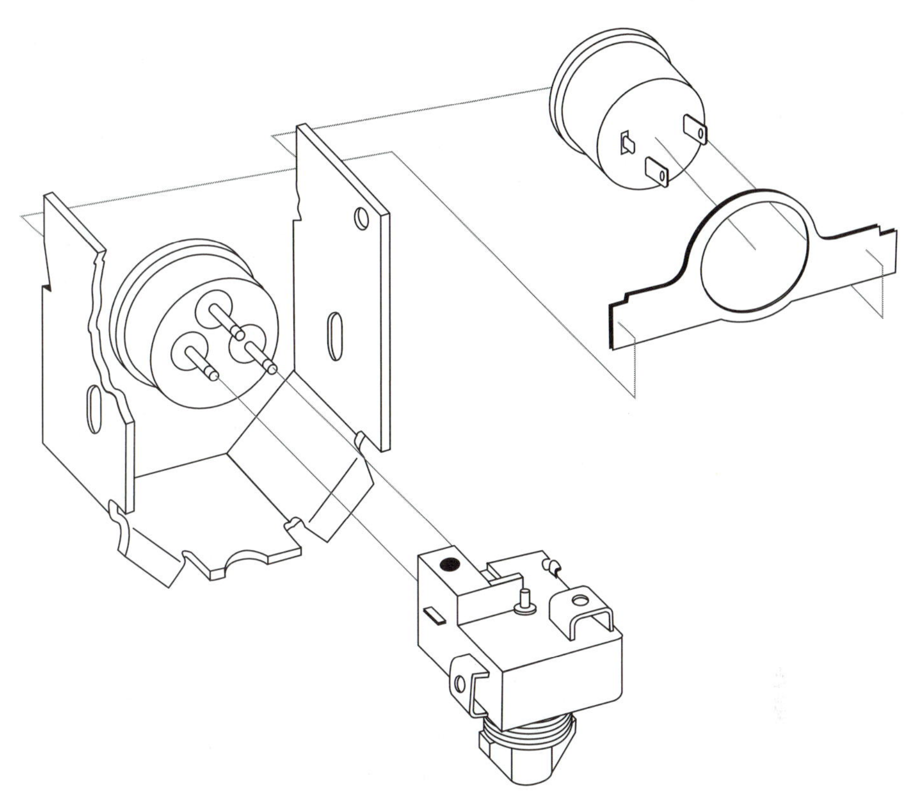

Figure 9–12 Push-on relay. *(Reprinted from Refrigeration and Air Conditioning, 3rd Ed. by Air Conditioning and Refrigeration Institute, Copyright © 1998, Prentice Hall, Inc. Reprinted by permission.)*

start pin. The operation of the circuit is not changed by this difference.

Potential Relay

The potential relay (figure 9–14) is used with a start capacitor and sometimes a run capacitor as well. The wiring for each is shown in figure 9–15. The coil of the potential relay is wired in parallel with the start winding. The potential relay switch is wired to take out the start winding and start capacitor, but unlike the switch in the current relay, it is normally closed. The potential relay coil will cause the switch to open when the voltage across it is slightly higher than the applied voltage. This is called the **pick-up voltage.** When power is applied to the motor circuit, both the start and run windings will be energized, and the motor will start. When the motor gets up to 75–80% of its rated speed, the voltage across the start winding actually exceeds the applied voltage, and the potential relay coil produces sufficient magnetic force to pull the switch open. With the switch open and the start winding out of the circuit, the start winding acts like a transformer. It is a coil of wire located close to a rotating magnetic field. The start winding produces a voltage in the same way as a transformer secondary winding. The voltage produced by the start winding is sufficient to produce enough current through the potential relay coil to hold the potential relay switch open. The potential relay coil consists of many turns of very thin wire. It can produce a significant magnetic field with a very low current flow.

Figure 9–13 Four-wire current relay.

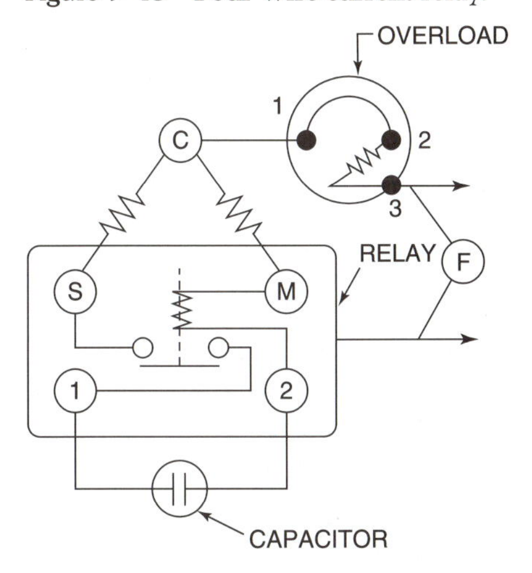

Figure 9–14 Potential relay.

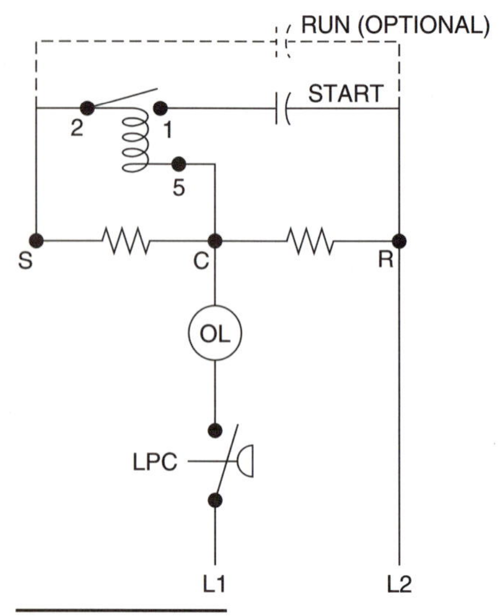

Figure 9–15 Wiring schematic—potential relay.

EXERCISE:
Using a compressor wiring trainer in which the wires can be easily moved, wire the compressor without a start relay. Bring Hot into both the start and run pins, and Neutral to the common pin. Energize the circuit. The compressor should start. Immediately remove the wire from the start pin (you are performing the same function as the switch in the start relay). The compressor continues to run. Measure the voltage across the start winding (between S and C pins on the compressor). You will find a voltage of 40–60 volts, even though there is no wire attached to the start pin! This is the same voltage that is applied to the potential relay coil. So long as this voltage remains higher than the **drop-out voltage** of the potential relay, the coil will hold the relay switch open.

Figure 9–16 shows the physical wiring diagram for the potential relay. Potential relays always have terminals numbered 1, 2, and 5, as shown in the previous schematic. Some potential relays also have a terminal 4. This is merely a **convenience terminal.** A convenience terminal is one that is provided only for the convenient connection of wires. On the potential relay, terminal 4 is not connected to anything inside the relay. It is only for the convenient connecting of three separate wires that would otherwise have to be placed in a wire nut, or all on the same terminal of the start capacitor. An easy way to remember how the wiring of the potential relay is done is to first bring the power into the run and common pins. This will be the running circuit. Then, add the potential relay by memorizing a simple poem, "5, 2, 1, common, start, run." However, you must also remember that 1 goes to the run pin, but through the start capacitor. Also, if a run capacitor is used, it is wired between the start and run pins of the compressor.

Figure 9–16 Physical wiring of potential relay.

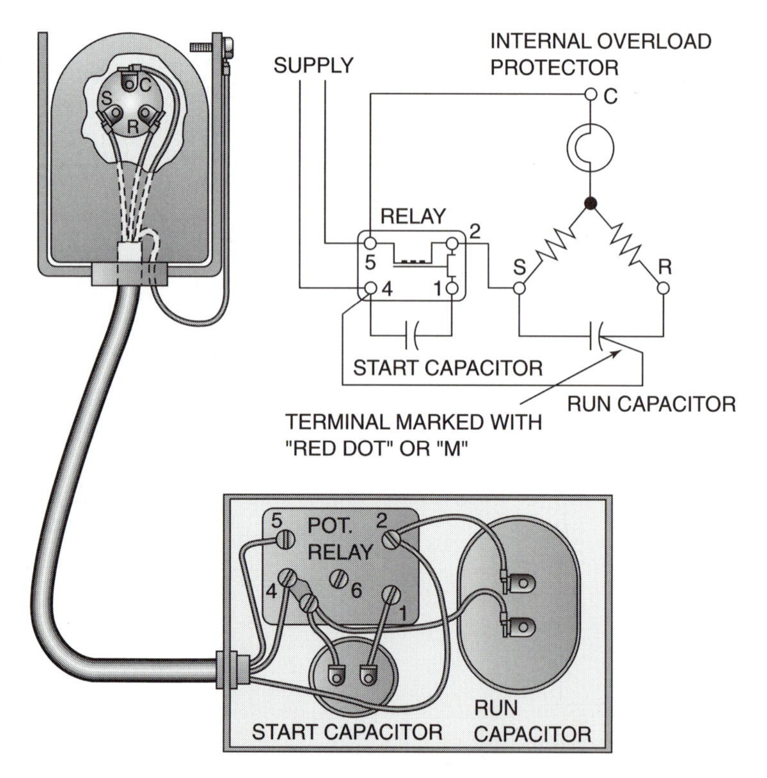

ADVANCED CONCEPTS

Run capacitors come with one terminal identified with a red dot, an M, or a (−) sign. This identifies the terminal of the capacitor plate that is internally located closer to the metal shell of the capacitor. If the capacitor shorts out to the casing, it is more likely this terminal will be the one that gets grounded. The **identified terminal** should be the one that is connected directly to the incoming power source. In this way, if the run capacitor becomes grounded, current will flow directly from the power source to ground, opening the circuit breaker. If the run capacitor is wired the opposite way, it will work fine, but in the event of a grounded capacitor, it can cause damage to motor and/or start relay by allowing the current to flow through the motor start winding to ground (figure 9–17).

EXERCISE:
Redraw the potential relay wiring diagram in figure 9–18 in the same format as figure 9–15. Note that in figure 9–18, where more than one wire is running in the same conduit, it is shown with a single line.

Solid State Relay ("Universal" Replacement Start Relays)

Because current relays and potential relays must be matched to the compressor, it presents a problem to the service technician. You must either stock a lot of different models on your truck, or you must make a trip to the wholesaler to purchase the correct replacement relay each time you need one. In recent years, manufacturers have introduced electronic "universal" start relays. Figure 9–19 shows one type of universal relay for use on compressors up to 1/3 or 1/2hp (depending on the brand of universal relay). It is simply a two-wire device that is wired between the start pin and the run pin on the compressor. If there is a start capacitor, it is wired in series with the relay. It doesn't matter if the capacitor is wired between the relay and the start pin or between the relay and the run pin as shown. If there is also a run capacitor, it is wired between the start pin and the run pin directly.

This type of relay device does the function of both sensing motor speed and switching out the start winding (and start capacitor). The relay consists of simply a piece of semiconductor material. This material has a very low resistance when it is at room temperature. When power is applied to the common and run pins, current flows through the run winding and the start winding (through the semiconductor material). Because the semiconductor has such low resistance upon start-up, it acts like a closed switch. However, after several seconds of carrying current to the start winding, the semiconductor material heats up and becomes a very high resistance. At that point, it acts like an open switch, and very little current is allowed to pass through the start winding.

There is not uniform agreement among technicians about how well this relay works. It has the following disadvantages:

1. It doesn't really sense motor speed. It really is only a time delay, and therefore is not as accurate as

Figure 9–17 Wiring of identified capacitor terminal.

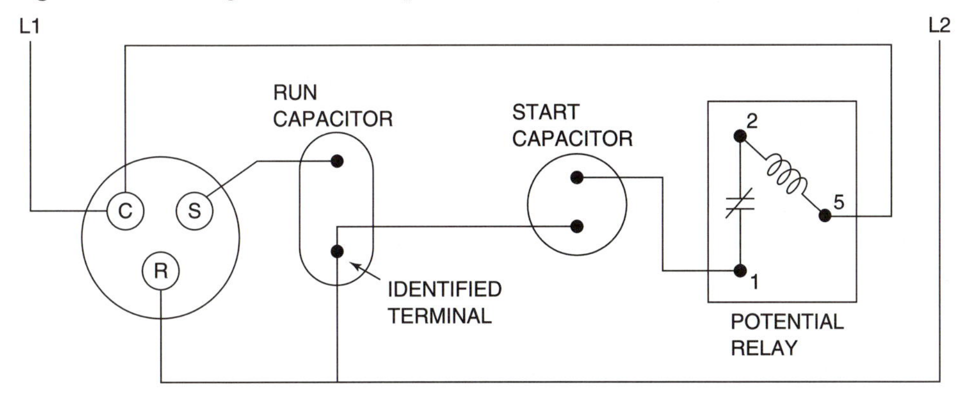

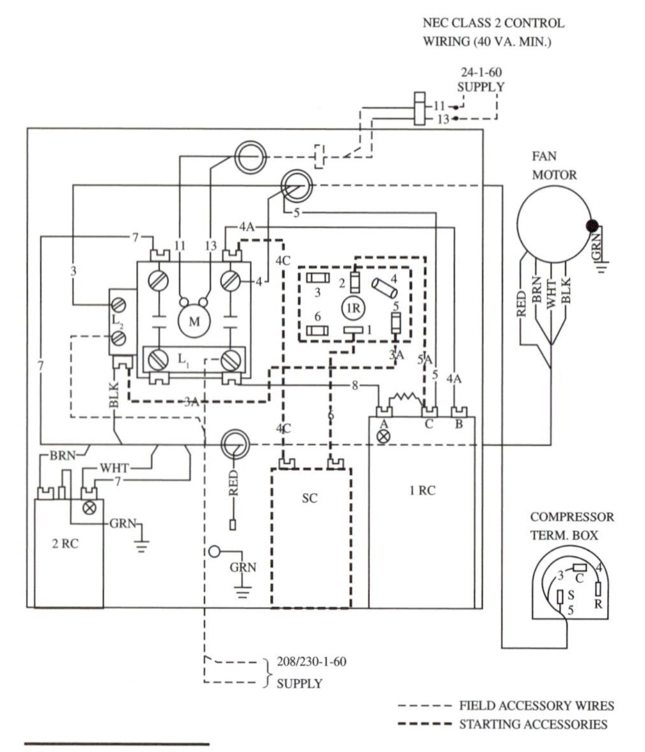

Figure 9–18 Exercise. *(Courtesy of York International Corp.)*

the original relay in taking out the start winding at the ideal instant.

2. The semiconductor must have sufficient time during the Off-cycle to cool, so that it will once again have low resistance at start-up.

Figure 9–20 shows another type of universal start relay. This can be used on compressors up to 5hp. It is truly an electronic circuit inside and may be wired to function as a current relay or a potential relay. The wiring methods are different, and the service technician must carefully follow the wiring diagrams supplied with the relay.

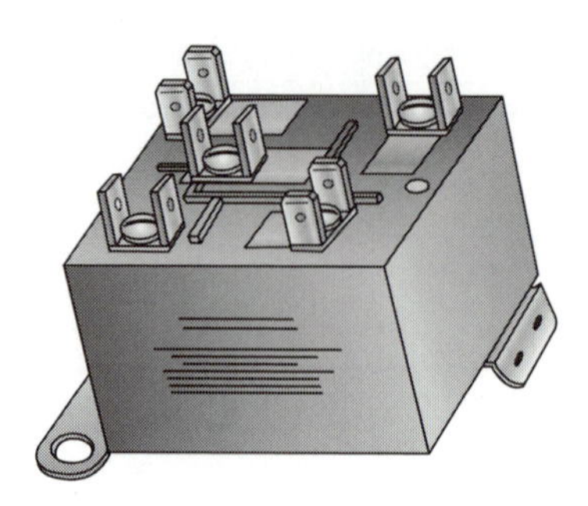

Figure 9–20 PR-90 Universal relay.

Another universal type of replacement relay is the adjustable potential relay shown in figure 9–21. The pick-up voltage of this relay is set to match the pick-up voltage of the failed potential relay. When the motor gets up to the appropriate speed so that the start winding generates the required pick-up voltage, the relay contacts will open, taking the start winding and start capacitor out of the circuit. If you don't know what the pick-up voltage is for the failed potential relay, you can probably set the adjustable potential relay for 190 volts for 115V compressors, and 370 volts for 208/230V compressors. If those settings are too low, the compressor may have difficulty starting, and you will have to increase the adjustment slightly. If these settings are too high, the internal circuitry of the adjustable relay will open the contacts 1 or 2 seconds after voltage is applied to the compressor. This prevents the start winding or start capacitor from failing due to remaining in the circuit too long.

PSC MOTORS

PSC (permanent split capacitor) motors are split-phase motors that usually have no start relay. The wiring of a PSC motor is shown in figure 9–22. The PSC motor is always used with a run capacitor to

Figure 9–19 Universal start relay 1/8 to 1/2hp.

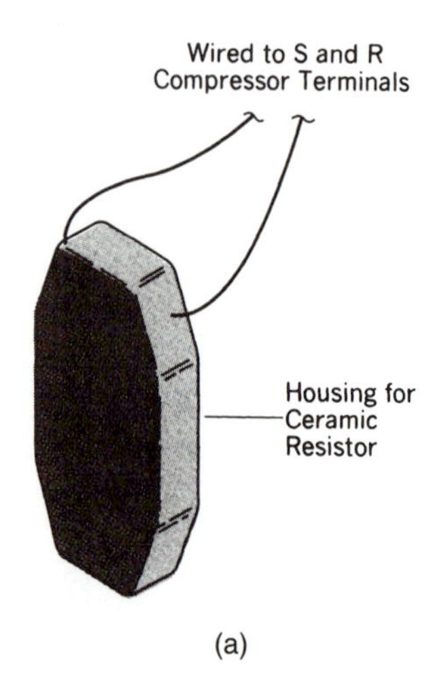

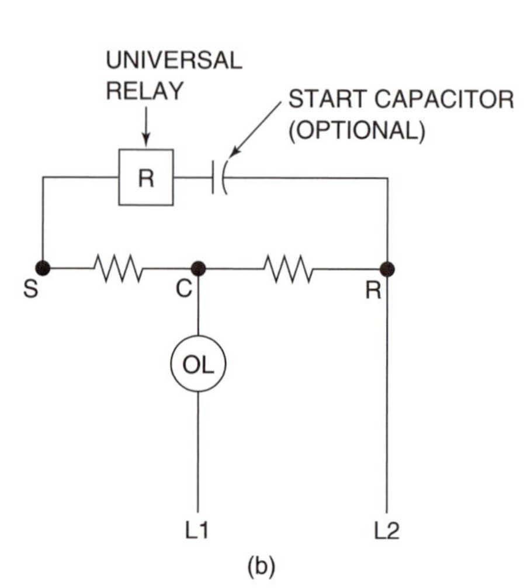

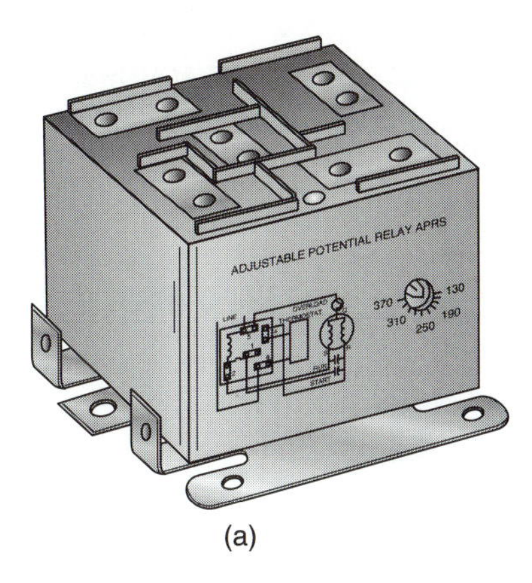

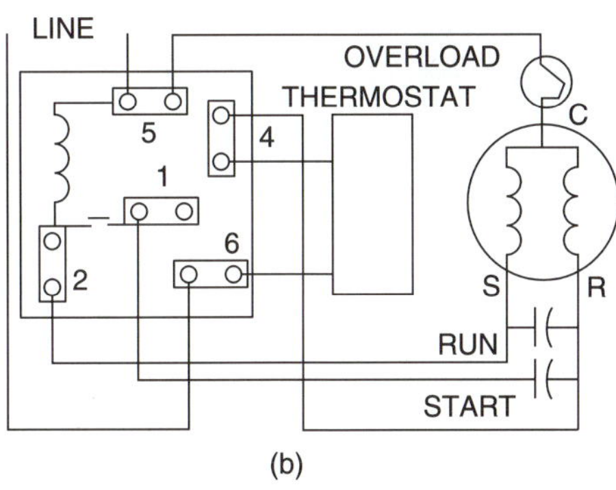

Figure 9–21 Adjustable potential relay.

prevent line voltage from being applied directly to the start winding. The permanent split capacitor motor is very efficient electrically, but it has very limited starting torque. It is most commonly used on residential air conditioning systems that use a capillary tube as the metering device. The capillary tube allows the high-side pressure and the low side pressure to equalize during the Off cycle, so the compressor doesn't have to start against a pressure difference. PSC motors are also commonly used as condenser fan motors.

Sometimes the PSC motor is furnished with two wires for the applied voltage (usually black and white) and two wires for the capacitor (brown and brown). Other times, there are only three wires: black, white, and brown. Figure 9–23 shows how the two configurations are really the same. If you are replacing a failed three-wire motor with a four-wire model, or a four-wire motor with a three-wire model, figure 9–24 shows how you would modify the wiring.

Figure 9–22 Wiring for a PSC motor.

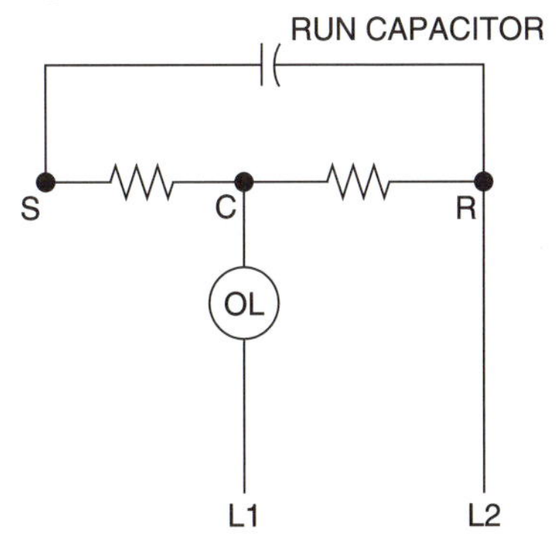

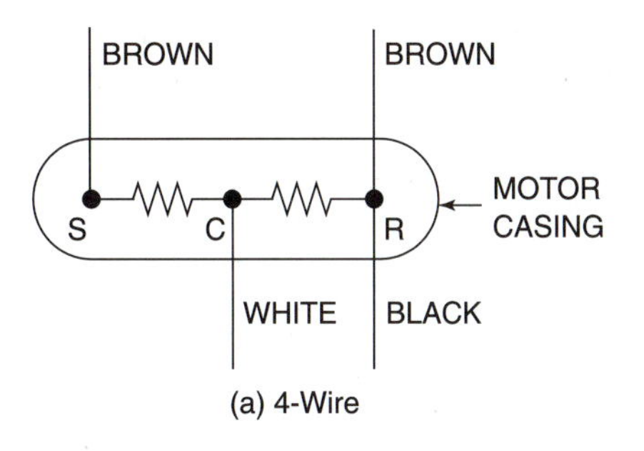

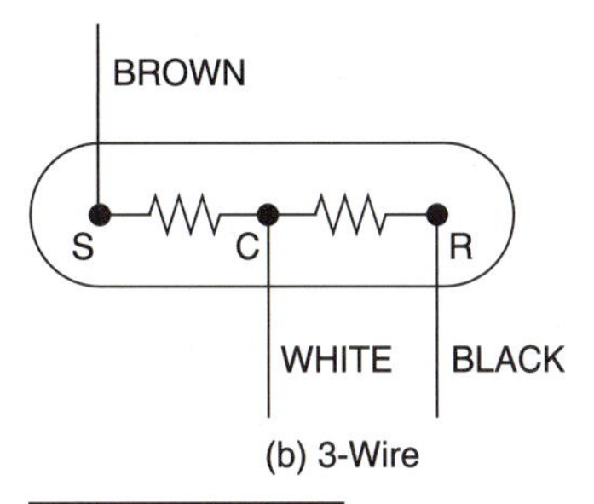

Figure 9–23 Comparison of 3-wire and 4-wire PSC motor. (a) 4-wire. (b) 3-wire.

PTC Device

Figure 9–25 shows a solid state **PTC device.** PTC stands for "positive temperature coefficient," and it means that as the temperature of the device increases, its resistance also increases. At normal ambient temperature, the resistance of the PTC device is very low, like a closed switch. The PTC is wired in parallel with the run capacitor, so that on start-up, the capacitor is bypassed, and the start winding gets full power. The starting current passes through the PTC, causing the PTC to heat up. The PTC resistance increases, and has no affect on the circuit after start-up. The run capacitor then takes over as if the PTC were not even there.

Hard Start Kit

Sometimes, even on new installations using a PSC compressor motor, the compressor will experience difficulty starting because there is insufficient starting torque to overcome the resistance in the compressor between the pistons and the cylinder walls. A **hard start kit** is a solid-state relay, prewired in

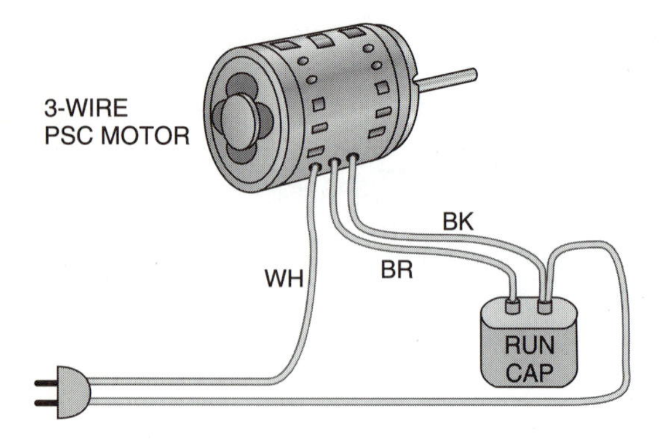

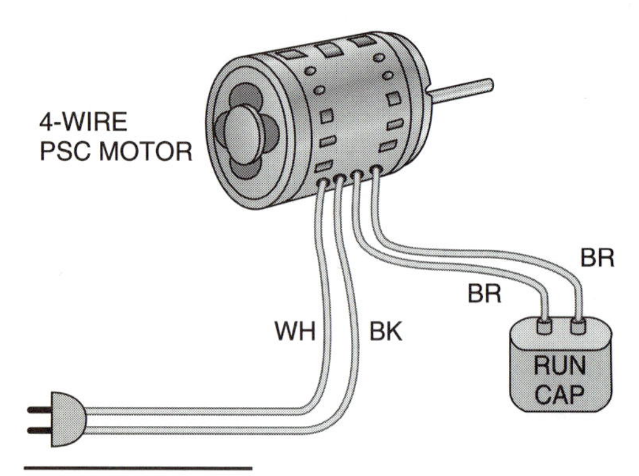

Figure 9–24 Wiring of 3-wire and 4-wire PSC motors.

Figure 9–25 PTC device.

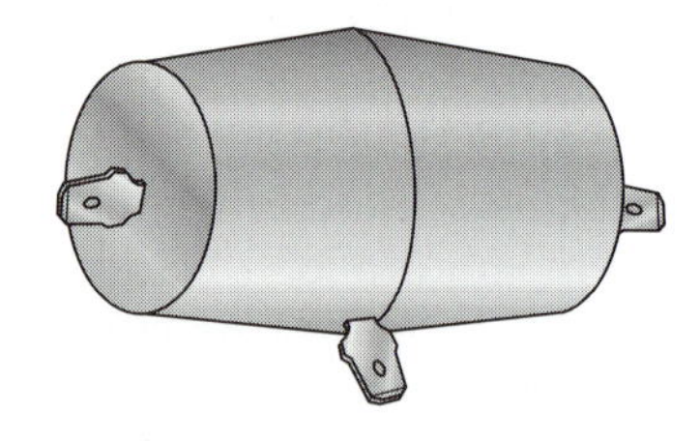

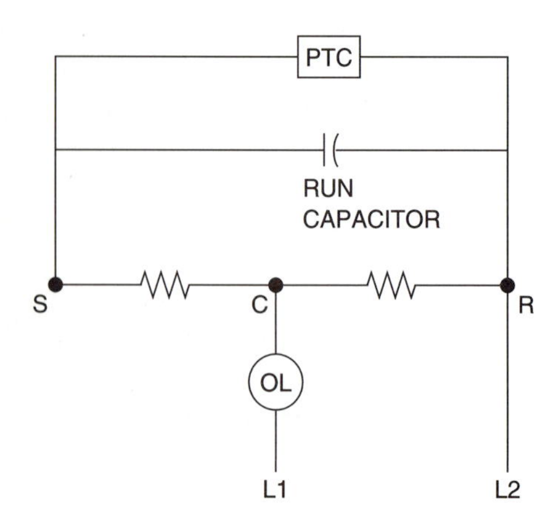

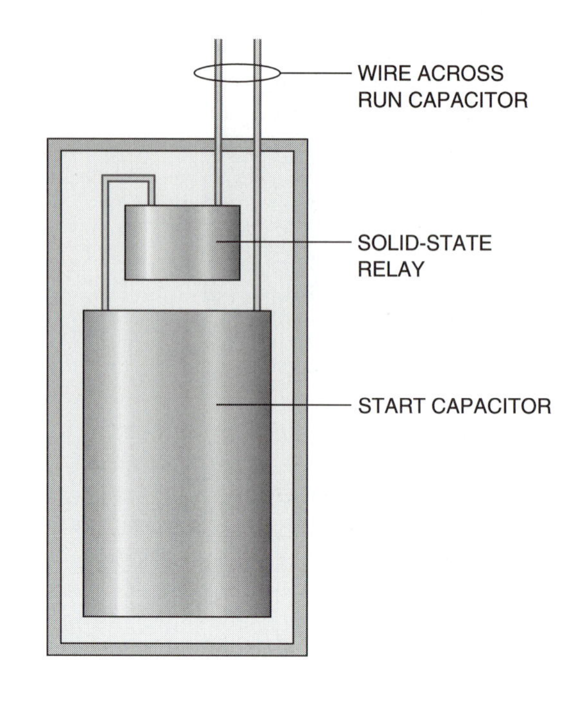

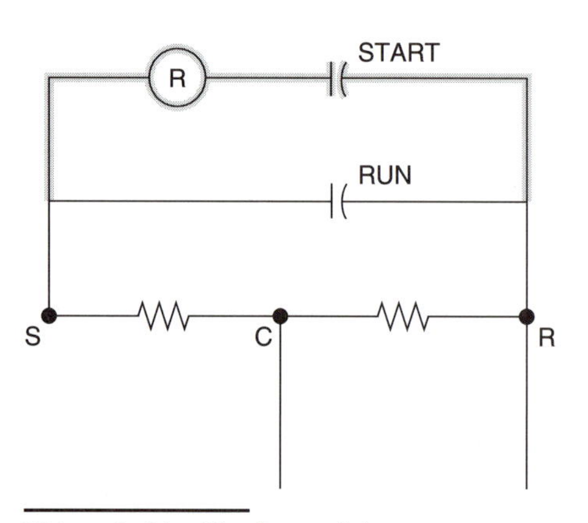

Figure 9–26 Hard-start kit.

series with a start capacitor in a single casing (figure 9–26). When wired across the terminals of the run capacitor (in parallel with the run capacitor), the start capacitor will add extra starting torque, and then the solid state relay will take the start capacitor out of the circuit.

For new installations, some technicians carry a large start capacitor with wires and insulated alligator clips attached to the terminals. With the disconnect off, the technician attaches the alligator clips to the run capacitor. Power is applied to the unit. When the compressor starts, the start capacitor is manually disconnected. After the compressor is allowed to run for a few minutes, the piston and cylinder will "work in" so that the next time, the unit will start without the start capacitor. If not, then the hard-start kit may be permanently wired into the system.

For old installations that have developed starting problems, the hard-start kit may be a temporary fix, but not a cure. Whatever caused the starting problem to develop (lubrication problem, turn-to-turn short in the start winding, etc.) will probably continue to deteriorate, and a compressor failure may result. Sometimes, if the compressor fails shortly after you have installed the hard-start kit, the customer will blame you for causing the failure. For this reason, use of the hard-start kit on old installations should be approached with caution.

MOTOR ENCLOSURES

Motor enclosures are designed for various environments where the motor may be used. Figure 9–2 shows an **open motor** with open-ventilation openings in the shell and end shields. The holes permit passage of external air over the windings. An **open drip proof motor** has ventilation openings that are placed so that droplets of liquid falling within an angle of 15 degrees from vertical will not affect the motor performance.

For applications where the motor windings must be protected from the atmosphere, a **totally enclosed motor** is used. It has no ventilation openings in the motor housing (but it is not airtight). The totally enclosed motor may be fan cooled (totally enclosed fan cooled, **TEFC**), non-ventilated (totally enclosed non-ventilated, **TENV**) or it may use an external source of airflow for cooling (totally enclosed air over, **TEAO**). The TEFC motor includes an external fan in a protective shroud to blow cooling air over the motor. The TENV motor operates without air cooling. It simply operates at a higher temperature, and heat is radiated from the enclosure. The TEAO motor is used to drive a fan. Whenever the motor is energized, the operation of the fan it is driving provides cooling for the motor which is located in the moving airstream. A broken belt on a TEAO motor can cause the motor to overheat.

Explosion proof motors are used in environments where the motor may be exposed to explosive atmospheres (flammable vapors or dust). They are designed to be air-tight, and to withstand an internal explosion without allowing it to propagate to the atmosphere. Explosion proof motors are a rarity because they are so expensive.

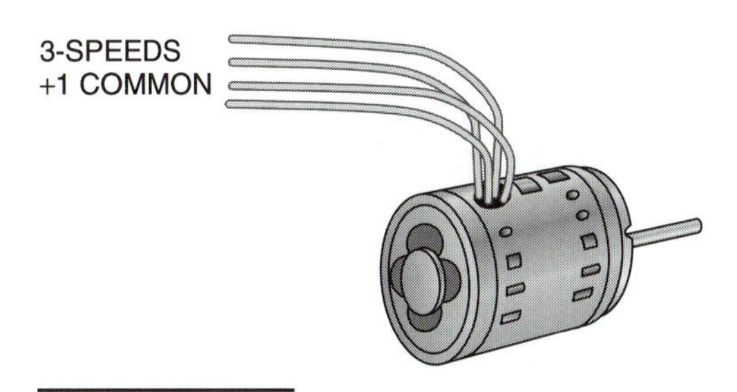

Figure 9–27 Multi-speed motor.

MULTI-SPEED MOTORS

Sometimes a motor will be supplied with multiple power leads (figure 9–27) to allow running the motor at different speeds. This is done for several reasons:

1. On heating-cooling systems, the fan sometimes runs at a higher speed on cooling and a lower speed on heating.
2. On single-speed systems, the manufacturer uses one size multi-speed motor for a range of cooling unit sizes, and then uses an appropriate speed to match the evaporator airflow to the unit size.

Multi-speed motors may be two, three, or four speeds. At most, only two speeds will be wired into the circuit. The NEMA standards for wire colors for the various speeds are given in figure 9–28 for 115V and 230V applications.

Sometimes the winding for one speed of a multi-speed motor will burn out, while another unused speed remains functional. It is usually acceptable to simply substitute the still-functioning

Figure 9–28 NEMA lead color codes.

	115 VOLTS	230 VOLTS
COMMON	WHITE (GROUNDED)	PURPLE (UNGROUNDED)
BLACK	HIGH	HIGH
YELLOW	MED-HIGH	MED-HIGH
BLUE	MED	MED
ORANGE	MED-LOW	MED-LOW
RED	LOW	LOW
CAPACITOR	BROWN OR BROWN W/WHITE TRACER	BROWN OR BROWN W/WHITE TRACER

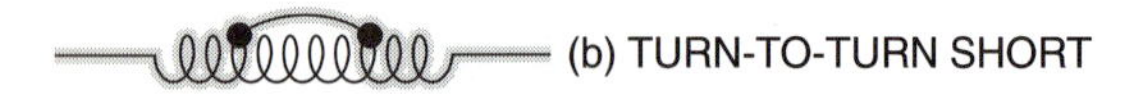

Figure 9–29 Turn-to-turn short due to failed winding insulation. (a) Normal winding. (b) Turn-to-turn short.

CLASS	MAXIMUM TEMP
A	105°C (221° F)
B	130°C (266° F)
F	155°C (311° F)
H	180°C (356° F)

Figure 9–30 Motor insulation classification.

motor speed instead of replacing the entire motor. However, there are two potential problems. If your substitute speed is slower than the original speed, the air conditioning system may tend to freeze up the evaporator coil due to insufficient air flow. Check your low-side pressure after you make the change. If it is above 60 psi (for an R-22 system), you should be OK. If the substitute speed is higher than the original speed, the increased air noise may be unacceptable to the occupant. Check with them on the noise level before you leave the job.

MOTOR INSULATION

The wire that forms the stator winding in a motor appears to be a bare copper wire, but it is not. Before the wire is wound into a winding, it gets a lacquer or enamel coating that acts as insulation. This keeps each individual turn of the winding electrically isolated from the neighboring turn. If this thin layer of insulation wears out, allowing contact between the turns of the winding, the motor has failed. This is called a turn-to-turn short in the motor. It has the effect of shortening the winding where the turn-to-turn short has occurred (figure 9–29). A motor in which the motor insulation has failed in this fashion is sometimes difficult to detect. Most times, you don't know what the actual resistance of the motor windings should be, unless you refer to the manufacturer's information (which may not be available to you while you are on the job). The symptoms can be difficulty in starting the motor if the start winding has a turn-to-turn short, or a high amperage draw if the run winding has a turn-to-turn short.

The insulation used on the motor winding is rated according to the maximum allowable operating temperature. The motor nameplate might tell you that it has class, for example, class B insulation. The classifications for motor insulation are shown in figure 9–30. The maximum temperature referenced in figure 9–30 is the motor temperature rise plus the maximum ambient temperature where the motor is located. Most common fractional horsepower motors are A and B rated. Generally, you should replace a motor with one of equal or higher temperature class. Replacement with one of lower temperature rating could result in nuisance tripping of the motor overload. Each 18°F rise above the temperature rating of a motor can reduce the motor life by one-half. A rise of 10°C is equivalent to a rise of 18°F.

DUAL-VOLTAGE MOTORS

Three-phase motors are commonly available with **dual-voltage** ratings. A single dual-voltage motor may be used on one of two different supply voltages.

Some common dual-voltage ratings are 115/230V for single-phase motors or 230/460V for three-phase motors. For single-phase dual-voltage motors, one pair of wires emerging from the motor is used for the lower voltage, and a different pair of wires is used if the motor is to be connected to the higher voltage. For three-phase motors, the dual-voltage motor is different from a single-voltage motor in the following ways:

1. There are six different windings inside the motor instead of just three.
2. Nine wires emerge from the casing instead of just three.

ADVANCED CONCEPTS

The reason that dual-voltage motors are manufactured is so that a single motor part can be stocked for two different applications. The added cost of making a motor capable of handling two different voltages is more than offset by the reduction in storage costs of having two different motors.

The nine wires are attached to the motor windings as shown in figure 9–31. Depending on which voltage is being used, you will connect the nine wires to each other and to the incoming three-phase power supply as shown. These same instructions will appear on the motor.

Figure 9–31 Dual voltage motor.

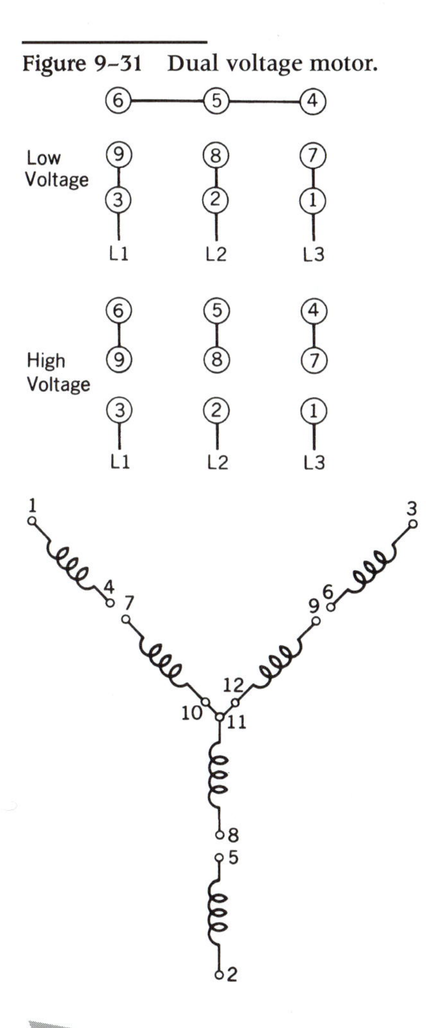

HORSEPOWER RATING

The **horsepower** rating of a motor is a measure of how much work it can do each second, minute, or hour. Some small fan motors will be rated in watts, rather than horsepower. (746 watts is equivalent to 1hp). The horsepower requirement for a motor is determined by the amount of work it must perform. For example, if a motor is driving a pump that moves 100gpm of water against a pressure difference of 30psi, the pump will require approximately 2.5hp of input to its shaft. The amount of work required to be done on a device like a pump, a fan, or a compressor is often called **brake horsepower (bhp)**. For the pump described above, you might find a 3hp motor, because motors are not commonly manufactured in a 2.5hp size. If the 3hp motor were to fail, you would have to replace it with another motor no smaller than 3hp. If you tried to use a 2hp motor, the pump would still try to draw 2.5bhp, and the motor would overload. If you used a 5hp motor, you would be wasting money, and the motor efficiency would be reduced slightly, but the pump would still only draw 2.5hp, and the pump performance would not change simply because you installed a higher hp motor (assuming the same pump/motor rpm). The horsepower rating of a motor is usually found on the motor nameplate.

Some motors are not rated in horsepower. For example, hermetic compressor motors will not show a horsepower rating on the nameplate. Other times, you will not be able to read the horsepower rating of the failed motor. In these cases, you can purchase a replacement motor that has a similar amp draw.

Motors can be rated for both their **full load amps (FLA)** and their **locked rotor amps (LRA)**. Sometimes full load amps are called **running load amps (RLA)**. Full load amperage is the current that the motor will draw when it is operating at its rated horsepower output. The chart in figure 9–32 shows what some average values are for the full load amperage of various motors. At part load, the motor amps will be roughly proportional to the load.

Locked rotor amps is the current that the motor draws upon start-up. This starting current is usually four to six times higher than the full load amps, but

ADVANCED CONCEPTS

You will notice that the different wiring schemes for the two different voltages each use a pair of windings to form a single winding connected to a single leg of the three-phase power. For the low-voltage wiring, the pair of windings is wired in parallel. For the high-voltage wiring, the pair of windings is wired in series. When the higher voltage is used, the motor winding resistance is higher because the loads forming each leg are in series. The motor produces the same horsepower as the lower voltage arrangement, but it consumes approximately half as much amperage. The operating cost is roughly the same, regardless of which voltage supply is being used.

		Motor HP	150	125	100	75	60	50	40	30	25	20	15	10	7½	5	3	2	1½	1	¾	½
Single Phase	115 V	Full Load Current												100	80	56	34	24	20	16	13.8	9.8
		Power Factor %												89	88.5	87.5	86	84	83	80	77	73
		Starting Current												575	460	322	195	138	115	92	80	56
	230 V	Full Load Current												50	40	28	17	12	10	8	6.9	4.9
		Power Factor %												89	88.5	87.5	86	84	83	80	77	73
		Starting Current												288	230	161	98	69	58	46	40	28
Three Phase	220 V	Full Load Current	353	293	223	180	144	120	103	75	64	52	40	27	22	15	9	6.5	5	3.5	2.8	2
		Power Factor %	91.5	91.4	91.2	91	90.8	90.6	90.4	90.2	90.1	90	89.5	89	88.5	87.5	86	84	83	80	77	73
		Starting Current	2118	1758	1338	1080	864	720	618	450	384	312	240	162	132	90	54	39	30	21	16.8	12
	440 V	Full Load Current	172	144	117	90	72	60	52	38	32	26	20	14	11	7.5	4.5	3.3	2.5	1.8	1.4	1
		Power Factor %	91.5	91.4	91.2	91	90.8	90.6	90.4	90.2	90.1	90	89.5	89	88.5	87.5	86	84	83	80	77	73
		Starting Current	1032	864	702	540	432	360	312	228	192	156	120	84	66	45	27	19.8	15	10.8	8.4	6
	550 V	Full Load Current	138	117	94	72	58	48	41	30	26	21	16	11	9	6	4	2.6	2	1.4	1.1	.8
		Power Factor %	91.5	91.4	91.2	91	90.8	90.6	90.4	90.2	90.1	90	89.5	89	88.5	87.5	86	84	83	80	77	73
		Starting Current	828	702	564	432	348	288	246	180	156	126	96	66	54	36	24	15.6	12	8.4	6.6	4.8

Figure 9–32 Average motor current ratings.

it only lasts for a few seconds. Once the motor gets up to speed, the amp draw diminishes.

If the nameplate is completely missing on the failed motor, you will not know the horsepower rating or the amp rating. In this case, you must look at the device that it is driving, and determine how much horsepower it requires. For example, say you have discovered a condenser fan motor that has failed. It drives a propeller-type fan blade. You can go to the wholesaler and look up the performance of a propeller type fan with the same number of blades, the same diameter, the same blade pitch, and the same rpm. Chances are that the motor requirements of two different fans with all of these same characteristics will have a similar bhp requirement.

SERVICE FACTOR

Some motor nameplates will show a **service factor (SF)**. It may be between 1.00 and 1.35. This number shows the percentage of full load that the motor can run at for short periods of time, without damaging the motor. For example, if a motor rated for 22FLA had a service factor of 1.15, it would be acceptable if the load, on occasion, loaded the motor to $22 \times 1.15 = 25.3$amps. When replacing a motor with an identical horsepower motor, the service factor of the new motor should be the same or higher than the failed motor.

FRAME SIZE

The frame size for a motor refers to its outside dimensions that have been standardized by the **National Electrical Manufacturers Association (NEMA)**. It is fortunate that NEMA has set standards for motor sizes, because if your replacement motor is the same frame size as the one it replaces, you know it will be the same physical dimensions. This includes the mounting holes, the diameter, the shaft size, the distance of the shaft from the mounting feet, motor length, and all those dimensions that are important if the new motor is to fit into the same space as the failed motor.

Simply matching the horsepower rating of the failed motor does not assure that it will be the same frame size. Refer to figure 9–33 which shows the various horsepower ratings that are available in each frame size. A 1/6 hp motor is available in either a 42 frame size or a 48 frame. Similarly, a 1/2hp motor is available in either a 48 frame size or a 56 frame.

Figure 9–33 Frame sizes.

NEMA Frame	Shaft Diameter	hp Range
42	⅜	1⁄20 1⁄15 1⁄12 1⁄10 ⅛ ⅙ ⅕
48	½	⅙ ¼ ⅓ ½ ¾
56	⅝, ⅞	1⁄12 ⅛ ⅙ ¼ ⅓ ½ ¾ 1 1½ 2
66	¾	1 1½
143T	⅞	1 1½
145T	⅞	1 1½ 2 3

SHAFT SIZE

Matching the shaft size of the failed motor is important because it must mate properly with the pulley or fan hub that the shaft fits into. Even though you match the frame size, sub-categories of the frame size may allow for different shaft diameters or shaft lengths. If the replacement motor has a shaft of a smaller diameter than the failed motor, you can use a ring between the shaft and the hub to effectively make the shaft size larger (figure 9–34). The shaft

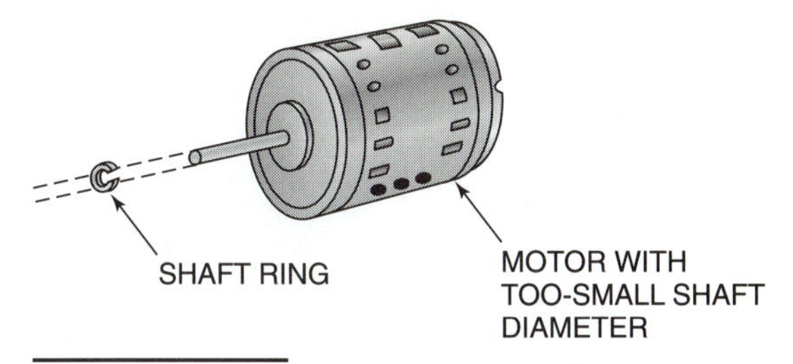

Figure 9–34 Shaft ring to increase shaft size.

ring is open in one place to allow a spot to fasten the hub to the shaft with a set-screw. If the shaft of the new motor is too long, it can be cut to length with a hacksaw. Shafts are made of soft material, and cutting them are not difficult.

Figure 9–35 shows the "flat" on a shaft. A fan or pulley may be fastened to the shaft by means of a set-screw. It is important when tightening a set-screw onto the shaft, that it bears down onto this flat. Otherwise, the device may quickly work its way loose after you leave the job. Some larger motors have two flats, 90 degrees apart, so that two set screws may be used for added strength.

MOTOR ROTATION

A replacement motor must be selected to rotate in the same direction as the failed motor. Single-phase motors may be designated as clockwise, counter-clockwise, or reversible. The motor will rotate in the designated direction, even if the power leads are switched. Usually (but not always), the direction of motor rotation is designated as the direction you would see when viewing the motor from the shaft end (figure 9–36). However, not all manufacturers follow this convention. When purchasing a replacement motor, you must confirm which convention that manufacturer uses. Some will designate the rotation direction with an arrow on the motor body. Others will say on the box something like CCWLE (counter-clockwise rotation when viewed from the lead end).

Figure 9–35 Flat portion of motor shaft.

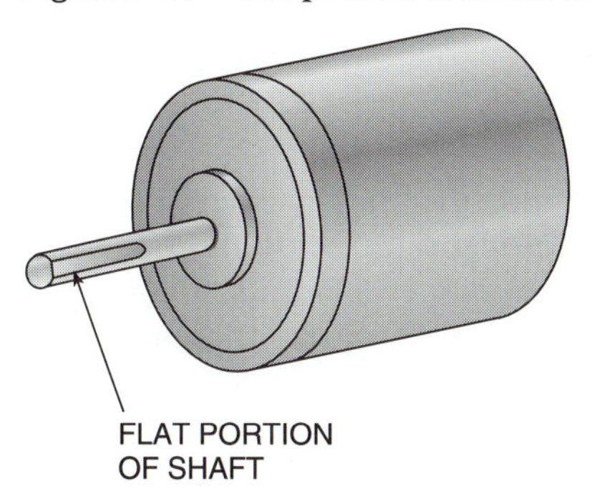

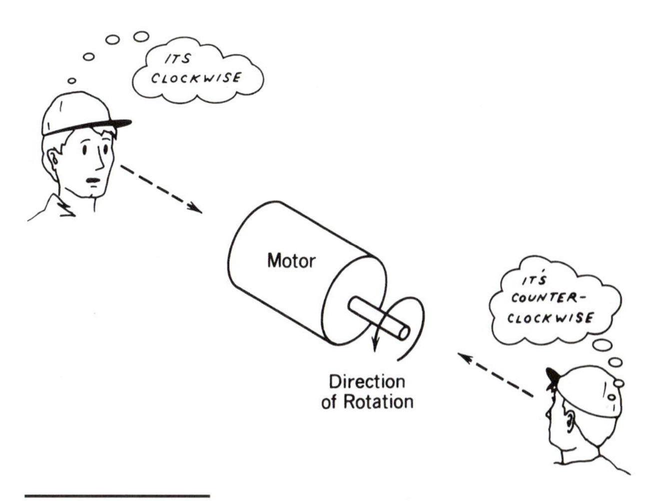

Figure 9–36 Motor rotation direction.

Some single-phase motors are reversible (figure 9–37). The reversing plug shown may be connected so that it connects purple-purple and yellow-yellow to give one direction of rotation. Or, it may be connected so that it connects purple-yellow and purple-yellow to give the other direction of rotation.

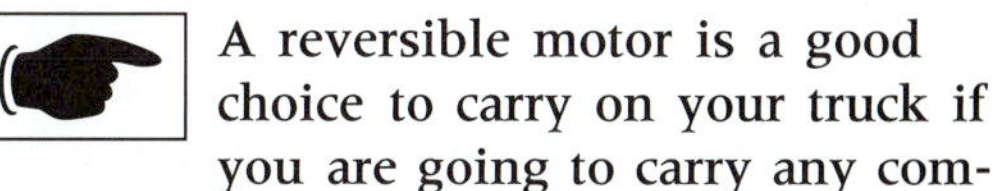

A reversible motor is a good choice to carry on your truck if you are going to carry any commonly used motors. It costs slightly more than a uni-direction motor, but you only need to stock one motor of a given horsepower for both clockwise and counter-clockwise applications.

Figure 9–37 Single-phase motor with reversing plug.

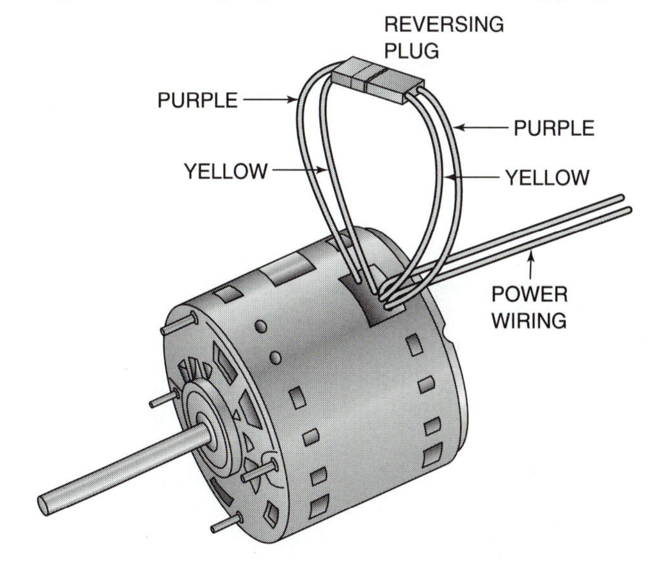

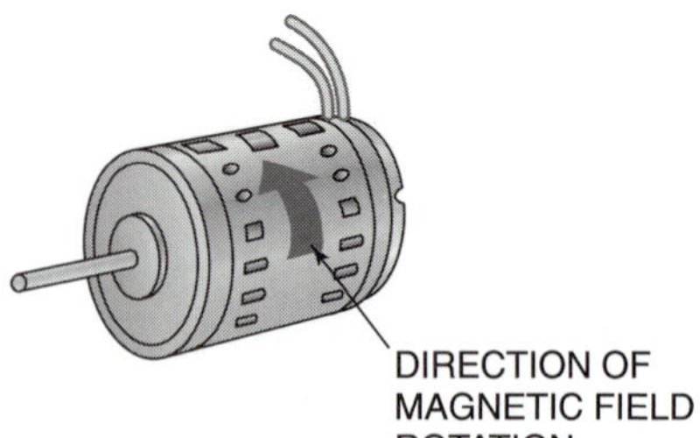

Figure 9–38 Mechanically reversing rotation.

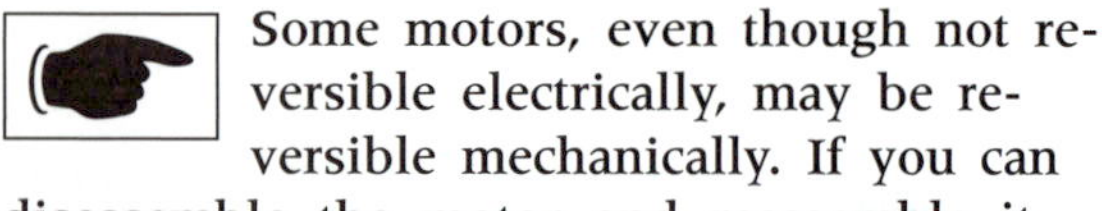

Some motors, even though not reversible electrically, may be reversible mechanically. If you can disassemble the motor and reassemble it so that the shaft emerges from the opposite end of the stator housing, the stator field direction will be unchanged, but the direction of the shaft will be reversed (figure 9–38).

For three-phase motors, the direction of the new motor will not be specified. That is because three-phase motors will operate in either direction. When you connect a three phase motor to the three power leads, you will not know which direction it is going to rotate. If, after you start it up, it rotates in the wrong direction, switching any two power leads will reverse the direction of rotation.

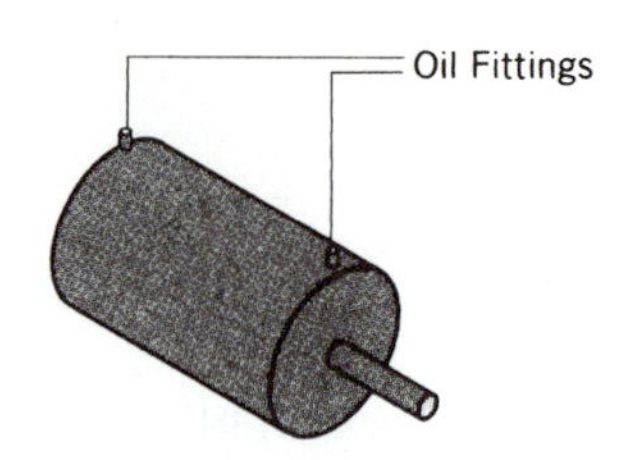

Figure 9–39 Motor oil fittings.

BEARINGS

The rotating portion of the motor is supported at each end by a **bearing** or a **bushing.** A bearing is similar to the ball bearings that separate a rotating wheel from a fixed axle on a car or a skateboard. A bushing is simply a smooth brass sleeve that supports a rotating shaft. Bearings are better than bushings from the standpoint of lower friction and therefore better motor efficiency. Bushings are superior to bearings from the standpoint of not needing as much (or any) lubrication. Motors with bushings are also quieter and less expensive than motors with bearings.

The bearing (or bushing) at the end of the motor where the shaft emerges is called the inboard bearing. The other end where the wiring is attached is called the outboard bearing.

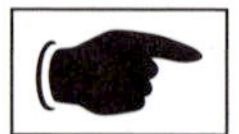

When lubrication ports are provided, the motor should be lubricated with a few drops of 30W oil once a year for most applications. When installing a new motor with oil ports, make sure that the motor is oriented so that they are above the elevation of the shaft (figure 9–39). Otherwise, it will be very difficult to lubricate the motor because the oil would have to defy the laws of gravity and flow uphill.

The primary causes of bearing failure in motors are lack of lubrication, overloading, and overheating. Overloading of a bearing is most commonly

ADVANCED CONCEPTS

The direction of rotation of any split-phase motor may be reversed by switching the leads on either the start winding or the run winding (but not both). However, unless the ends of one of the windings is brought outside the motor enclosure, you are unable to access them to reverse the direction of rotation.

ADVANCED CONCEPTS

It is possible to predict the direction of rotation that a three-phase motor will rotate by using a special meter to measure the phases of the power supply. But most technicians find that it is quicker to just attach the three power wires randomly, and then switch two if it turns out wrong. Over the long haul, attaching the wires in random order will be correct half the time.

caused on belt-drive applications where an overzealous technician makes the belt too tight. The sideways pull on the bearing will result in premature failure. On direct-drive fans, an out-of-balance fan wheel can cause the same result. Overheating can be prevented by maintaining proper motor ventilation by keeping the ventilation ports clean.

Figure 9–40 4-pole and 6-pole motors.

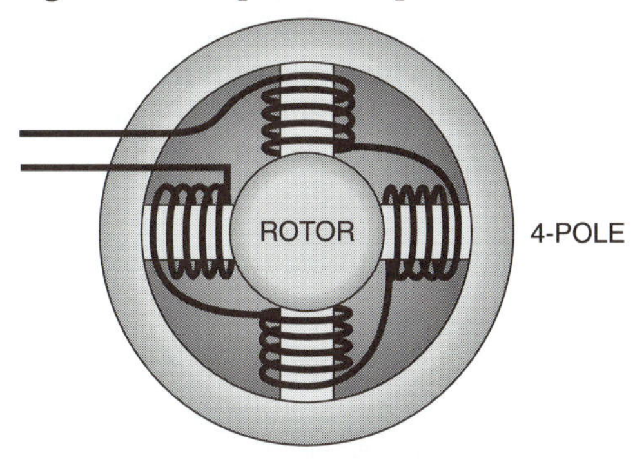

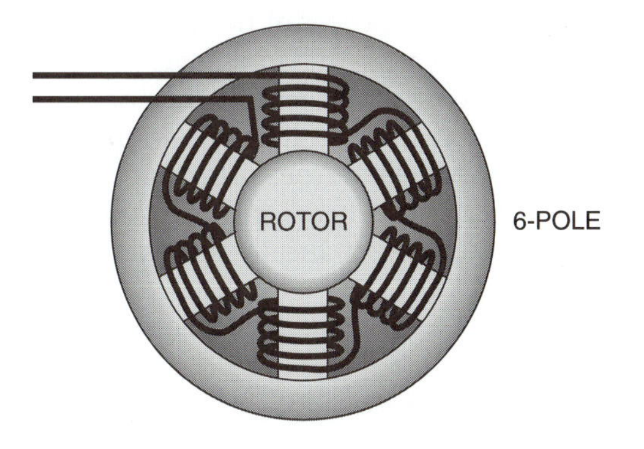

MOTOR SPEED

The rotation speed of a motor is determined by the number of poles in the stator winding. Figure 9–40 shows a four-pole motor and a six-pole motor. On a four-pole motor, the magnetic field rotates at 1800 revolutions per minute. This is called the **synchronous speed.** However, the motors used in heating, air conditioning and refrigeration work generally run at a speed slightly slower than synchronous speed. For example, a four-pole motor is usually rated at 1725 to 1750 rpm. A six-pole motor has a synchronous speed of 1200 rpm, but is rated at 1050 to 1075 rpm. The difference between the synchronous speed and the actual speed is called **slip.**

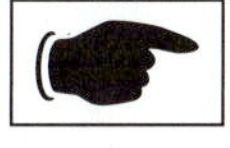

Sometimes, you find a failed motor, and you can't determine from the nameplate what the rated speed is. You can determine the correct speed for the replacement motor by disassembling the failed motor and counting the poles. If there are six poles, buy a 1075 rpm motor. If there are four poles, replace it with a 1725 rpm motor.

AUTO RESET VS. MANUAL RESET OVERLOAD

Some motors have built-in overload protection. Others rely on some external method of sensing and preventing overload. Most motors that have internal overload protection use some sort of thermostat

ADVANCED CONCEPTS

The synchronous speed of any motor may be calculated as:

RPM = 7200/number of poles

The minimum number of poles is two. Therefore, the highest rpm motor (without gears to increase the speed) is 3600 rpm.

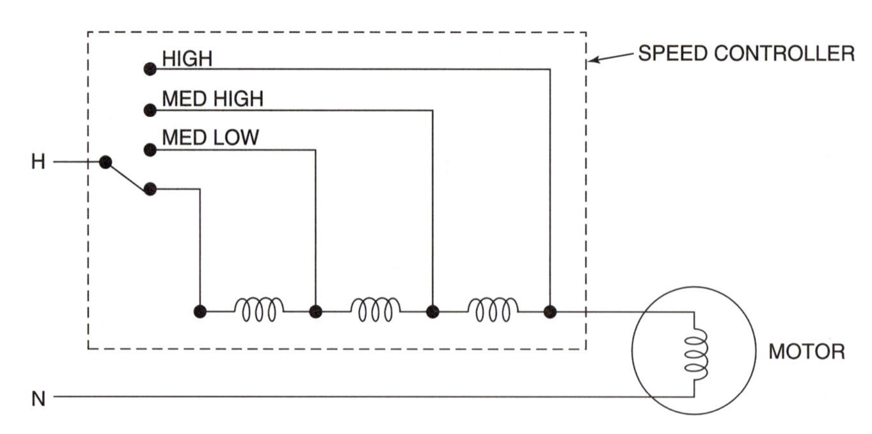

Figure 9–41 Motor speed selector.

imbedded in the stator windings. When the motor draws more than its rated current, the motor windings get warmer than normal and the thermostat opens, shutting down the motor. When the motor then cools, the thermostat re-closes, and the motor can attempt to start again. This is called **automatic reset.**

Other motors, particularly those used to drive the auger on an ice flaker, are **manual reset.** When these overloads sense an abnormal condition, they operate a switch located on the motor shell. The switch will stay open until it is manually pressed to reset it. When you push on this reset button, if the motor was actually out on overload, you will feel a slight "click."

VARIABLE SPEED MOTORS

Small **variable speed motors** are used in some high-efficiency furnaces (blowers) and condensing units (condenser fans) in order to match the fan performance with the load requirements. They are usually either shaded-pole motors or PSC motors, and they use a separate speed controller.

ADVANCED CONCEPTS

The solid-state speed controller uses a device called a **triac.** The triac operates to "chop off" a portion of the voltage signal coming to the motor. As the knob on the triac is adjusted, it chops off more or less of the voltage signal. As more of the input voltage signal is dropped, the amperage through the motor winding is decreased, reducing motor speed. The triac does *not* act like a resistor, and does not generate any significant amount of heat. The adjustment of the triac may be accomplished from some mechanical sensing device, such as a bellows that senses head pressure. The condenser fan motor speed can then be adjusted automatically in response to the sensed head pressure.

MORE ADVANCED CONCEPTS

A true Variable Frequency Drive varies both the voltage and the frequency of the output voltage supplied to the motor. These devices are more expensive, but result in very precise control of motor speed. They also optimize power consumption and motor efficiency. The lower-cost motor speed controls generally found in A/C equipment usually vary the line voltage to the motor but not the frequency. This results in somewhat lower overall motor efficiency at reduced loads, but this is of little concern on small fan motor applications.

One method of controlling the speed is to insert a coil of wire in series with the motor winding. The coil does not have a high resistance, so it does not generate much heat or waste energy. But the back EMF that it creates reduces the current flow through the motor winding, therefore reducing its torque, and increasing its slip. Figure 9–41 shows how a rotary speed switch can be used to insert varying amounts of coil in series with the motor coil. As the selector switch moves to lower speed settings, the voltage actually applied to the motor decreases. The same effect can also be achieved with a device called an **autotransformer.** It has multiple taps that provide different voltages to the motor.

Infinitely variable speed can be achieved by using a solid-state speed controller.

TRICKLE HEAT CIRCUIT

Figure 9–42 shows a circuit that is sometimes used on compressor motors to take the place of a crankcase heater. There is a **trickle heat** capacitor that allows a small amount of current to pass through the motor windings, even when the contactor is open. This produces a small amount of heat to prevent refrigerant vapor from condensing in the compressor and diluting the oil.

When the contactor switch closes, the trickle heat circuit has no effect.

Figure 9–42 Trickle heat circuit.

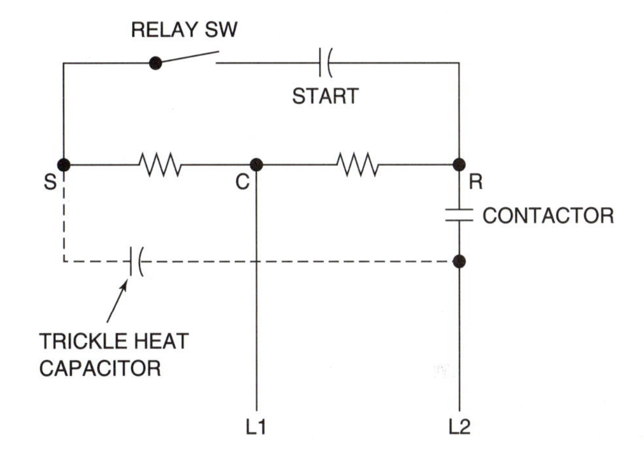

Power Wiring

Utility Power Distribution

The utility company distributes electricity from the power-generating station at a very high voltage of 120,000V or higher. The reason for such a high transmission voltage is to reduce power loss and to minimize the required transmission line sizes. Both the wire size and the power loss ($I^2 \times R$) vary with the amperage. Therefore, the power company distributes power at high voltage and low amperage.

The high voltage power is supplied to a **substation** that contains transformers that reduce the power down to a usable voltage. For residential use, additional transformers are supplied by the power company to provide single-phase power. For commercial or industrial applications, the customer may provide their own transformers to take the high voltage supplied by the utility company.

This chapter describes the various types of transformers that can be used to step-down the utility power. Figure 10–1 shows a transformer used to deliver 115V and 230V single-phase power. This type of transformer is usually provided by the utility company to supply residences. Three wires are supplied to the individual residence. After passing through the meter, the power comes into a circuit breaker box. Individual circuits may use either hot leg with a neutral leg to provide 115V, or an individual circuit may use two different (out-of-phase) 115V legs through a pair of circuit breakers ganged together to provide 230V (see figure 10–2). Each 115V circuit has a neutral wire that comes back to the circuit breaker box ground to complete the circuit. The 230V circuit is supplied from two different circuit breakers whose trip levers are mechanically connected together. If either circuit breaker trips, it will also pull the other breaker switch open.

Figure 10–3 shows the primary and secondary windings of a utility **three-phase transformer** called a **delta transformer** (it looks like the Greek letter Δ). This would be used for commercial and industrial accounts requiring three-phase power. Four wires are supplied to the consumer, who may then select the appropriate wires in order to obtain either three-phase (L1, L2, and L3), 230V single-phase (between any two of the hot legs L1, L2, or L3), or 120V single-phase power between either the L2 or L3 hot legs and ground. Note that the L1 terminal opposite the center tapped leg is at a voltage of 208V when compared to ground. This leg is referred to as the **high leg,** the **wild leg,** or the **stinger leg.** This type of transformer is used when the primary requirement of the consumer is for 230V three-phase power, with only a minor requirement for the 120V single-phase power.

Figure 10–4 shows a different type of utility transformer called a **wye transformer.** It is used where the customer's primary requirement is for 110V single-phase, and a minor requirement for three-phase power. Of the four wires supplied, 110V single-phase may be taken from any of the hot legs and ground. Between any pair of hot legs,

ADVANCED CONCEPTS

Suppose a power company delivers 60kW (60,000W) of power to consumers one mile away over a power distribution line that has 9 ohms of resistance. If they send the power at 120,000V, the amps that the wire will need to carry is given by:

$$\begin{aligned} \text{Amps} &= \text{Watts/Volts} \\ &= 60{,}000\text{watts}/120{,}000\text{volts} \\ &= 0.5\ \text{amps} \end{aligned}$$

The voltage drop through the mile-long wire will be:

$$\begin{aligned} \text{Volts} &= \text{Amps} \times \text{Resistance} \\ &= 0.5\text{amps} \times 9\text{ohms} \\ &= 4.5\text{V} \end{aligned}$$

And the power dissipated due to heat formed in the power lines will be:

$$\begin{aligned} \text{Watts} &= \text{Volts} \times \text{Amps} \\ &= 4.5\text{V} \times .5\text{A} \\ &= 2.25\text{W} \end{aligned}$$

However, for comparison, suppose that the power company delivered the same 60kW at 30,000V instead of 120,000V.

The current would be 60,000W/30,000V, or 2 amps. The voltage drop through the line would be 2A × 9 ohms, or 18V, and the power dissipated due to heat would be 18V × 2A or 36W. By transmitting at 120,000V instead of 30,000V, the transmission voltage drop is 75% less, and the power loss is 94% lower.

208V are available. Although 208V are suitable for most motors rated at 230V, they are not as desirable. This is near the low end of the acceptable motor supply voltage, and will tend to cause somewhat higher amp draws on motors. During periods of peak utility usage, the voltage supply from the utility may drop slightly. If this happens, the voltage available from the wye transformer may drop below 208V and cause motors to trip out on overload.

Figure 10–5 shows a wye connection transformer that is popular in many industrial applications. It provides 480V three-phase power and 277V single phase for fluorescent lighting. There is no way to obtain either 115V or 230V single-phase power with this system. Separate transformers must be provided if either of these two voltages is required.

SIZING CONDUCTORS

Power wiring is the business of a qualified electrician. Heating, refrigeration, and air conditioning service technicians will only do power wiring that is incidental to the installation of equipment. The HVAC/R technician must also be able to recognize when a problem rests with the power wiring and not with the HVAC/R equipment. Because power wiring can carry such large amounts of current, the laws that govern its installation are primarily concerned with two potential hazards:

1. The size of the conductor must be large enough in cross section to carry the required amount of amps.

2. The physical installation of the power wiring must be mechanically sound to prevent power wiring from shorting together.

The ability of a wire to carry amperage is analogous to the ability of a pipe to carry water. If you try to push too much water through a pipe, its velocity will be very high, and there will be a high friction rate between the water and the inside of the pipe wall. This will cause a high pressure drop in the pipe. Similarly, when electrical wiring is too small to carry the required load, there is a voltage drop in the wire.

Wire size is defined according to its **AWG** (American Wire Gauge). The largest wire is 0000 (4/0). As the numbers get larger, it denotes a smaller wire diameter. The smallest standard wire is number 50. Figure 10–6 gives information for some of the more commonly used wire sizes. The wiring used to supply power to a unit must have an **ampacity** (amp

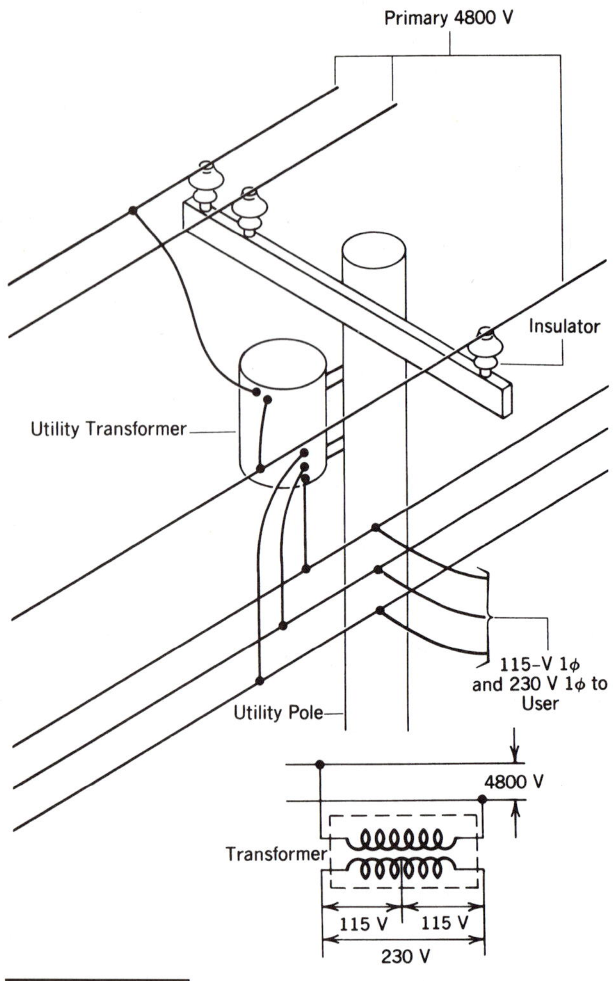

Figure 10–1 Utility transformer to deliver 115V and 230V single-phase power.

carrying capacity) greater than the current draw of the equipment being supplied. Where exceptionally long runs of power wiring are required, larger wire sizes may be required. The resistance values shown in figure 10–6 can be used to determine how much voltage drop will occur in a wire.

EXAMPLE

Select the proper wire size and determine the voltage drop for a copper wire to supply a 35-amp load at 440V. The length of the power wiring is 60 ft.

SOLUTION: A number 10 wire can only carry 30 amps; therefore we would choose a number 8 wire with an ampacity of 40 amps. The resistance of this wire is .628 ohms per 1000 ft. Therefore, the resistance of a 60-ft. length would be

$$0.628 \text{ ohms} \times 60/1000 = 0.038 \text{ ohms}.$$

The voltage drop in the wire is

$$\begin{aligned} \text{volts} &= \text{amps} \times \text{resistance} \\ &= 35 \text{ amps} \times 0.038 \text{ ohms} \\ &= 1.33\text{V}. \end{aligned}$$

■

Copper is the most popularly used conductor. Aluminum is sometimes used due to its low cost. Aluminum wiring cannot carry as much amperage as copper for a given AWG size. Note also that the ampacity values in figure 10–6 are based on a 140°F rating on the wire insulation. For higher temperature ratings, the ampacity would be higher.

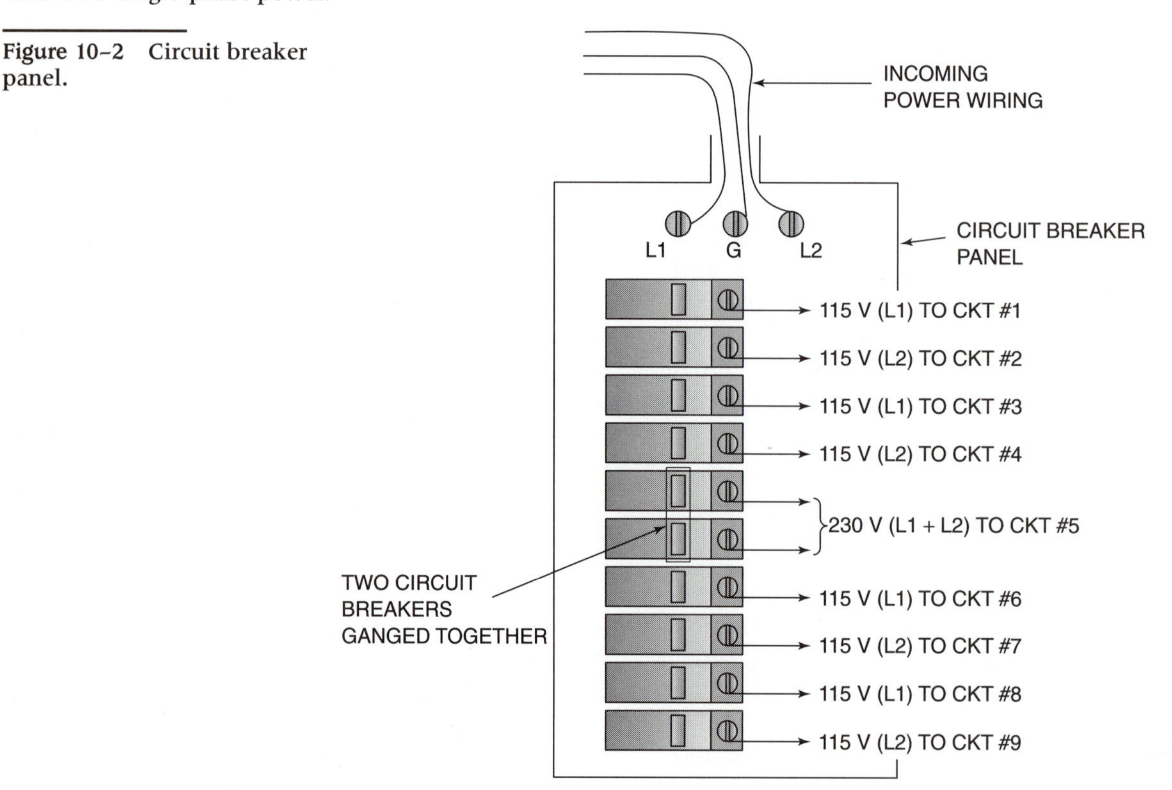

Figure 10–2 Circuit breaker panel.

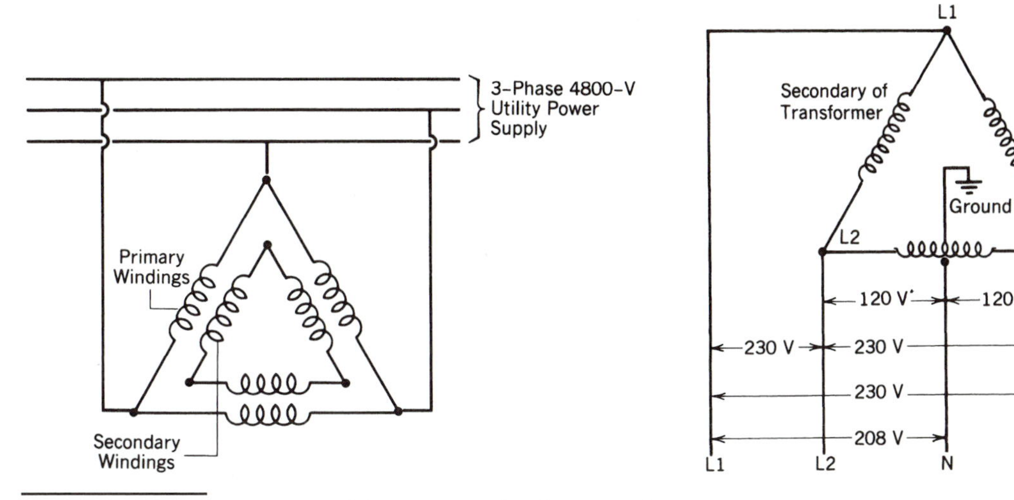

Figure 10-3 Voltages from a delta-connected utility transformer.

Oversizing wire diameter causes no operational problems. It is just a waste of money. However, undersizing wire can cause overheating, voltage drop, and motor burnout due to low voltage supply.

Diagnosing Undersized Power Wiring

On construction trailers and other temporary structures, it is not uncommon to find an air conditioner that blows fuses from time to time, for no apparent reason. One possibility is that the temporary power wiring to the air conditioner is undersized. This makes it difficult for the compressor to start, and on start-up, the compressor will draw higher than FLA for a longer period. Sometimes this will be long enough to blow the fuse.

Start your diagnosis with the disconnect to the air conditioner open. Measure the voltage available at the unit. This will be equal to the voltage being supplied from the other end of the power wiring because there is no current flow and therefore no voltage drop through the wire. If the available voltage is within the acceptable voltage rating of the air conditioner (nameplate voltage plus or minus 10%), continue. Using an analog voltmeter, read the voltage at the disconnect when you close it and the compressor motor tries to start. Immediately, it will drop because of the LRA being supplied through the power wiring, and the voltage drop that is caused by this high draw. After the compressor starts, the amp draw diminishes, and the voltage available at the disconnect should once again rise to almost the same voltage that you measured with the unit off. If the voltage available at the air condi-

Figure 10-4 Voltages from a wye-connected utility transformer providing 110V and 208V.

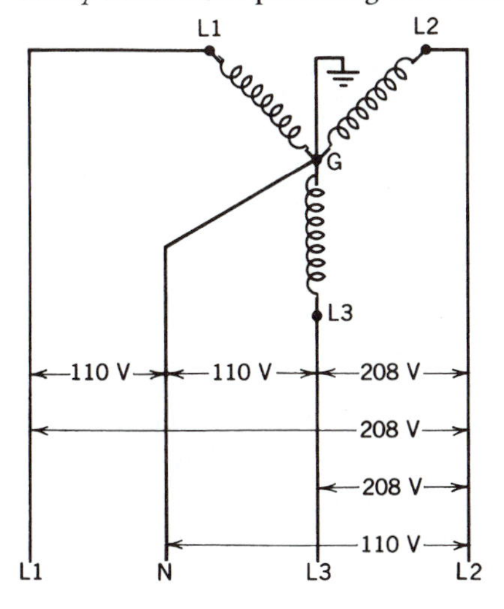

Figure 10-5 Utility wye transformer providing 480V 3Φ and 277V 1Φ for lighting.

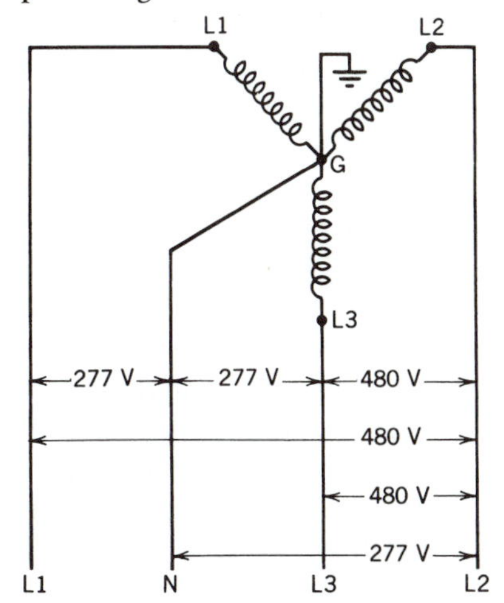

AWG	DIA (IN.)	OHMS/000 FT		AMPACITY	
		COPPER	ALUMINUM	COPPER	ALUMINUM
0000	.460	.049	.080	195	150
4	.204	.248	.408	70	55
8	.128	.628	1.03	40	30
10	.102	1.00	1.64	30	25
12	.081	1.59	2.61	20	15
14	.064	2.52	4.14	15	—

Figure 10–6 Amp carrying capacities of wire.

tioner remains significantly lower during normal operation than it was with the unit off, it indicates either undersized power wiring or a poor connection in the power wiring that is causing a voltage drop.

If you are using a digital volt meter, you will not be able to read how low the voltage drops as the unit is starting. This is because the digital meter takes a few seconds to "zero in" on the correct reading. If the voltage being measured is changing, the digital meter cannot read it accurately as it is changing.

ALUMINUM WIRING

In the 1970s many cities changed their building codes to allow the use of aluminum wiring due to its lower cost. However, there are two characteristics of aluminum wiring that caused some unexpected and undesirable results. Those properties are:

1. The oxide coating that forms on the surface of aluminum wiring does not conduct electricity (unlike copper oxide which is a fine conductor).
2. The coefficient of thermal expansion for aluminum tubing is quite high. As its temperature increases, its size changes considerably.

These two characteristics combined to cause a number of fires. Here's how. When the aluminum wiring was tightened under a connector lug, the copper oxide coating between the wire and the connector would present a slight resistance to the flow of electricity. This caused the generation of a slight amount of heat (due to the I × R drop of the aluminum oxide coating), and a slight amount of expansion of the aluminum wire. Over a period of years, the continuing cycle of expansion and contraction would cause the mechanical connection to loosen slightly. This, in turn, allowed air to the aluminum, caused more aluminum oxide to form, creating more electrical resistance and more heat to be generated, more expansion, more loosening of the connection, and an ever-accelerating march toward a failure of the connection due to heat.

Most cities outlawed the use of aluminum wire in buildings when the failure mode was discovered, but there are still millions of feet of aluminum wire out there, lurking in wait to cause you problems. The most common problem you will encounter is a failure of the connection of aluminum wire to a circuit breaker. When you find a loose connection, turn off the power and remove the wire. Clean the oxidation from the surface of both the wire and the connector lug. You can apply a special compound (you'll probably have to go to an electrical wholesaler) to the wire to inhibit the reformation of the oxide coating. Then the connection can be remade. Make sure that you get it *really tight.*

WIRE INSULATION

In the previous section, you saw that wire size was determined only by the current carrying requirement of the wire. It had nothing at all to do with the voltage in the conductor. Selection of the wiring insulation is just the opposite. It depends only upon the voltage inside and the potential physical abuse to which the insulation is likely to be exposed. Some of the different wire insulation designations are:

1. Type RH is a heat-resistant rubber insulation used in dry applications where the temperature will not exceed 167°F.
2. Type RHH is similar, except that it is rated up to 194°F.
3. Type RHW is a heat-and-moisture-resistant rubber used in dry and wet locations at temperatures not to exceed 167°F.
4. Type TW insulation is a moisture-resistant thermoplastic for wet and dry locations up to 140°F.
5. Type THHN is a heat-resistant thermoplastic for dry locations up to 194°F. It has an outer covering consisting of a nylon jacket.

6. Type THW is a moisture- and heat-resistant insulation for wet and dry applications up to 167°F.

CONDUIT

To protect power wiring from mechanical damage, it is run inside a protective pipe called **conduit.** Conduit can be threaded like a water pipe or it may be thin-wall tubing using special connectors. The thin-wall conduit is the most popular due to its low cost. It is referred to as **EMT (electro-mechanical tubing).**

Wires are pulled through the conduit from the power source to the load. Any wire connections that are required are made with wire nuts inside a junction box (figure 10–7). You will hear technicians refer to this as a J-box.

Conduit may be rated for indoor or for outdoor service. Outdoor conduit connections are gasketed to keep rain from getting inside to the wiring. Conduit supplying power to walk-in freezer boxes can present a unique challenge. The box can cool the conduit enough so that the moisture inside the conduit condenses, resulting in an ice-ball forming on the junction box.

Conduits are available in sizes starting at 1/2-inch diameter. The size of the conduit required is determined by the National Electric Code, according to the number and diameter of the wires. The maximum number of conductors allowed in conduit or EMT is given in figure 10–8. The maximum number of conductors allowed in an electrical box is given in figure 10–9.

Figures 10–10(a) and 10–10(b) show **flexible conduit** as opposed to **rigid conduit.** This may also be rated for indoor or outdoor service. Codes govern the maximum allowable lengths of flexible conduit that may be used. Rigid conduit is used for the long, straight runs. At the end near the air conditioning or refrigeration unit, flexible conduit is used to simplify the installation. Indoor flexible conduit is commonly called **armored cable** or **BX cable.** Outdoor flexible conduit is commonly called **Sealtite,** although this brand name is only one of the many brands available.

Figure 10–7 Junction box.

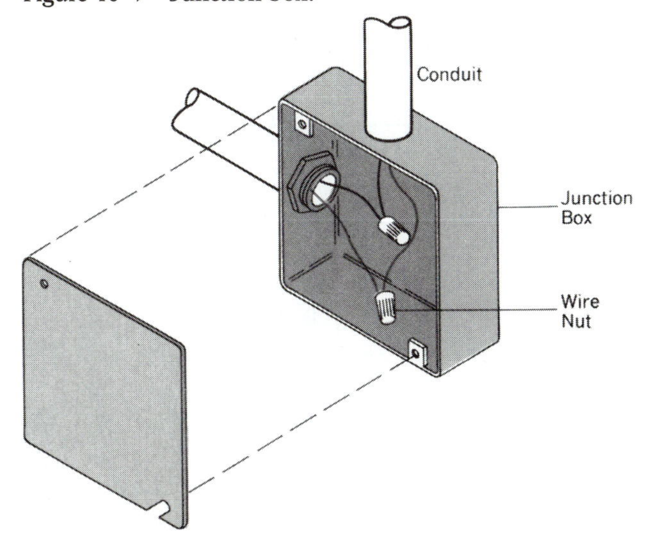

DISCONNECT

Figure 10–11 shows a **disconnect switch.** It is usually located on or near the equipment that it controls. It is basically a set of knife switches that open all the legs (two or three, depending on whether the incoming power is single-phase or three-phase) of the incoming power supply to a load. When the handle is in the down position, the knife switches are open (Off), and when the handle is in the up position, the switches are closed (On). Disconnect boxes are provided with holes to accommodate a padlock, to allow locking the handle in either the On position or the Off position. A service technician might lock the disconnect in the Off position when s/he is working on a unit. An owner might lock the disconnect in the On position to prevent it from being turned off by mischievous children.

Many local codes require the installation of a disconnect switch within a line of sight from the unit it controls. This is a safety precaution, so that if the service technician shuts down the unit, s/he can make sure that the unit will not be turned back on by another person. (Of course, the safest way is to lock the disconnect switch open.)

On construction jobs, **locking out** the disconnect is common, where there may be several tradespeople working at the same time. Following are some important safety considerations about lockouts:

1. If you lock out a unit, have a tag on your lock that clearly identifies you as the owner of the lock. The tag must bear sufficient identification so that anyone who sees the tag will be able to locate you.

2. If there is already another person's lock attached to the disconnect, do not assume that this is sufficient protection for you. Add your own lock.

3. Never take it upon yourself to cut off another person's lock, even if you are sure that they have mistakenly left it on after going home for the day. Advise the owner's representative, foreman, or other responsible person in charge. There may be a reason for the switch to remain locked.

TYPE OF WIRE	CONDUCTOR SIZE AWG KCMIL	CONDUIT SIZE (INCHES)					
		1/2	3/4	1	1 1/4	1 1/2	2
TW, XHHW (14 THRU 8)	14	9	15	25	44	60	99
	12	7	12	19	35	47	78
	10	5	9	15	26	36	60
	8	2	4	7	12	17	28
RHW AND RHH (WITHOUT OUTER COVERING), THW	14	6	10	16	29	40	65
	12	4	8	13	24	32	53
	10	4	6	11	19	26	43
	8	1	3	5	10	13	22
TW, THW, FEPB (6 THRU2), RHW AND RHH (WITH-OUT OUTER COVERING)	6	1	2	4	7	10	16
	4	1	1	3	5	7	12
	3	1	1	2	4	6	10
	2	1	1	2	4	5	9
	1		1	1	3	4	6
	1/0		1	1	2	3	5
	2/0		1	1	1	3	5
	3/0		1	1	1	2	4
	4/0			1	1	1	3
	250			1	1	1	2
	300			1	1	1	2
	350				1	1	1
	400				1	1	1
	500				1	1	1
	600					1	1
	700					1	1
	750					1	1
THWN, THHN, FEP (14 THRU 2) FEPB (14 THRU 8) XHHW (4 THRU 500 KCMIL)	14	13	24	39	69	94	154
	12	10	18	29	51	70	114
	10	6	11	18	32	44	73
	8	3	5	9	16	22	36
	6	1	4	6	11	15	26
	4	1	2	4	7	9	16
	3	1	1	3	6	8	13
	2	1	1	3	5	7	11
	1		1	1	3	5	8
	1/0		1	1	3	4	7
	2/0		1	1	2	3	6
	3/0		1	1	1	3	5
	4/0		1	1	1	2	4
	250			1	1	1	3
	300			1	1	1	3
	350			1	1	1	2
	400				1	1	1
	500				1	1	1
	600				1	1	1
	700				1	1	1
	750					1	1
XHHW	6	1	3	5	9	13	21
	600				1	1	1
	700					1	1
	750					1	1

Figure 10–8 Maximum number of conductors in conduit.

BOX DIMENSIONS, INCHES TRADE SIZE OR TYPE	MAXIMUM NUMBER OF CONDUCTORS			
	14 AWG	12 AWG	10 AWG	8 AWG
$4 \times 1\frac{1}{4}$ RND. OR OCTAG.	6	5	5	4
$4 \times 1\frac{1}{2}$ RND. OR OCTAG.	7	6	6	5
$4 \times 2\frac{1}{8}$ RND. OR OCTAG.	10	9	8	7
$4 \times 1\frac{1}{4}$ SQUARE	9	8	7	6
$4 \times 1\frac{1}{2}$ SQUARE	10	9	8	7
$4 \times 2\frac{1}{8}$ SQUARE	15	13	12	10
$4\frac{11}{16} \times 1\frac{1}{4}$ SQUARE	12	11	10	8
$4\frac{11}{16} \times 1\frac{1}{2}$ SQUARE	14	13	11	9
$4\frac{11}{16} \times 2\frac{1}{8}$ SQUARE	21	18	16	14
$3 \times 2 \times 1\frac{1}{2}$ DEVICE	3	3	3	2
$3 \times 2 \times 2$ DEVICE	5	4	4	3
$3 \times 2 \times 2\frac{1}{4}$ DEVICE	5	4	4	3
$3 \times 2 \times 2\frac{1}{2}$ DEVICE	6	5	5	4
$3 \times 2 \times 2\frac{3}{4}$ DEVICE	7	6	5	4
$3 \times 2 \times 3\frac{1}{2}$ DEVICE	9	8	7	6
$4 \times 2\frac{1}{8} \times 1\frac{1}{2}$ DEVICE	5	4	4	3
$4 \times 2\frac{1}{8} \times 1\frac{7}{8}$ DEVICE	6	5	5	4
$4 \times 2\frac{1}{8} \times 2\frac{1}{8}$ DEVICE	7	6	5	4
$3\frac{3}{4} \times 2 \times 2\frac{1}{2}$ MASONRY	7	6	5	4
$3\frac{3}{4} \times 2 \times 3\frac{1}{2}$ MASONRY	10	9	8	7

Figure 10–9 Number of conductors in an electrical box.

Figure 10–10 Flexible conduit. *(Courtesy of Anamet Electrical, Inc.)*

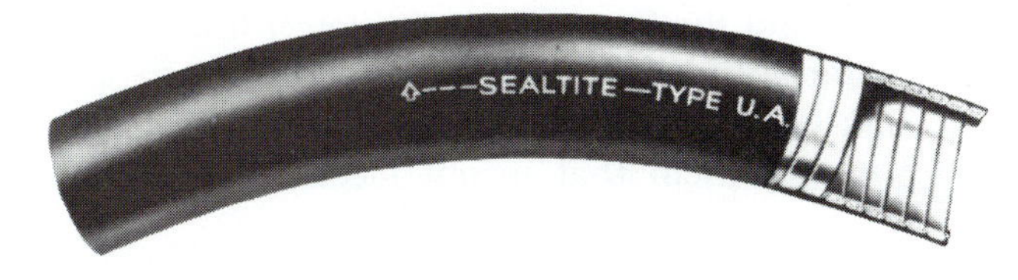

The disconnect switch can be either fused or non-fused. **Cartridge-type fuses** (figure 10–12) are used in the fused disconnect switch. One size disconnect switch may accommodate several different size fuses. For example, the physical size is the same for 15-, 20-, 25-, and 30-amp fuses. A different size is used for 40-, 45-, 50-, and 60-amp fuses. Where the fuses are the same physical dimensions, you can replace one fuse with a fuse of a different amp rating. But be sure that if you are replacing a fuse with a higher-rated fuse, that you do not exceed the amp rating given on the nameplate of the unit. The standard ratings for the disconnects are 30, 60, 100, and 200 amps (and higher).

EXAMPLE

You have arrived at a job site and found that the 45-amp fuse has blown. You can find no short or other apparent cause. Can you replace it with a 50-amp fuse to prevent another nuisance trip-out?

SOLUTION: Physically, the 45-amp and the 50-amp fuses are the same dimensions. So long as the "Maximum Fuse Size" rating on the nameplate is 50 amps or higher, you can use the 50-amp fuse. ■

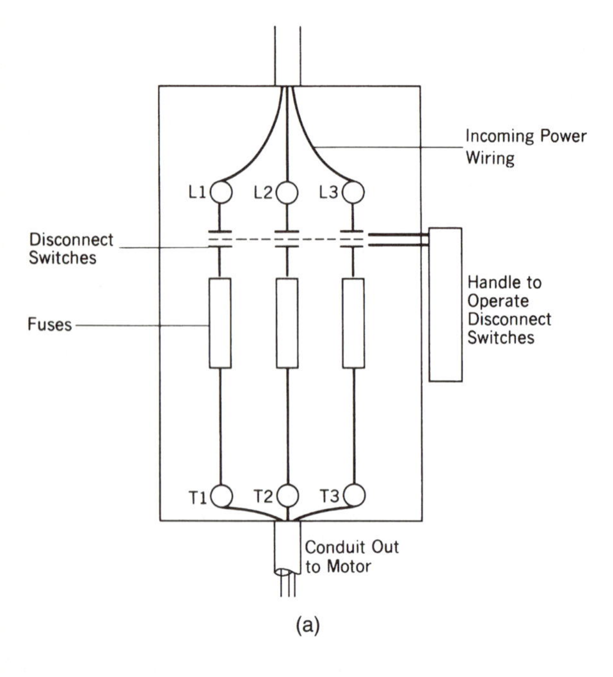

(a)

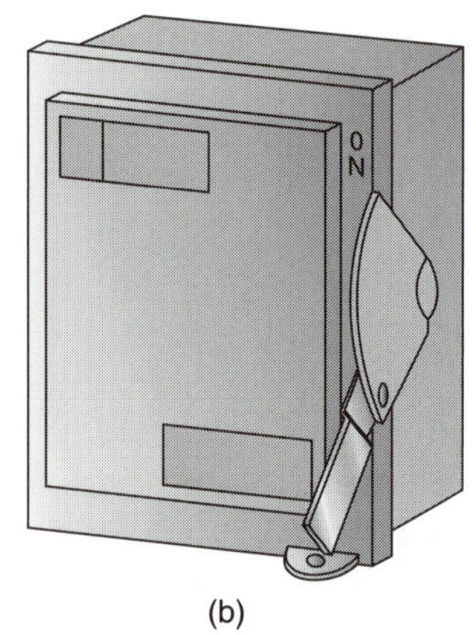

(b)

Figure 10–11 Disconnect switch.

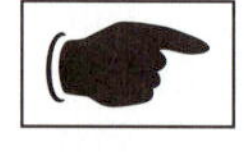

The maximum fuse size on the unit nameplate will sometimes be identified as the HAC or HACR rating. *Do not* use Locked Rotor Amps as an indicator of the maximum allowable fuse size. The wiring inside the unit can handle the LRA for only a few seconds. The maximum permissible fuse rating will be significantly lower than the LRA rating.

Figure 10–13 shows another type of fused disconnect. It is probably a safer design, because the fuses can be removed by simply pulling on the handle and removing the entire fuse block. The fuse block may be returned upside down to the receptacle, and this will provide the same protection as having turned the handle of the disconnect in figure 10–13 to the Off position.

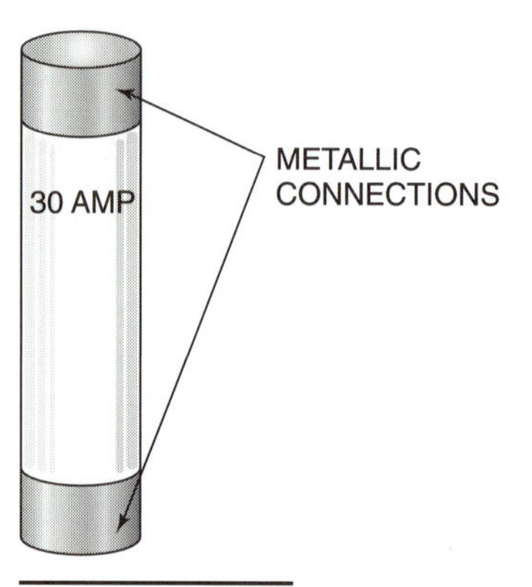

Figure 10–12 Cartridge fuse.

FUSES

Figure 10–14 shows various types of **fuses.** A fuse is a thin wire that carries current to a load. The wire inside can only carry a limited voltage without overheating and breaking. In this way, the fuse senses the amount of current that the wire is carrying, and acts like a switch that opens if the current rating is exceeded.

A common misconception is that fuses are used to protect a load such as a motor. This is not true. A fuse is sized to protect the downstream wiring. For example, suppose you have a motor rated to run at 26 amps. The wiring that supplies the motor can handle 30 amps, and a 30-amp fuse is provided. Suppose the motor starts drawing 28 amps. Eventually, the motor will fail because of the overload, but the fuse will never detect it until after the motor failure results in a short circuit.

Fuses that protect motor circuits have an application problem that is not faced by fuses that

Figure 10–13 Pull box.

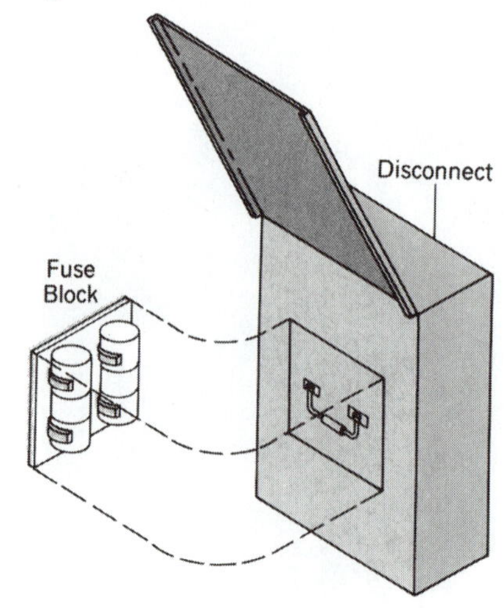

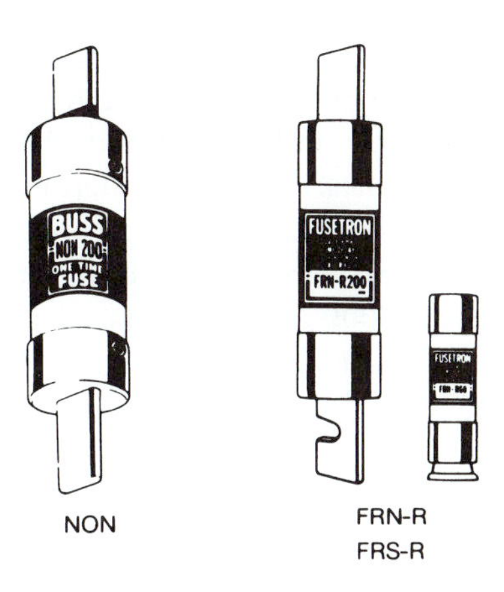

Figure 10–14 Fuse types. *(Courtesy of Cooper Industries.)*

protect circuits with resistive loads. The motor will draw locked rotor amps on start-up, but will drop down to its normal operating amperage within a few seconds. The wiring inside the unit is sized to supply the running load amps. In order to prevent blowing the fuse during the few seconds of locked rotor amps, we use a **dual-element** or **slow blow** or **time delay** fuse. This type of fuse will allow the momentary surge of power at start-up (4 to 6 times the running load amps) without blowing. If this current persists for more than a few seconds, the fuse will blow. In the event of a short-circuit condition, there will be no time delay in the operation of the fuse.

Figure 10–15 Plastic fuse pullers. *(Courtesy of Cooper Industries.)*

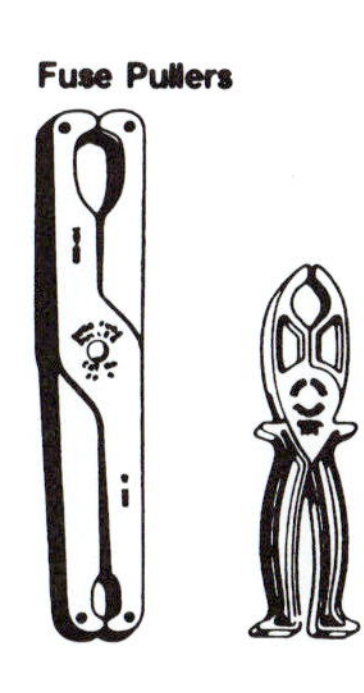

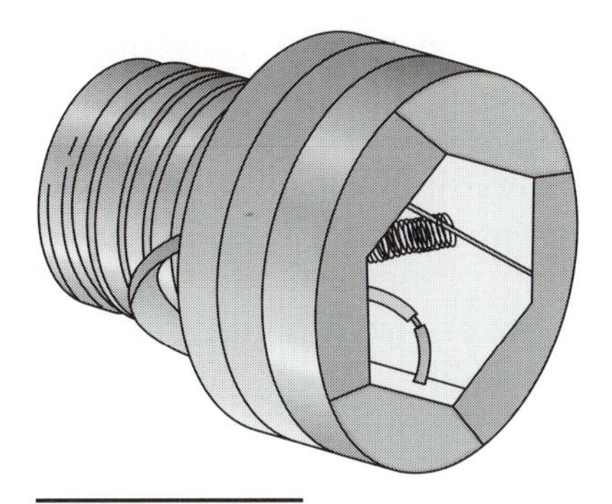

Figure 10–16 Plug-type fuse.

When changing a fuse, the power supply should first be turned off. In the case of cartridge fuses, they can be pulled using a plastic fuse puller as shown in figure 10–15.

Technicians have been killed (electrocuted) using screwdrivers, pliers, fingers, and other inappropriate tools for removing cartridge fuses. Do not do this. This is no idle safety warning.

For circuits of less than 30 amps, sometimes **plug-type fuses** are used (figure 10–16). They are sometimes called **Edison-base** fuses, because years ago they all had base sizes that are the same as an ordinary socket. However, newer plug-fuses each have a different size base for the different amp ratings.

CIRCUIT BREAKERS

Circuit breakers (figure 10–17) serve the same purpose as fuses. They protect the downstream wiring. They have the advantage of being able to be reset

Figure 10–17 Circuit breakers.

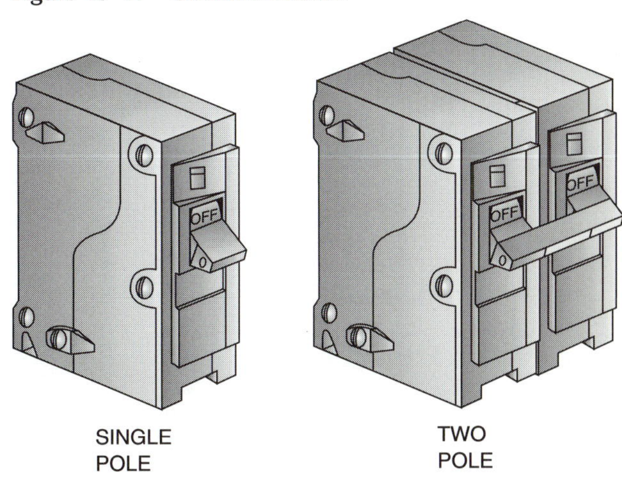

after they have tripped. Besides the convenience, the possibility of personal injury that is present when changing a fuse is avoided. However, they probably provide a slightly lower level of protection for the following reasons:

1. When a circuit breaker switch has not been operated for a long period of time (one year or more), the contacts tend to become somewhat fused together. It will still provide adequate short-circuit protection, but for a gradually added overload, the circuit breaker may be able to carry far more than its rated amperage.

2. The current interrupting capacity of a circuit breaker is approximately 10,000 amps, compared to a fuse short-circuit rating of 100,000 amps.

CHAPTER ELEVEN

Testing and Replacing Motors and Start Relays

Checking Compressor Motor Windings (Single-Phase)

If a compressor motor is receiving the correct voltage but it refuses to run, or it trips a fuse or a circuit breaker, the problem might be that the motor windings have failed. The most common types of motor winding failures are:

1. **Open winding**
2. **Grounded winding**
3. **Shorted winding**

An open winding is one in which the wire that forms the winding has separated or broken. A grounded motor is one in which the insulation around the winding wire has broken down, and the copper wire itself is making electrical contact with some part of the motor casing. A shorted winding is one in which the winding insulation has failed, and one part of the winding is making unintended contact with another part of the winding. These conditions are illustrated in figure 11–1.

Four ohm measurements are required to check the motor windings in a compressor. An ohm measurement must be taken between each pair of pins emerging from the casing (three measurements shown in figure 11–2). If all show continuity (as they should), then take another ohm measurement between any pin and the compressor casing. It should read infinite.

The pin-to-pin readings should be something like 2, 6 and 8 ohms, or 4, 11 and 15 ohms. The highest reading should equal the sum of the other two because it is actually a reading of the two other windings in series. The second highest reading should be two to four times higher than the lowest reading (the start winding normally has two to four times as much resistance as the run winding). If one winding is open, two of the readings will be infinite. If any reading is zero, it indicates that a winding is shorted.

If the winding resistances are all reasonable, then you must take an ohm reading between any one of the pins and the casing ground. You'll need to scrape off a bit of paint on the casing to assure that you're making good electrical contact between your ohm meter probe and the metal of the casing. You'll also need to make sure that your ohm meter is set on the highest ohm range available. If you get an infinite reading, the motor is not grounded to the casing. It is not necessary to check the other pins to casing, because you have already determined that the pins are continuous with each other. If any winding is grounded, all three pins will show a less-than-infinite reading to the casing.

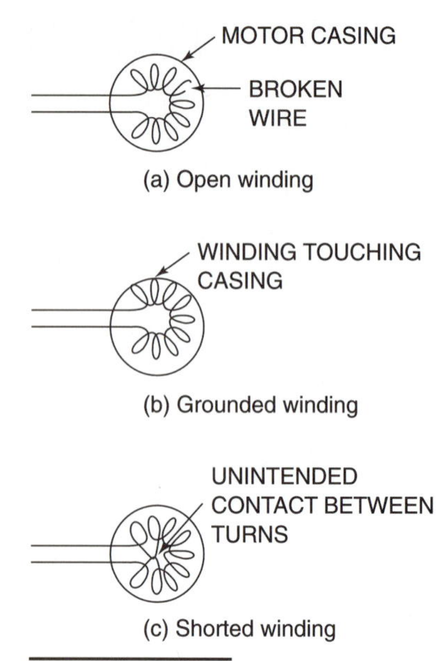

Figure 11–1 Motor failure modes. (a) Open winding. (b) Grounded winding. (c) Shorted winding.

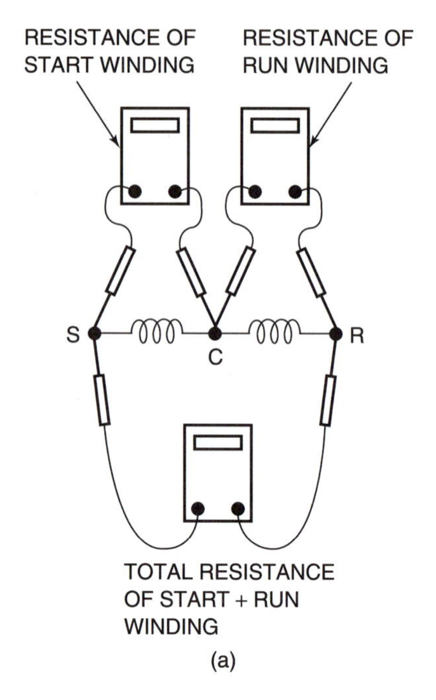

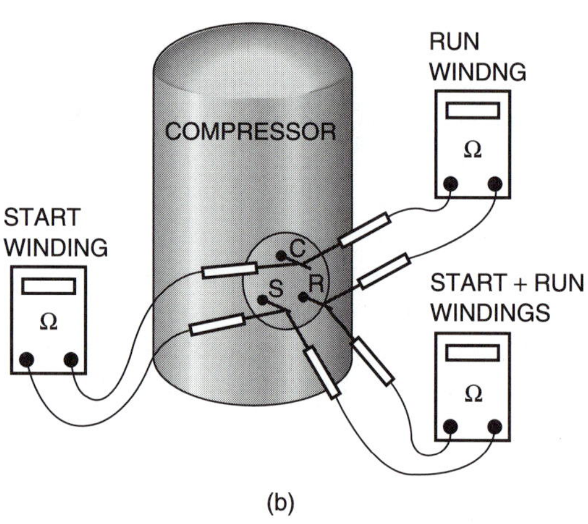

Figure 11–2 Measuring motor winding resistances.

CHECKING COMPRESSOR MOTOR WINDINGS (THREE-PHASE)

Ohming out the windings on a three-phase motor is even easier than on a single-phase motor. The pin-to-pin resistances should all be equal. They may be quite low (maybe one-half ohm or less). Be careful not to jump to the conclusion that just because the winding resistance is very low, that the windings are shorted.

The three-phase motor windings are checked to the casing ground in the same way as a single-phase motor.

CHECKING MOTOR WINDING INSULATION

In critical applications where a motor failure would cause catastrophic loss to the owner (usually the shutdown of a facility), as a preventive measure, maintenance personnel routinely check for an impending failure of motor winding insulation. As you know, if you measure resistance between a motor winding and the casing of the motor, it should be infinite. Well, that's not quite true. In fact, the resistance of the insulation that separates the motor winding from the casing is extremely high, but it's not quite infinite, even though most ohm meters will indicate that it is. However, if a meter with a high enough range is used, the actual less-than-infinite resistance of the motor winding insulation can be read. A record of the readings is kept, and compared from year to year. If the resistance is lower than the historical resistance, the owner may want to consider buying a replacement motor so that it will be available in the event of a future failure.

Ohm meters capable of reading very high resistances are called **megohm meters.** For years, a megohm meter included a hand-operated crank that would generate a sufficiently high voltage across the insulation to get a reading. However, modern megohm meters (figure 11–3) look much like a standard VOM. Instead of relying on a high voltage produced by a hand-crank, the electronic circuit is sufficient to detect the slightest current across the insulation, using only standard batteries. The megohm meter in figure 11–3 is capable of reading

Figure 11-3 Digital Megohmmeter. *(Courtesy of AEMC Instruments.)*

up to 2000MΩ. There are also analog models of megohm meters available (figure 11–4).

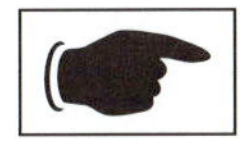

You may hear a technician use a phrase "megger the motor." This simply means to measure the motor winding insulation resistance with a megohm meter.

Figure 11-4 Analog megohm meter. *(Courtesy of TIF Instruments, Inc.)*

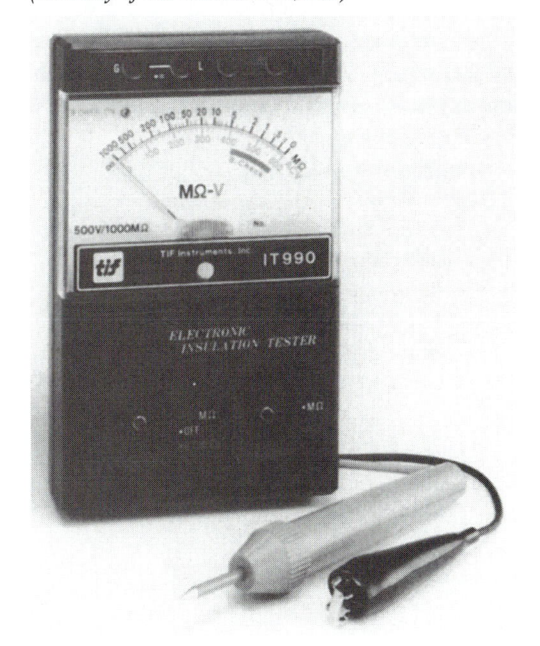

TESTING THE CURRENT RELAY

Figure 11–5 shows the wiring of the current relay. Figure 11–6 shows a manufacturer's diagram using a current relay.

EXERCISE:
Redraw the manufacturer's diagram in figure 11–6 in the same format as figure 11–5. Identify all line colors. Draw it one time without the start capacitor and a second time with the start capacitor and the jumper removed.

Some of the wires in figure 11–6 are identified as "red tracer" or "white w/red tracer." A "tracer" refers to a colored stripe on the wire. It simply serves as another color for identification.

There are four ways that the current relay can fail:

1. The coil is open—the motor will not start, hum, or anything else.
2. The switch is welded closed—the compressor will start, but because the start winding is not being taken out of the circuit, it will trip out on overload within a few seconds after starting.
3. The switch is mechanically stuck open—the run winding will hum when the operating control closes, but the motor will not start.
4. The coil is shorted—the run winding will hum when the operating control closes, but the motor will not start.

CHECK THE SWITCH

With the current relay removed from the system and oriented in its normal operating position, check the resistance between L and S. It should be infinite. If it is zero, it means that the switch is closed when it

Figure 11-5 Current relay.

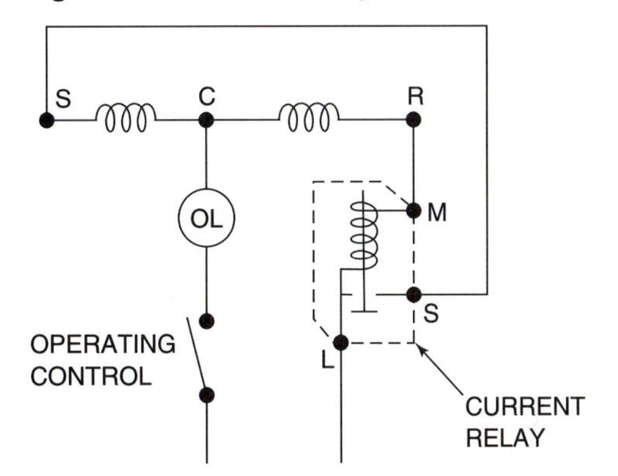

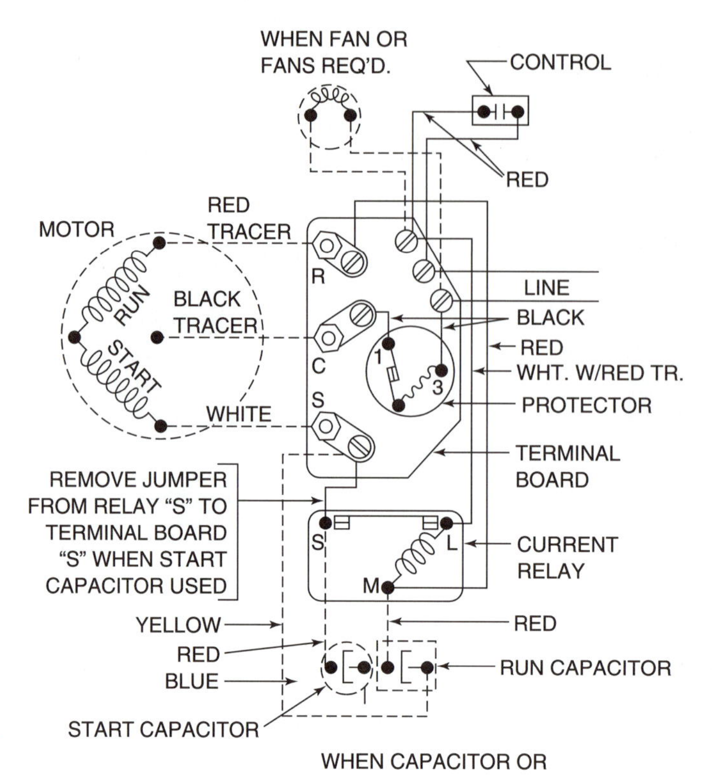

Figure 11–6 Manufacturer's diagram—current relay.

should be open, and the current relay must be replaced.

Turn the relay upside down so that gravity can pull the switch closed. Measure resistance between L and S. This time, it should be zero. If there is any significant resistance, it means that the switch contacts are pitted. An infinite reading indicates that the switch is mechanically stuck open.

Check the Coil

Measure the resistance between L and M. It should be continuous. Because of the large diameter wire, the resistance should be almost zero. If the resistance is infinite, the coil is open and the start relay must be replaced.

If the switch or the coil show that it has failed, the current relay must be replaced. The new current relay should be an exact replacement for the old.

Testing the Potential Relay

Figure 11–8 shows the wiring for a potential relay. Figure 11–9 shows a manufacturer's diagram.

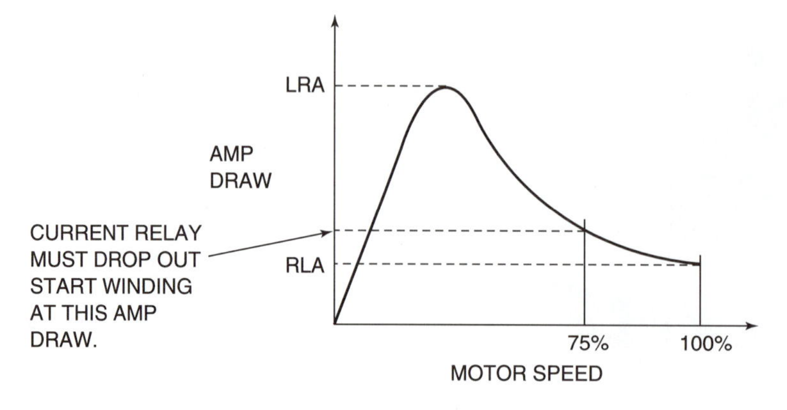

Figure 11–7 Motor amp draw vs. motor speed on start-up.

ADVANCED CONCEPTS

Various current relays will drop out the start winding at different motor amperages (figure 11–7). If you replace a current relay with one that drops out the start winding at a too-high amperage, the start winding will be de-energized before the motor is up to a sufficient speed for the run winding to take over on its own. The motor speed will begin to diminish, the amps will increase, the current relay will pull the switch closed, and the cycle will repeat rapidly, causing the switch to chatter.

If the current relay is replaced with one that has a too-low amperage rating, it will drop out the start winding too late, or if the rating is below the RLA, it will not drop out the start winding at all, and the compressor will trip out on overload.

EXERCISE:
Redraw the manufacturer's wiring diagram in figure 11–9 in the same format as the wiring diagram in figure 11–8. Show all wire colors.

The potential relay will generally fail in one of three ways:

1. The coil is open—the compressor will start, but will go out on overload several seconds later.

2. The switch contacts are pitted and not making good electrical contact—the compressor will try to start but it won't. It will hum for a few seconds and then go out on overload.

3. The switch contacts are welded closed. The compressor will start, but will go out on overload after running for a few seconds.

CHECK THE COIL

Measure the resistance between terminals 2–5. It should be continuous. Because the potential relay coil has many turns and the wire gauge is very small, the resistance will be several thousand ohms. If the resistance is infinite, the coil is open and the start relay must be replaced.

CHECK THE SWITCH

The switch in the potential relay, unlike the current relay, is normally closed when there is no power applied to the coil. With the potential relay removed from the system, measure the resistance across terminals 1–2. It should be zero, indicating a closed switch. If there is significant resistance or infinite resistance, the switch has failed, and the potential relay must be replaced. When measuring re-

Figure 11–8 Potential relay.

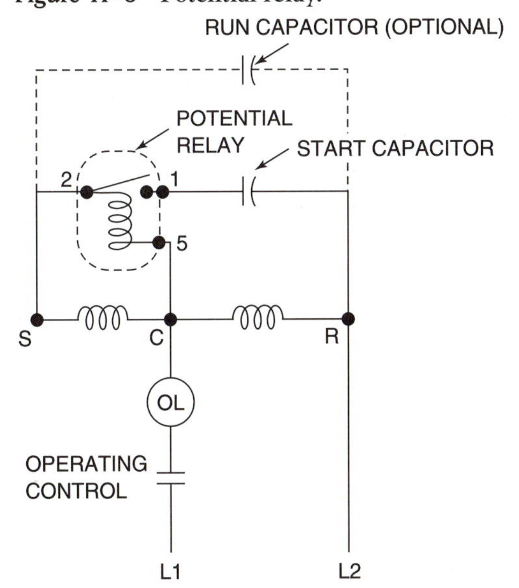

Figure 11–9 Potential relay wiring.

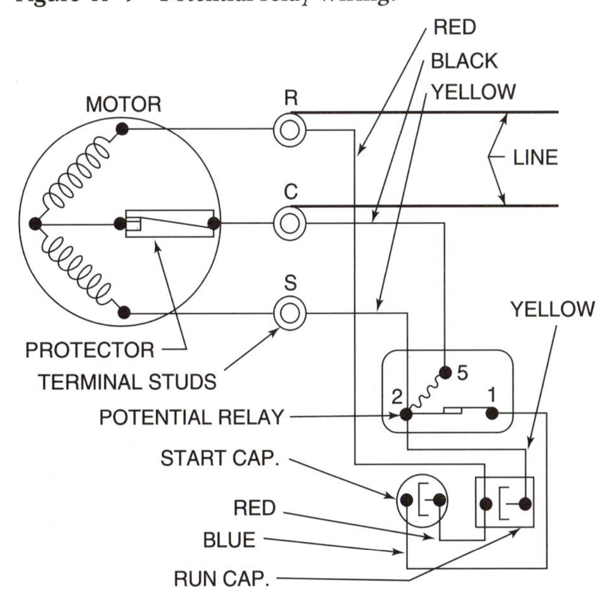

sistance of the potential relay switch, the relay may be oriented in any position. Unlike the gravity-operated switch in the current relay, the switch in the potential relay is spring-operated to close.

Check the Operation

Even if the coil and the switch check out electrically, that is not a guarantee that the switch is opening when the motor gets up to speed. If the compressor was starting and then going out on overload, reinstall the potential relay. As soon as the compressor starts, pull the wire off terminal 1 (be very careful, and use a well-insulated needle-nose pliers to pull the wire). If the compressor continues to run, then you can be sure that the compressor motor and the overload are functioning properly, and replacing the potential relay is all that is required.

HERMETIC ANALYZER

Suppose you discover a failed current relay or potential relay. You quote a price to the customer to repair the system. Then you go to the wholesaler to purchase the proper replacement relay, return to the job and install the new relay, only to find that there is some mechanical compressor failure that caused the relay to fail in the first place. How can you avoid this embarrassing situation?

Figure 11–10 shows a tool called a **Hermetic Analyzer.** It allows you to remove all of the starting equipment (relay and capacitor), and replace it with manually operated switches. If you cannot get the compressor motor to run using the hermetic analyzer, don't bother to replace any starting relays or overloads or capacitors. On the other hand, if you *do* get the motor running with the hermetic analyzer, then you know that the compressor motor is probably good.

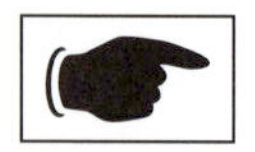

You can first try to start the compressor connecting the hermetic analyzer wiring as shown. If the compressor starts, then rewire it so that the overload is in the circuit between the common pin and the black wire of the hermetic analyzer. If it runs again, then the overload is also good. If the overload trips, test the amp draw of the compressor to determine if the compressor is drawing too many amps or if the overload is too weak to carry the proper compressor amps.

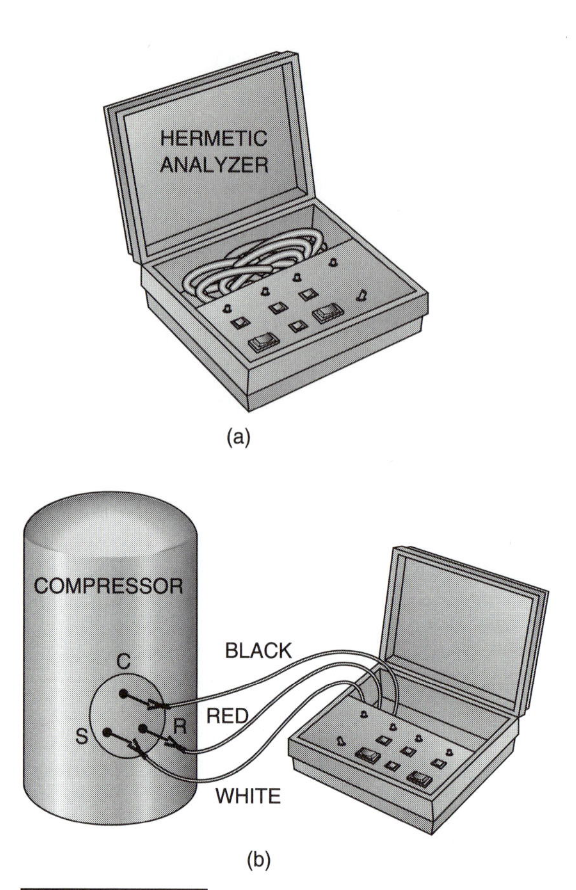

Figure 11–10 Hermetic analyzer.

IDENTIFYING COMMON, START, AND RUN PINS

In order to hook up the hermetic analyzer to the compressor, you must first know which pin on the compressor shell is Common, which is Start, and which is Run. You can always identify the pins using ohm readings between the pins. But before you do that, here are some helpful hints.

1. If there is an external overload (Klixon overload), one side of the overload will be wired to one side of the power supply, and the other side of the overload will be wired to the Common pin.

2. The wires connected to the compressor pins are usually different colors. Most usually (but not always), the pin with a black wire is the Common pin, the pin with the white wire is Start, and the pin with red is Run (just remember red is to **R**un).

3. The physical arrangement of the pins is usually Common, Start, Run as viewed from left to right and top to bottom. See figure 11–11 for examples.

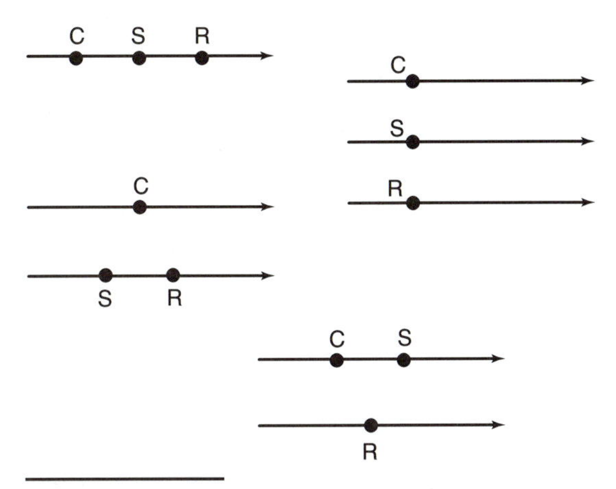

Figure 11–11 Usual arrangements for single-phase hermetic compressor pins.

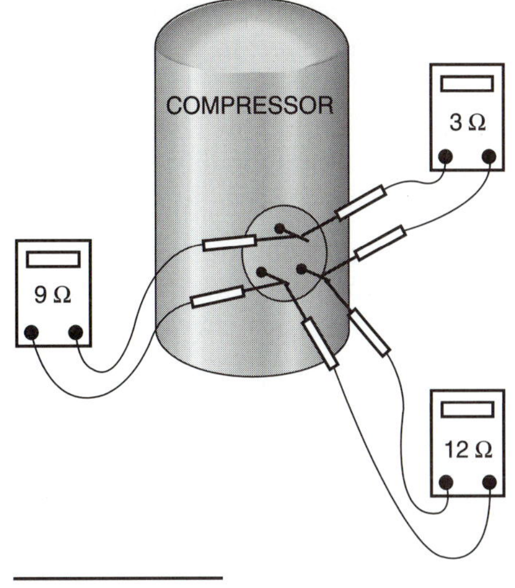

Figure 11–12 Exercise.

Ohm readings are required to confirm if your initial guess from the above guide is correct. Take ohm readings between each pair of pins (three readings total). The sum of the two lowest readings will equal the highest reading. Once you have taken the three readings:

1. Note which two pins produce the highest ohm reading. The pin that is not involved with that reading is the Common pin.

2. Note which two pins produce the lowest ohm reading. One of them is the Common pin (which you have already identified). The other pin involved with the lowest ohm reading is the Run pin.

3. The remaining unidentified pin is the Start pin.

> **EXERCISE:**
> For the compressor winding measurements shown in figure 11–12, label each of the three pins as either C, S, or R. Does the physical arrangement match what you expected?

> **EXERCISE:**
> Locate several single-phase hermetic compressors in your shop. Take ohm readings on the pins of each. Identify each pin as Common, Start, and Run. Does each compressor follow the guidelines above regarding wiring colors and pin arrangements?

MAKING YOUR OWN HERMETIC ANALYZER

Figure 11–13 shows a tool that you can make for manually starting compressors. Use it as follows:

1. Identify the compressor pins.

2. Connect the black wire to the Common pin, the white wire to the Start pin, and the red wire to the Run pin.

3. Plug into a 115V circuit. (If the compressor is a 230V compressor, you may want to make an

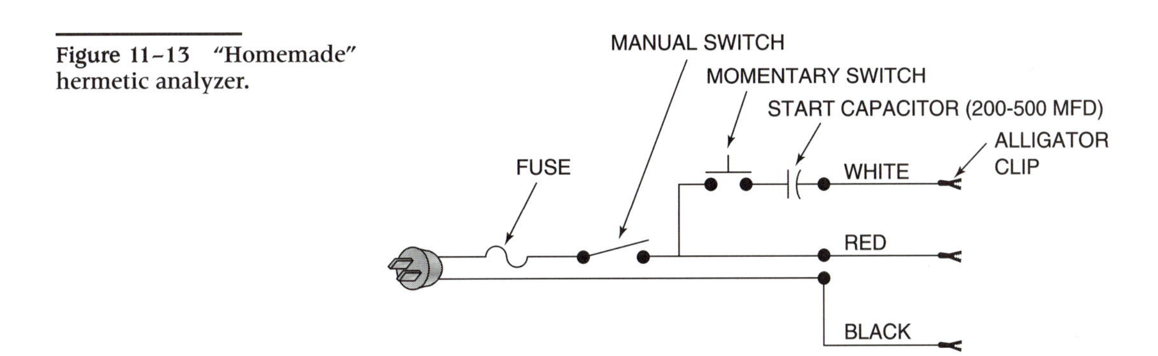

Figure 11–13 "Homemade" hermetic analyzer.

adapter from a 115V receptacle to another pair of alligator clips that can be attached to a 230V power source.)

4. Turn the manual switch to On. This will energize the Run winding. Immediately following, press the momentary switch for two seconds.

This will energize the Start winding as well, and the compressor should start. If it does not, the compressor has failed. If it does, then check your starting components.

CHECKING INTERNAL MOTOR OVERLOADS

Compressor motors and other motors are sometimes furnished with an internal thermal overload. It is a switch, or a thermistor that simulates a switch, that senses the temperature of the stator windings. If the windings get too hot, the switch will open. On a single-phase compressor motor, the switch may be wired as shown in either figure 11–14(a) or 11–14(b). In figure (a), the overload is line duty. That is, when the overload switch is closed, it carries the total current that flows through the compressor. In figure (b), the overload is pilot duty. It carries only the current that is consumed by the compressor contactor.

In figure 11–14(a), you cannot tell from looking at the compressor that there is an internal overload. However, if you measure voltage between C and R, and the compressor does not run, you might suspect that the compressor windings have failed. When you ohm out the compressor pins, if you find normal resistance between the **S** pin and the **R** pin, while the other two readings are infinite ohms, you would then realize that there is probably an internal overload that is open. If the compressor is cold, you

Figure 11–14 Line-duty and pilot-duty overloads.

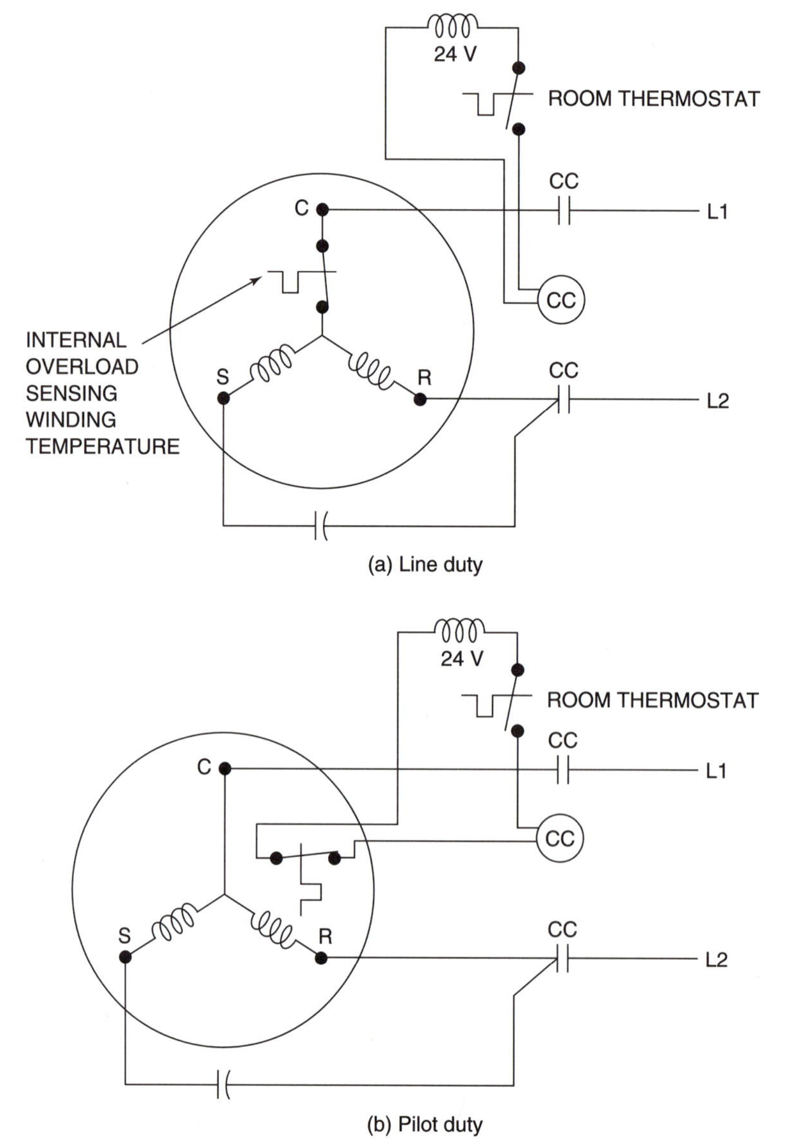

would replace it. If the compressor was warm, you would give it time to cool (maybe a couple of hours). If the overload did not re-close after that time, then you would condemn the compressor.

In figure 11–14(b), you would know from looking at the compressor that there is an internal overload, because you would see two extra wires emerging from the compressor shell. You can simply isolate the wiring to the overload, and measure the resistance. If it is infinite, the internal overload is open. The compressor must be replaced, or the internal overload may be bypassed. If appropriate, you can replace the overload function with an external overload.

CHAPTER TWELVE

How Motors Work

(This chapter was prepared and is reprinted with permission from Dirk DeKreek of DeKreek Technical Services.)

OVERVIEW

Motors, as well as transformers, generators, and alternators, are devices that make use of the properties of magnetism to achieve a desired effect. In the case of motors, this desired effect is to cause the shaft of the motor to rotate.

There are many different types of motors, each designed for specific purposes. The most common type of motor used in the air conditioning and refrigeration industry is called an **induction motor**. The name reflects the fact that current is induced in the rotating part of the motor (called the rotor).

MAGNETICS

The easiest way to develop an understanding of induction motors is to first explore the magnetic principles of generators and transformers. We will then find that an induction motor is actually just a variation of a simple transformer.

The important magnetic principles are:

1. An electric current flowing in a wire creates a magnetic field, commonly called flux, which wraps around the wire along its length.

2. Magnetic flux "flows" most easily through iron and will always seek the path of least resistance. In the absence of iron or other materials, it will flow through air or a vacuum.

3. A wire exposed to *changing* magnetic flux will develop an induced voltage.

A SIMPLE GENERATOR

Let us begin by using the second and third principles to build a simple electric generator. Figure 12–1 shows a simple bar magnet. This type of magnet is called a permanent magnet. The magnetic field is constant (i.e., unchanging). The magnetic flux "flows" from the "North" end through the surrounding air to the "South" end. This magnetic flux would much rather flow through iron or certain other magnetic materials than though air.

Figure 12–2 shows the bar magnet inserted in an iron core. Now almost all of the flux flows through the iron, which offers a much lower resistance path than the surrounding air. The flux is still constant, since the magnet is not moving. The natural tendency of magnets is to align themselves with their surroundings in such a way as to create the easiest path for the magnetic flux. This results in the mutual attraction between magnets and ferro-metallic objects. The magnet in figure 12–2 will pull itself into the position shown and attempt to stay there. Some external force will be required to move it out of this position.

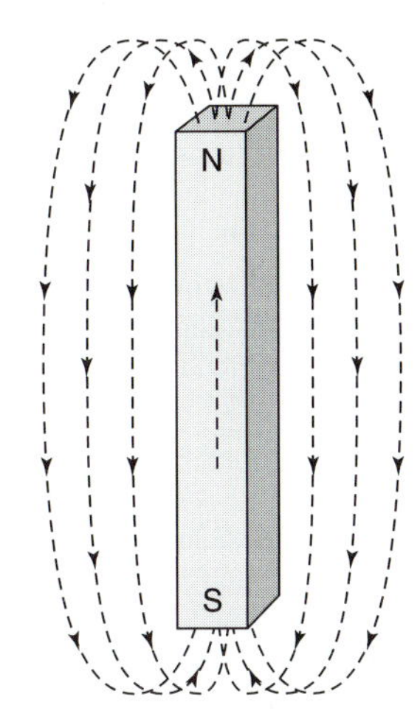

Figure 12–1 Magnetic flux lines in a bar magnet.

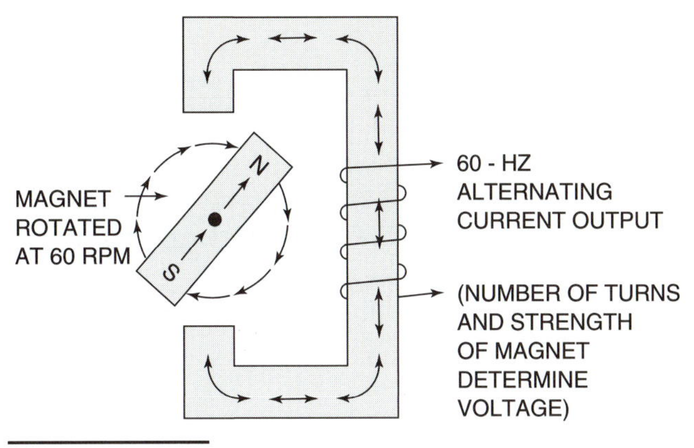

Figure 12–3 Changing magnetic field causes AC voltage to appear.

Figure 12–3 shows the permanent magnet and core with some turns of wire wound around the iron core. If the bar magnet is now forced to rotate as shown, the resulting *changing* magnetic field will cause a voltage to appear in the windings. There will not be any voltage induced in the wires unless the magnet is moved, because a changing magnetic flux field is required to induce voltage. When the magnet is rotating, the voltage produced in the windings will alternate in polarity with each half-turn of the magnet, thus generating AC (alternating current). This type of generator is referred to as an **alternator,** a word composed from the contraction of the expression alternating current generator.

The simple alternator is commonly used for low-power applications such as bicycle head lamps. The voltage produced in the windings depends on the strength of the magnet, the speed at which it is rotated, and the number of turns of wire wound on the iron core. If a load is connected to the alternator, the amount of current drawn by the load will also affect the output voltage.

A more practical alternator can be constructed by using an **electromagnet** instead of a permanent magnet. Automotive alternators are designed this way. The strength of the rotating electromagnet (called the field winding) can be varied by regulating the amount of DC current flowing through it. This allows the output voltage of the alternator to be controlled. An electronic device called a voltage regulator is used in most automotive alternators to perform this function. Automotive alternators also use a full-wave rectifier to convert the output alternating current into direct current, since automotive systems generally operate on DC.

Figure 12–2 Flux lines follow iron core.

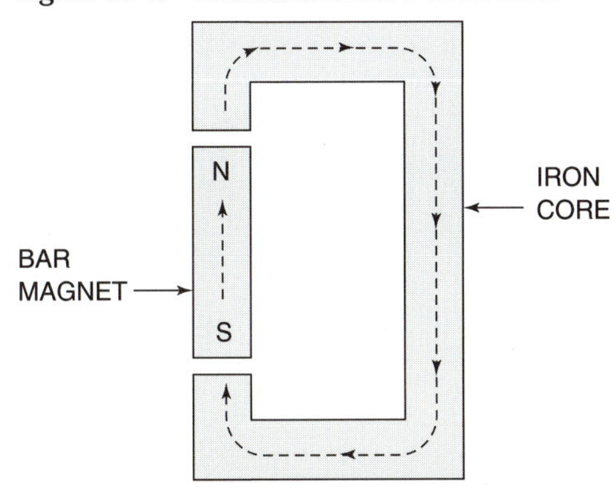

A SIMPLE TRANSFORMER

Another way to create the necessary changing magnetic flux field is to use a stationary electromagnet connected to a source of alternating current. Figure 12–4 shows this arrangement. Note that the secondary winding is identical to the previous alternator example. The essential difference between an alternator and a transformer is the way the changing magnetic field is created.

The primary of the transformer is connected to a 120V AC source. The resulting current flow creates a magnetic field which will follow the path of least resistance, which is the iron core of the transformer. This magnetic field is constantly changing in direction and intensity since the primary voltage is alternating. Therefore, in accordance with the third principle, a voltage is produced in the secondary.

At this point a few rules about transformers are in order:

1. Transformers cannot function on direct current since a *changing* magnetic field is required to generate electricity.

2. Transformers conserve power, which means that the wattage (volts × amps) in the primary will

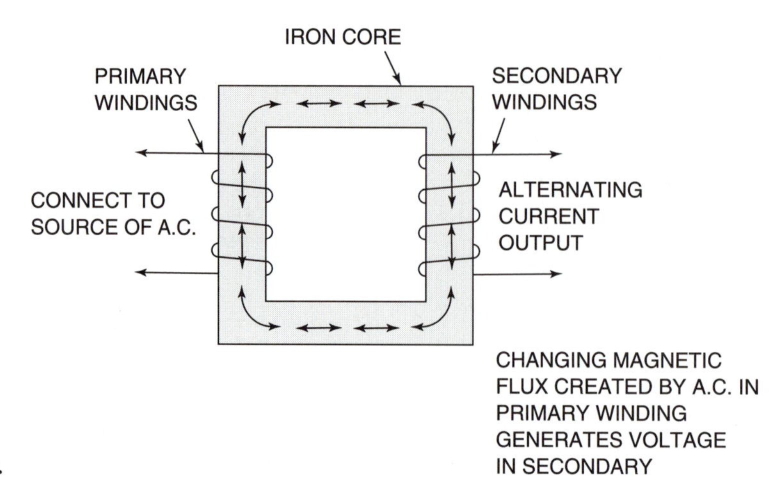

Figure 12–4 Transformer.

be very nearly equal to the wattage in the secondary. Instead of watts, transformers are generally rated in VA (volts × amps) or KVA (thousands of VA).

3. The voltage ratio will equal the turns ratio. A transformer with half as many turns of wire on the secondary as on the primary will develop a secondary voltage equal to half the primary voltage.

THE SHORTED-SECONDARY TRANSFORMER

The current draw in the primary will increase as the current in the secondary increases, in accordance with rule #2 above. In the extreme case, if the secondary leads of a transformer are shorted together, as shown in figure 12–5, the primary side behaves as though it were also shorted. This often causes some bewilderment, since there is no direct connection between the two windings. Since this phenomenon also occurs in induction motors, and is important to the understanding of both transformers and motors, we will take the time to explain it now.

The reason for this "reflection" of secondary load current back to the primary side is that the current flowing in the secondary windings generates a magnetic flux of its own, which also is confined to the iron core of the transformer. This flux is opposite in polarity to that created by the primary windings, and has the effect of weakening the flux field created by the primary windings. It is the resultant combined flux field that actually determines the current flow into the primary windings, using a property called inductance.

About Inductance

Contrary to popular belief, it is not the resistance of the wire that limits the current flow in the primary of a transformer or the windings of an induction motor. A quick calculation using Ohm's Law will verify this. A typical transformer may have a primary resistance of only 5 ohms. Applying 120V to a

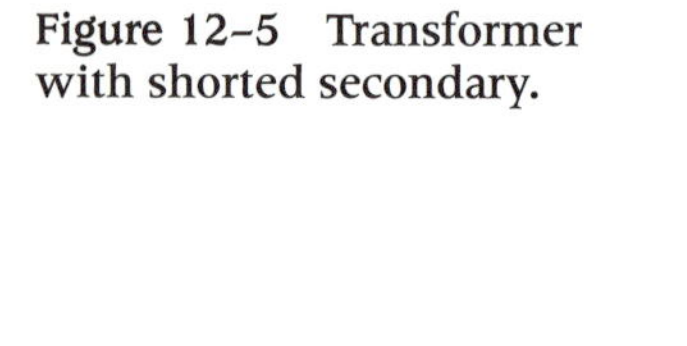

Figure 12–5 Transformer with shorted secondary.

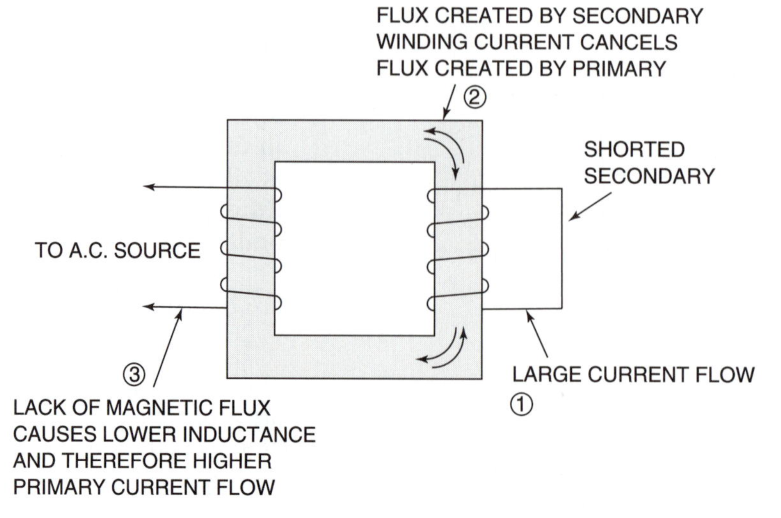

5-ohm resistive load would result in a current flow of 24 amps, yet the actual primary current is found to be almost zero. The reason for the lower current flow is that the transformer winding also exhibits a property called **inductance.**

Inductance limits current flow because the magnetic flux field created by the primary windings self-generates a voltage in these primary windings which is in direct opposition to the applied line voltage. This has the effect of "pushing" back against the applied line voltage, effectively limiting the current flow to a very small amount. The current flow increases to a higher value only when some of the magnetic flux energy is consumed by other windings, such as the secondary of the transformer.

Inductance behaves much like resistance with one very important difference: it exists only in alternating current applications. When direct current is applied to a circuit, the inductance is always zero, so the resulting current flow is limited only by the resistance of the circuit. It is for this reason that AC and DC inductive devices (i.e. relays and solenoids) are not interchangeable, even though their voltage ratings may be identical.

MOTOR TYPES

The following sections describe various types of motors. The first three—permanent magnet, shaded pole, and brush-type—are not induction motors. The remaining types—split-phase, RSIR, CSIR, PSC, and CSCR—are all induction motors.

A SIMPLE PM MOTOR

Figure 12–6 shows a very simple PM (permanent-magnet) motor. Note the resemblance to the transformer in figure 12–4 and the generator in figure 12–3. The permanent magnet in the rotor always seeks to align itself with the magnetic field created by the windings on the stator. Applying an AC voltage to the stator windings will create a magnetic field between the poles of the iron stator which reverses each time the current flow reverses. This will tend to drive the rotor one-half revolution each AC half-cycle. The resulting motor speed would be 60 revolutions per second, or 3600 revolutions per minute (RPM).

A SIMPLE SHADED-POLE PM MOTOR

Note that the PM motor in Figure 12–6 has no particular inclination to favor clockwise or counterclockwise rotation. In fact, it is usually necessary to manually start these motors before they will run, and they run equally well in whichever direction they were started. This problem can be solved by adding a second set of poles (figure 12–7), positioned just slightly clockwise or counterclockwise from the primary poles, and electrically *shaded* so that their alternating magnetic cycle slightly lags behind that of the primary poles.

This shading is achieved by means of a small, heavy copper ring wrapped around each shaded pole. These rings behave exactly like a one-turn transformer secondary winding with the leads shorted together. One turn generates only a tiny voltage, but since there is virtually no resistance, a large amount of current flows in the rings. These rings create additional inductance, causing a slight delay in the changing of the magnetic field at these poles. This delay is just enough to cause the motor to favor one rotational direction over the other and it will start rotating by itself when power is applied.

Figure 12–6 Permanent-magnet motor for positive and negative half-circles of applied AC power.

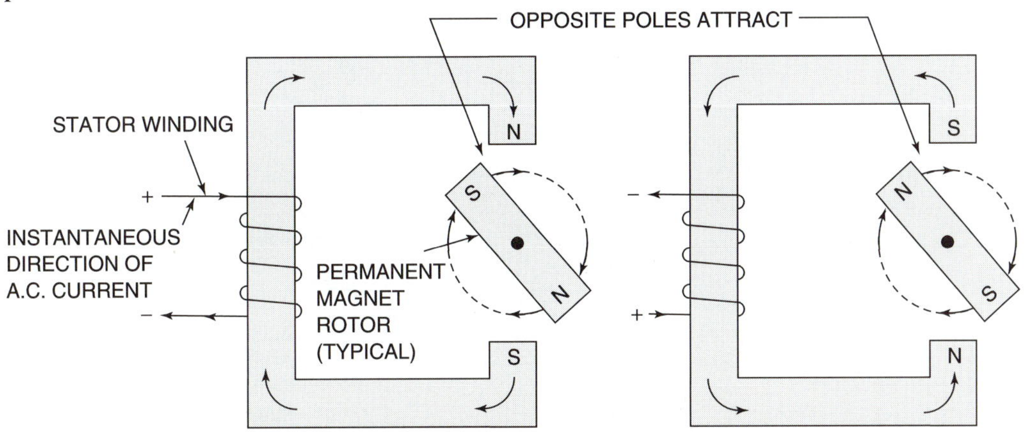

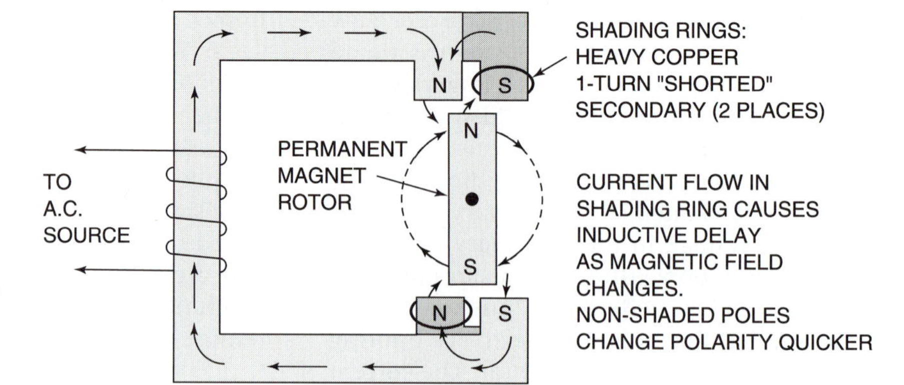

Figure 12–7 Shaded-pole motor: shaded poles are slower to reverse polarity when AC voltage reverses. Motor favors rotation in direction shown.

Shaded-pole PM motors are ideal for small loads such as small evaporator fans. They are extremely reliable. They are also **synchronous**, which means they "lock on" to the frequency of the AC power line and rotate at exactly that speed. Their only limitation is that they have very low torque and come only in very small horsepower sizes (typically 1/40 to 1/10hp) due to the limited strength of available permanent magnets in practical motor sizes.

BRUSH-TYPE MOTORS

PM motors larger than about 1/10 horsepower would require the use of prohibitively large (and expensive) permanent-magnet rotors, so most motors above this size use electromagnet rotors instead. Electromagnets can be made very strong using relatively small amounts of iron and wire. However, since the rotor moves, power wires cannot be connected directly to the rotor. One common solution is to use **brushes** and a **commutator** to make the electrical connections to the rotor. Figure 12–8 shows this arrangement.

Brush-type motors have very high starting torque and can develop high horsepowers while remaining very small. Portable electric drills and saws almost always use brush-type motors. These motors are relatively inexpensive, but the brushes wear out and need replacement after just a few hundred hours of operation. This is fine for intermittent-use items such as power tools, but not for fan motors and air conditioning compressors, which must run hundreds of hours each season.

A SIMPLE INDUCTION MOTOR

Fortunately, there is a way to cause current to flow in the rotor windings of the motor without using brushes. The method used is called **induction** (not

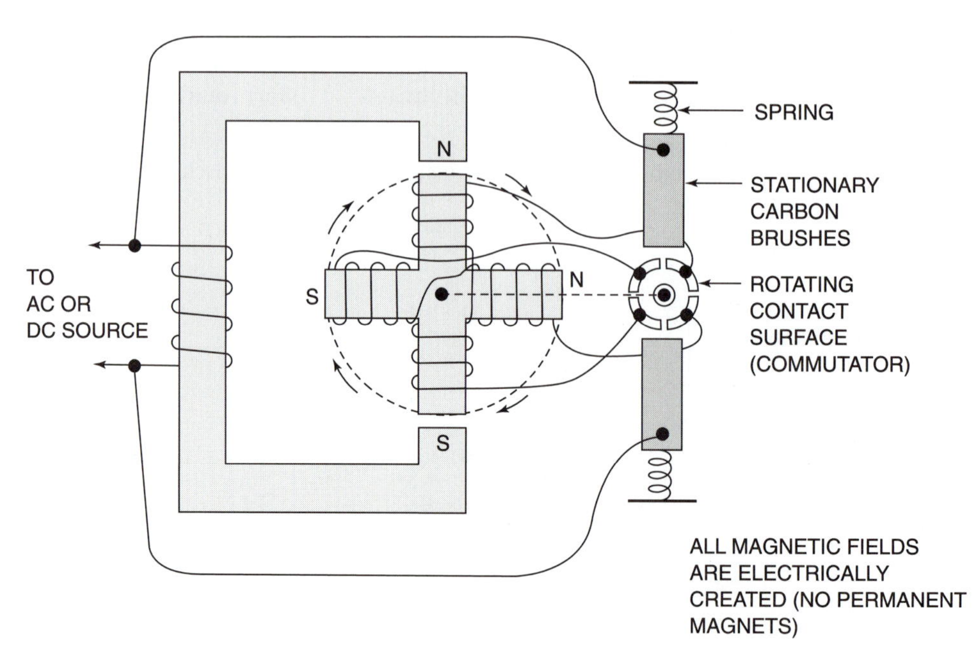

Figure 12–8 Brush-type motor with 4-pole rotor. Arcing of brushes makes this type of motor unsuitable for most HVAC applications.

to be confused with the word "inductance" used earlier). Figure 12–9 shows a very simple induction motor. Note that the stator (iron frame and windings) is very similar to the simple PM motor in Figure 12–6, but that the permanent magnet in the rotor has been replaced with a piece of iron with shorted windings.

The key to understanding induction motors lies in recognizing that this arrangement looks very much like the shorted-secondary transformer of figure 12–5, with the iron core cut to allow the secondary section to rotate. When AC voltage is applied to the primary of the transformer (or the **run winding** of the induction motor), the resulting alternating magnetic field will cause current to flow in the shorted secondary windings of the transformer (or the shorted rotor winding of the induction motor). If the secondary (rotor) is kept from turning, large currents will flow in both windings. The current flowing in the rotor winding will cause the rotor to behave like a powerful electromagnet, much stronger than a permanent magnet of similar size.

When the rotor is turning, it tends to follow the alternating magnetic field created by the stator, much the same as the permanent magnet motor did. However, induction motors are not synchronous, and the rotor must always run slightly slower than the synchronous speed of the AC line voltage. If the rotor were able to "catch up" completely with the rotating magnetic field created by the stator, there would no longer be any current flowing in the rotor since it would not be "seeing" a changing magnetic field (remember Rule #1 about transformers).

Some current flow is always required in the rotor to keep it magnetized, so induction motors always run slightly slower than the synchronous speed of the applied AC line voltage. This speed difference between the magnetic field and the rotor is called **slip.** The greater the load on an induction motor, the greater the slip. A typical 2-pole induction motor will operate at approximately 3525 rpm under full load, and 3550 to 3575 rpm unloaded. Since the synchronous line voltage creates a 3600 rpm magnetic field, the motor is said to have a slip of 75 rpm at full load. Actual slip varies from motor to motor and depends primarily on the electrical resistance of the rotor winding.

A PRACTICAL INDUCTION MOTOR

The simple induction motor of figure 12–9 exhibits the same shortcoming as the simple PM motor of figure 12–6, which is that the motor will not start by itself. One solution would be to add shaded poles as we did for the PM motor. This would work on very small induction motors but would be of no help on larger motors, since shaded poles exert only a very small force. A better starting method must be used for induction motors, since they range in size from about 1/4 up to several hundred horsepower.

Figure 12–10 shows the arrangement most commonly used for single-phase induction motors. Instead of shaded poles, this type of induction motor uses a completely separate set of poles with a dedicated **start winding.** This arrangement is often referred to as a **split-phase** motor. There are several variations of the split-phase induction motor, but they all operate on the same principle, namely that the magnetic cycle in the start circuit leads or lags that of the run circuit.

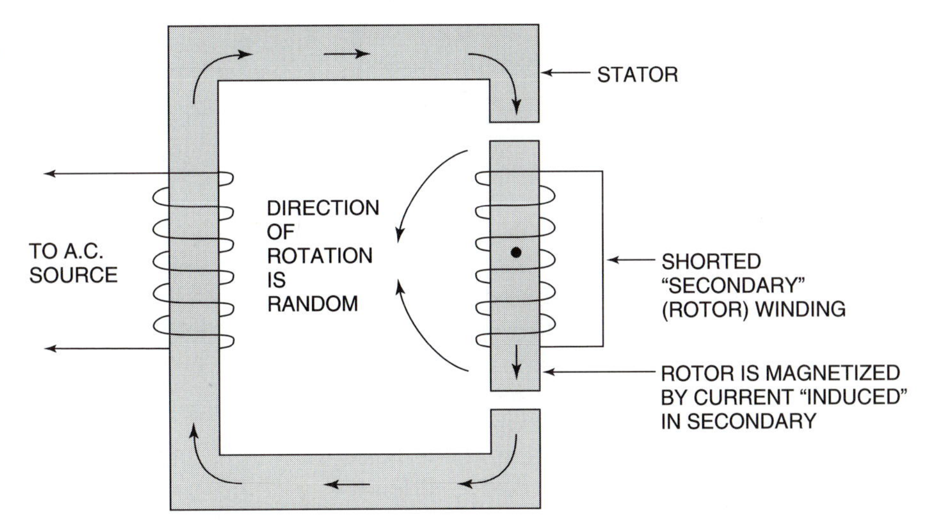

Figure 12–9 Simple induction motor. Uses no brushes or permanent magnets.

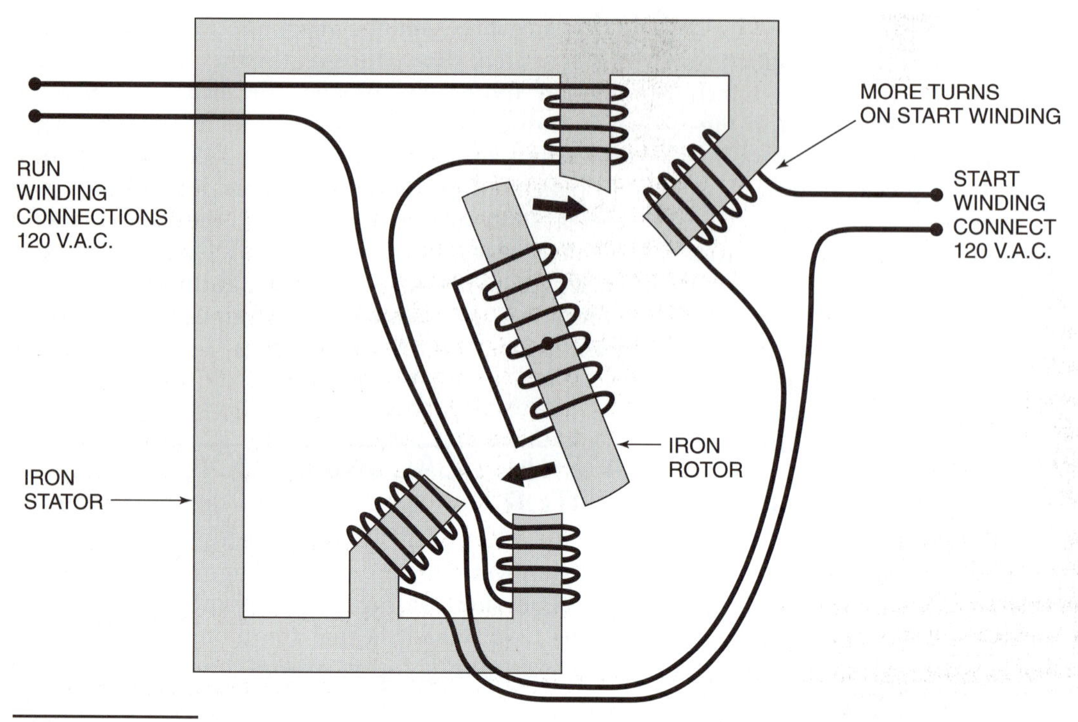

Figure 12–10 Simplified split-phase induction motor. Start winding current "lags" slightly behind run winding due to extra inductance.

THE BASIC SPLIT-PHASE MOTOR

The easiest way to create a lag in the start winding is to simply wind more turns of wire on the start winding than on the run winding. The extra inductance of the start winding causes the alternating current flow to lag behind that in the run winding, which has fewer turns and therefore less inductance (and also less DC resistance). The actual phase shift (lag) obtainable in this way is limited, so these motors have low starting torque. The efficiency is also rather poor, since the start winding creates a fair amount of heat. Basic split-phase motors are commonly found on small furnace blowers in sizes up to about 1/4 horsepower.

RSIR (RESISTIVE-START, INDUCTION-RUN) MOTOR

The RSIR motor is simply an improved split-phase motor. The significant improvement is in the operating efficiency of the basic split-phase motor. It is achieved simply by switching the start windings out of the circuit once the motor is running. Figure 12–11 shows the schematic equivalent of the basic split-phase motor, with a switching device added to disconnect the start winding once the motor reaches operating speed. The following two common methods are employed.

CENTRIFUGAL SWITCH

Open-frame motors often use switch contacts operated by a centrifugal weight and a spring-loaded slide mounted on the shaft. When the motor reaches running speed, the weights move outward and push the slide against a cam mounted on the end plate of the motor. The cam opens a set of contacts which disconnect power to the start winding.

Figure 12–11 "RSIR." Basic split-phase motor with automatic switch added to disconnect start winding once motor reaches speed.

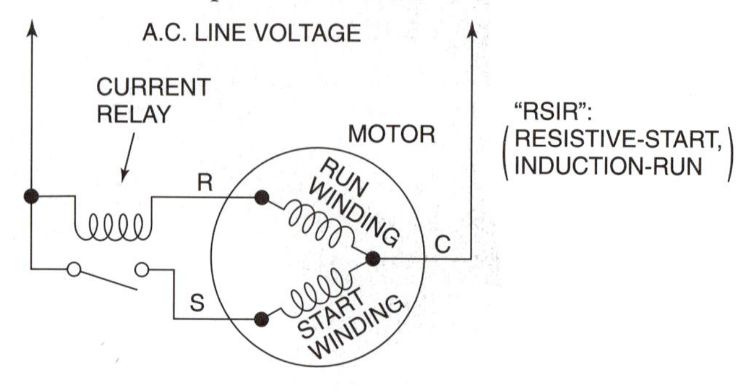

Current Relay

Current relays are specially wound relays that use the motor's run winding current. Since motors will draw more current when stalled than when running, these relays have normally open contacts which will close when power is first applied to the motor. Once the motor comes up to speed, the amperage will drop to normal and the relay contacts will open, disengaging the start windings.

THE CSIR (CAPACITIVE-START, INDUCTION-RUN) MOTOR

A significant improvement in **starting torque** can be achieved by increasing the angle between the start winding poles and the run winding poles inside the motor. This requires a corresponding increase in the time delay (phase lag) between the two windings. The simplest way to achieve this phase lag is to insert a **capacitor** in series with the start winding (figure 12–12).

The capacitor has the effect of "cancelling" the induction of the start winding, allowing the current flow to exactly follow the line voltage, while the run winding continues to "lag" about 90 degrees, or 1/4 cycle. This allows the two poles to be spaced about 90 degrees apart, and enables the motor to develop maximum starting torque. A capacitor which is used only to assist the starting of a motor, but is switched out once the motor is running, is called a **start capacitor.**

About Capacitors

Capacitors are probably the most misunderstood components in air conditioning equipment. They are extremely simple devices, consisting of nothing more than two conductive "plates" or surfaces (usually tinfoil) separated by a thin layer of insulating material (called **dielectric**). The "sandwich" of tin foil and insulation is usually rolled up into a cylindrical or oval shape and installed in a plastic or metal housing for protection.

A capacitor is able to store a limited amount of electric charge on its plate surfaces. The more surface area, the more charge can be stored. Once the plates are fully charged, no more current can flow into the capacitor unless the voltage is further increased. This ability to store limited amounts of electric charge can be very useful in two different ways:

1. Filters: Capacitors connected directly across DC voltage sources provide a ready reservoir of charge to smooth out variations in voltage. Note that this works only for DC circuits, not AC.

2. Inductive cancellation: Capacitors very effectively cancel out inductive lag in AC circuits. (See section above to review induction.) The means by which capacitors cancel out induction is explained below. The most important thing to realize about this use of capacitors is that the size of the capacitor is very important, and varies with the amount of inductance present. Since inductance is determined by the number of turns of wire in the motor and the load on the motor, it follows that different motors require different capacitor sizes for optimum performance.

Imagine an alternating current applied to the motor circuit shown in figure 12–13. Initially, current builds up slowly in the direction that the AC voltage happens to be pushing at that instant. (The current does not instantly reach full intensity because the induction of the motor windings resists sudden changes in current flow.)

The capacitor initially looks like a conductor, with current flowing into one side and out of the other. However, the capacitor quickly charges up and starts looking more and more like a resistor as

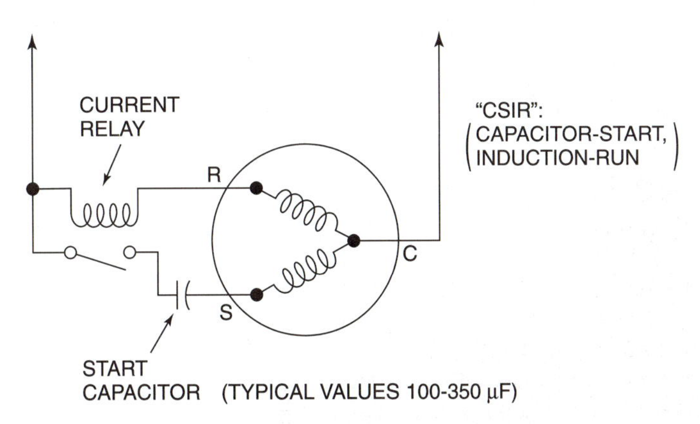

Figure 12–12 "CSIR." RSIR and start capacitor.

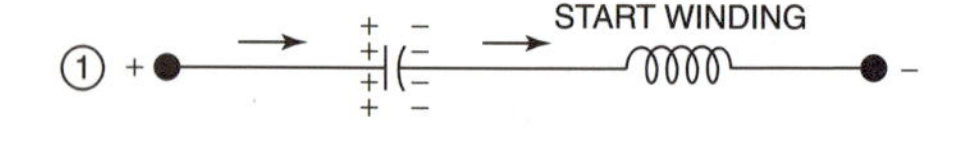

1. CAPACITOR IS INITIALLY NOT CHARGED. CURRENT FLOWS INTO CAPACITOR, CHARGING LEFT PLATE.

2. CURRENT FLOW IN START WINDING STOPS WHEN CAPACITOR IS FULLY CHARGED.

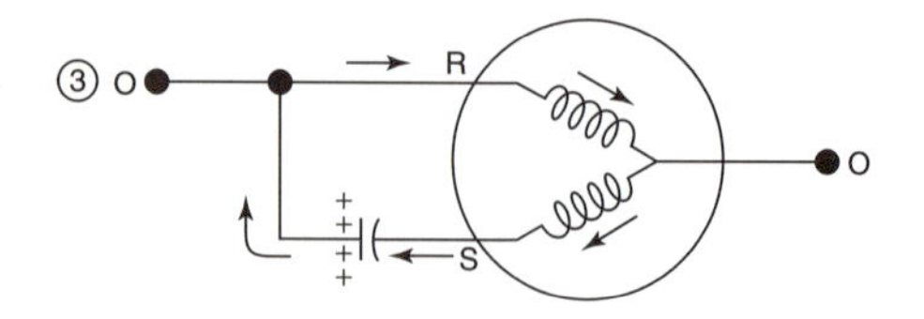

3. AS AC LINE VOLTAGE DROPS TO ZERO, RUN WINDING CURRENT, CONTINUES TO FLOW DUE TO INDUCTIVE LAG, WHILE START CAPACITOR DISCHARGES THROUGH START WINDING. THIS PRODUCES THE DESIRED PHASE DIFFERENCE.

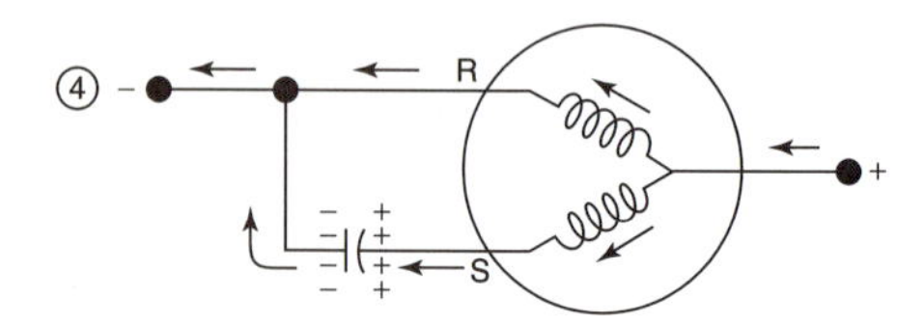

4. CYCLE REPEATS IN OPPOSITE DIRECTION FOR NEXT HALF-CYCLE.

Figure 12–13 Capacitor Basics.

the charge increases. This has the effect of "slowing down" the current flow into the start winding well before the line voltage starts to turn around for the next half-cycle of the AC sine wave. The capacitor is now sitting, fully charged, waiting for an opportunity to discharge. As soon as the line voltage begins to turn around, the capacitor starts discharging through the start winding and back in the opposite direction through the run winding, which is still carrying current due to the inductive effects. This entire cycle then repeats in the opposite direction for the next half-cycle.

The "size" of a capacitor is measured in farads. Most capacitors are small enough to be more conveniently rated in microfarads (millionths of farads). Microfarads simply measure the charge-storing capabilities of a capacitor.

THE PSC (PERMANENT SPLIT CAPACITOR) MOTOR

The presence of a capacitor in the start winding circuit not only improves the starting torque of a motor, it also enhances the running efficiency and power factor (see section below) of most induction motors. Sometimes the running efficiency is more important than the starting torque. A common example is for air conditioning compressor and fan motors. For these applications, the capacitor is sized differently and left in the circuit continuously, as shown in figure 12–14.

A capacitor installed in this way is called a **run capacitor.** Run capacitors are generally much smaller, electrically speaking, than start capacitors, and are built differently. Since they remain in the circuit continuously and therefore must be able to dissipate any heat which builds up due to the constant charging and discharging of the plates, run capacitors generally have a steel case and are physically larger than the equivalent microfarad-size start capacitor would be. Start capacitors, on the other hand, are designed for very short intermittent use and thus can be built much more compactly. Start capacitors are generally packaged in round plastic or phenolic cases.

THE CSCR (CAPACITOR-START, CAPACITOR-RUN) MOTOR

The ultimate single-phase induction motor employs both a start capacitor for optimum starting torque and a run capacitor for optimum efficiency and

Figure 12–14 PSC motor: run capacitor, connected in series with start winding, remains in circuit at all times.

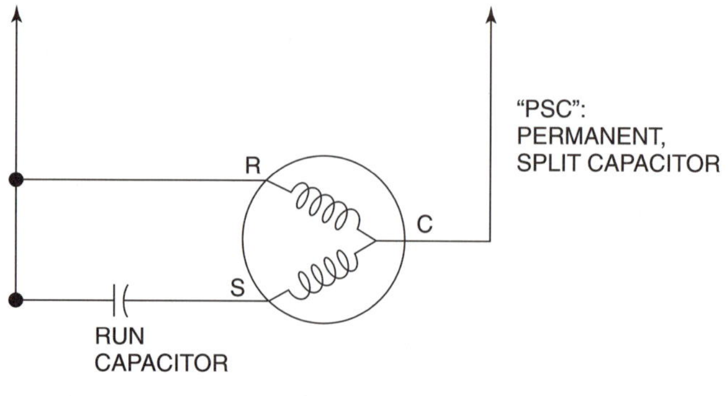

(TYPICAL VALUES: 5–60 μF)

PSC MOTOR: RUN CAPACITOR, CONNECTED IN SERIES WITH START WINDING, REMAIN IN CIRCUIT AT ALL TIMES.

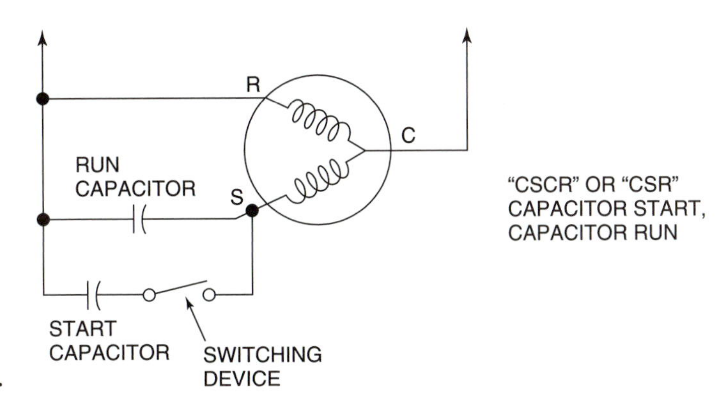

Figure 12–15 CSCR motor.

power factor. These motors are called **CSCR motors**, sometimes abbreviated to **CSR** (figure 12–15). Since the optimum capacitor value is different when starting than during normal running, two separate capacitors are used. The start capacitor is switched out of the circuit once the motor reaches running speed, while the run capacitor remains in the circuit at all times. Note that both capacitors are connected in series with the start winding.

The device most commonly used to switch the start capacitor out of the circuit in CSCR motors is called a **potential relay** (figure 12–16). A current relay or centrifugal switch could probably be used also, but CSCR motors are generally large enough to make current relays impractical, since all of the run winding current would have to flow through the coil of the current relay.

About Potential Relays

Potential relays make use of the "motor-generator" effect of induction motors. Simply stated, the run winding causes the rotor to become magnetized and rotate. The spinning rotor acts on the start winding in much the same way as the spinning magnet in the generator from figure 12–3, or the transformer in figure 12–4, which means that a voltage is induced in the start winding which increases as the rotor speed increases. By connecting the coil of a specially designed relay to the two leads from the start winding (terminals "S" and "C"), the relay can detect when the motor reaches operating speed and switch a set of contacts to switch the start capacitor out of the circuit. The voltage generated by the start winding is commonly referred to as **back EMF** (for Back Electro-Motive Force) and varies from motor to motor, necessitating the use of the proper potential relay for a given motor. It is possible for the back EMF to exceed the line voltage applied to the motor.

Figure 12–16 Potential relay and CSCR motor.

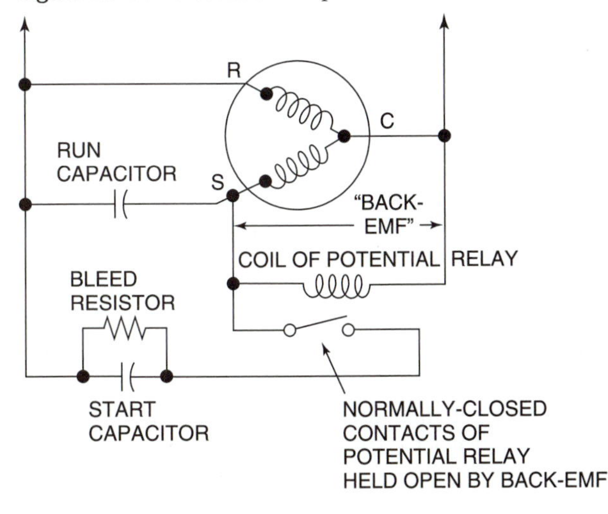

Note that the contacts on a potential relay are normally closed, since the back EMF is very low when first starting a motor. The back EMF increases as the motor speed increases. Also note that potential relays can only be used in conjunction with CSCR motors. A potential relay cannot be used in place of a current relay for RSIR or CSIR motors, because the back EMF is directly connected to the incoming line voltage and therefore cannot vary with motor speed.

About Bleed Resistors

Another commonly misunderstood component is the **bleed resistor.** Bleed resistors are commonly connected across large start capacitors to allow any stored charge to slowly dissipate. A charged capacitor can give a nasty electrical shock to a careless service technician, but this is not the reason why bleed resistors are installed. The real reason is to prevent welding the contacts in the potential relay when the motor stops and the back EMF drops to zero. At this time, the contacts revert to their normally closed position, creating a series circuit consisting of the run capacitor, the start capacitor, and the contacts of the potential relay (figure 12–17). If the start capacitor and run capacitor happen to both be charged in the same direction at the instant the contacts close, an extremely large momentary surge of current will flow as the capacitors discharge through the relay contacts. Since this situation will occur roughly 50% of the time, and since most of the charge is stored in the start capacitor, it becomes necessary to provide a means for the start capacitor to discharge itself during the time the motor is running. The bleed resistor accomplishes this very nicely. Repeat failures of potential relays can often be traced to missing or disconnected bleed resistors.

POWER FACTOR

Often confused with efficiency, **power factor** is actually a separate characteristic of all non-resistive loads and has little or nothing to do with efficiency. Simply stated, power factor is a measure of how much lag exists between the time when peak voltage is reached on a sinusoidal AC power line vs. the time when peak current flow occurs. Pure resistive loads such as light bulbs or electric heaters have no time lag at all, and are said to have a power factor of one. Motors and transformers typically exhibit some time lag, and have power factors of less than one. Typical motor power factors range from .4 to .9 for single-phase induction motors. Power factor can be measured by connecting a true wattmeter to a motor, along with a volt meter and an amprobe. The governing formula for single-phase motors is:

Power Factor = Watts/VA

Most motors exhibit poor power factor under light loads, and improve as the motor approaches full load. This is the reason why a motor with no shaft load still appears to pull significant power. In reality, such a motor operating at no-load conditions may draw nearly 50% of the nameplate rated full load amperes, but because the power factor is so low at no-load conditions, the actual watts consumed are very low. Useable power is measured in watts, which may differ from VA (volts times amps) in any reactive load.

Figure 12–17 Discharge path of start capacitor at end of cycle if bleed resistor is omitted.

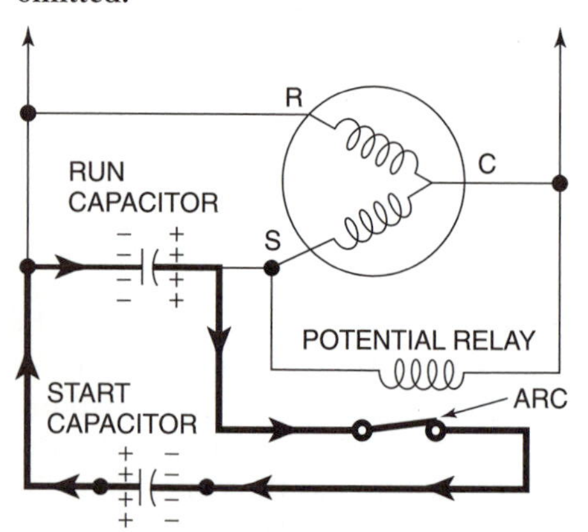

Electric utilities are concerned about low power factor loads because the out-of-phase current flow does not register on normal electric meters (which measure watt-hours, not amp-hours) but still causes current to flow in the power company's distribution wiring, creating voltage drops and IR losses. Industrial users typically have a special meter to measure this out-of-phase current (called a KVAR meter, for Kilo-Volt-Amps-Reactive), and pay a surcharge on their electric bill if overall power factor falls below a specified level.

Power factor may be corrected by adding a carefully sized capacitor in parallel with the motor leads of three-phase motors. Single-phase motors should be corrected by carefully sizing the run capacitor to match the motor's actual load.

Service technicians often ask why the identical compressor is equipped with a given size run capacitor in one manufacturer's air conditioner, while another manufacturer uses a slightly different value of run capacitor. The reason for this is because the actual load on the compressor, determined by suction and discharge pressures and refrigerant type, will vary slightly from one model of air conditioning unit to another.

CONCLUSION

This chapter was intended to provide some deeper insight into the actual electro-magnetic principles underlying common HVAC motors. While most motor failures are relatively easy to diagnose and correct, occasionally a particularly troublesome repeat failure condition occurs. Often these can be traced to low line voltage or mechanical overload conditions, but on occasion an intermittent start component or improperly sized run capacitor is found to be the cause. Technicians armed with a good understanding of the role of capacitors and start relays in single-phase induction motors are far less likely to overlook or misdiagnose the problems that these devices sometimes cause.

There are many more types of motors than those described in this chapter, but since they are rarely applied to HVAC equipment they were omitted from this discussion. Those with the curiosity to explore further may find stepper motors, reluctance motors, and direct-current motors particularly interesting.

CHAPTER THIRTEEN

Low-Voltage Room Thermostats

STANDARD HEATING-COOLING

There are many different types and styles of low-voltage room thermostats that will be encountered. Generally, they all share the same characteristic of using a bimetal element that changes shape when its temperature rises or falls. This mechanical movement is used to operate one or more switches which, in turn, operate various components in a heating or air conditioning system. Figure 13–1 shows a very common arrangement. The bimetal element is coiled around a fixed center. As its temperature changes (indicating a change in room temperature), the coil either winds tighter because the bimetal wants to curve more, or unwinds because the bimetal wants to unwind. A **mercury bulb** is attached to the bimetal. A mercury bulb is a sealed bulb with mercury inside. As the mercury rolls from one end to the other, it will make contact between different pairs of wires.

Not all thermostats use a mercury bulb as the switch. Figure 13–2 shows several arrangements of switches with **open contacts,** as compared to the **sealed contacts** of the mercury bulb. The sealed contacts are used in higher-quality thermostats because their contacts are much less prone to get pitted or contaminated than open contacts. However, in applications where vibration of the thermostat is a problem (i.e., in a shipping office next to a railroad loading dock), the open-contact thermostat may actually be better because it has a more positive (spring-loaded) action.

Figure 13–3 shows a mercury bulb that acts as an SPDT switch. If the mercury rolls to the right, it makes contact between wires R and Y. If the mercury rolls to the left, it makes contact between wires R and W. As the bimetal moves due to a change in temperature, the mercury bulb tilts, and makes R-W or R-Y, bringing on the heating or cooling equipment.

Figure 13–4 shows a thermostat with and without a **subbase.** The subbase (figure 13–5) is the part that screws to the wall and contains the screws that hold the thermostat wires that emerge from inside the wall. When the thermostat is screwed to the subbase, it makes electrical connections between the thermostat and the subbase. In figure 13–6, the connections that are made when the thermostat is screwed to the subbase are R-R1, Y-Y1, and W-W1. The subbase contains two manual switches. One is called the Fan switch. It may be set to either On or Auto. The other switch is called the System switch. It may be set to either Heat, Cool, or Off. The circles at the manual switches in figure 13–6 show the different connections that are made when the switches are set in their different positions. Note that when the System switch is set in the Heat position, it will make R-W on the subbase when the mercury bulb tilts to the left, calling for heat, but it will not energize any terminals if the mercury bulb tilts to the right (calling for cooling). If the System switch is set on Cool, it will make R-Y and R-G when the mercury bulb calls for cooling, but it will not energize

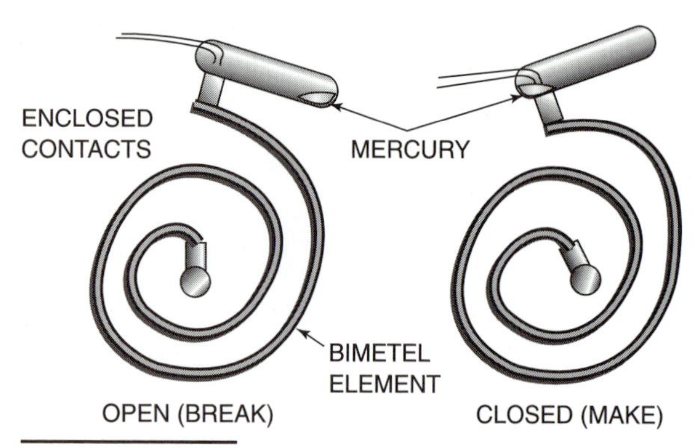

Figure 13–1 Thermostat using coiled bimetal element and mercury switch.

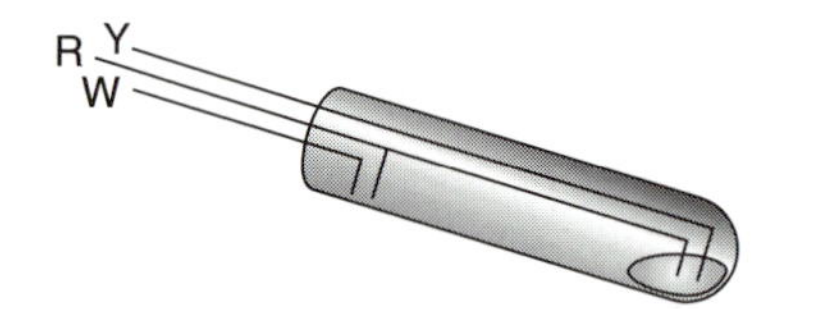

Figure 13–3 SPDT mercury switch.

any terminals on the subbase if the mercury bulb calls for Heat. This arrangement prevents the thermostat from calling for heating, then cooling, then heating, etc., as the mercury bulb tilts to one side and then the other. Therefore this thermostat can be used to control a heating and cooling system.

The terminal identifications used on the thermostat in figure 13–6 are very common and should be memorized by the service technician. Red is the terminal that receives power from the transformer. It can be switched out of the thermostat through white, green, or yellow. White is most commonly connected to the wire that will energize the gas valve, or other heating device. Green will be connected to the wire that energizes the blower relay. Yellow will be connected to the wire that energizes the compressor contactor. On a call for heating, the thermostat will make R-W. On a call for cooling, the thermostat will make R-Y and R-G. Note that the thermostat does not normally bring on the fan during heating. When the Fan switch is placed in the On position, it will make R-G without regard to the position of the System switch or the temperature being sensed by the thermostat.

Although the terminal identifications discussed above are common, they are not universal. White-Rodgers and Honeywell use this convention. A less-used convention used by some manufacturers such as General Controls is the following:

V = Voltage (similar to R)
H = Heating (similar to W)
C = Cooling (similar to Y)
F = Fan (similar to G)

Thermostats with R, W, Y, and G terminals may also have a B terminal and an O terminal. The thermostat will make R-B whenever the System switch is set to Heat, and it will make R-O whenever the system switch is set to Cool. For these two terminals, it does not matter what temperature is being sensed by the thermostat. The B or the O terminal can be used to energize damper motors in either the heating or the cooling mode. But in practice, the B and O terminal are rarely used on thermostats that are not heat-pump thermostats.

Figure 13–5 shows a thermostat subbase with an R_C and an R_H terminal. Normally, the red wire coming from the transformer will be wired across both of these terminals, making them into a single terminal. The reason that these terminals are sometimes provided is for an add-on cooling system. Consider a system that has a heating-only furnace, and you are going to add on a rooftop packaged cooling system. The existing furnace has a transformer and a gas valve. The existing thermostat has only R and W terminals. You can replace the existing thermostat with one with a subbase as in figure 13–5. The existing two wires that were attached to the old thermostat will be connected to R_H and W. The thermostat wires from the new rooftop packaged cooling system (with its own transformer) will be connected to R_C, Y, and G. The heating circuit will be electrically completely independent from the cooling circuit. The advantage of this type of thermostat is that it avoids having one thermostat for heating and a separate thermostat for the new

Figure 13–2 Thermostats with open contacts.

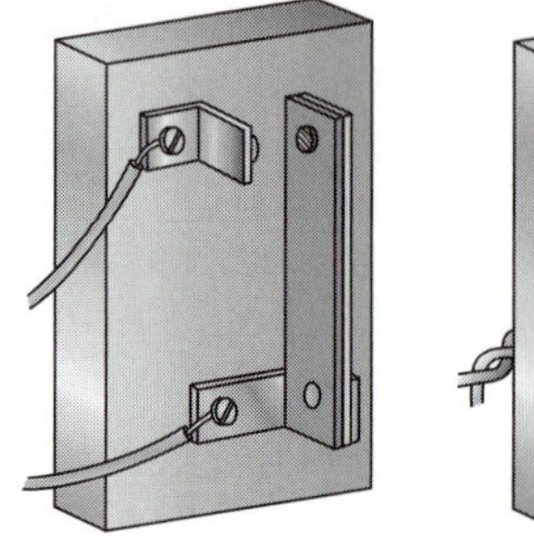

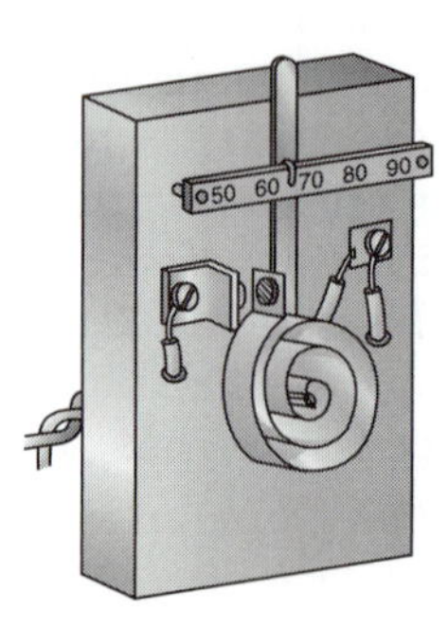

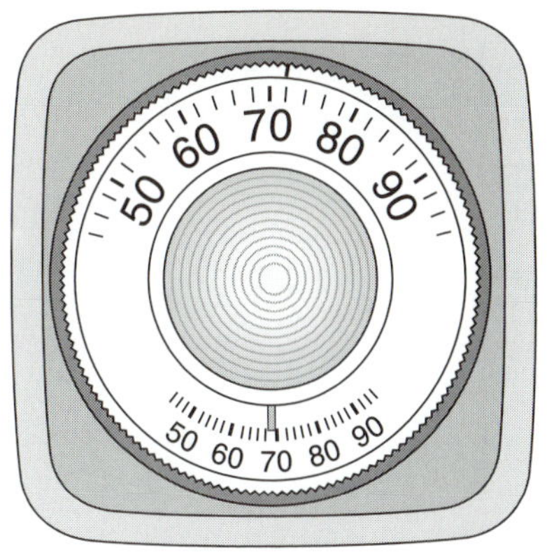

(a) Without sub base

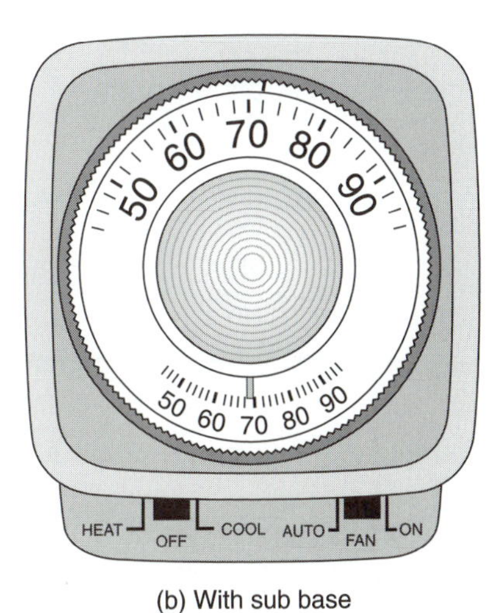

(b) With sub base

Figure 13–4 Thermostat with and without subbase. (a) Without subbase. (b) With subbase.

cooling unit. Using two thermostats simultaneously on the same system has the disadvantage of potentially having the heating system and the cooling system run at the same time if the set point on the heating system thermostat is higher than the set point on the cooling system thermostat. Figure 13–7 shows a ladder diagram for an add-on cooling system.

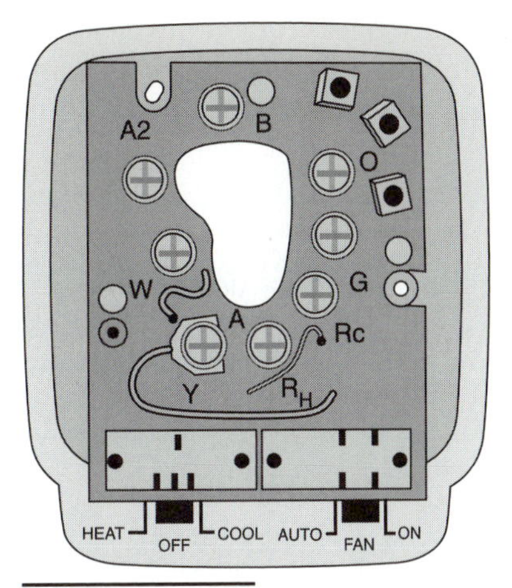

Figure 13–5 Thermostat subbase.

HEATING AND COOLING ANTICIPATORS

On a heating application, once the thermostat is satisfied, the R-W contacts open. However, on most applications, this does not stop heat from being provided to the room. Typically, when R-W opens, it de-energizes the gas valve, but the fan continues to blow warm air into the room until the heat exchanger has cooled. This will cause the room temperature to "overshoot" the temperature at which the thermostat was actually satisfied. It would be nice if we could somehow anticipate when the thermostat was going to be satisfied, and somehow open the R-W contacts a little sooner than they would otherwise have opened, to prevent overshooting the temperature actually desired.

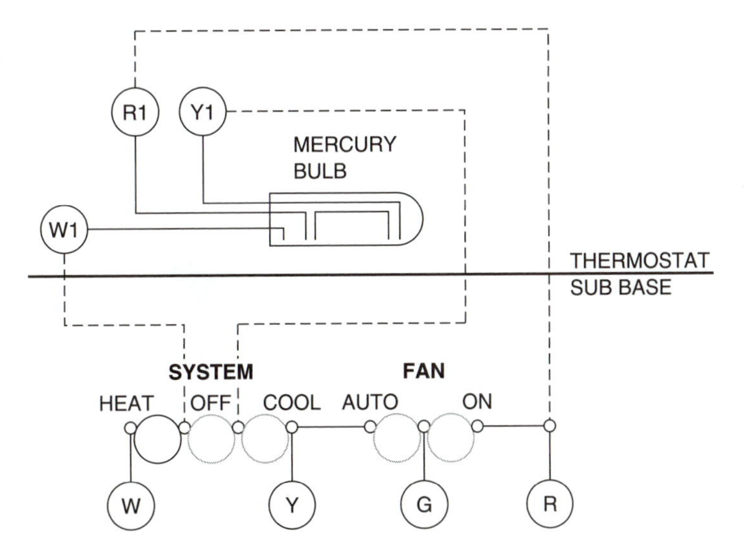

Figure 13–6 Dotted wiring shows connections that are made when thermostat is attached to subbase.

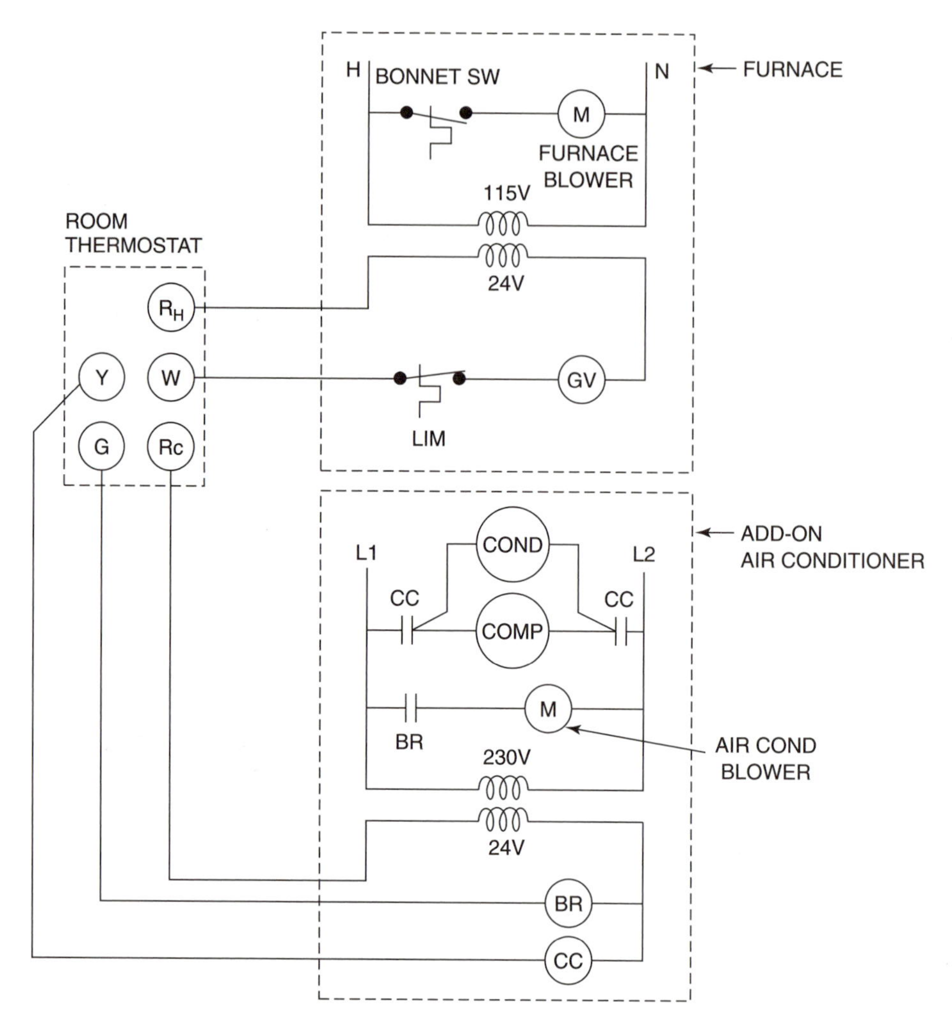

Figure 13–7 Add-on cooling schematic using R_C and R_H thermostat terminals.

The **heat anticipator** does exactly that. It is a small, adjustable resistor located in the thermostat, very close to the bimetal element (figure 13–8). It is wired in series with the R-W switch, so it is energized whenever the thermostat is calling for heat.

Figure 13–8 Heat anticipator.

The current passing through the anticipator generates a small amount of heat that tends to warm the bimetal element slightly quicker than the room itself would otherwise warm the bimetal. This causes the bimetal to open the R-W switch a little sooner than it would have without the anticipator.

The amount of resistance used in the heat anticipator will determine how much heat it produces. If more of the anticipator element is used, the cycles will be shorter; using less of the heat anticipator element will produce longer cycles. The correct amount of heat anticipator element to be used must be selected by the installing service technician.

To set the heat anticipator, remove the thermostat from the subbase, and place a 10-turn multiplying loop jumper wire between R-W. This jumper wire will cause the heating system to come on. Using a clamp-on ampmeter, measure the current being carried by the thermostat (figure 13–9). Because of the 10 turns on the multiplying loop, the magnetic field produced

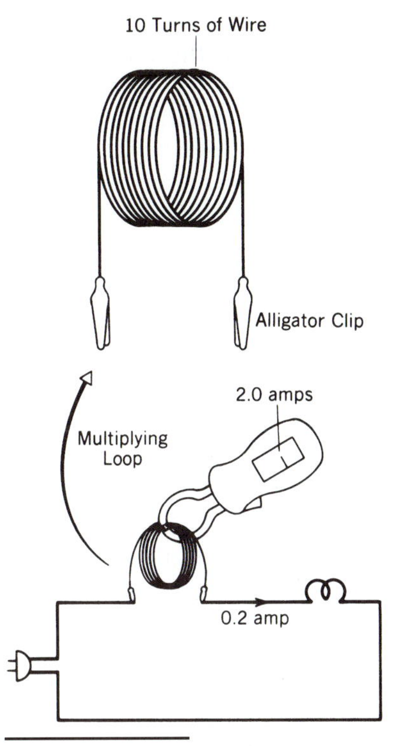

Figure 13–9 Multiplying loop.

is 10 times the field produced by one wire, and the amp reading obtained will be 10 times the actual amps being carried by the thermostat. Divide your reading by 10 to determine the actual amps. Then set the heat anticipator to a number that matches the actual amps that flow through the thermostat on the heating cycle.

On cooling, there is a similar problem. When the room gets warm, the thermostat will close R-Y and R-G, bringing on the air conditioning. However, it will take a minute or two for the refrigeration system to get down to operating temperatures, and to cool the duct work in the attic that has warmed during the Off-cycle. It would be desirable to turn the air conditioning system on prematurely, just as it was desirable to turn the heating system off prematurely.

The **cooling anticipator** (figure 13–10) is a fixed resistor in the thermostat that is wired between the R and Y terminals. It is energized whenever the R-Y switch is open (whenever the air conditioning system is off). It tends to heat the bimetal and close the R-Y and R-G earlier than would otherwise happen from the rising room temperature alone. It is not adjustable, and the service technician does not have to do anything about the cooling anticipator.

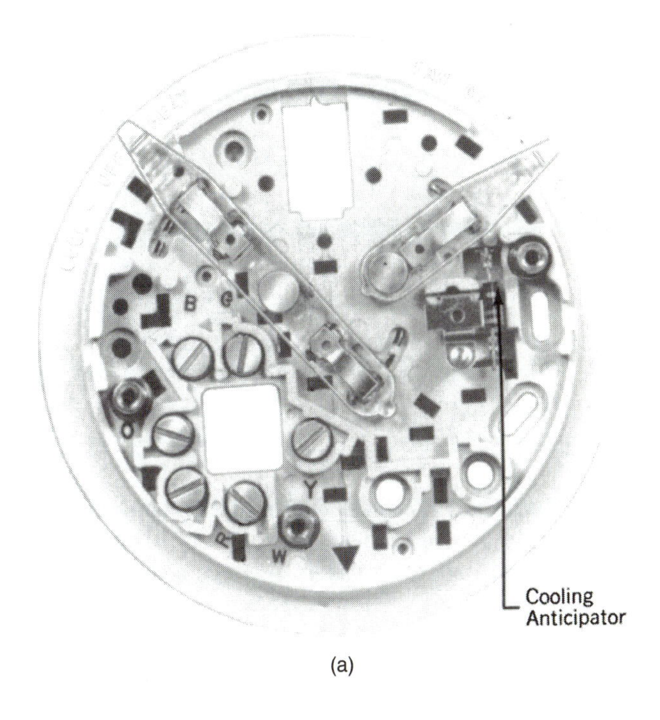

(a)

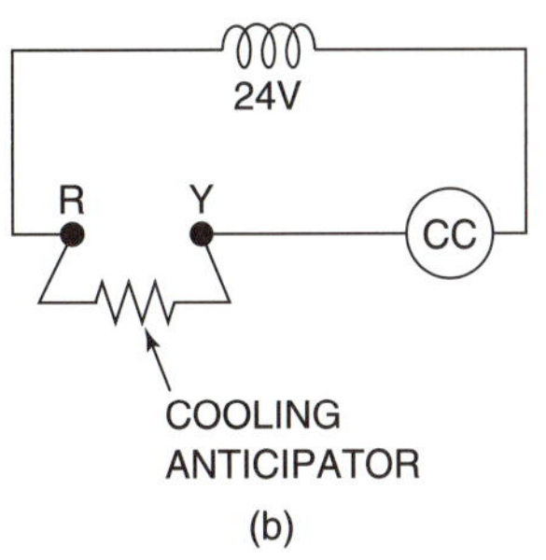

(b)

Figure 13–10 Cooling anticipator. *(a. Courtesy of Honeywell, Inc.)*

QUESTION:
Why must the heating anticipator be adjustable, while the cooling anticipator is non-adjustable?

QUESTION:
What will be the customer complaint if the heat anticipator is burned out (open)? What will be the customer complaint if the cooling anticipator is burned out?

AUTO-CHANGEOVER THERMOSTAT

The most common heating-cooling thermostats have a System switch that allows three different selections: Heat, Cool, and Off. The system is always in either the heating mode or the cooling mode, but

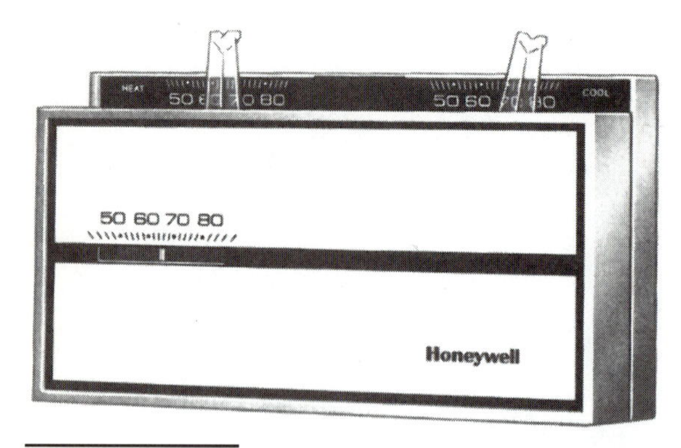

Figure 13–11 Auto-changeover thermostat.

cannot switch to the other unless the System switch is manually operated. This presents some problems in office environments. Often, setting of the thermostat can be a source of disagreement among office occupants. Owners will sometimes place a locked box around the thermostat so that it cannot be adjusted by the office personnel. However, if it is set for Heating, and there is an exceptionally warm day on which cooling is needed to maintain the set temperature, there will be a problem.

Figure 13–11 shows an **auto-changeover** thermostat. The System switch has four selections: Heat, Cool, Auto, and Off. When the System switch is set in the Auto position, the thermostat can provide heating when the room temperature is below set point and cooling when the room temperature is above set point. Note that there are two different set point levers on the auto-changeover thermostat. The lower set point is for heating, and the higher set point is for cooling. Typically, the heating set point will be color coded red, and the cooling set point will be color coded blue. They may be set independently, but a mechanical separation between them prevents them from being set so there is less than 2°F between them. They may be set as far apart as desired, for energy savings. For example, if the red lever is set to 70°F and the blue lever is set for 75°F, the thermostat will make R-W bringing on the heating whenever the room temperature is below 70°F, and the thermostat will make R-Y and R-G whenever the room temperature is above 75°F. If the room temperature is between 70°F and 75°F, neither heating nor cooling will be used to change the room temperature.

SETBACK THERMOSTAT

Figure 13–12 shows a thermostat that looks similar to the auto-changeover thermostat because it also has two set point levers. But the function of these two set points is different from the auto-changeover thermostat. This is a **setback** thermostat. It has a clock, and colored pins (red and blue) that may be placed on a 24-hour wheel at a place that corresponds to a particular time of day. Whenever it is a time corresponding to a blue pin, the thermostat set point will correspond to the set point indicated by the blue lever. That will be the thermostat set point until the time corresponding to the next red pin occurs. Then, the set point of the thermostat will automatically change to the temperature indicated by the red lever. It is common to use four pins, set up to provide a schedule that uses the higher set point starting at 6 a.m., or shortly before everyone gets up in the morning. Another is set at 7:30 a.m. when everyone has left for work or school, the lower set point. Then, at 4 p.m. when people start returning

Figure 13–12 Setback thermostat.

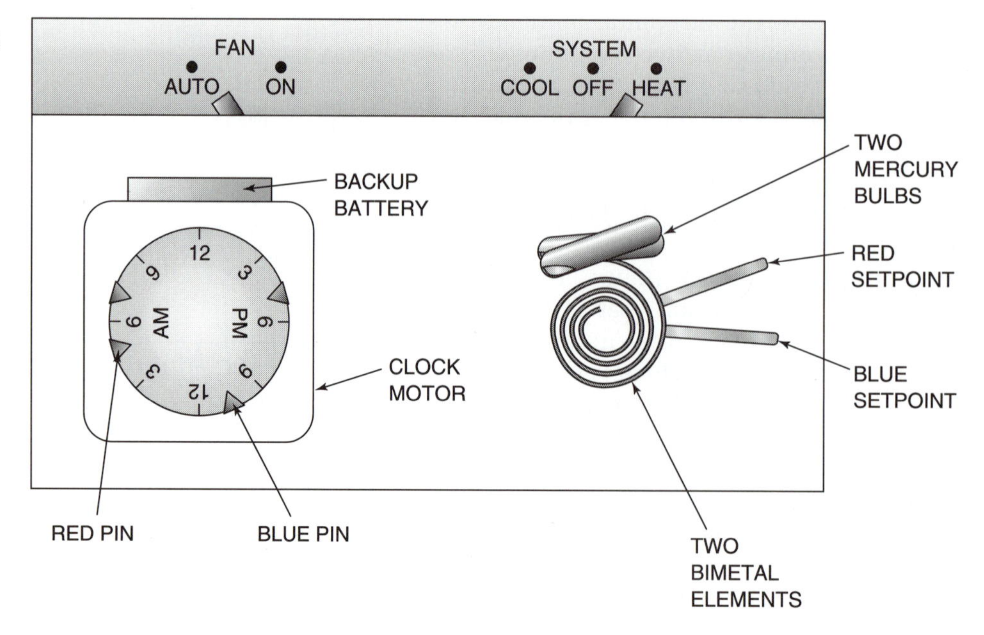

home, the higher set point is used, until 11 p.m. when everybody has gone to sleep. Of course, in an office or commercial application, you would only need two programs, not four. Only two pins would be used, and the extra pins would be stored in a holder on the thermostat.

The setback thermostat must have a clock, and the clock is a load that must receive power. Several methods are employed to provide this power to the clock:

1. Older setback thermostats used an 18V clock motor. There were two clock terminals on the subbase, and two extra thermostat wires went back to the furnace where they were connected to a 120V to 18V transformer that was provided for the sole purpose of providing power to the clock. This was separate from the 120V to 24V transformer that was used as control voltage.

2. Newer thermostats use a clock that operates on 24V. The power source for the clock is the same transformer that provides 24V control voltage. Some of these thermostats have a terminal labelled C on the subbase (for Common). This terminal is connected to the other side of the transformer (the one opposite from the side connected to the red terminal on the thermostat).

It is important that you never attach a jumper wire to this Common terminal. It can cause a short circuit that can ruin the thermostat in a "flash."

Some "smarter" thermostats with a 24V clock circuit do not need a Common terminal. They recognize that there is always at least one open switch in the thermostat, and there will be 24V available across that switch. The thermostat can use that 24V potential to power the clock.

MICROELECTRONIC THERMOSTAT

The mechanical setback thermostat described above has a number of shortcomings:

1. While there can be four different programs, there can be only two different set points. For example, if you wanted a minimum of 66°F when you were sleeping, 62°F when nobody was home during the day, you could not do it with the mechanical setback thermostat.

2. The mechanical setback thermostat must have the pins reset for different times when you change from heating to cooling. In heating, you want the warmer set point to be used during the occupied periods of the day, but for cooling, you want the warmer set point to be used only when there is nobody home.

3. There is no provision to have a different program for the weekends when the occupancy pattern is different from weekdays.

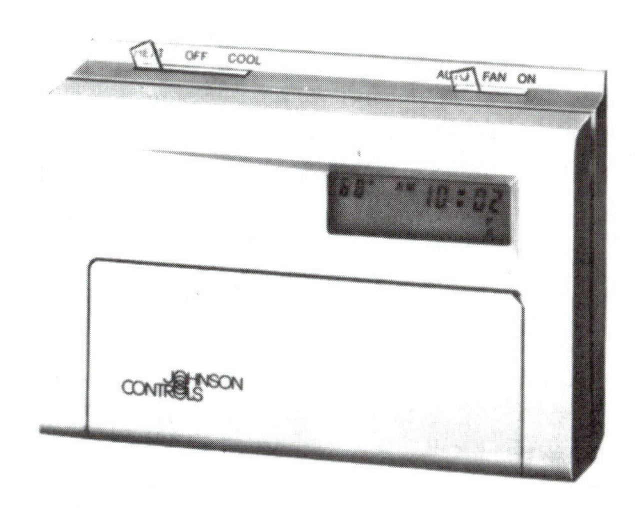

Figure 13–13 Microelectronic thermostat. *(Courtesy of Johnson Controls, Inc.)*

The **microelectronic** thermostat in figure 13–13 addresses all of these problems, and more. It has the same terminal identifications on the subbase, but the temperature sensing and switching is all based on microprocessors similar to those found in a computer. Many microelectronic thermostats have the following features:

1. A different set point can be selected for each of the four daily programs.

2. A different program (times and/or temperatures) can be used for each day of the week.

3. Separate programs are used for heating mode and cooling mode.

ADVANCED CONCEPTS

Using all of the above options, you can see that there are potentially 56 different set points that might be programmed into the microelectronic thermostat (4 programs per day times 7 days per week times 2 modes of operation = 56).

The microelectronic thermostats have a battery that is used to store the program in case there is a power outage. The method of programming each thermostat varies, depending upon the manufacturer.

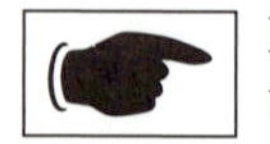

Before attempting to install one of these thermostats for a customer, you should review the instructions and practice programming until you can do it quickly and easily. Then, when the customer tells you what his or her daily routine is like, you can program it quickly and easily, as well as teach the customer how the programming is done.

TWO-STAGE THERMOSTATS

Figure 13–14 shows a thermostat that has two mercury switch elements attached to a single bimetal element. They are mounted slightly out of parallel, so that as the bimetal tilts, it will close one mercury switch, and then, if the bimetal continues to tilt more in the same direction, it will close the other mercury switch. In this way, we can provide two stages of heating, or two stages of cooling, or both. In a system with two stages of heat, instead of there being a W terminal on the subbase, there will be a W1 and a W2. If there are two stages of cooling, instead of a Y terminal, there will be a Y1 and Y2.

Here's how a two-stage cooling thermostat might work. A rooftop air conditioner has two completely separate refrigeration circuits, but with a single evaporator fan and a single condenser fan (figure 13–15). The thermostat set point is 73°F. When the room temperature rises to 74°F, the thermostat makes R-Y1, bringing on the first compressor, and R-G, bringing on the blower motor. Most times, this will be sufficient to cool the room back down to 72°F, and the thermostat will become satisfied and open R-Y1, turning off the first stage of cooling. However, on a warm day, the first stage of cooling may not be sufficient to stem the rise of room temperature, and even after the first stage compressor turns on, the room temperature continues to rise. When the room temperature rises to 75°F, the thermostat makes R-Y2 (in addition to R-Y1 which has already been made). Then, the second stage of cooling operates along with the first stage. The second stage of cooling is the second to come on, but when the room temperature drops, it is the first to go off.

Figure 13–14 Two-stage thermostat element.

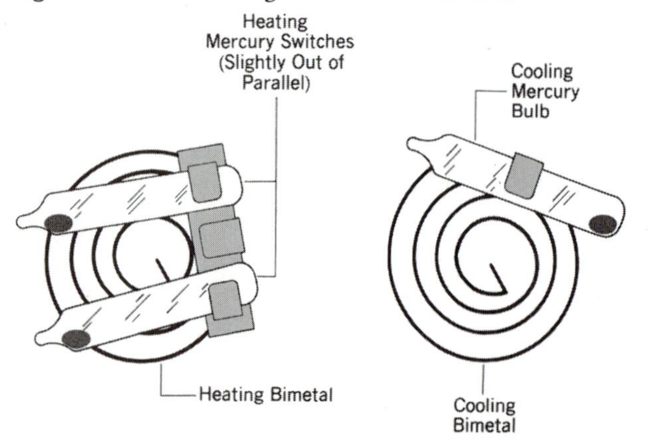

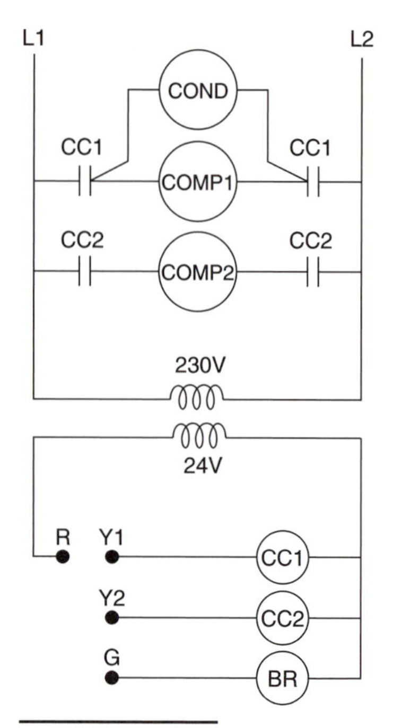

Figure 13–15 Packaged rooftop air conditioner with two stages of cooling

HEAT PUMP THERMOSTATS

With the rising popularity of heat pumps, thermostats have been designed to facilitate the wiring of heat pumps. Heat pump thermostats are different from the standard heating-cooling thermostats in two respects:

1. On heating, the heat pump thermostat makes R-Y and R-G, compared to a standard thermostat that makes R-W, and uses a separate furnace switch to bring on the indoor fan.
2. The B or the O terminal is always used (to operate the reversing valve), whereas on the standard heating-cooling thermostat, it is almost never used, and often is not even supplied.
3. The heat pump thermostat will normally have a second stage of heating available, because auxil-

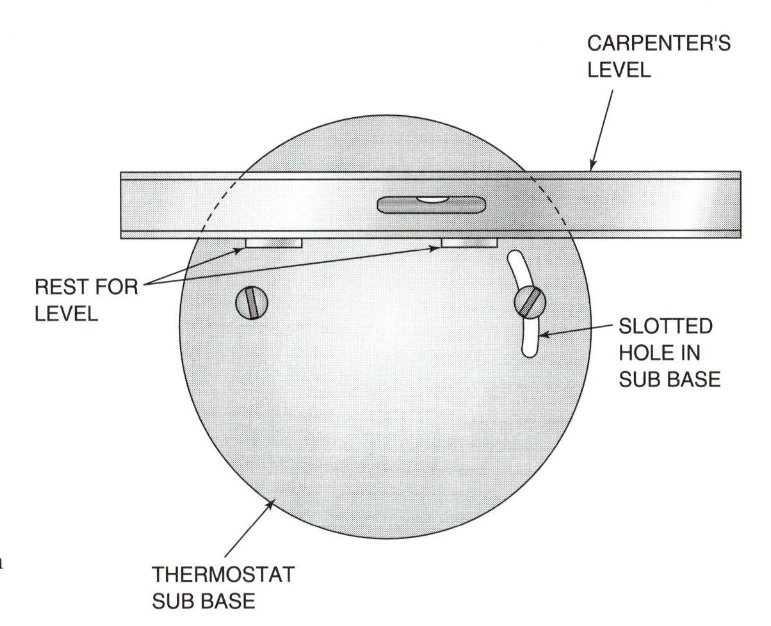

Figure 13–16 Leveling a thermostat.

iary heat will be required in all but the mildest climates. The second stage heat terminal may be labelled W2. There is no W1. The first stage of heating is accomplished by bringing on the compressor.

ELECTRIC HEAT THERMOSTATS

An electric furnace consists of a blower that circulates room air over one or more electric resistance heating elements. Electric heat is different from gas heat in that we cannot tolerate the time-delay for the operation of the furnace fan. The electric heat thermostat makes R-G at the same time that it makes R-W. The R-G switch will bring on the furnace fan at the same time that R-W is bringing on the electric resistance heating elements. On cooling, R-G is made at the same time as R-Y, just as with a standard thermostat.

MILLIVOLT THERMOSTATS

Millivolt systems use a pilot generator that provides an operating voltage of between 185 and 600 millivolts (1000 mV = 1V). Because of these very low voltages, the millivolt thermostat is designed to provide less resistance than a standard 24V thermostat. Even the slightest resistance on contact points or resistance of a heat anticipator can render a millivolt system inoperable. When replacing a millivolt thermostat, make sure that the new thermostat is rated for millivolt service.

CALIBRATING THERMOSTATS

There can be two calibrations that you might need to do on a room thermostat:

1. When the thermostat lever is set for 72°F, the thermostat may actually be controlling the room at a different temperature. The customer complaint will be something like, "we need to set the thermostat at 78°F in order to maintain the room at 72°F."

2. The thermometer on the thermostat (not technically part of the thermostat) reads an incorrect temperature.

The first problem may be caused if you have a mercury bulb thermostat that is not mounted in a level position. There is usually one slotted screw hole for the screw holding the subbase to the wall. This allows you to move the thermostat and hold it in a level position (using a carpenter's level) while you tighten the screw (figure 13–16). If the thermostat is level, and it is still controlling around a different temperature than the set point lever, the bimetal element must be rotated. Sometimes a special thermostat tool is required.

If you don't have the tool necessary to rotate the bimetal element on its mounting, you may be able to compensate for an out-of-calibration thermostat by actually mounting it out-of-level to compensate. If the calibration is not too out-of-line, this may be acceptable, especially if the thermostat is round.

CHAPTER FOURTEEN

Electronic Ignition Gas-Fired Furnaces

In the 1980s, energy shortages around the country moved manufacturers to start designing furnaces that did not use a standing pilot flame. The theory was that keeping a pilot flame lit when heat was not required was wasting a lot of energy. Several methods were developed to accomplish ignition using a spark instead of a pilot flame. Generally, these methods can be classified as either **intermittent pilot** or **direct spark** systems. A description of the operation of each follows.

INTERMITTENT PILOT SYSTEMS

Intermittent pilot systems use two valves in one body (figure 14–1). The first valve, when opened, will allow pilot gas to flow through the pilot tubing to a pilot burner. It is called the **pilot valve** and is usually abbreviated in wiring diagrams as PV. The second valve, when opened at the same time that the pilot valve is already open, will allow gas to flow into the main burner where it can be ignited if there is a flame at the pilot burner. This second valve is the **main valve**, and is usually abbreviated MV in wiring diagrams.

The general sequence of operations for these systems follows:

1. The room thermostat calls for heat, closing its contacts.

2. The pilot valve is opened, supplying gas to the pilot burner. At the same time, an electronic device provides a spark between a wire and the pilot burner hood. This will cause the pilot gas to ignite.

3. The pilot flame will be sensed by one of several different methods. If the pilot flame has, in fact, lit, then the main valve will be energized, allowing the main flame to come on.

4. If the pilot flame were not lit, or if it were not sensed, some systems will allow the pilot valve to remain energized and the spark to keep sparking. These are called non-100% shut-off systems. Other systems are 100% shut-off. They will only allow the trial for ignition to continue for a fixed time (maybe 30 seconds). After that, they will go into **lock-out.** That means that the pilot valve will be de-energized, the spark will stop, and the system will not try to relight again until it is reset. Most systems can be reset from a lock-out condition by simply turning the room thermostat down so that its contacts open. When the thermostat is then returned to a higher setting, the light-off sequence will be initiated again.

DIRECT SPARK SYSTEMS

Direct spark systems differ from intermittent pilot systems in the following ways:

1. There is no pilot burner.

2. The spark is made at the main burner. The spark lights the main flame directly, without using the intermediate pilot flame.

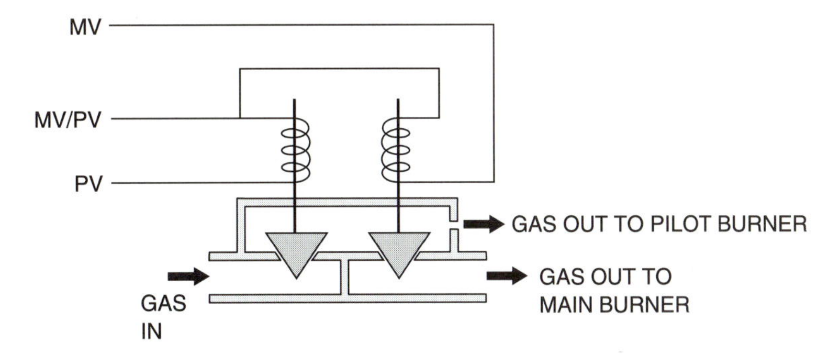

Figure 14–1 Gas valve for an intermittent pilot furnace.

3. Instead of sensing pilot flame, the system senses main flame.

4. All direct spark systems must be 100% shut-off. While a steady flow of unburned pilot gas in an intermittent pilot system would simply vent itself through the flue stack, a much larger steady flow of unburned main gas is too unsafe to be tolerated. A direct spark system will go into lock-out if the main flame is not established within a very short time (maybe 10 seconds).

INTERMITTENT PILOT FURNACE WIRING

Figure 14–2 shows a typical furnace manufacturer's wiring diagram for an intermittent pilot system. Some of the features of this wiring scheme are:

1. There is an ignition control (IC) that performs the functions described above.

2. A **vent fan** assists with the removal of flue gas. In modern furnaces, the heat exchangers are more efficient, but have higher pressure drop. Sometimes gravity venting of the products of combustion is not sufficient by itself. The vent fan is operated through a relay (RVF) that is energized on a call for heat. When the centrifugal switch on the vent fan motor closes, then the ignition sequence is allowed to proceed.

3. There is a **roll-out switch** (identified in the diagram as S_{FR}). This is common on furnaces that use a vent fan for the products of combustion. If the vent fan does not properly remove the products of combustion, the flame may roll out of the heat exchanger and endanger the wiring on the front panel of the furnace. The roll-out switch senses if the flame reaches outside the heat exchanger where it doesn't belong. It the roll-out switch opens, the ignition sequence is interrupted.

EXERCISE:
When you approach the above furnace, you hear the vent fan running, but there is no spark, no flame, and no indoor fan. Which of the following voltage readings would make sense to take as your first reading? For each, state why it would or would not make sense:

a. Voltage across the vent fan relay switch.
b. Voltage across the vent fan relay coil.
c. Voltage across TH-TR on the ignition control.
d. Voltage across the vent fan centrifugal switch.

SOLUTION:

a. It would not make sense to measure across the vent fan relay switch. You already know that it is closed, because the vent fan is running.

b. It would not make sense to measure across the vent fan relay coil. Its job is to close the vent fan relay switch, and the vent fan relay switch is already closed.

c. A measurement at TH-TR makes sense. A 24V reading would steer you to a combustion control problem (maybe a failed ignition controller or sparker wire), while a 0V reading would indicate that maybe one of the three switches between the room thermostat and TH is open.

d. A measurement across the vent fan centrifugal switch makes sense, as would any other measurement across a switch between the room thermostat and the TH terminal. ■

DIRECT-SPARK FURNACE WIRING

Figure 14–3 shows the wiring for two different Honeywell direct-spark modules. The difference between them is that the S87 module lights the main flame using an **ignitor-sensor** that creates a spark to the pilot burner hood (figure 14–4). The S89 module lights the main flame using a **hot surface ignitor** (figure 14–5). The Alarm terminal is only used on some models. It provides power to an alarm to notify occupants if the module has gone into lock-out. The vent-damper plug is also available only on some

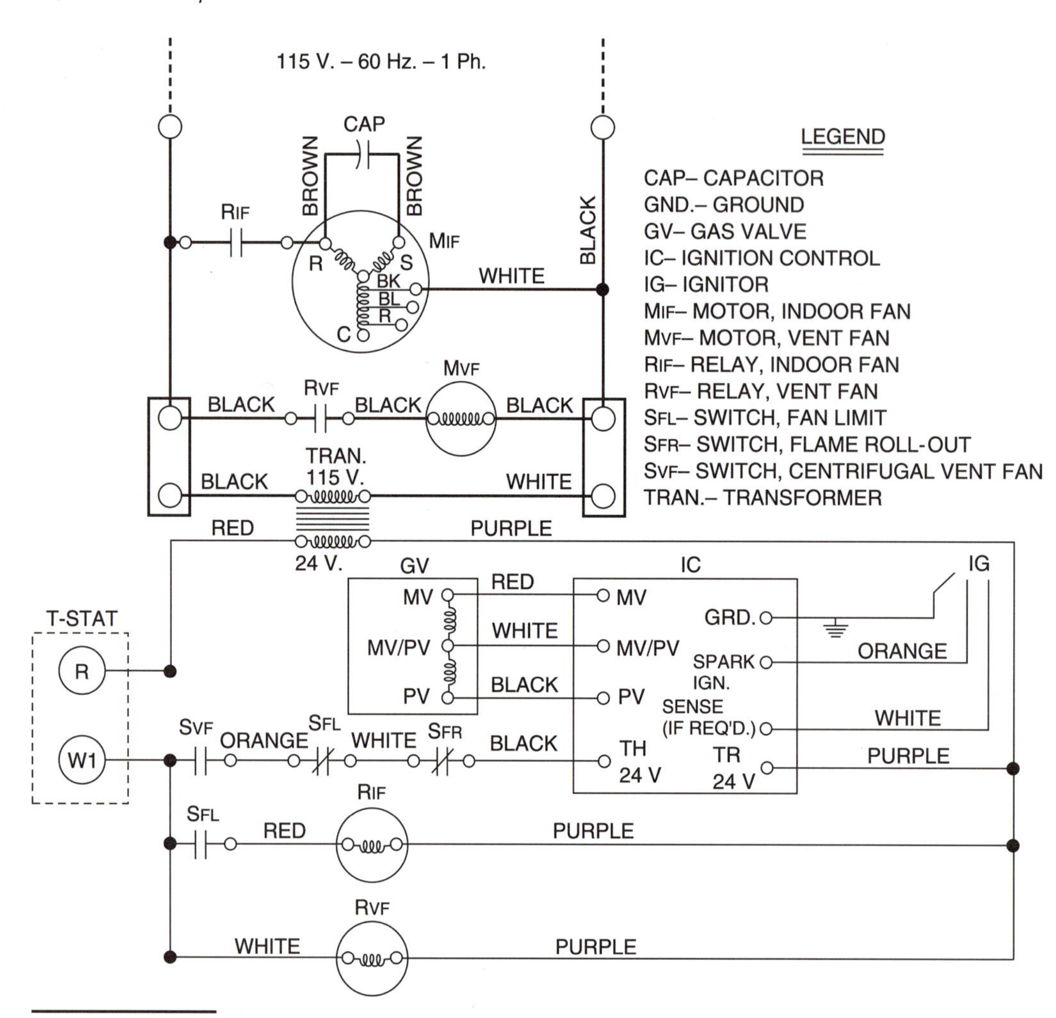

Figure 14–2 Intermittent pilot system.

models. It provides an output signal to a 24V motor that closes a damper in the flue stack to conserve heat energy during the off-cycle.

SENSING METHODS

Standing pilot systems usually use a thermocouple to prove the existence of a pilot flame before main gas is allowed to be supplied to the burners. Intermittent pilot systems, depending on the manufacturer of the ignition system, use either a **bimetal pilot safety switch** (figure 14–6), a mercury-filled bulb that operates a switch (figure 14–7), or an electronic means of sensing such as **flame conduction** or **rectification** (figure 14–8). The bimetal and mercury bulb systems are easy to understand. When the pilot flame is established, expansion of the bimetal or expansion of the mercury in a sensing bulb causes an SPDT switch to operate. For the bimetal switch, the common terminal is yellow, the normally closed terminal (when the bimetal is cool) is green, and the normally open terminal is white.

However, to understand the principles of flame conduction and rectification, you must understand what happens in a gas flame. The flame is actually a series of small explosions that cause the immediate atmosphere to become ionized. This ionization causes the atmosphere to become conductive. The flame can be thought of as a switch. This "switch" is located between the pilot burner tip and the flame sensor. When there is no flame between the pilot tip and the flame sensor, the switch is open. When a flame is in contact with the pilot tip and the flame sensor, the switch is closed.

JOHNSON CONTROLS ELECTRONIC IGNITION

The G-60 (figure 14–9) is the first generation of electronic ignition controls manufactured by Johnson Controls. It is available in 24V and 115V models, 100% or non-100% shutoff, and with an optional vent damper plug. The G65 and G66 are

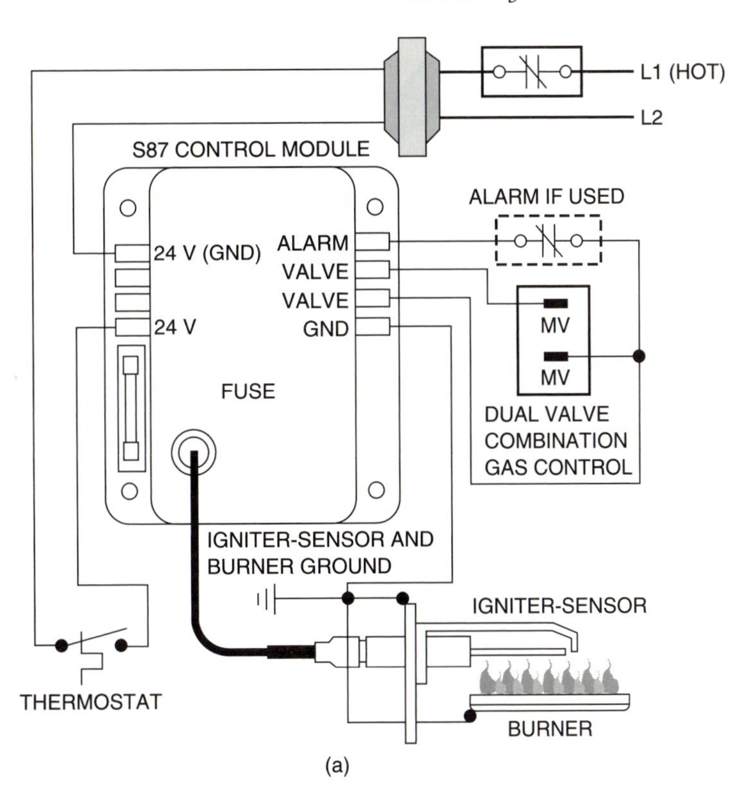

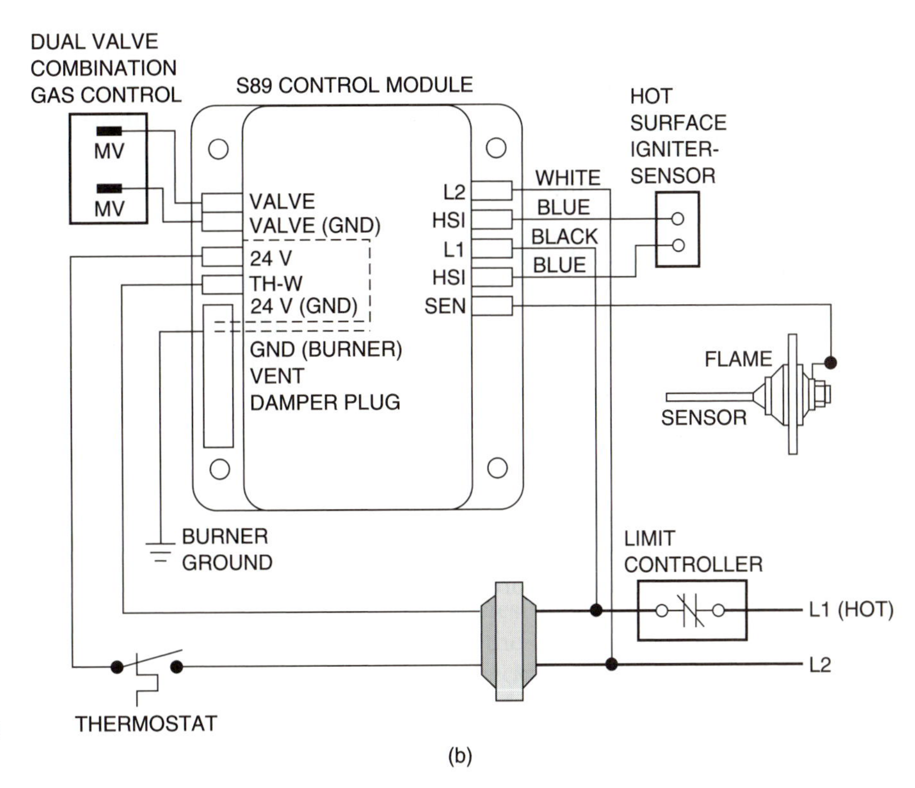

Figure 14–3 Honeywell direct spark systems.

newer models. The sequence of operation is similar for all of these controls:

1. When the temperature control calls for heat, the ignition control's spark transformer and pilot valve are energized.

2. The spark lights the pilot.

3. The flame sensor proves the presence of the flame. The ignition control then shuts off the spark. At the same time, the main burner valve is opened. Some models permit the spark to continue for a

HONEYWELL AND ROBERTSHAW CONTROL BOXES

The Honeywell and Robertshaw boxes (figure 14–10) are similar in operation to the Johnson Controls G60 box. The differences are as follows:

1. On some models, there is no sensor wire. The same wire is used to send the spark and to sense the pilot flame. This eliminates one step of troubleshooting, because if the pilot flame has lit, you don't need to check the sensor wire for continuity.

2. Instead of the valves being wired to the ground connection, a terminal called MV/PV is used. The pilot valve is wired between the PV and PV/MV terminals, and the main valve is wired between the MV and PV/MV terminals.

Except for the minor differences noted, the troubleshooting of the Honeywell and Robertshaw boxes is the same as for the G60 box.

PICK-HOLD SYSTEM

Furnaces manufactured by Carrier, Bryant, Day/Night, and Payne, have their own method of accomplishing the intermittent pilot sequence. The pick-hold gas valve is unique. It has five terminals. It is not interchangeable with any other type of furnace control system. It consists of two solenoid coils that operate the pilot valve, and a small heater (called a **heat motor**) that opens the main valve (after a time delay of approximately 10 seconds after it is energized). The PICK coil is designed so that when it is energized, it picks up the pilot valve off its seat. The HOLD coil, when energized, is not strong enough to open the pilot valve by itself, but if the pilot valve has already been opened (by the PICK coil), the HOLD coil is strong enough to hold the pilot valve open. The PICK coil performs the same function as your thumb when you manually open the pilot valve in a standing pilot system. The HOLD coil performs the same function as the thermocouple in the standing pilot system.

The Pick-Hold system uses a sparker as shown in figure 14–11. When 24V are applied to the terminals, the box will put out a 10,000V spark between the sparker wire and the pilot burner hood. If the pilot does not light, the sparker will keep going for as long as it is supplied with 24V. Therefore, this is a non-100% shut-off system. If the pilot flame is established, it will be sensed through the sparker wire, and the sparking will stop.

A simplified wiring diagram for the combustion portion of the Pick-Hold system is shown in figure 14–12. When the room thermostat calls for heat, it makes the connection between the R and W terminals. The HOLD coil is energized directly from the thermostat, and the PICK coil and the sparker are energized through the normally closed contacts of the bimetal pilot safety switch. Pilot gas and spark should both be present. When the pilot flame is established, the sparker turns itself off, and the bimetal element in the pilot safety switch heats up. First, the normally closed contact opens. Some time later (maybe 15–30 seconds), the normally open contact will close, energizing the main gas valve, and the main flame will ignite. The same signal that energizes the main valve also sends a signal to the fan circuit which will be discussed later.

Figure 14–10 Honeywell and Robertshaw electronic ignition control boxes.

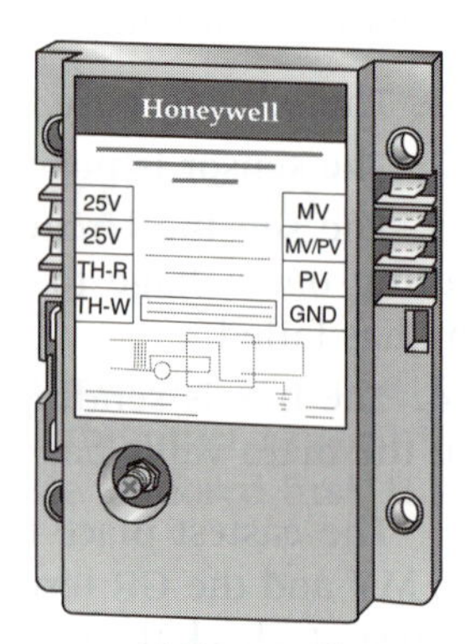

(a) Honeywell

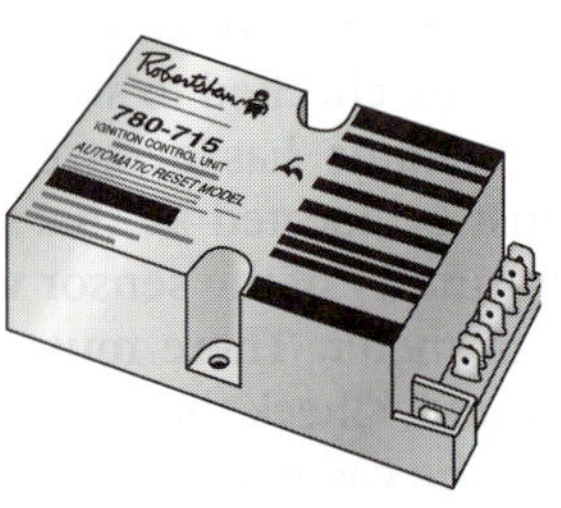

(b) Robertshaw

Figure 14–11 Pick-hold sparker box.

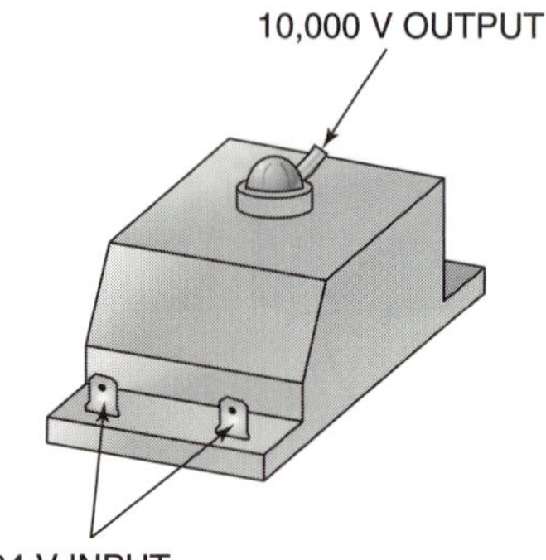

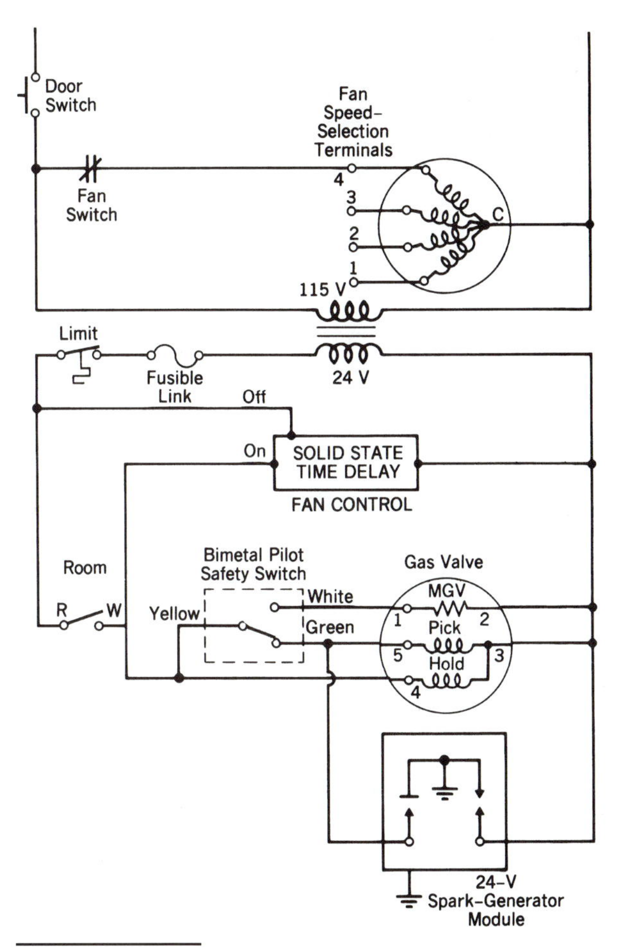

Figure 14–12 Simplified pick-hold wiring.

Troubleshooting: Spark Present, But No Pilot Flame

We know that the room thermostat has closed, and the normally closed contact of the bimetal pilot safety switch is closed. Potential causes are:

1. No gas available to the furnace.
2. Plugged pilot gas orifice.
3. Spark positioned badly.

Use a match to try to light the pilot. If the pilot gas lights, then change the position of the sparker. If the pilot doesn't light, confirm the availability of gas to the furnace by loosening the supply gas piping. If gas is available, check the availability of pilot gas by loosening the pilot gas tubing. If no gas is available, the pilot valve has failed. If gas is available, the pilot orifice must be plugged.

Troubleshooting: No Spark, No Flame

Check across the R and W terminals to make sure that the room thermostat is closed. Then check between W and 5 on the gas valve to see if the normally closed contact in the bimetal pilot safety switch is closed. If you get 24V across the switch (indicating it's open), it must be replaced. Sometimes a sharp rap on the switch with a screwdriver will restore it to operation, but it should still be replaced.

Troubleshooting: Pilot Flame Cycles On and Off

The HOLD coil has failed. After the pilot flame is established, the normally closed contact opens. As soon as it does, the pick coil is de-energized, and the pilot valve closes. The bimetal pilot safety switch closes, and the pilot lights again. The cycle continues. The gas valve must be replaced.

QUESTION:
Why does the Pick-Hold system use a hold coil? Wouldn't it be simpler to have the Pick coil energized directly from the room thermostat?

SOLUTION: If the Pick coil were energized directly from the thermostat, it would work fine, except for a safety problem. Consider a situation where the furnace is operating, and there is a momentary power outage. The main valve and pilot valve would both close, and the flame would go out. When the power is restored, the Pick coil will be energized, opening the pilot valve. And because the normally closed contacts of the bimetal pilot safety would still be warm, the Main coil will also be energized. We would therefore have main gas being admitted to the furnace before the existence of a pilot flame has been proven.

In figure 14–12, when power is restored, the pick coil would not be energized until after the bimetal pilot safety switch had a chance to cool. Even if the main valve opens, no main gas will flow so long as the pilot valve is closed. ■

Figure 14–13 shows the completed wiring diagram for the pick-hold system. The operation of the fan deserves further explanation. The control of the fan is accomplished electronically by a chip on a circuit board. There is no bonnet-sensing fan switch or mechanical time-delay switch. The time-delay control gets a constant 24V signal from the transformer. This energizes relay 2A coil, opening the 2A contacts, and turning the fan off! Yes, the blower relay switches are normally closed, and a 24V signal is required to keep the fan off. This results in a unique problem with the pick-hold furnace. If either the transformer has failed, the fusible link or the limit switch is open, or if the blower relay coil has failed, the fan will run continuously!

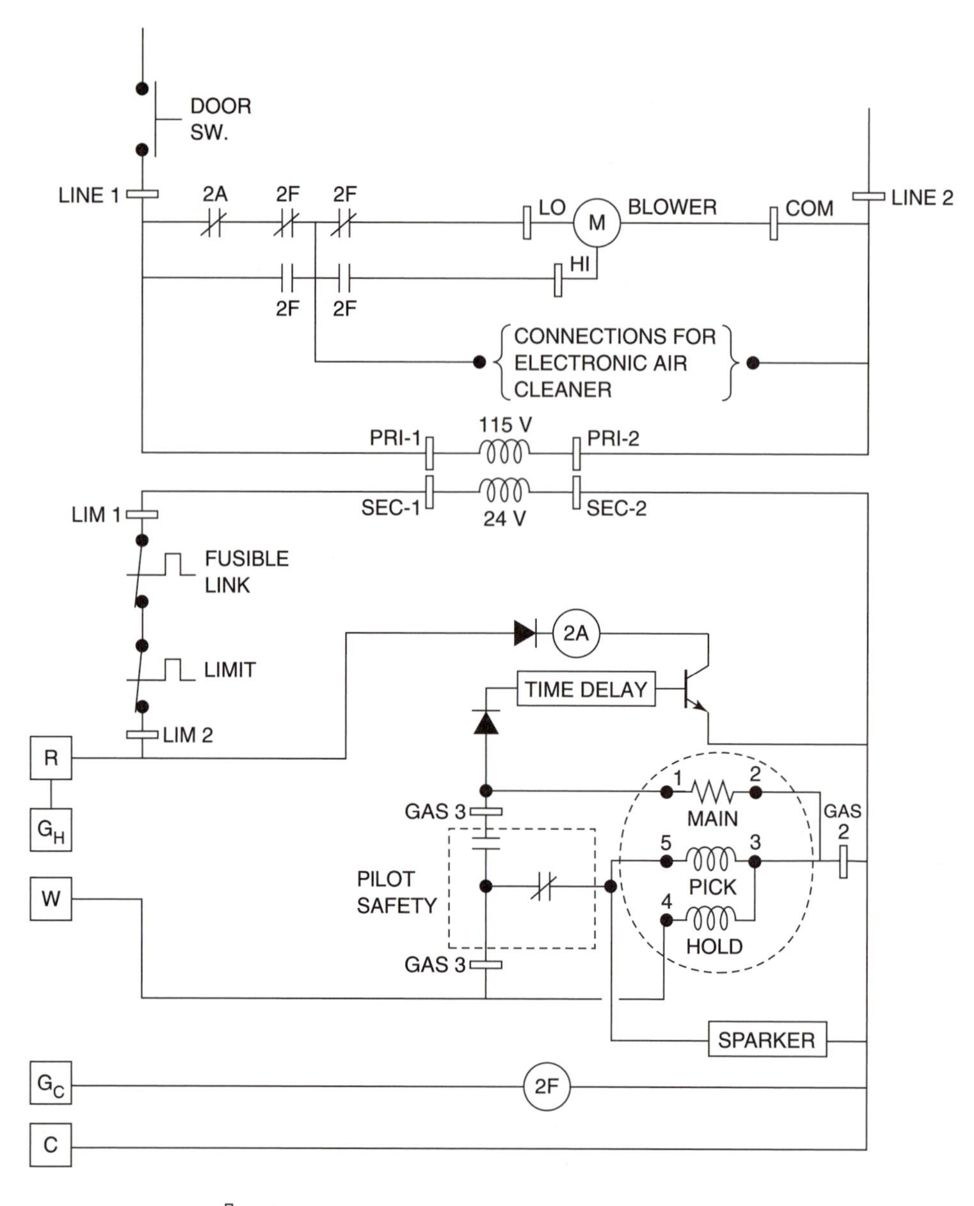

Figure 14–13 Pick-hold furnace.

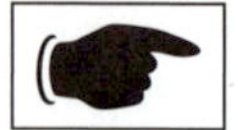

A fusible link is shown in figure 14–14. It is used in various applications where it is desired to open a circuit if a predetermined limit temperature is reached. In the pick-hold furnace, it is located on the front of the heat exchanger, and is clearly visible when the front panel of the furnace is removed. It will only reach its opening temperature if the flame, for some reason, escapes from the confines of the inside of the heat exchanger (for example, if the flue stack is blocked). In this application, it is sometimes called a **roll-out switch.** The fusible link is also commonly found on electric heating systems where it operates as a high-temperature limit switch.

For normal blower operation, the 2A relay coil is energized immediately when the furnace is plugged in. On a call for heat, when the bimetal pilot safety operates, it energizes the MGV, and at the same time, provides 24V to the timer circuit, which some seconds later, emerges as an "off" signal which defeats the constant 24V to the 2A coil. This de-energizes the 2A coil, and the normally closed switch contacts then bring on the blower motor.

ADVANCED CONCEPTS

The time delay circuit includes two symbols that have not been previously explained, a **diode** and a **transistor** (figure 14–15). The service technician will never be called upon to troubleshoot down to this level.

A diode is the electronic equivalent of a mechanical check valve. It will only allow electrons to flow in one direction. In order to test a diode you can use an ohm meter. It should show continuity across the diode in one direction, but if you reverse the meter lead locations and measure in the opposite direction, it should indicate no continuity.

A transistor is the electronic equivalent of an electric valve, which opens when an external voltage is applied. For the transistor shown, current will not flow from the collector lead to the emitter lead (like a closed valve) until there is first a small current that flows from the base lead to the emitter lead. The current from the base to the emitter can be quite small compared to the current that can flow from the collector lead to the emitter lead.

Diodes are also sometimes arranged in a circuit such as shown in figure 14–16, a **rectifier.** Even though the current in the supply voltage is continuously changing direction, the flow of electricity between terminal D and the load will always be in one direction only, to the right.

When the furnace is used for cooling as well as heating, there is an extra relay (2F). When the room thermostat makes R-G, it energizes the 2F coil, causing the blower motor to run on High speed.

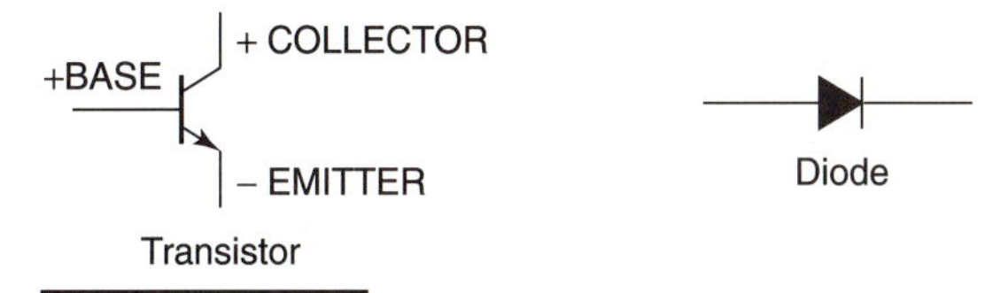

Figure 14–15 Diode and transistor symbols.

Pick-Hold with an Inducer Fan

Figure 14–17 shows the same furnace, but with the addition of an inducer fan to help pull the products of combustion through the heat exchanger. When the inducer fan is not running, the pressure switch (7V) is in the "up" position. The pressure switch senses operation of the inducer fan, and will move to the "down" position whenever the inducer fan runs.

On a call for heating, voltage is supplied through the W terminal to the 2D inducer fan motor relay. Its contacts close, energizing the inducer fan motor 3A. Another set of 2D contacts also close in the 24V circuit, "sealing in" the normally closed contacts of the pressure switch. Even though the pressure switch will open as soon as it senses the operation of the inducer fan motor, the 2D coil will remain energized.

At the same time that the 2D coil is energized, the Pick coil is also energized through the bimetal pilot safety switch 6H. Its operation is also "sealed

Figure 14–14 Fusible link.

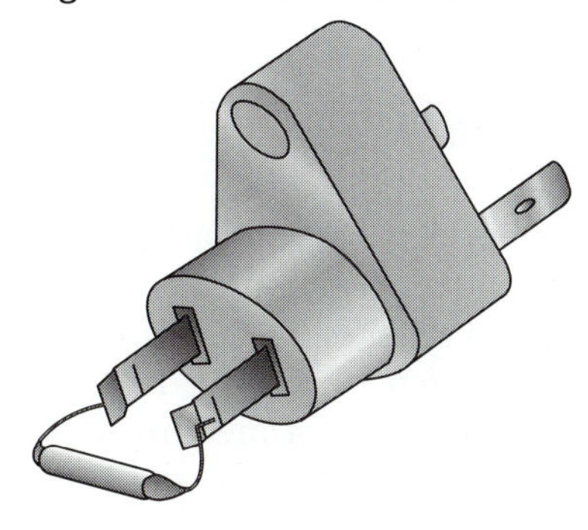

Figure 14–16 Rectifier.

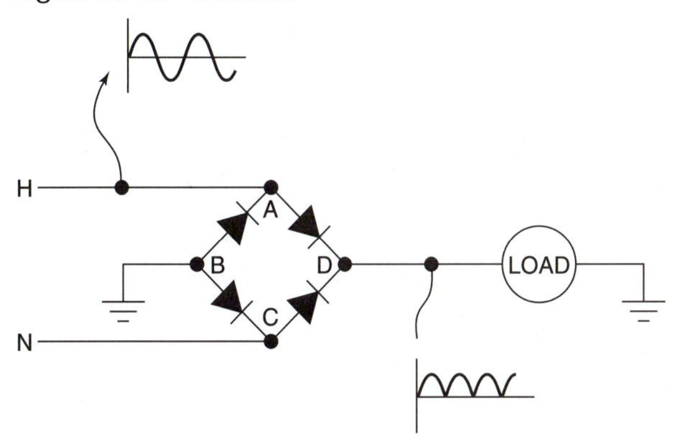

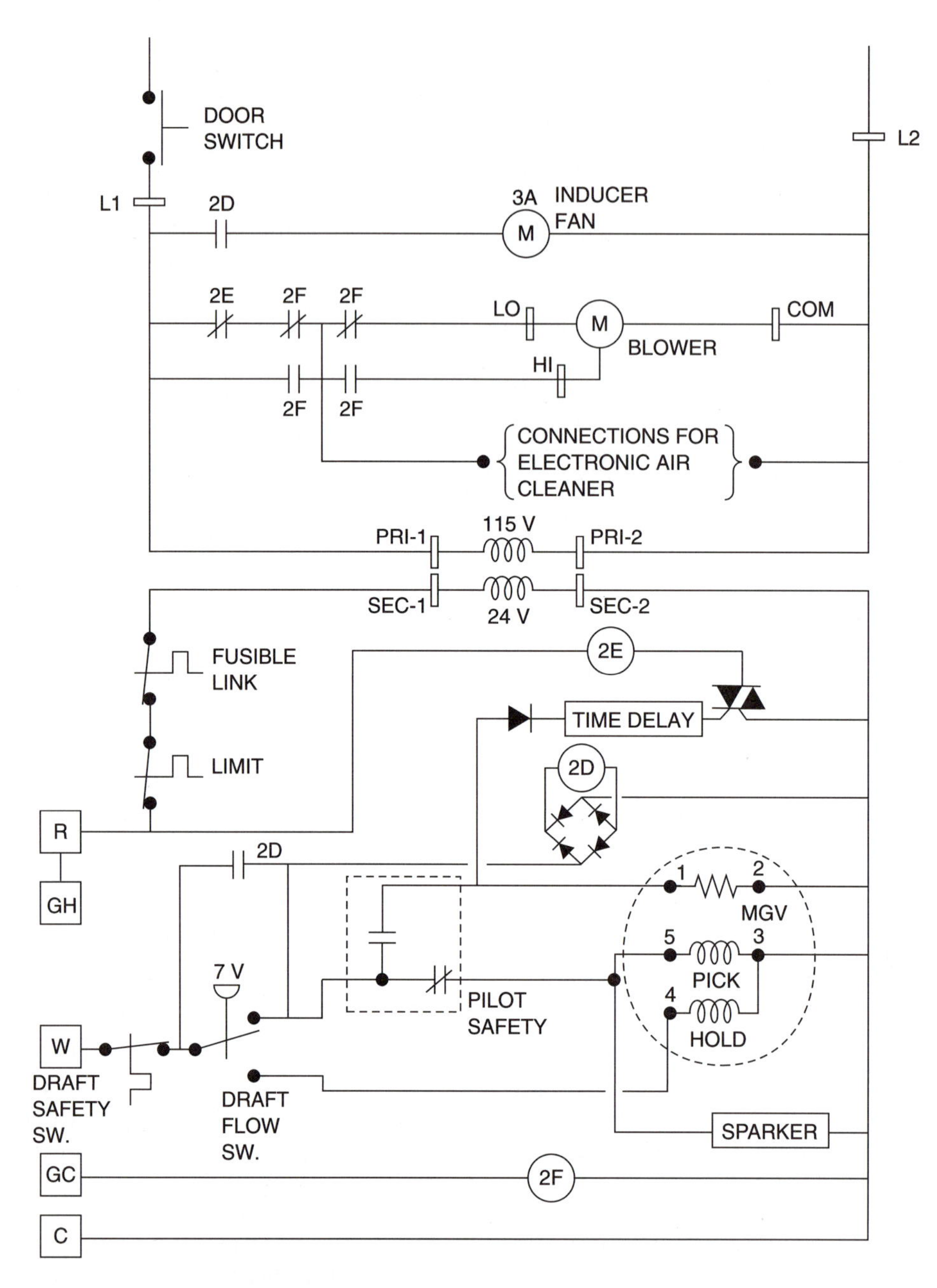

Figure 14–17 Pick-hold furnace with inducer fan.

in" around the 7V pressure switch, through the operation of the 2D contacts in the 24V circuit.

> **QUESTION:**
> The inducer fan is running, but the pilot flame cycles on and off every few seconds. One of two components has likely failed. Which ones?

MERCURY BULB SYSTEM

The White-Rodgers system (figure 14–18) involves the use of a special gas valve. It has a main valve and a pilot valve (similar to other valves), but it also has an internal pressure switch (figure 14–19), and a place to plug in the terminals of a mercury-bulb pilot safety switch. In figure 14–20, the mercury-

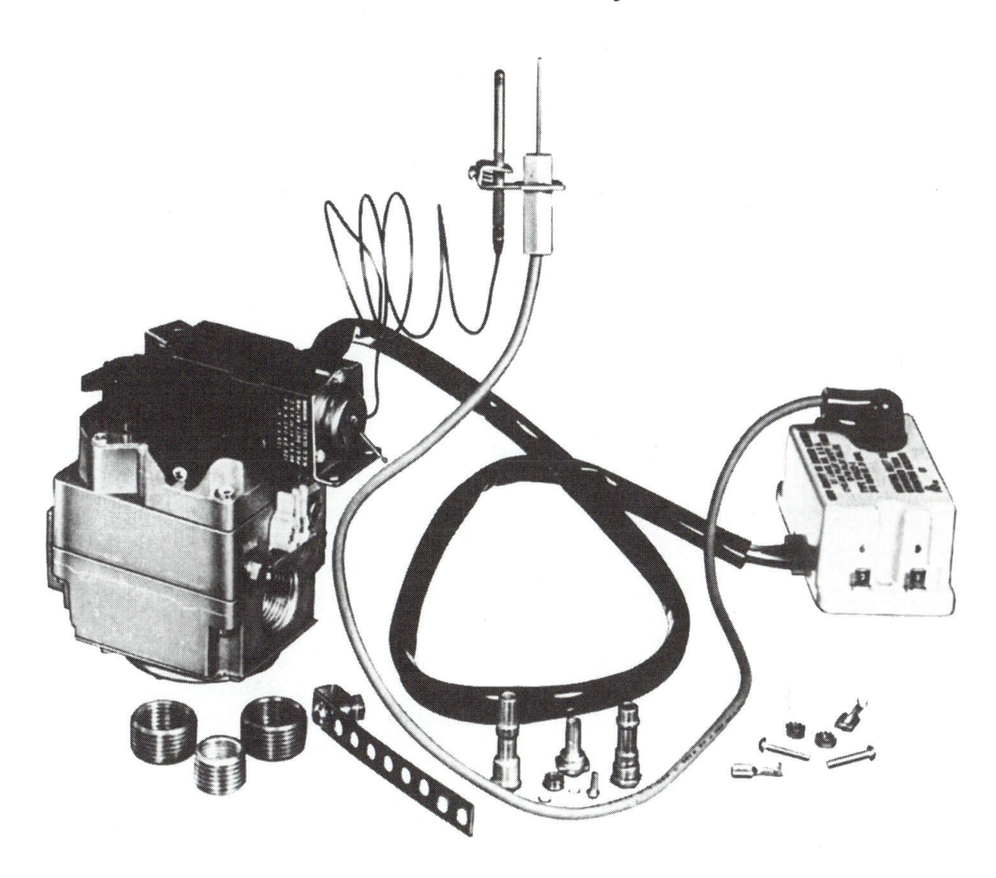

Figure 14–18 White-Rodgers mercury-bulb system. *(Courtesy of White-Rodgers Division, Emerson Electric Company)*

bulb pilot safety switch is shown as an SPDT switch in a dotted box separate from the gas valve because it is a remote switch. The pressure switch is located in the chamber of the gas valve between the gas inlet and the gas outlet. It is a normally open switch, and it closes when the pilot valve opens.

When the mercury bulb is cool, the switch makes between terminals 4–3. When the room thermostat calls for heat, a circuit is completed from the transformer, through the room thermostat, the pilot safety switch and the pilot valve. A parallel circuit is completed through the sparker. The sparker ignites the pilot gas, and the pilot flame heats the pilot safety valve. The switch between 4–3 opens, and some seconds later, the switch between 4–2 closes. This completes the circuit to the main gas valve. The main gas is ignited by the pilot flame.

Figure 14–19 White-Rodgers gas valve with internal pressure switch.

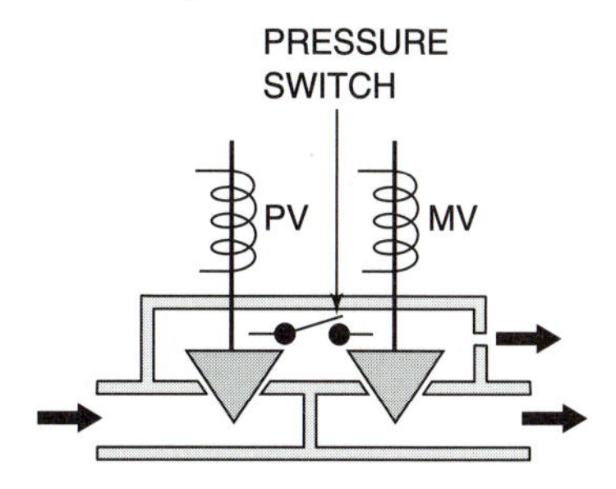

When the switch between 4–3 breaks, the pilot valve remains energized because the internal pressure switch closes as soon as the pilot valve opens. The pressure switch "seals-in" the circuit around terminals 4–3.

QUESTION:
If the pressure switch fails (won't make contact), what symptom will you observe?

SOLUTION: The pilot flame will cycle on and off every few seconds. The main flame will not light. Each time the pilot flame is established, the pilot safety switch will warm and open 4–3. The mercury bulb will cool and reclose 4–3, repeating the cycle.

If the pressure switch fails, the technician might be tempted to bypass it by simply cutting the two wires leading to it, and wire-nutting them together. The technician might think that the pilot safety switch alone is enough protection to make sure that there is a pilot flame proved before opening the main gas. Don't do it! Bypassing the pressure switch will allow the unit to operate normally, except for one situation. Suppose that the unit is in its normal heating cycle, with the Pilot valve and

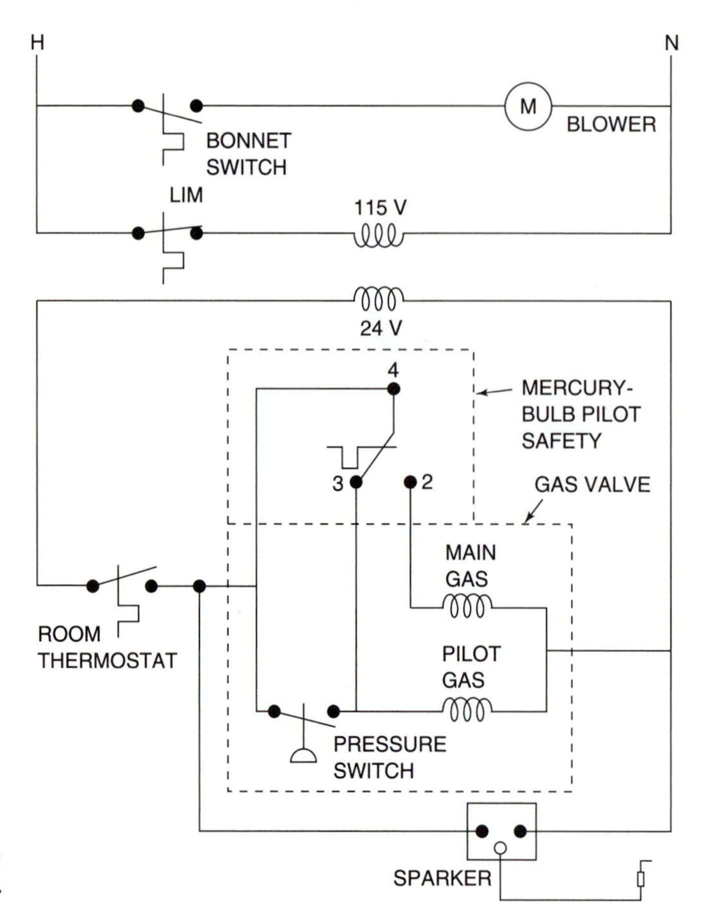

Figure 14–20 Wiring for White-Rodgers gas valve.

the Main valve energized and the mercury bulb hot. Now, suppose there is a momentary power interruption, either because of a utility interruption, or an occupant who moves the room thermostat down and then back up. When the power is restored, the main gas valve, the pilot gas valve, and the sparker are all energized at the same time. It creates the unsafe situation of allowing main gas into the furnace without having first proved the existence of a pilot flame.

EXERCISE:

Even with the pressure switch properly in the circuit, the above situation will allow the main gas valve to be energized when power is restored. Explain how the start-up sequence will be safer with the pressure switch properly in the circuit.

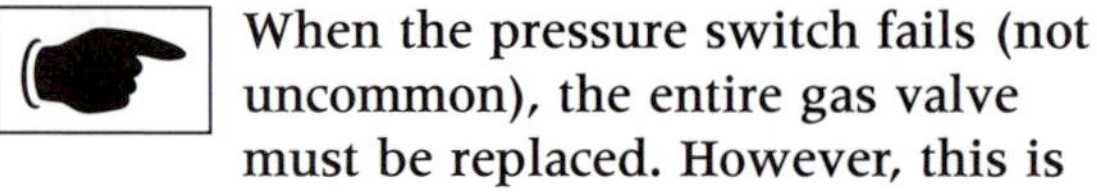

When the pressure switch fails (not uncommon), the entire gas valve must be replaced. However, this is expensive. White-Rodgers has a "fix" avail-

Figure 14–21 Repairing a gas valve with a failed pressure switch.

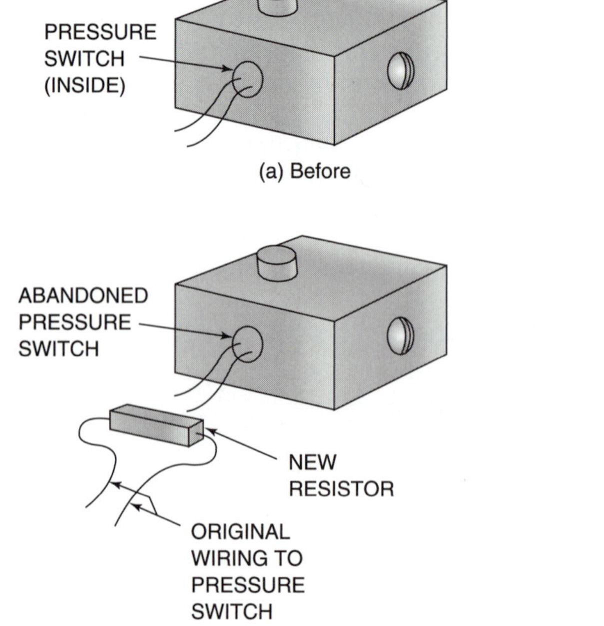

able that is safe and inexpensive. Instead of bypassing the failed pressure switch, a resistor (of a specific resistance) is wired external to the valve in place of the failed pressure switch (figure 14–21). During start-up, with the mercury bulb switch closed from 4–3, the pilot valve receives 24V and the resistor has no effect. However, when 4–3 opens, the added resistance in series with the pilot valve reduces the current through the pilot valve circuit. There is sufficient current to hold the pilot valve open, but if there is a momentary loss of power, the reduced current is insufficient to actually pull the pilot valve open from a closed position.

QUESTIONS:

1. What symptom would you observe if the Pilot valve coil failed open?
2. What symptom would you observe if the Main valve coil failed open?
3. What symptom would you observe if the Sparker failed? How would you confirm this diagnosis?

RETROFIT KIT

Retrofit kits are available that can be used to convert a standing pilot furnace into an electronic ignition furnace. This conversion is usually not justified unless the standing pilot gas valve has already failed. Then, the incremental cost of retrofitting to electronic ignition over the cost of simply replacing the standing pilot valve can sometimes be justified. These retrofit kits may use either a flame rectification sensor or a mercury bulb sensor. They are designed so that the sparker and sensor can be mounted using the hole in the pilot burner assembly that was used to hold the thermocouple (the thermocouple is removed when converting to electronic ignition).

LENNOX PULSE FURNACE

The Lennox system of electronic ignition is unique. It uses a pulse combustion process in which pressure pulses from the ignition process operate flapper valves inside a specifically designed gas valve. It uses a flame rectification sensor similar to other electronic ignition furnaces. A spark plug is used to provide the spark for initial ignition. A separate spark plug is used as an ignition sensor. The two spark plugs are not interchangeable due to different thread diameters. However, the control box for this furnace (figure 14–22(a) and (b), two different brands) has other functions as well.

On a call for heat, prior to starting the combustion sequence, this control box sends a 115V signal to a small blower (**purge blower**) that circulates outside air through the combustion chamber (pre-purge). The same blower is energized after each heating cycle (post-purge). The sequence of operation after a call for heat is as follows (times are approximate and will vary depending on the control module manufacturer):

1. Purge blower is energized.
2. At 30 seconds, the gas valve and ignition spark are energized for 5 seconds.
3. When ignition occurs and is sensed by flame rectification, the spark and purge blower are de-energized.
4. When the room thermostat is satisfied, the gas valve is de-energized and the purge blower is energized.
5. Post purge continues for 30 seconds.

If no ignition is sensed in step 3 above, the purge blower continues to run. After an additional 30 seconds, step 2 above is repeated (another 5-second trial for ignition). This sequence will continue to be repeated until normal ignition occurs. If no ignition is sensed after five trials, the control goes into lock-out and must be reset at the room thermostat.

A graphical representation of a successful and an unsuccessful ignition sequence is shown in figure 14–23.

All times shown may vary slightly, depending on the manufacturer of the particular control box being used.

Figure 14–24 shows the wiring diagram for a Lennox Pulse furnace.

1. Line voltage is supplied to the unit through the door interlock switch. The door interlock switch assures that if the door to the fan compartment is not properly closed, nothing will be energized. Door interlock switches are provided by manufacturers as a safety device and to prevent lawsuits. But they also provide a steady stream of service calls from customers whose furnaces will not work after they have changed the air filters. Many owners fail to replace the access panel cover properly, and the door switch does not close.
2. The transformer provides 24V control voltage. Note that the "common" side of the transformer is grounded at terminal T. A good ground is required for the electronic ignition primary control to operate properly.

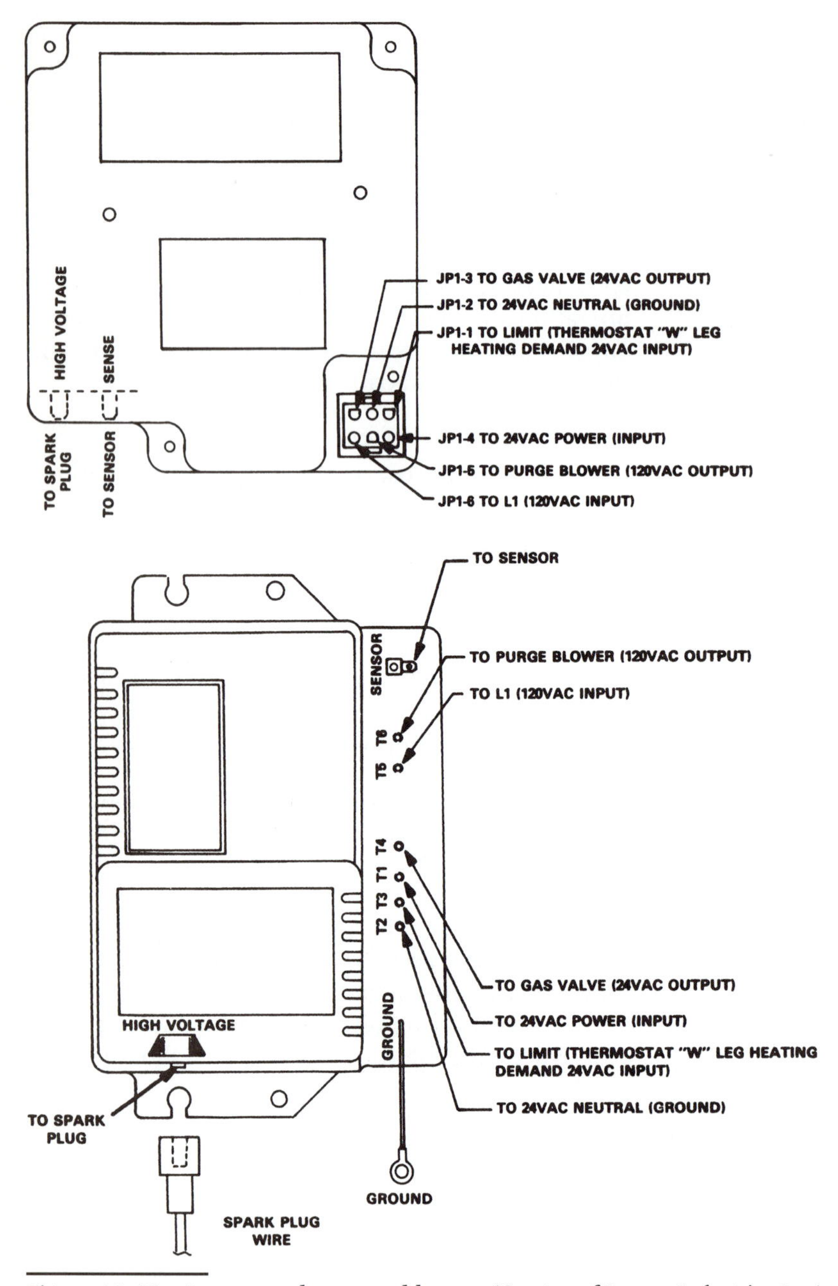

Figure 14–22 Lennox pulse control boxes. *(Courtesy of Lennox Industries, Inc.)*

3. The room thermostat closes, making R to W.

4. The signal from the room thermostat feeds through two pressure switches and a limit switch. The first pressure switch will open if there is excessive pressure in the flue gas exhaust pipe. The second pressure switch will open if there is excessive vacuum in the combustion air intake pipe. These conditions can occur if there is a blockage in either pipe, such as from a bird's nest.

5. The primary control is energized, and the furnace goes through its sequence of control.

Note the symbols that are labelled JP1-1, JP1-2, JP1-3, etc. JP1 refers to a quick-connector, and the -1, -2, -3, etc., tell you which pin on the connector is being used to complete the connection. Also note that the indoor blower relay (terminals 1, 2, and 3) will operate the furnace blower motor on low speed (red lead) for heating, and a higher speed (black lead) for cooling. A separate set of terminals on the indoor blower relay (4, 5, and 6) are used to provide 24V to an accessory such as an electronic air cleaner when the fan runs on either heating or cooling.

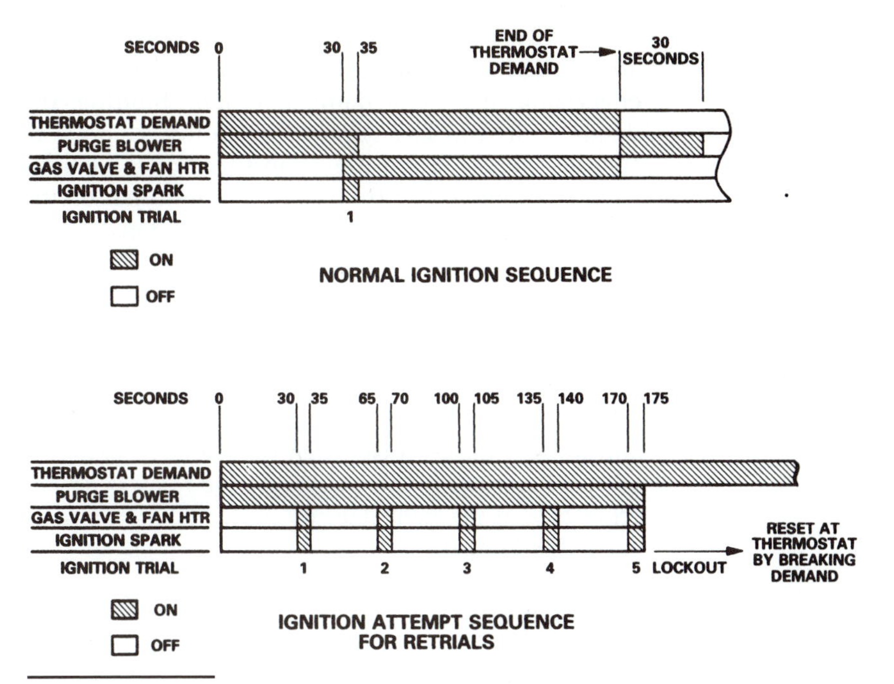

Figure 14–23 Timing of functions—pulse furnace. *(Courtesy of Lennox Industries, Inc.)*

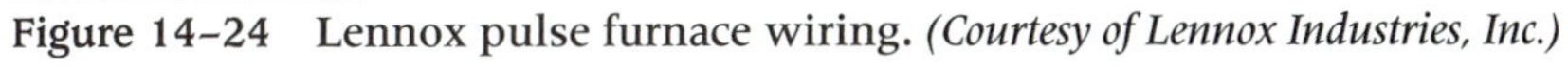

Figure 14–24 Lennox pulse furnace wiring. *(Courtesy of Lennox Industries, Inc.)*

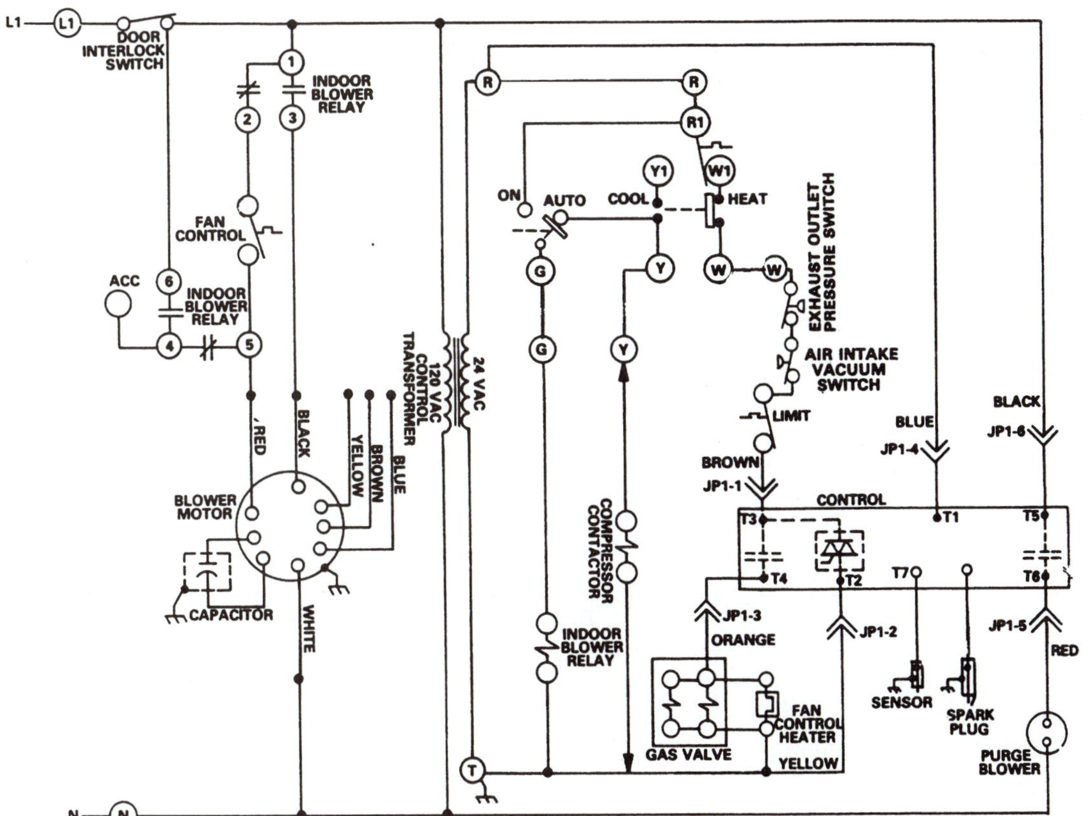

QUESTION:
The furnace will not run. You found a reading of 115V between terminal T6 on the primary control and Neutral. What problem is indicated?

QUESTION:
The furnace will not run. You found a reading of 24V between the white thermostat wire and terminal T, and 0V between T3 on the primary control and terminal T. What are the possible problems?

QUESTION:
Why does the blower relay have two switches instead of just one?

VENT DAMPER

When a furnace (or boiler) turns off, there is still some residual heat in the heat exchanger that becomes lost through the flue stack during the furnace off-cycle. As an energy conserving measure, some furnaces use a damper in the flue stack that closes during the off-cycle. Figure 14–25 shows the typical wiring. The unique feature of the vent damper is that it is spring-loaded to open, motor-operated to close. This way, if the motor should fail, the spring will keep the damper in the open (safe) position.

When the furnace is ready for operation, 24V are applied to terminals 1 and 2 of the vent damper, energizing the vent damper motor, holding the damper closed. The furnace or boiler is now in standby mode. When the room thermostat calls for heat, it provides 24V from the TH-R terminal on the pilot control to terminal 4 on the vent damper, energizing the relay in the vent damper. The relay contacts open, de-energizing the vent damper motor. The springs open the vent damper. When it is fully open, the damper pushes on the end switch (a microswitch), closing it. This completes the circuit from TH-R to TH-W on the ignition control box, and normal light-off can then proceed.

When the thermostat opens, the relay is de-energized, and the vent damper motor is energized to close the damper.

QUESTION:
1. The furnace has no pilot flame and no main flame, even though you have jumpered across the terminals on the thermostat subbase. What potential problems would you investigate?
2. What symptom or customer complaint would there be if the motor in the vent damper failed?

EXERCISE:
Explain the sequence of operation of the vent damper in figure 14–25.

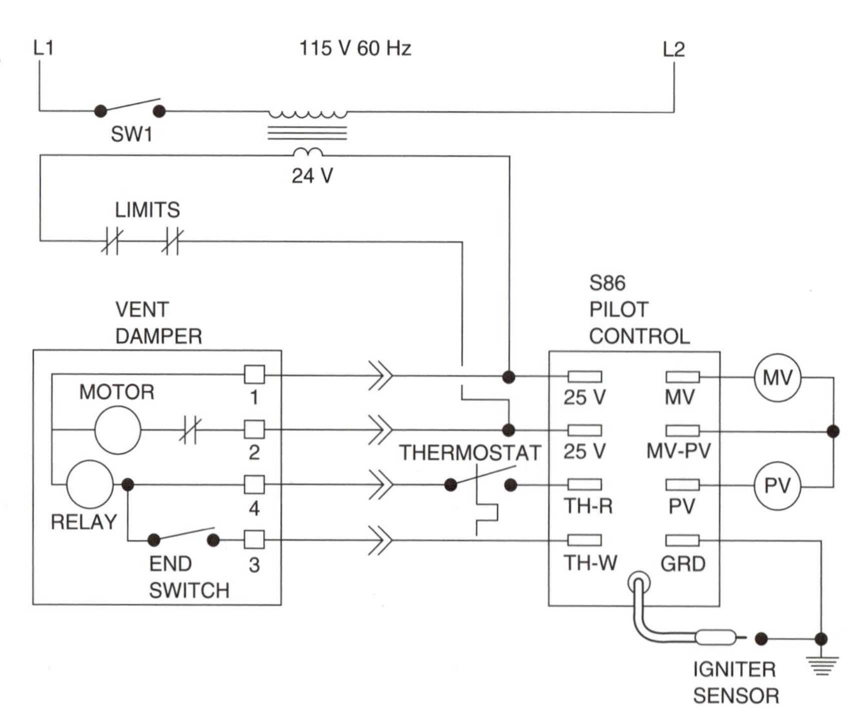

Figure 14–25 Vent damper.

CARRIER IGC BOARD

The Carrier IGC (Integrated Gas Unit Controller) board is one of the more modern electronic control schemes used on rooftop furnaces that controls not only the ignition, but the blower motor and inducer fan motor as well. Figure 14–26 shows the terminals that are available on this control board. Also built onto the board is a combustion relay, a blower relay, a gas valve relay, and all of the fan and safety control logic.

All external components are wired to the appropriate terminals on the IGC board. The room thermostat is wired to terminals W, G, and R. A roll-out switch is wired between the two terminals RS, the limit switch is wired between the two LS terminals, and the gas valve is wired to terminals GV and GV. 24V from the transformer secondary are wired between terminals R and C.

On a call for heat, the board detects continuity between R-W, energizing the on-board combustion relay which closes a switch between L1 and the CM terminal. This energizes the inducer fan. The operation of the inducer fan can be proved in one of three ways:

1. Pressure switch.
2. Centrifugal switch.
3. Hall Effect sensor.

Once the inducer fan motor operation is proved, the board sends a 10,000V spark through the ignition wire, and at the same time, energizes the GV terminals to supply gas. The flame rectification sensor (terminal FS) proves the flame, and 45 seconds later the BR terminal is energized, bringing on the blower motor. If the flame is not established within 5 seconds after the sparking starts, the gas valve will be de-energized, and the inducer fan mo-

Figure 14–21 Carrier IGC logic board.

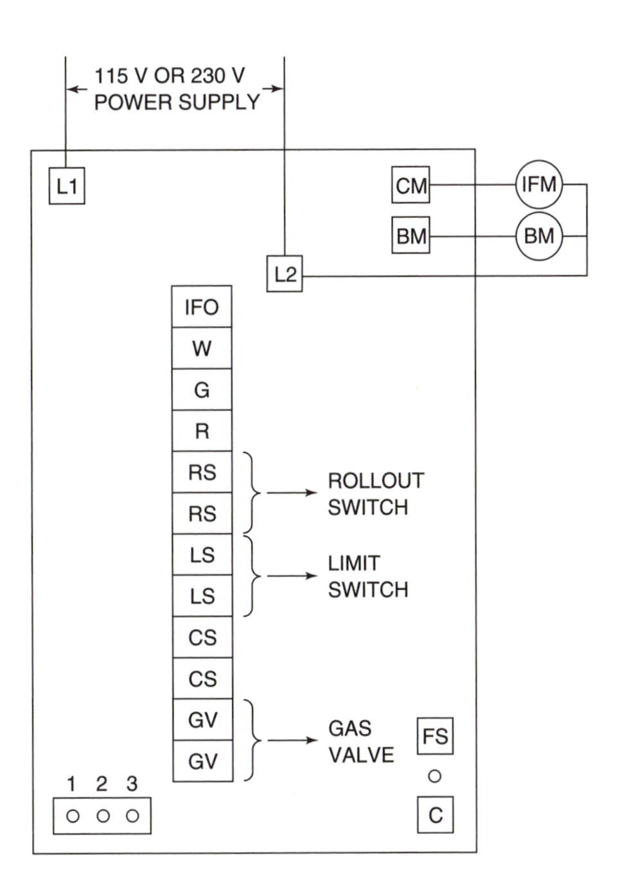

ADVANCED CONCEPTS

The Hall Effect sensor is a magnetic device that can be mounted on the inducer fan motor. When the motor is rotating at the correct speed, a current is induced. When the Hall Effect sensor is used to prove operation of the inducer fan motor, its wiring is connected to terminals 1, 2, and 3 in the lower left corner of the IGC board.

tor will run for another 20 seconds to purge out the heat exchanger. Another ignition attempt follows. If, after 15 minutes of trying, no ignition is proved, the unit goes into lock-out.

Diagnostic Signals

A single LED mounted on the control board is illuminated continuously during normal operation. If it is off, there are probably no 24V between terminals R-C. It can also give 8 different error signals, as signaled by the number of flashes between pauses:

1 flash: "Fan On/Off Delay Modified" indicates that the limit switch has opened (although it may be closed now). This signal will continue until power is disconnected. You should investigate to determine why the limit switch may have opened (i.e., low air flow).
2 flashes: "Limit Switch Fault" is open now.
3 flashes: "Flame Sense Fault" is indicating a flame without the gas valve being open.
4 flashes: "4 Consecutive Limit Faults" indicates that the furnace has tried to start four times, but each time, the limit switch was open. The furnace is in lock-out.
5 flashes: "Ignition Lock-out Fault" indicates that the furnace has tried unsuccessfully for 15 minutes to light off. It is in lock-out.
6 flashes: "Induced draft Motor Fault" indicates that there have been 24V at the W terminal for more than 60 seconds and the control board has not gotten a signal proving that the inducer fan was running. The combustion relay remains energized, and if the board receives proof of inducer fan rotation, ignition will be attempted.
7 flashes: "Rollout Switch Fault" indicates that the rollout switch has opened, and the unit is in lock-out.
8 flashes: "Internal Control Fault" indicates that there is most likely a problem in the IGC board. If disconnecting power and reconnecting does not clear the fault, replace the IGC board.

If more than one fault exists, they will all be indicated, in sequence, separated by a three-second pause between each.

Make sure you read the LED so that you are aware of all problems before you disconnect power, which will cancel and reset the error messages.

QUESTIONS:

During normal heating operation, what voltage should you read between each of the following terminals? Assume 115V input voltage.

1. R-W?
2. RS-RS?
3. LS-LS?
4. GV-GV?
5. CM-L1?
6. CM-L2?
7. BM-L1?
8. BM-L2?

CHAPTER FIFTEEN

Oil Heat

Oil is used as a fuel in both oil-fired furnaces (figure 15–1) and oil-fired boilers (figure 15–2). This chapter will deal only with the combustion portion of the oil-fired systems. For oil-fired furnaces, the controls used to operate the fan are the same as for gas-fired furnaces discussed in Chapter three. For oil-fired boilers, the control of pumps, dampers, and other accessories is the same as for gas-fired boilers, discussed later in Chapter 17.

OIL BURNER

Oil is far more difficult to burn than gas. If you were to have a dish of heating oil at room temperature, you would not be able to ignite it, even if you applied heat from a propane torch. The reason the oil won't easily burn is that at room temperature, there is insufficient oil vapor above the liquid to ignite. In order to burn the oil, it must be finely atomized. This increases the total surface area of the oil, and increases the rate of oil vapor formed off the surface. An oil burner (figure 15–3) includes a pump to increase the pressure of the oil so that it can be pushed through an atomizing nozzle, a burner motor that drives the oil pump and a fan that supplies combustion air to mix with the atomized oil, a transformer that produces approximately 10,000V, electrodes that produce a 10,000V spark, a sensor that determines if a flame has been established, and a primary control that monitors the combustion process.

BURNER MOTOR

The **burner motor** (figure 15–4) may be either split-phase or capacitor start. Residential burners will most commonly use either a 1/6 or a 1/8hp split-phase. Manual overload protection is required, with the reset button provided on the motor housing. The motor is usually flange mounted to the burner housing. The combustion air blower is fastened directly to the motor shaft. The fuel oil pump may be either directly coupled to the opposite end of the motor shaft, or it may be belt driven. The motor is usually 3450 rpm when the pump is belt driven, and 1725 rpm where the pump is direct driven.

PRIMARY CONTROL

The controller for the oil burner is an electronics "black box" called a **primary control.** One type is mounted on the flue gas stack, and is sometimes called a **stack control** (figure 15–5). Another type of primary control is mounted on the oil burner (figure 15–6). It differs from the stack control in that it uses a remote sensing device to sense whether combustion is taking place, rather than the integral temperature sensor on the stack controller that senses stack temperature.

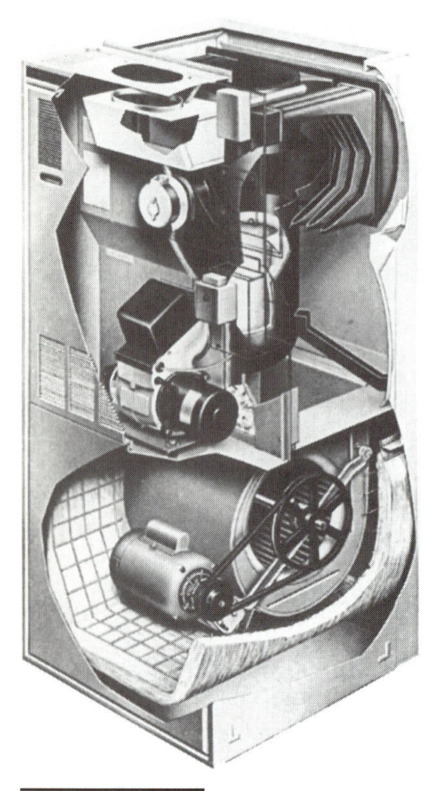

Figure 15-1 Oil furnace. *(Courtesy of Lennox Industries, Inc.)*

Figure 15-2 Oil-fired boiler. *(Courtesy of Mestek, Inc.)*

The function of the primary control is:

1. On a call for heat from the thermostat, start the burner motor and energize the ignition transformer.

2. When the oil flame has been established, some models will turn off the ignition transformer (intermittent ignition). On other models, the transformer remains energized, and the spark continues to fire throughout the call for heat.

3. If the oil flame is not established within a specified time (usually around one minute), shut down the burner motor. If unsuccessful, the control

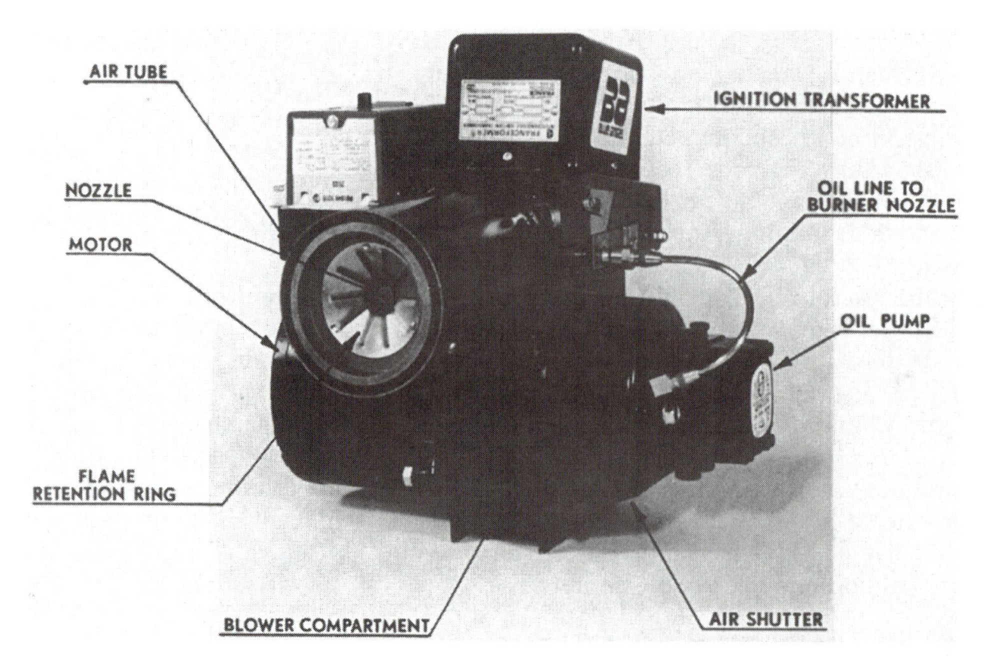

Figure 15-3 Oil burner. *(Courtesy of Wayne Combustion Systems.)*

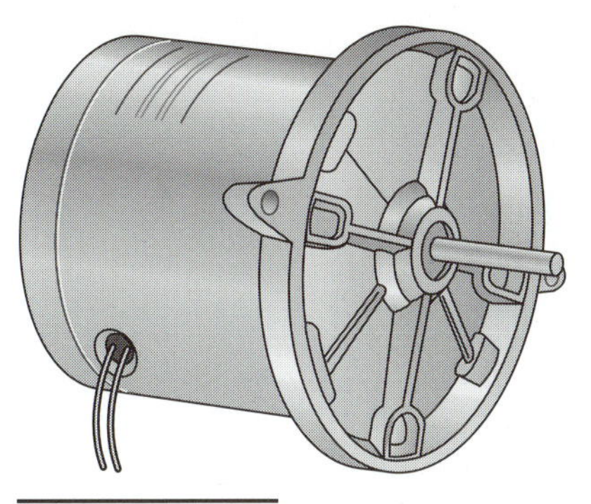

Figure 15–4 Flange-mounted oil burner motor.

will go into lock-out. The reset button on the primary control must be reset before a new trial for ignition may proceed.

4. When the thermostat is satisfied, the primary control turns off the burner motor and the ignition transformer (for constant ignition models).

Some primary controls provide **non-recycling control.** They will attempt to restart a burner immediately upon a loss of flame. Ignition attempt will continue until control locks out. Others are **recycling** controllers. They will shut down the burner immediately on loss of flame, then attempt to restart the burner once before locking out.

Figure 15–5 Stack controller. *(Courtesy of Honeywell, Inc.)*

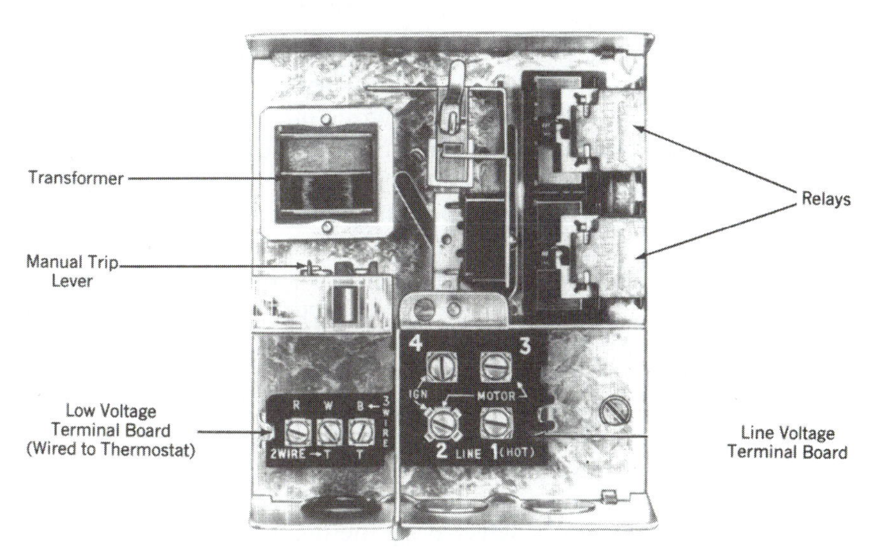

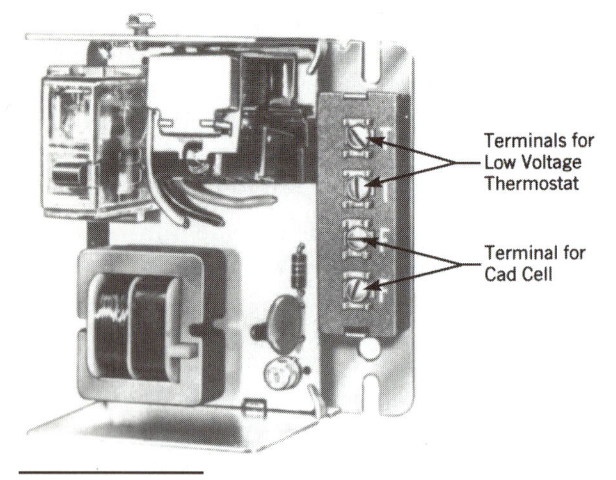

Figure 15–6 Burner-mounted primary control. *(Courtesy of Honeywell, Inc.)*

FLAME SENSING METHODS

There are several ways in which the primary control determines whether the trial for ignition has been successful. The stack control has its built-in temperature sensing element that is inserted into the flue stack when the controller is mounted. For remote-mounted primary controllers in residential and small commercial use, the combustion is sensed either with a remote temperature sensing switch mounted on the stack (**stack switch**, figure 15–7), or a cadmium sulfide cell (**cad cell**, figure 15–8). The cad cell is a light-sensitive device that is mounted where it can "see" the oil flame. A cad cell has a very high resistance when it senses darkness, but its resistance becomes very low when it sees light. It therefore behaves as if it were a switch in the primary control circuit, open when there is no flame, and closed when there is flame. If the burner flame is properly adjusted, the cad cell resistance will be in the range of 300 to 1000 ohms when the burner is operating. If the resistance is higher, either the cad cell needs cleaning or re-aiming, or the burner flame may need adjustment. The two wires that emerge from the cad cell will be connected to the terminals F-F on the primary control (flame) or the S-S terminals (sensor).

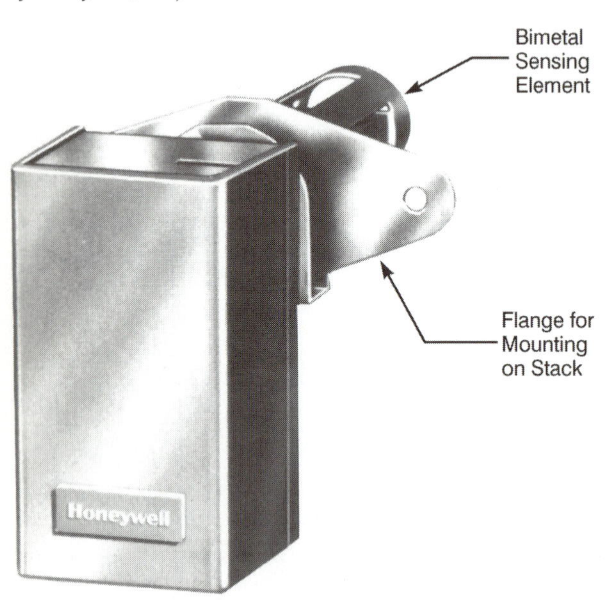

Figure 15–7 Stack switch. *(Courtesy of Honeywell, Inc.)*

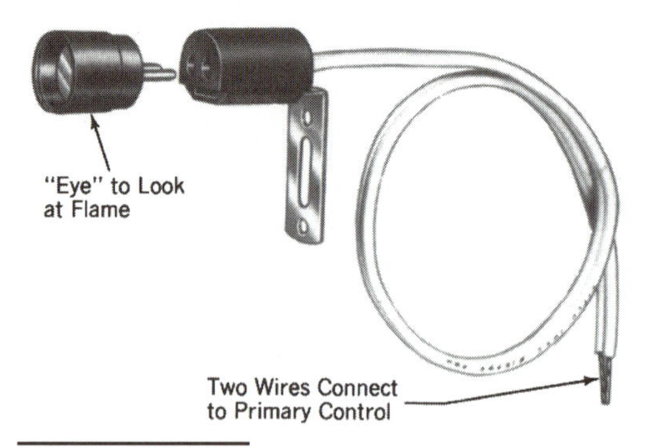

Figure 15–8 Cad cell.

The stack detector switch (figure 15–7) is either a two-wire or three-wire device. On the two-wire model, the switch is normally closed. When the sensor detects heat, the switch opens. On the three-wire model, there is an SPDT switch. In the cold starting position, the switch between terminals R and B is closed. On a temperature rise, R-B will open, and R-W will close.

WIRING DIAGRAMS—PRIMARY CONTROL

The wiring diagrams that follow all show the internal wiring of the primary control. Some are just electro-mechanical, while others have electronic devices such as thermistors, diodes or triacs. It is not necessary for the service technician to understand exactly what goes on inside the primary control. As with other "black box" controllers, you simply need to know the required inputs (thermostat signal and flame signal) that will produce a desired output (a voltage output to the burner motor and transformer). The general sequence of operation is as follows. When the burner is started, a bimetal operated safety switch (inside the primary control) starts to heat. This switch will open, and shut down the burner unless an oil flame is established. If the flame is detected by the flame sensor, the circuit to

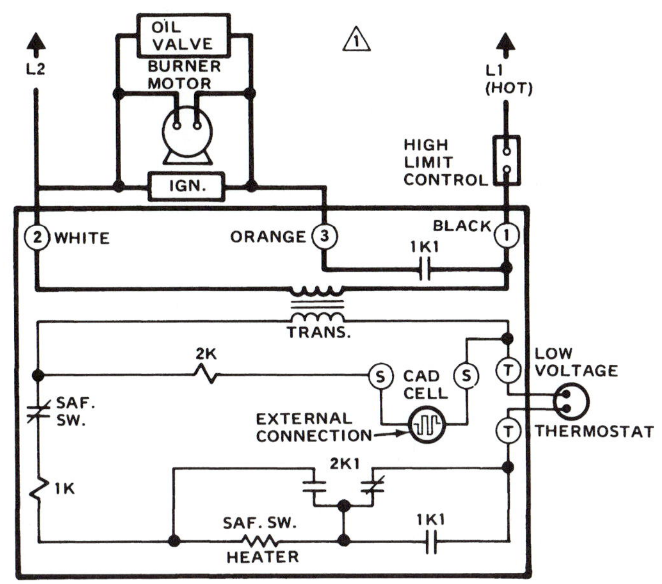

Figure 15–9 Wiring of primary control with low-voltage room thermostat. *(Courtesy of Honeywell, Inc.)*

the safety switch element is broken, the safety switch will stop heating, and the burner will be allowed to continue to run.

Figure 15–9 illustrates the typical internal wiring for a primary control using a low-voltage thermostat. When the thermostat calls for heating, a circuit is completed through the 1K relay coil, safety switch heater, and the normally closed contacts of 2K1. The 1K contacts complete the line voltage circuit through the ignition transformer, the burner motor, and the oil solenoid valve. If a flame is established and sensed by the cad cell, the 2K relay coil will become energized. This will create a short around the safety switch heater, causing it to stop heating. If the cad cell circuit is not completed within the predetermined time, the safety switch heater will open the 1K coil circuit.

Figure 15–10 shows the internal wiring diagram for a stack controller using a line-voltage thermostat.

Figure 15–10 Wiring of stack-mounted primary control with line-voltage room thermostat.

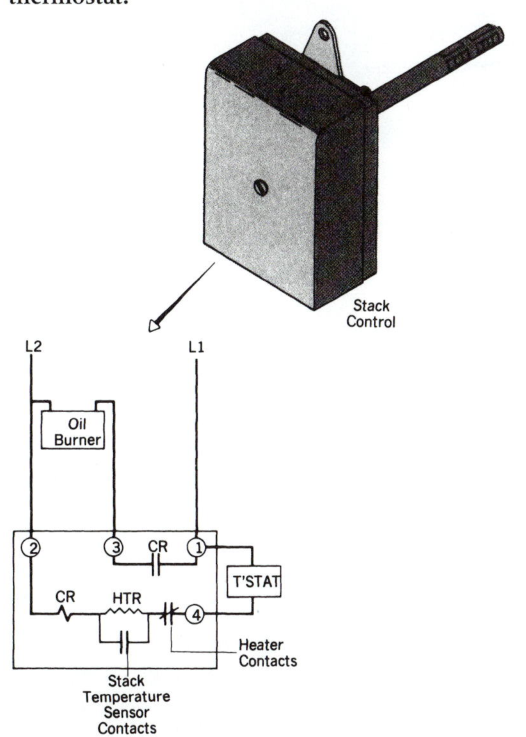

EXERCISE:

Write a sequence of operation for the stack controller.

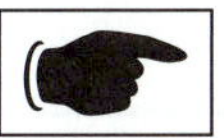

If a cad cell system burner locks out on safety after flame is established, the problem could be the primary control or the cad cell. Push the manual reset button on the primary control to attempt ignition. Quickly, jumper the S-S (or F-F) terminals. The burner should continue to run. If it does, the primary control is good. If the control works with the jumper in place but fails when the cad cell is reconnected, the cad cell has either failed, or become covered with soot. [Note: do not jumper the S-S (or F-F) terminals prior to calling for heat. The primary control will think there is already a flame, and it will not allow ignition to happen.]

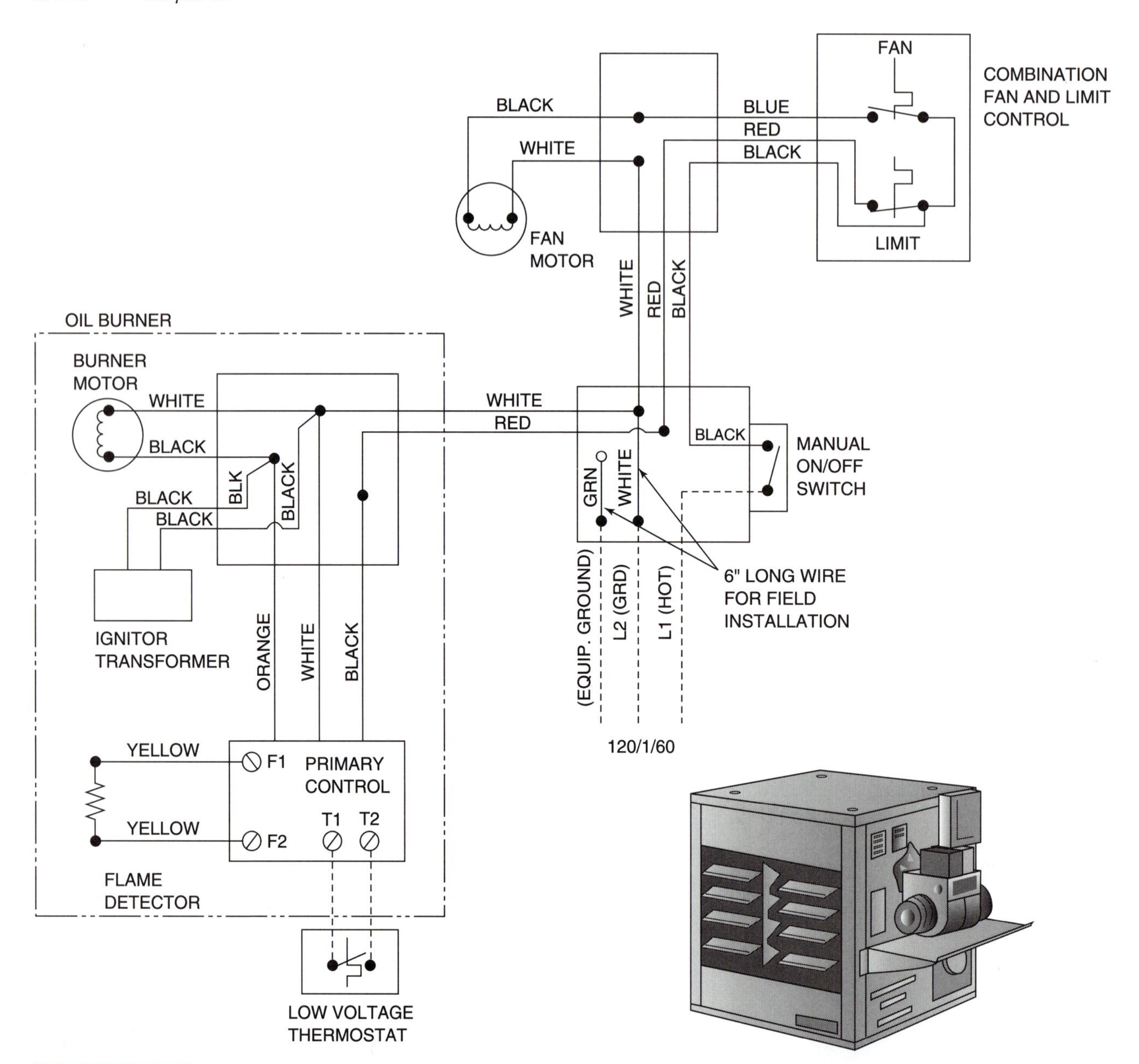

Figure 15–11 Oil-fired unit heater.

OIL-FIRED UNIT HEATER

Figure 15–11 shows the interconnection diagram for a unit heater that has an oil-fired burner. The sequence of operation is similar to that of the gas-fired furnaces discussed in earlier chapters. When the room thermostat closes, the primary control causes the burner motor and the ignition transformer to start. Assuming that the flame detector properly senses the flame, the heat exchanger will begin to warm. When the heat exchanger reaches approximately 130°F, the fan switch will close, energizing the fan motor. If, for any reason, the heat exchanger exceeds approximately 200°F, the safety limit switch will open, shutting down the entire system.

EXERCISE:
Redraw the interconnection diagram as a ladder diagram.

EXERCISE:
Figure 15–12 shows a manufacturer's wiring diagram with unusual symbols. Redraw the diagram as a ladder diagram, using conventional symbols.

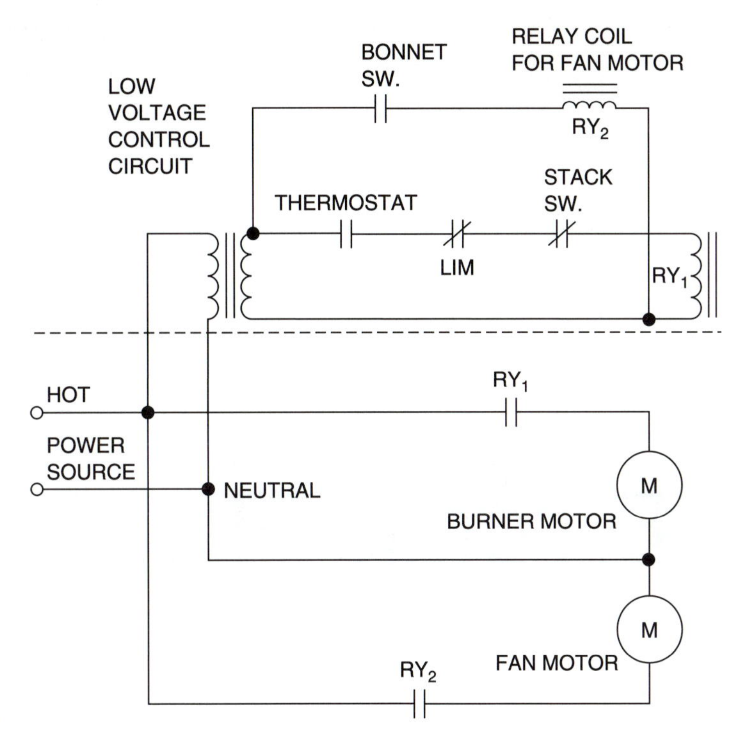

Figure 15-12 Exercise.

CHAPTER SIXTEEN

ELECTRIC HEAT

RESISTANCE HEATING

When an electric current is passed through a resistor, the electrical energy is converted into heat energy. The principle of **resistance heating** is used in toasters, electric ovens, electric water heaters, and electric furnaces and heaters. All of the electrical input is converted into heat, at the rate of 3415 Btu/hr for each kW of input. It is 100% efficient because of this complete conversion, but it is very expensive to operate because the energy input in the form of electricity is three to four times as expensive as energy in the form of fossil fuel. Resistance heat may be designed either to stand alone (without a fan), as a resistance heater with a fan (electric furnace or self-contained electric heater), or as an insert into the ductwork of a system that already has a fan (duct heater).

ELECTRIC FURNACE

Figure 16–1 shows a cutaway view of an upflow electric furnace. Instead of a heat exchanger, the blower sends the room air through an electric heating element that operates on 230V. When the thermostat calls for heat, it energizes both the heating element and the blower relay.

Figure 16–2 shows the heating element. It is a long spiral-wound wire, supported by insulators. It is made of a nickel and chromium alloy commonly called **Nichrome.** A furnace will use several of these elements, each drawing between 3 and 8kW. The furnace control panel will have a 20 to 40amp cartridge fuse on each leg of each heater. Therefore, an electric furnace with three heating elements would have six cartridge fuses.

In addition to the large fuses that protect the furnace wiring against a shorted heating element, there is also a thermal cutout and a limit switch mounted on the heater element. The limit switch senses the air temperature around the heating element and will de-energize the element if the limit set point is reached (usually between 140° to 190°F).

The **thermal fuse** will also open if it detects an abnormally high temperature around the heating element. The primary cause of a potentially too-high air temperature is restricted or lost airflow across the heating element. The heating element will continue to produce the same amount of heat, regardless of the airflow. If the airflow is too low, it will get too hot. The resistance heating element will also get hotter than normal, and if left unchecked, can get hot enough to burn out the heating element.

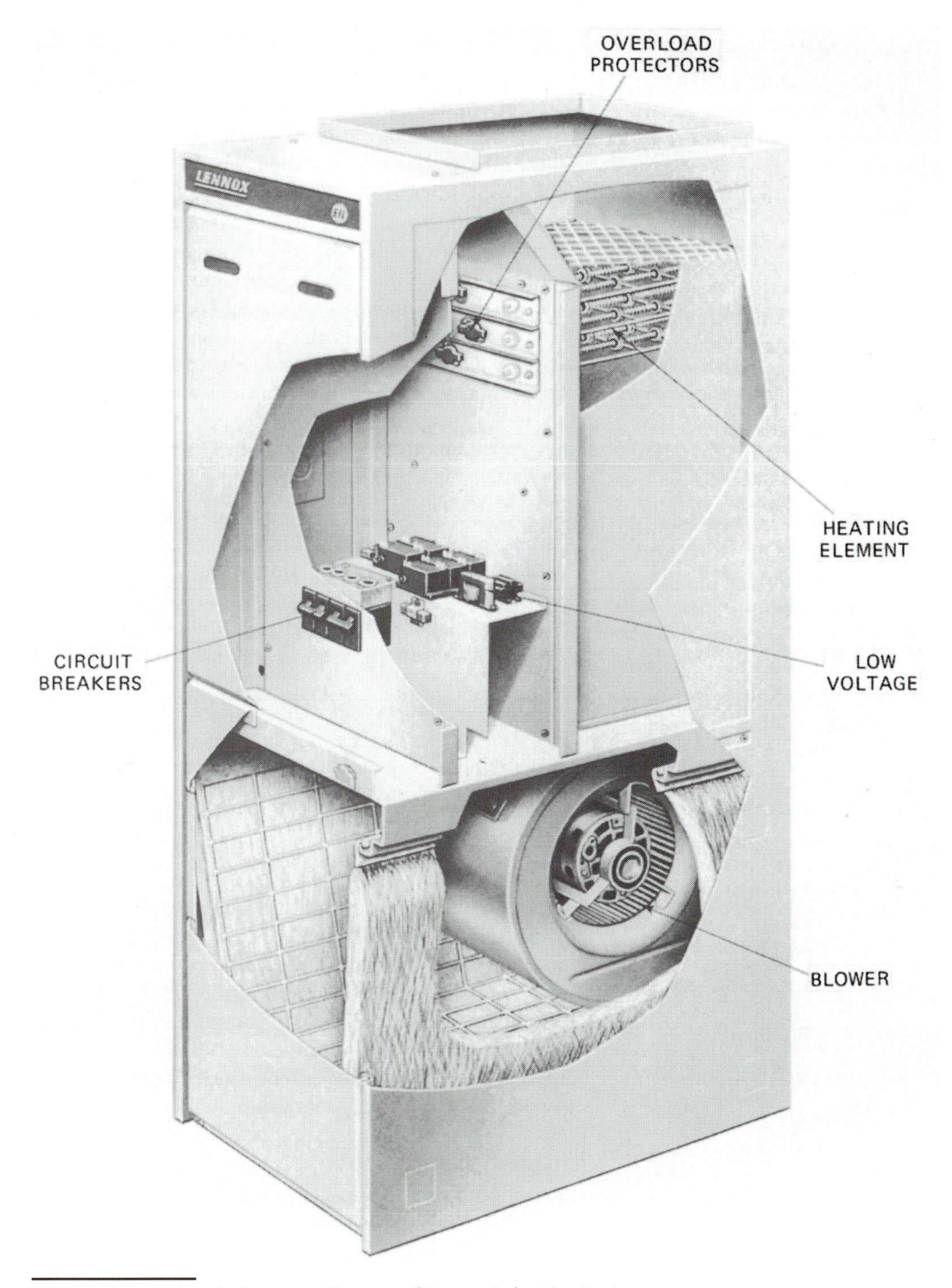

Figure 16–1 Electric furnace. *(Courtesy of Lennox Industries, Inc.)*

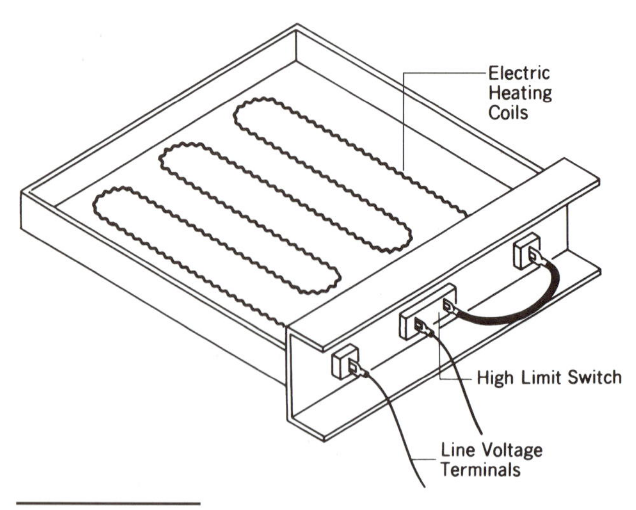

Figure 16–2 Electric heating element.

Most electric furnaces (as well some other types of units) have a door latch that is operated by a solenoid coil that is energized whenever power is applied to the unit. There is nothing more embarrassing to a service technician than not being able to open the front panel, as the customer watches. The only way to release the latch is to turn the power off to the unit from the circuit breaker or disconnect. After the door is opened, the power may be switched on again.

Figure 16–3 Electric duct heater. *(Courtesy of Tutco, Inc.)*

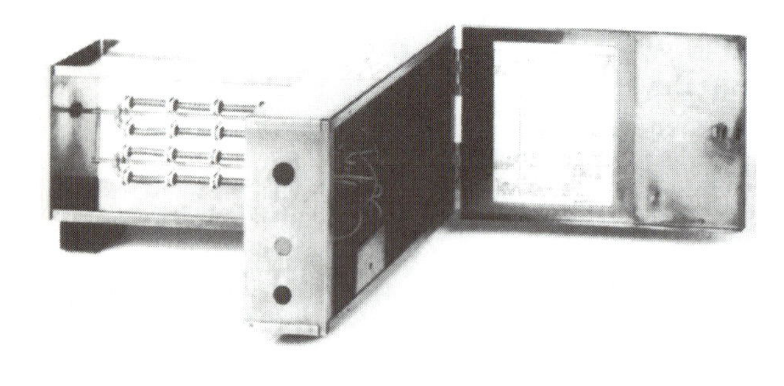

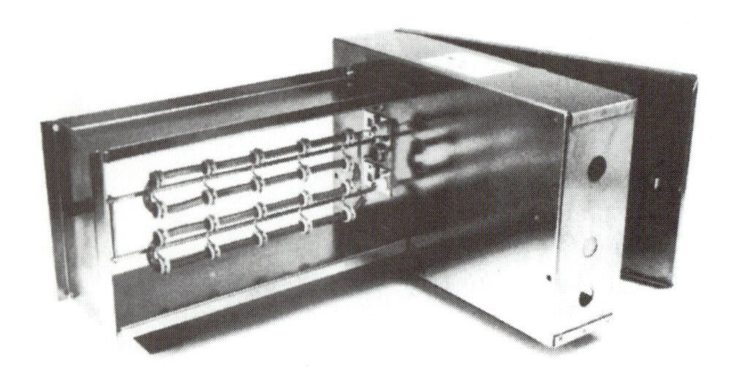

DUCT HEATER

A **duct heater** is an electric heater element that is installed in a run of duct. The element slides inside the duct, and an attached control compartment hangs outside the duct (figure 16–3). Duct heaters require a minimum airflow across them in order to keep the temperature of the elements below an acceptable level. Turns in the ductwork can cause uneven airflow distribution, which can result in a portion of the electric element not getting sufficient airflow. This can be a difficult-to-diagnose reason for elements burning out. Sometimes a sail switch is used in the duct to prove that airflow is present before the electric heating element is allowed to operate.

EXERCISE:
Create a ladder wiring diagram for a large duct heater, controlled by a low-voltage room thermostat and a sail switch. Add whatever other components you need.

PACKAGED COOLING WITH ELECTRIC HEAT

Figure 16–4 shows a wiring diagram for a conventional air conditioning system with an electric resistance heater. The low-voltage thermostat connected to the unit's terminal board is designed for electric heat. On a call for heat, it makes R-W, energizing the heat sequencer. The heat sequencer has a pilot-duty heating element inside that causes a bimetal line-duty switch to close. When that switch closes, it energizes the heating element and the fan motor (through the normally closed contacts of the blower relay). On other heaters, a specially designed electric-heat thermostat is used. On a call for heat, it makes both R-W and R-G. The R-W circuit operates only the heater, while the R-G circuit energizes the blower relay.

SEQUENCING THE HEATING ELEMENTS

When there are two or more heating elements, they are sometimes staged or sequenced to come on one at a time. There are two ways that this can be accomplished:

1. Two-stage thermostat.
2. Heat sequencers.

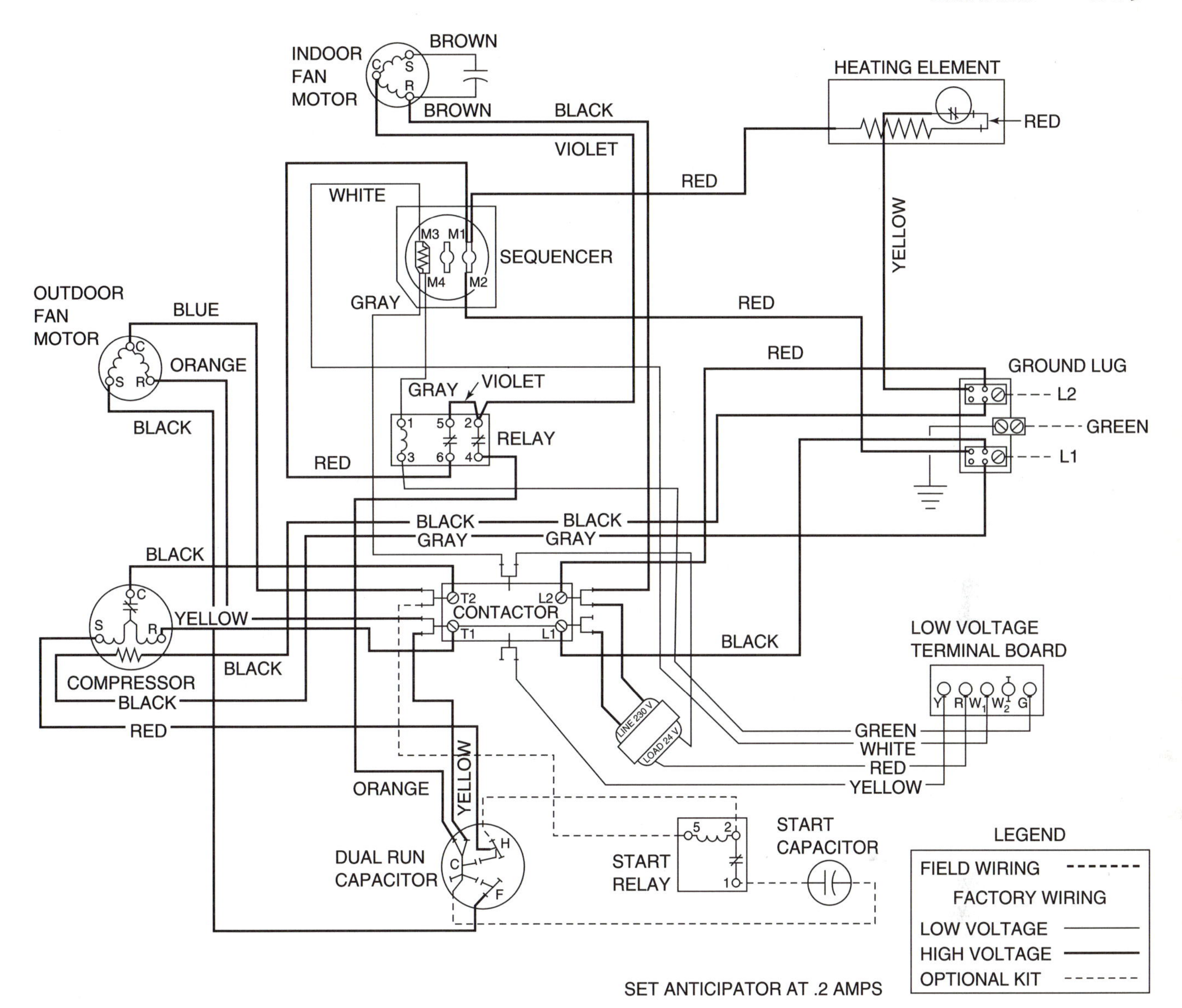

Figure 16–4 Air conditioner with electric resistance heat.

Figures 16–5 and 16–6 show the comparison of these two methods. In 16–5, when the room temperature drops below the setting of the first stage set point of the room thermostat, it makes R-W1, energizing heat sequencer #1, and some time later, element #1. If this does not provide enough heat to satisfy the room, the room temperature will continue to fall until the thermostat makes the second stage switch, R-W2. This will energize heat sequencer #2, and then element #2.

In figure 16–6, the heat sequencer is slightly different. Instead of the internal heater element operating a single bimetal switch, there are two internal bimetal switches, identified on the diagram as A and B. A heat sequencer can have three or even more bimetal switches. Here, unlike with the two-stage thermostat, all of the heating elements are going to be energized, even if only one would be sufficient to maintain the room temperature.

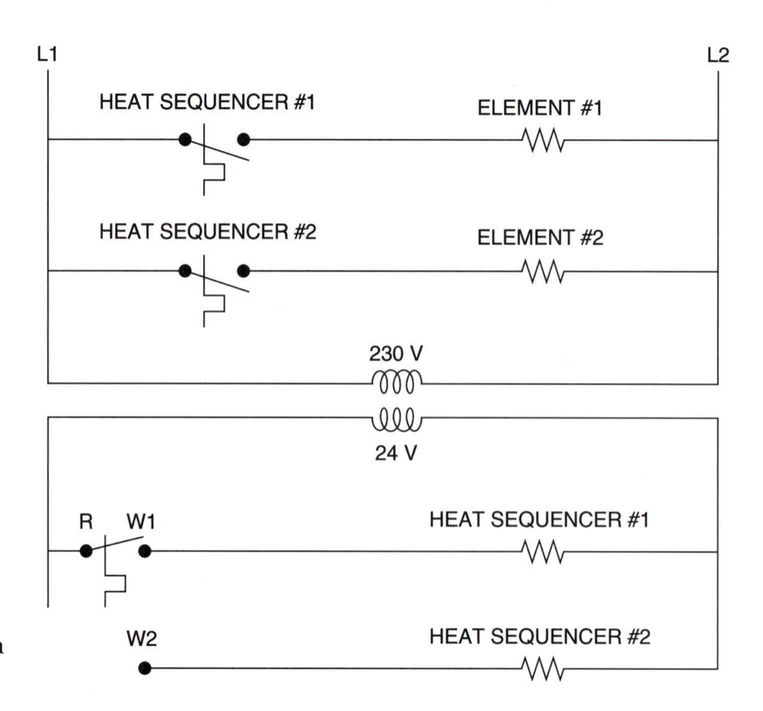

Figure 16–5 Electric heat with 2-stage thermostat.

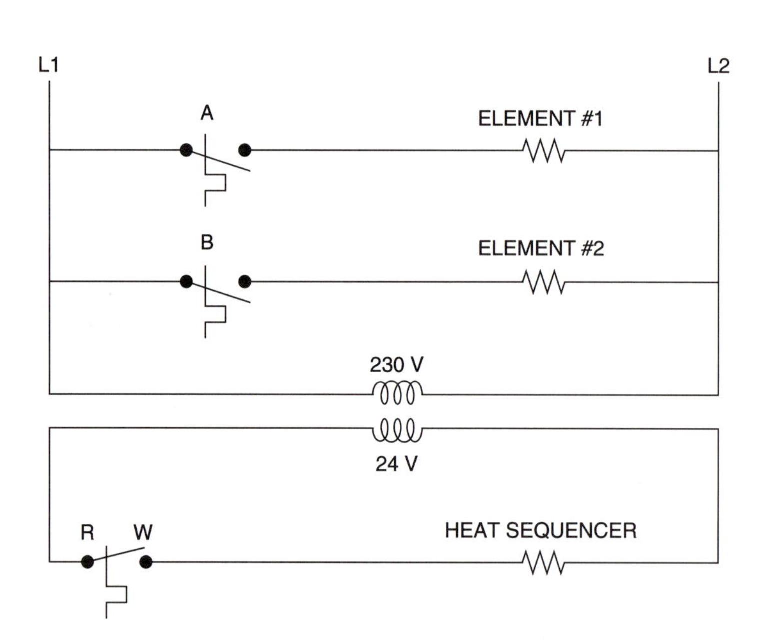

Figure 16–6 Two stages of heat controlled by a single-stage thermostat.

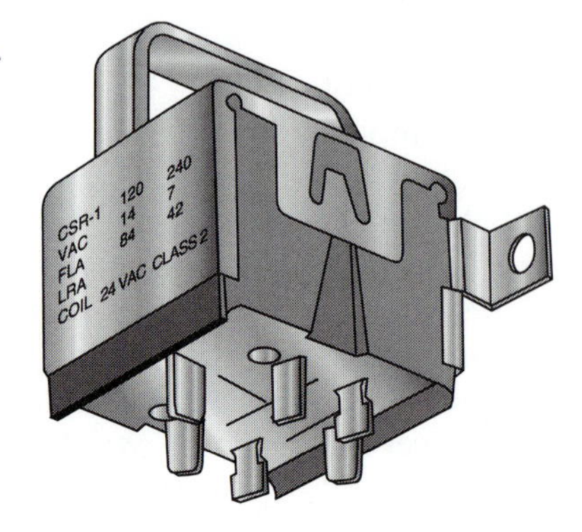

Figure 16–7 Current-sensing fan switch.

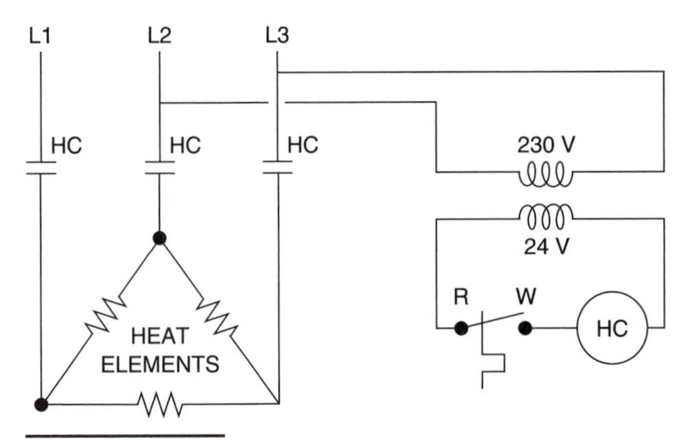

Figure 16–8 Three-phase electric heat.

QUESTION:
Assume that two identical occupancies are being maintained at the same temperatures, one by the system in figure 16–5 and the other by the system in figure 16–6.

1. Over the course of a winter, which system will operate element #2 for a greater number of hours?
2. Which system will cost more to operate? Explain.

EXERCISE:
Redraw figure 16–4 as a ladder diagram. It is only necessary to show those components that are involved with the heating mode. Redraw your ladder diagram, showing how it can be simplified by using an electric heat thermostat.

QUESTION:
In figure 16–4,

1. What type of overload protection is provided for the compressor?
2. If you removed the yellow wire from the compressor contactor, would you be able to safely work on the compressor wiring? Explain.

BLOWER CONTROL

Another method of operating the blower on an electric heat system is by using a switch that senses when current is flowing through the electric heating element (figure 16–7). The wiring to the element is passed through the current-sensing loop of the blower control. Whenever the element is energized, the current-sensing loop will cause a set of normally open contacts in the control to close, energizing the fan motor.

THREE-PHASE ELECTRIC HEAT

All electric heat is single phase, but electric heat is used on three-phase systems. Figure 16–8 shows how three single-phase heaters can be applied to a three-phase system. This diagram also illustrates the use of a contactor (HC = heating contactor) to energize the heaters.

SCR CONTROLS

An **SCR (silicon-controlled rectifier)** is an electronic device that can be used to regulate the flow of electricity to an electric heater. The SCR can turn the electric heating element on and off many times each second. By modulating the time on and the time off, the SCR acts as a modulating control for the heating element.

ADVANCED CONCEPTS

When connected in an AC circuit, the SCR acts as a rectifier, and its output is DC. The symbol for an SCR is shown in figure 16–9. When the current flowing through the gate is high enough (as modulated by a temperature-sensing device), it will turn the SCR on, allowing current to flow through the anode-cathode path, and through an electric heating element. Figure 16–10 shows how the first portion of the AC wave-form is "chopped off," because the rising voltage is not high enough to cause the gate to open the current flow to the heater. When the voltage gets high enough to cause the turn-on current to flow through the gate, the remaining portion of the wave form is allowed to flow through the heater. When the downside of the applied voltage wave form signal drops to zero, the SCR turns off again, until the turn-on current is reached once again. By adjusting the turn-on current setting, the amount of the wave-form that can be allowed to pass through the heater can be adjusted. By connecting two SCRs in tandem, one for each direction of current flow, it becomes easy to control AC current flow. Each half-cycle is "chopped" at low-power settings, and the full AC sine wave can be passed through when full power is required.

Low-power devices use a **triac** which is essentially the same as two back-to-back SCRs in a single semiconductor package. Higher-power capability can be achieved by using two discrete SCRs as described above. Often the user is unaware of the difference, since most manufactured power controllers are prepackaged by the manufacturer.

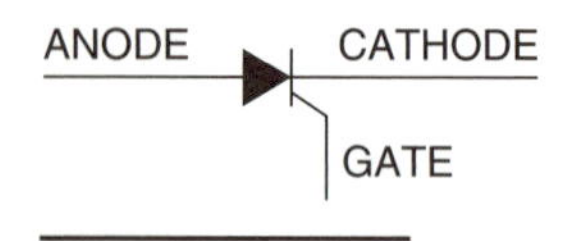

Figure 16–9 SCR.

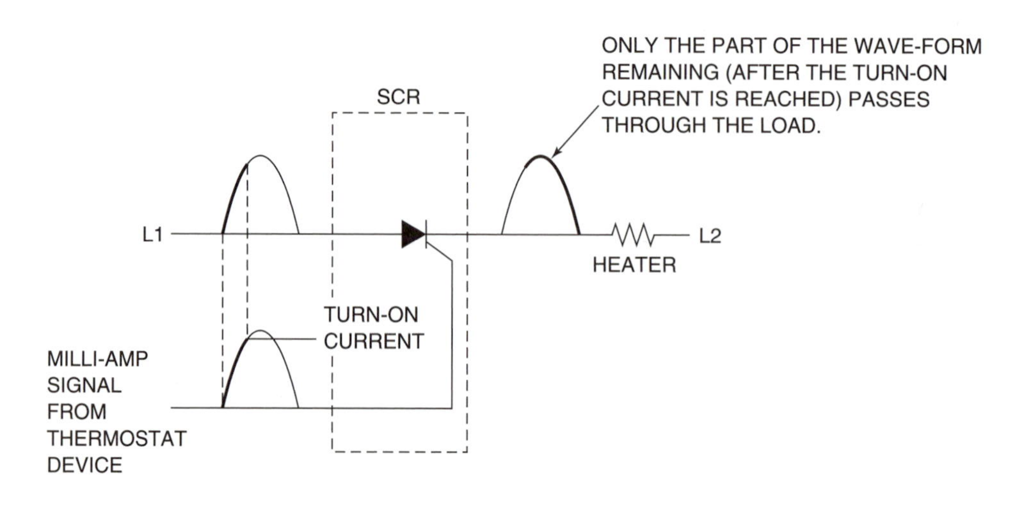

Figure 16–10 "Chopped off" wave form to control heater capacity.

CHAPTER SEVENTEEN

Boilers

The control scheme for a boiler ranges from a very simple millivolt-operated gas valve on a pool heater to a very complex large boiler with control for combustion air fans, dampers, air-fuel mixtures, modulating controls, purge cycles, multi-step start-up sequences, and belt-with-suspenders safety controls. The boiler control scheme may be thought of as two separate functions.

The first function is combustion, either gas-fired or oil-fired. This is usually controlled to maintain a constant boiler condition, either hot water temperature or steam pressure. It may be done with on-off control, or with modulating controls. The boiler condition that is being maintained might be constant year-round, or it may be automatically reset depending on the outside air temperature. When it is warm outside, the need for heating in a building is reduced, and the temperature of the boiler hot water may be adjusted downward for energy savings and improved control.

ADVANCED CONCEPTS

Sometimes the heaters that are receiving hot water are controlled with modulating valves to modulate the amount of water flowing through the heater. When the temperature is mild outside, the modulating valves will be operating at very close to a fully closed position. This makes control less accurate, and can cause the valve seat to wear out quickly. By resetting the hot water temperature downward, in order to provide the same amount of heat, the control valve will open more, allowing more flow of the lower-temperature water.

The second function of the control scheme is to deliver the heat to the occupied space, or in some cases, to a heat exchanger that will use the hot water to heat city water for domestic uses (sinks and showers). Often there are multiple **"zones,"** such as upstairs and downstairs, or living quarters and sleeping quarters, or interior offices and perimeter offices. A zone is defined as any area that has its own thermostat that can maintain the area at a selected temperature, not dependent upon what the temperatures are in the rest of the building. Zone control can be accomplished by having a thermostat adjust either the flow rate or the temperature of the hot water or steam being delivered.

RESIDENTIAL HOT-WATER BOILER

Figure 17–1 shows a boiler of the type commonly used in a residence. A simple circuit for a boiler (not the one shown in figure 17–1) is shown in figure 17–2. A hot water thermostat (commonly referred to as an **aquastat**) controls the water temperature in the boiler at a constant 180°F (sometimes as low as 160°F or as high as 200°F). When the room thermostat calls for heat, it energizes the hot water pump (figure 17–3) through the control relay.

> **QUESTION:**
> The control relay coil in the above circuit has failed. Will this cause the boiler to trip out on high limit? Explain.

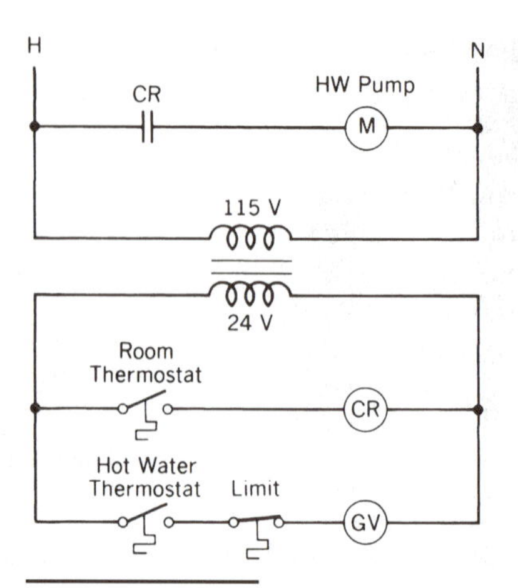

Figure 17–2 Single-zone hot water boiler schematic.

ZONE CONTROL FOR SMALL BOILERS

Zone Control Valves

Figure 17–4 shows a boiler system that has two different zones of control, operated by two different thermostats. It would be common to find the upstairs of a house as one zone, and the downstairs as the second zone. Each zone has its own zone valve (figure 17–5). When the thermostat closes, it energizes a clock-type motor inside the valve that slowly opens the valve, allowing hot water to flow through the heater coil in that zone. When the valve reaches the end of its travel, an end-switch is mechanically closed. The end-switches for the zone valves are wired in parallel, so that if either zone needs heat,

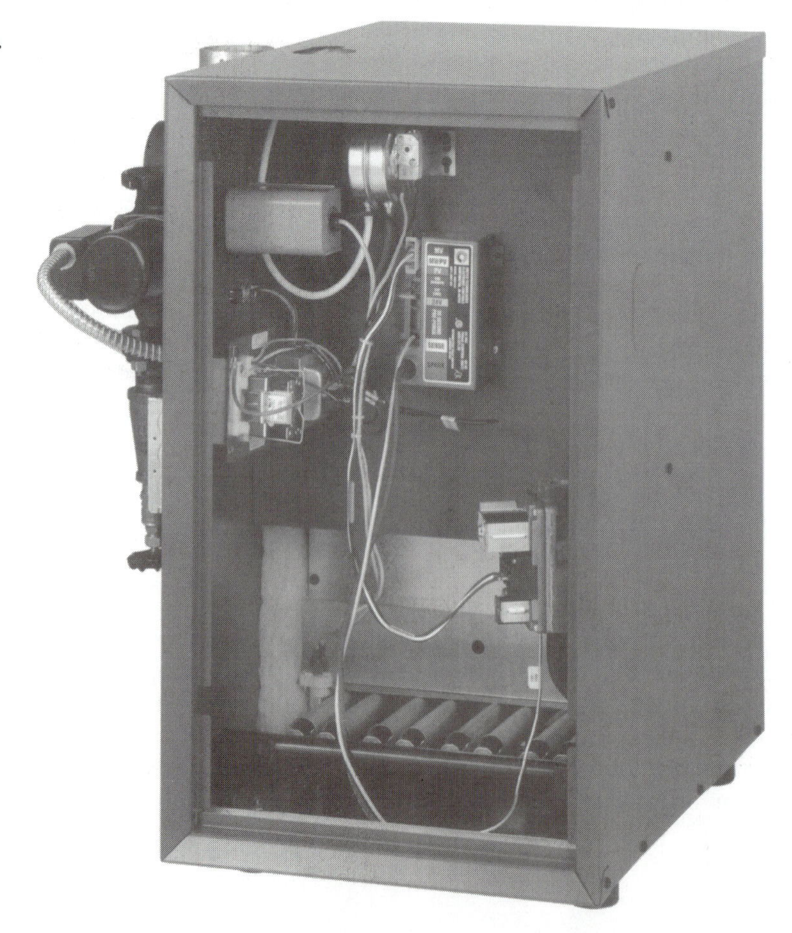

Figure 17–1 Hot-water boiler. *(Courtesy of Weil-McLain Co.)*

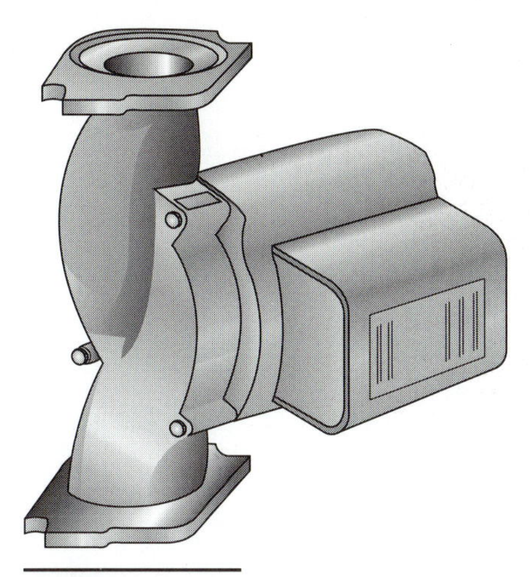

Figure 17–3 Hot water pump for a residential boiler.

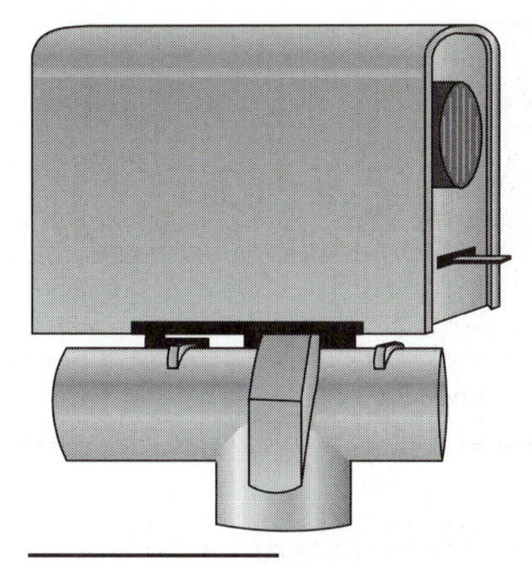

Figure 17–5 Zone valve.

the hot water pump will run. The pump will only shut down when both thermostats are satisfied.

The water inside the boiler is maintained at a constant temperature by an aquastat.

MULTIPLE PUMPS

Instead of each zone thermostat operating a zone valve, each zone can be provided with its own circulating pump (figure 17–6). Each room thermostat uses a room thermostat to energize a control relay. The control relay contacts are wired in parallel so that either zone can energize the gas valve (or oil burner). In this way, during the summer months when the boiler does not need to be maintained at 180°F, it will be allowed to cool to room temperature.

Figure 17–4 Boiler with zone valves.

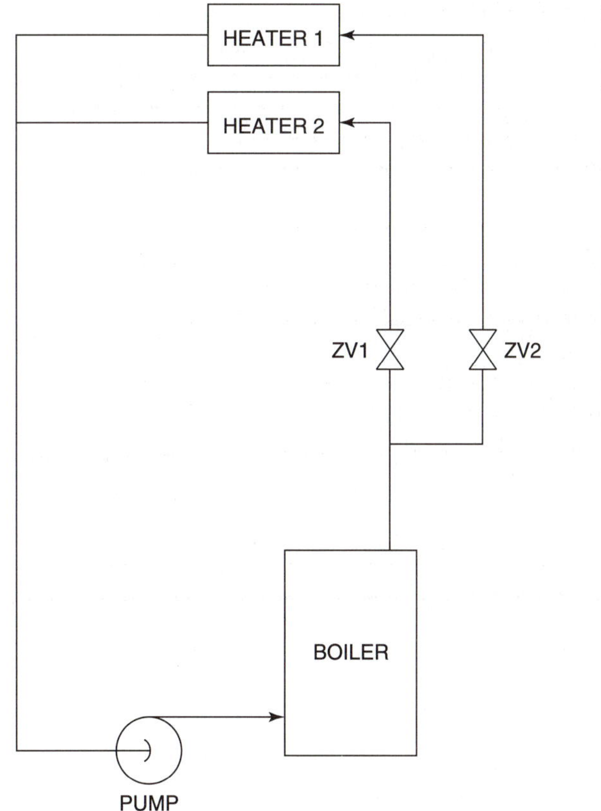

(a)

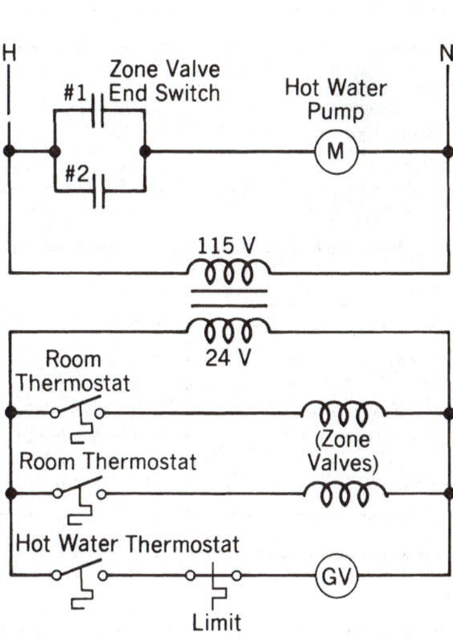

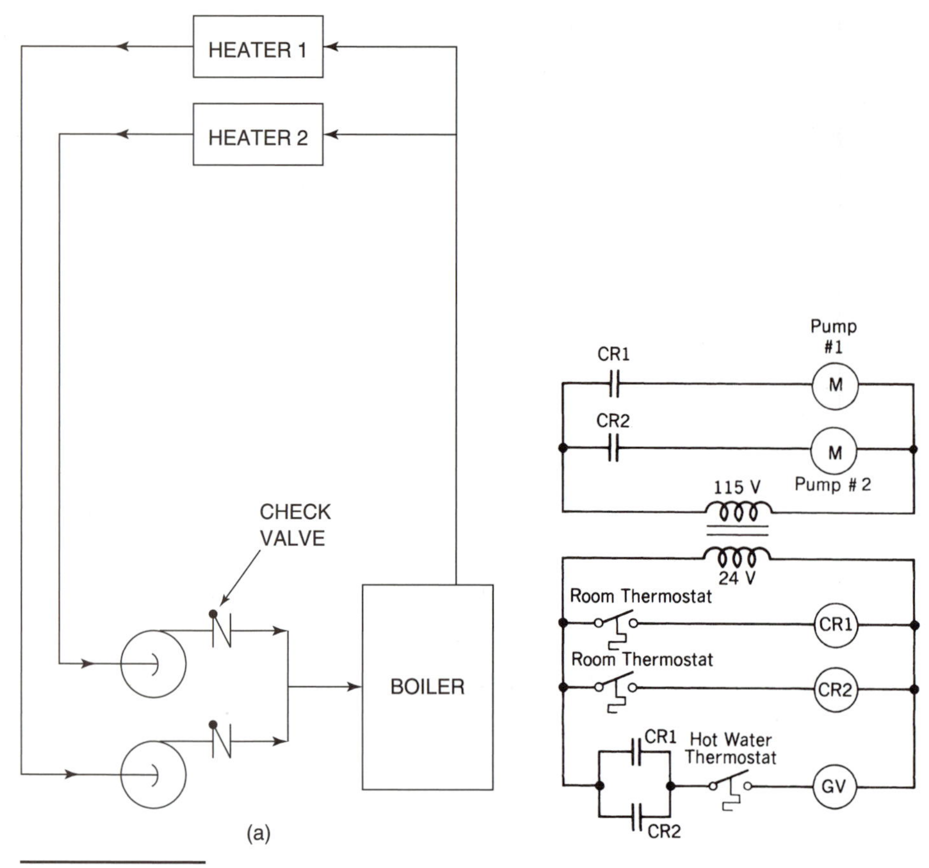

Figure 17–6 Boiler with multiple pumps.

Allowing a small boiler to cool to room temperature is fine, but extreme caution must be exercised if a large boiler is to be allowed to cool. The same applies to start-up of a large boiler. It must be done *very* slowly (maybe over a two or three-day period) to prevent the boiler from being damaged from expansion/contraction.

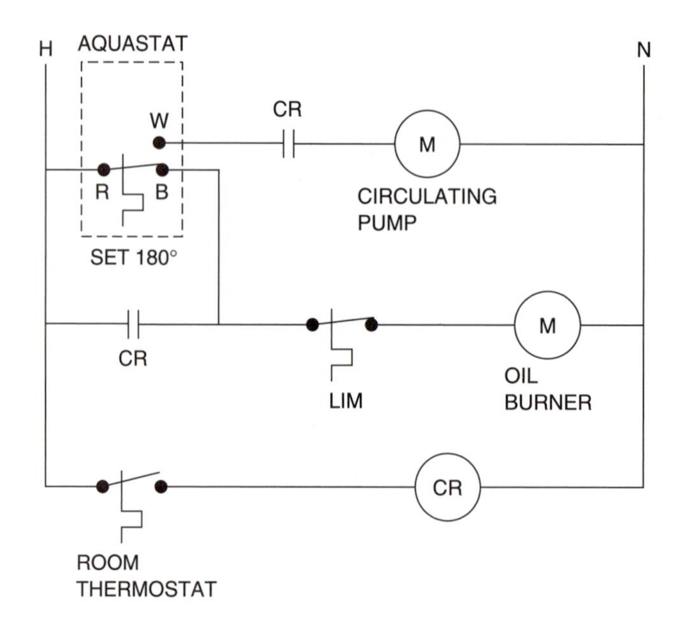

Figure 17–7 Boiler with tankless heater.

Figure 17–8 Electronic boiler controller. *(Courtesy of Erie Controls.)*

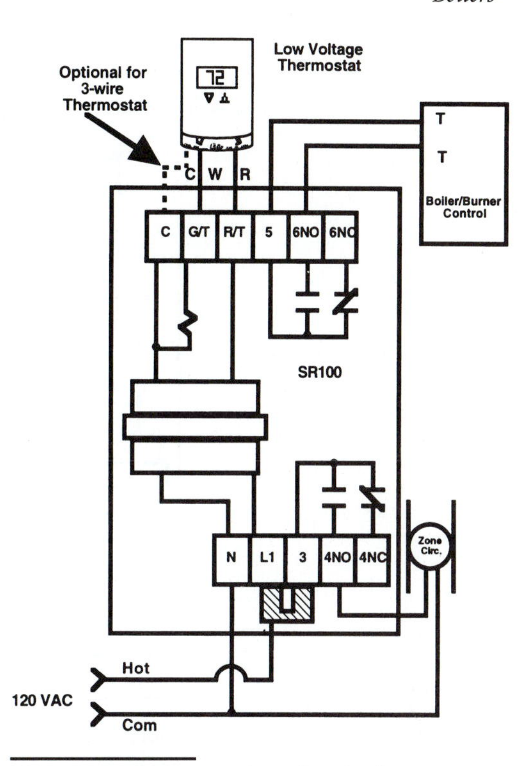

Figure 17–9 Connections for single-zone operation. *(Courtesy of Erie Controls.)*

BOILER WITH TANKLESS HEATER

Some boilers use an auxiliary heat exchanger located inside the boiler to supply **domestic hot water** (water for sinks, baths and showers). Figure 17–7 shows a wiring diagram for this system. It uses a three-wire aquastat in order to give the domestic water zone priority over heating. For example, suppose that the demand is so great that the boiler cannot maintain 180°F. The aquastat will make R-B and break R-W, turning off the circulating pump, but letting the oil burner continue to run. Only after the aquastat reaches 180°F again will the heating system be allowed to draw heat from the boiler by turning on the circulating pump.

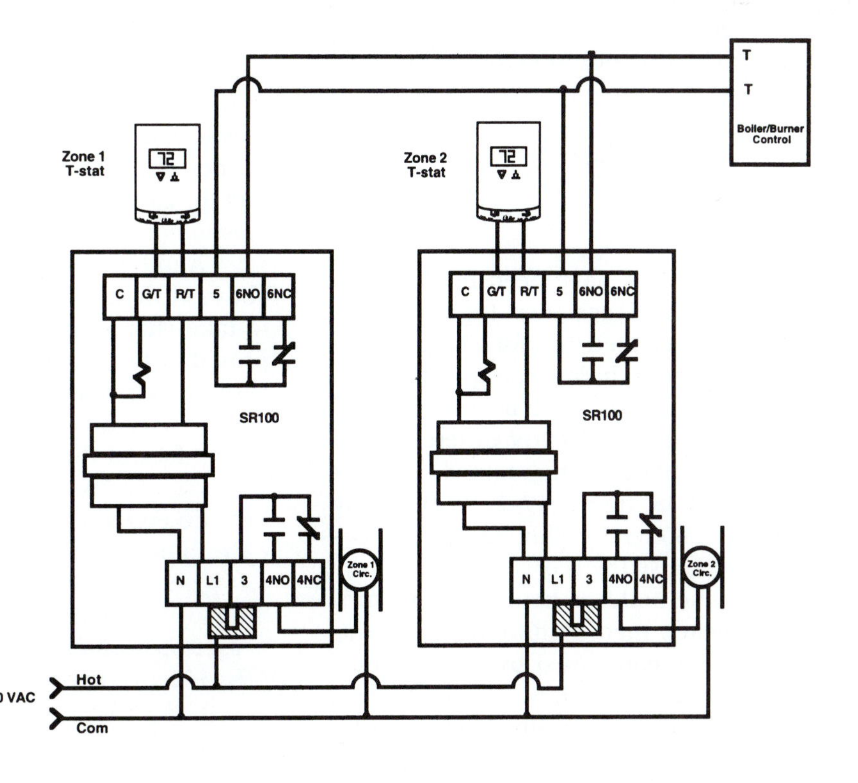

Figure 17–10 Zone control with water temperatures allowed to fall between cycles. *(Courtesy of Erie Controls.)*

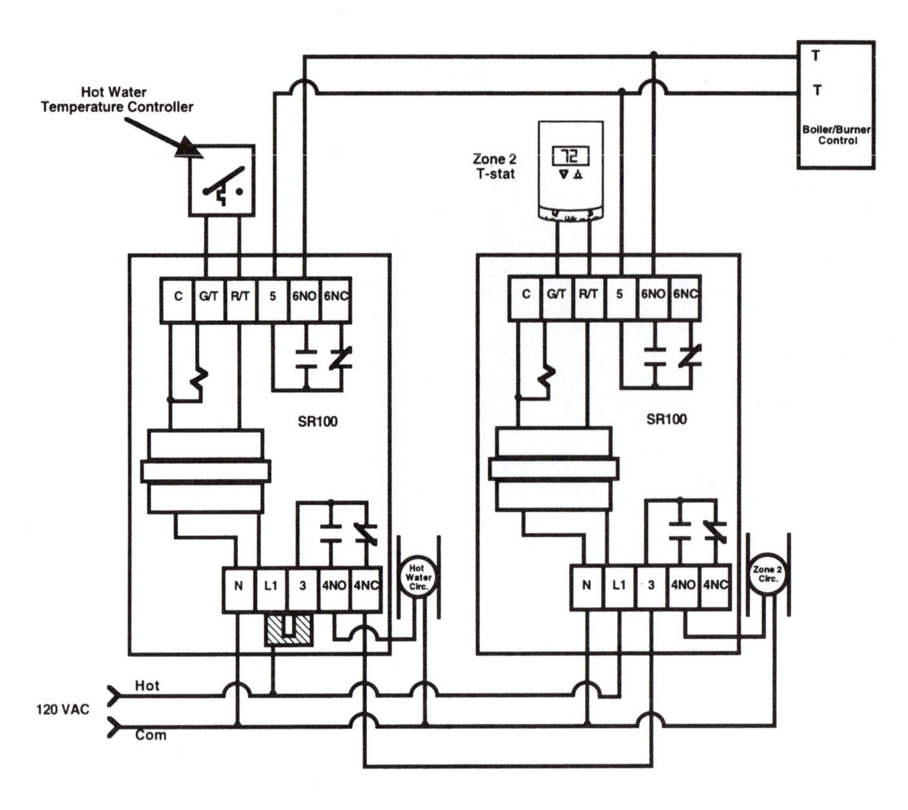

Figure 17–11 Zone control with domestic hot water priority. *(Courtesy of Erie Controls.)*

EXERCISE:
You have just changed the circulating pump, and reconnected the boiler. The room thermostat is calling for heat. What voltage would you measure between each of the following pairs of terminals?

1. Across the CR coil?
2. Across a CR switch?
3. Across R-B on the aquastat?
4. Across R-W on the aquastat?
5. Across the circulating pump motor?

ELECTRONIC BOILER CONTROLLER

Modern electronic controls have been marketed to replace the above control schemes (figure 17–8). Figure 17–9 shows the connections for a single-zone operation where the room thermostat operates both the circulating pump and gas valve or primary controller on the boiler. This is slightly different from the scheme where the boiler is maintained at a constant temperature as in figure 17–2. With many modern boilers, the boiler is allowed to cool between each cycle, because the recovery time is very quick. However, if the boiler is being used to supply domestic hot water, then a minimum temperature of 160°F must be maintained by controls outside the electronic boiler control.

The electronic boiler controllers may also be used in multiples to provide zone control (figure 17–10), or to provide domestic water priority (figure 17–11).

Figure 17–12 shows multiple-zone controllers that can incorporate up to six double-pole single-throw zone relays to control up to six circulators and a boiler operating control (only three zones are shown). There is a field-selectable priority for zone 1, which eliminates the need for an additional relay to provide domestic hot water priority. When priority is selected, zone 1 (which is the domestic hot-water) has priority control over zones 2 through 6. If any zone is calling for heat and a demand for domestic hot water is detected, zone 1 turns on the burner and domestic hot water circulator. Circulators for zones 2 through 6 are disabled until the domestic hot water demand is satisfied.

Another feature that may be used with the above system is that it only allows the priority zone to take charge for a maximum of one hour. This is provided to prevent freeze-up of the heating zones if there is a malfunction in the priority zone. This feature may be enabled or bypassed with a switch on the face of the controller.

QUESTION:
The boiler in figure 17–12 won't run. You read 24V between terminals X1 and X2. Identify all of the following components that may be the culprit:

a. Transformer
b. Electronic controller
c. Gas valve
d. Room thermostat

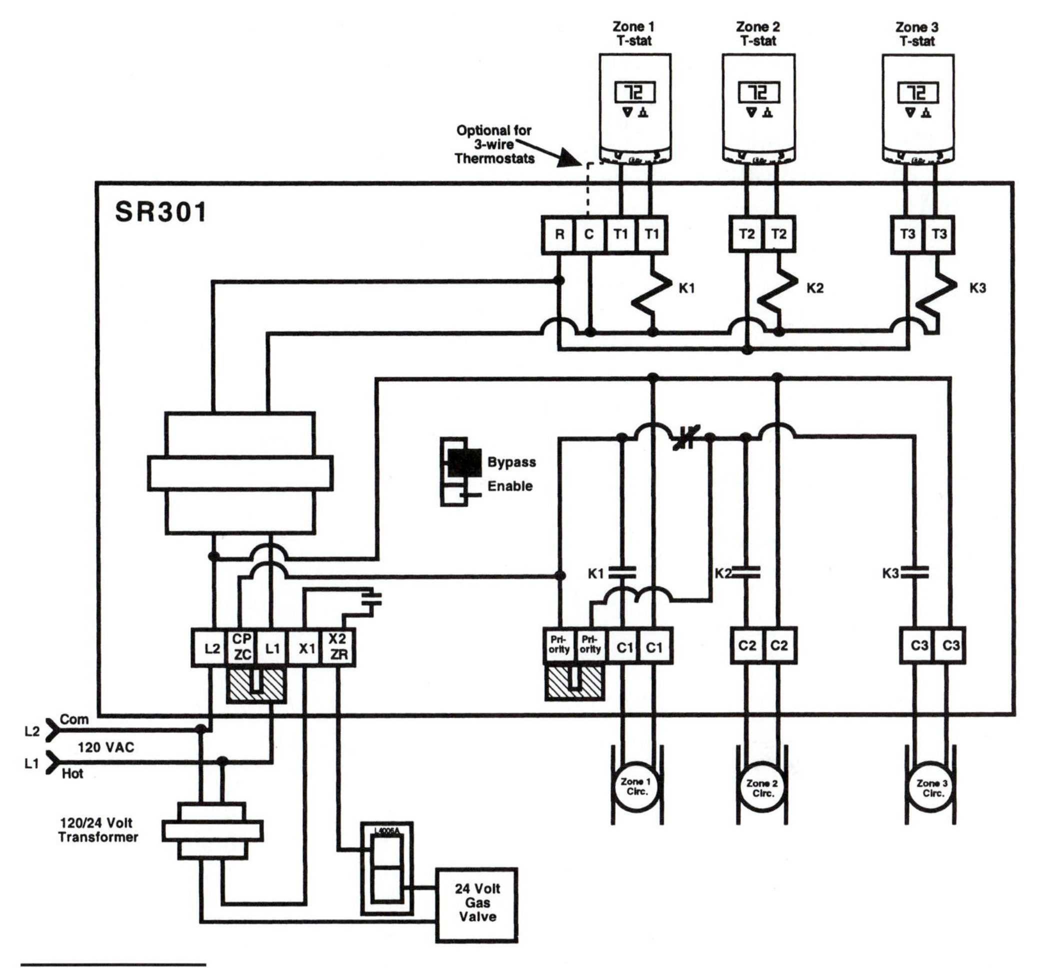

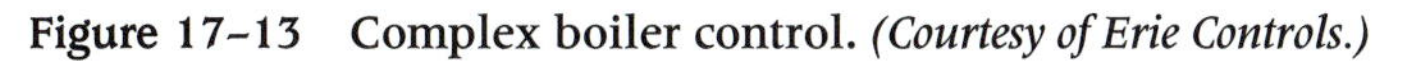

Figure 17–12 3-zone field-selectable priority. *(Courtesy of Erie Controls.)*

Figure 17–13 Complex boiler control. *(Courtesy of Erie Controls.)*

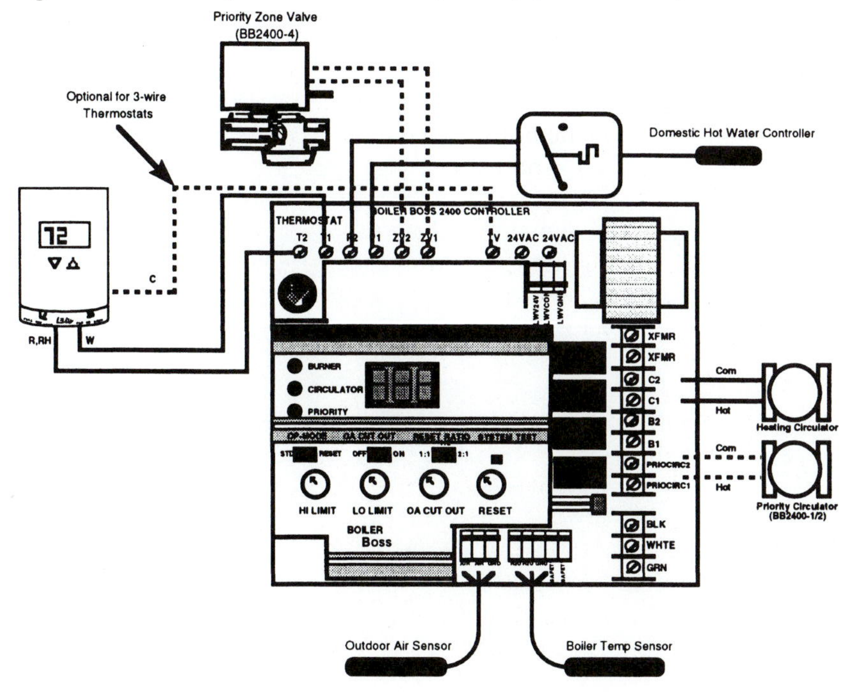

MULTI-FUNCTION ELECTRONIC CONTROLLER

The controller shown in figure 17–13 represents another step in control complexity and sophistication. It provides the following control features:

1. High and low limit protection.
2. Control of the circulating pump.
3. Automatic reset of the boiler temperature with outside air temperature.
4. Warm weather shutdown.
5. Thermal safety cut-off.
6. Constant operation of the circulating pump in the event of a burner failure.
7. Automatic disabling of the priority domestic hot water zone in the event of a domestic hot water failure.

When control becomes this complex, you often are not even given a wiring diagram that shows how it is accomplished. The most you might have is a diagram such as figure 17–13 that shows the inputs and the outputs to the electronic "black box." For this system, the inputs include the room thermostat, the outdoor air sensor, the boiler temperature sensor, and the domestic hot water controller. The outputs are to the circulating pumps (or zone valves), and to the status-indicating **LEDs**. LED stands for "light emitting diode." You can think of it as a small light in an electronic system. The numbers that are formed on digital displays (even including your wrist watch) may be formed using individual LEDs.

CHAPTER EIGHTEEN

Heat Pumps

The operation of the heat pump is the same as an air conditioner, except for the addition of a **reversing valve** and a defrost cycle.

REVERSING VALVE

The reversing valve is shown in figure 18–1. A cutaway view is shown in figure 18–2. Hot gas from the compressor enters the top port, and is directed out to either the indoor coil or the outdoor coil, depending on the position of the piston inside the barrel. Whichever coil is receiving the hot gas from the compressor acts as the condenser. The other coil acts as the evaporator. Figure 18–3 shows the direction of refrigerant flow for heating and for cooling.

When the position of the reversing valve changes, the operation of the unit changes from cooling to heating. In the cooling mode, the indoor coil is cold, and absorbs heat from the indoor air. The outdoor coil is warm, and the heat is rejected to the outside air. In the heating mode, the outdoor coil is cold, and it absorbs heat from the outside air. The indoor coil is warm, and the heat is rejected into the room. Some manufacturers energize the solenoid coil on the reversing valve to put it into the heating position. Other manufacturers energize the solenoid coil in the cooling mode. There is no standard. Some manufacturers believe it is more important to provide heat in the event of a solenoid coil failure, while others believe that it is more important to provide cooling in the event of a solenoid coil failure. It will be up to the service technician to determine which method is being used.

There are two general schemes for controlling the reversing valve. They are shown, in general terms, in figure 18–4. SW1 is a switch that is operated by the operation of the room thermostat. In figure 18–4(a), it may be a normally open switch that is closed when the thermostat calls for cooling (usually through a control relay called the reversing valve relay), or it may be a normally closed switch that is opened when the room thermostat calls for heating. SW2 is a switch that is normally open, and closes when the unit needs to go into defrost cycle (usually through a control relay called a defrost relay). The load, L, may be the reversing valve solenoid coil, or a control relay coil whose contacts will operate the reversing valve. In figure 18–4(a), the reversing valve will be energized when the room thermostat calls for cooling or when the defrost control says that it's time for a defrost. When the reversing valve is de-energized, the heat pump will deliver heat. In figure 18–4(b), if the room thermostat needs cooling, it opens SW1. If the defrost control wants a defrost, it opens SW2. Either of these two switches opening will cause L to become de-energized. Therefore, L will be energized only in the heating mode.

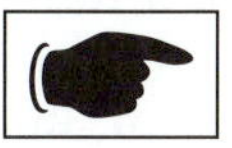

In most cases, if the reversing valve relay switch and the defrost relay switch are wired in parallel, the re-

Figure 18–1 Reversing valve. *(Courtesy of Ranco North America)*

versing valve is de-energized on heating. If the reversing valve relay switch and the defrost relay switch are wired in series with the reversing valve, the reversing valve is energized on heating.

QUESTION:
You are replacing a damaged reversing valve. A label with the diagram in figure 18–5 is attached to the new reversing valve. You have determined that when you set the room thermostat to Cool and set the temperature to 60°F, there is no voltage at the reversing valve solenoid, but when you set the room thermostat to Heat and set the temperature to 85°F, there are 24V at the reversing valve solenoid. Which port on the reversing valve (A, B, C, or D) will you connect to the indoor coil, and which port will you connect to the outdoor coil?

SOLUTION: The diagram tells you that when the reversing valve solenoid is energized (lightning bolt), it will send the hot gas entering port A out port D. You know that for your system, the reversing valve is energized on Heating, so you will want to connect port D to the indoor coil to receive the hot gas from the compressor during the Heating cycle. ■

HEAT PUMP THERMOSTAT

The room thermostat used to control a heat pump might be of the same type that is used to control a standard gas-heating electric-cooling unit. But more likely, it is a special thermostat designed specifically for heat pump service. Some of the differences

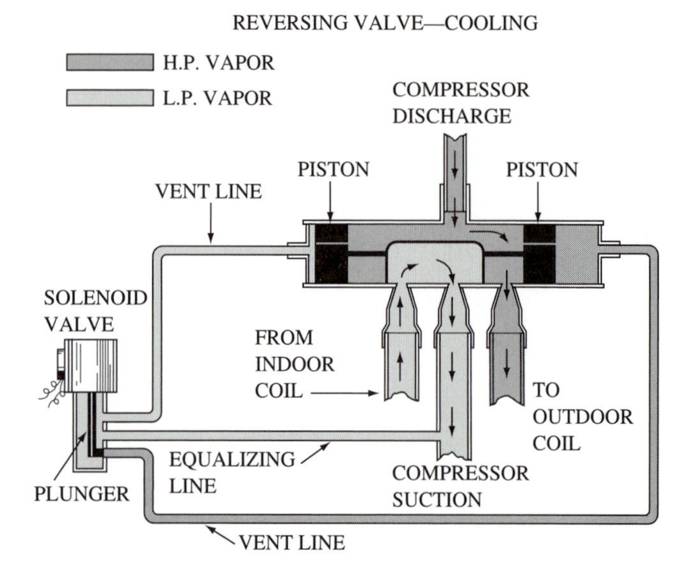

Figure 18–2 Reversing valve cutaway. *(Reprinted from Refrigeration and Air Conditioning, 3rd Ed. by Air Conditioning and Refrigeration Institute, Copyright © 1998, Prentice Hall, Inc. Reprinted by permission.)*

between a heat pump thermostat and a standard gas-heating electric-cooling thermostat are as follows:

1. The heat pump system uses the B or the O terminal on the thermostat subbase. While a standard thermostat may have these terminals, they are usually not used. The thermostat makes R to B when the system switch on the thermostat subbase is set to Heat. It makes R to O when the system switch is set to "Cool" (to help you remember . . . R to **O** on **C**o**o**l).

2. The standard thermostat makes R to Y only on a call for cooling to energize the compressor contactor. The heat pump thermostat makes R to Y on a call for heating as well as on a call for cooling. This

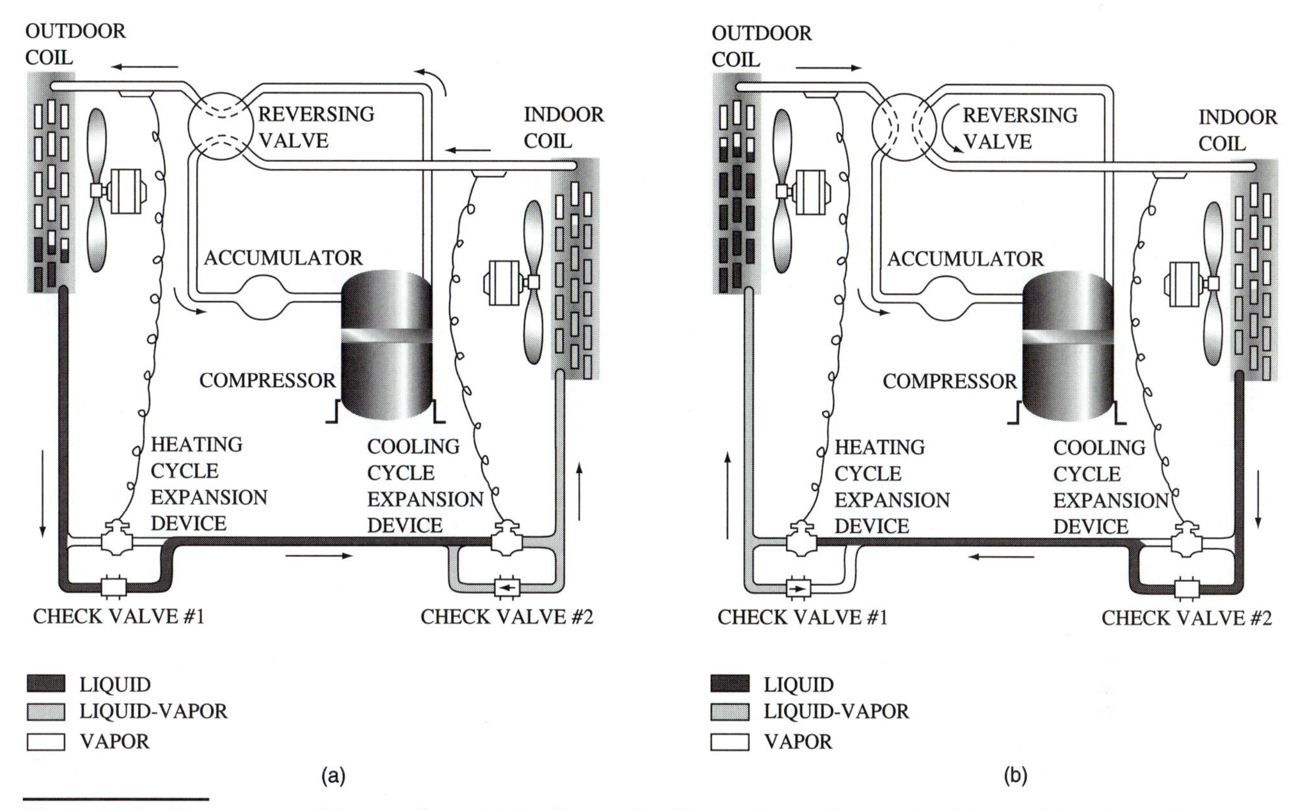

Figure 18–3 Heat pump refrigerant flow. (a) Cooling cycle. (b) Heating cycle. *(Reprinted from Refrigeration and Air Conditioning, 3rd Ed. by Air Conditioning and Refrigeration Institute, Copyright © 1998, Prentice Hall, Inc. Reprinted by permission.)*

is necessary because the heat pump system needs to run the compressor for heat.

3. The standard thermostat makes R to G (energizing the blower relay) only on a call for cooling, or when the fan switch is moved from Auto to On. The heat pump thermostat makes R to G on a call for heating, as well as on a call for cooling. This is necessary because on heating, the indoor fan acts as the condenser fan. It must start at the same time as the compressor, or the head pressure will go too high.

Figure 18–4 Two schemes for controlling the reversing valve.

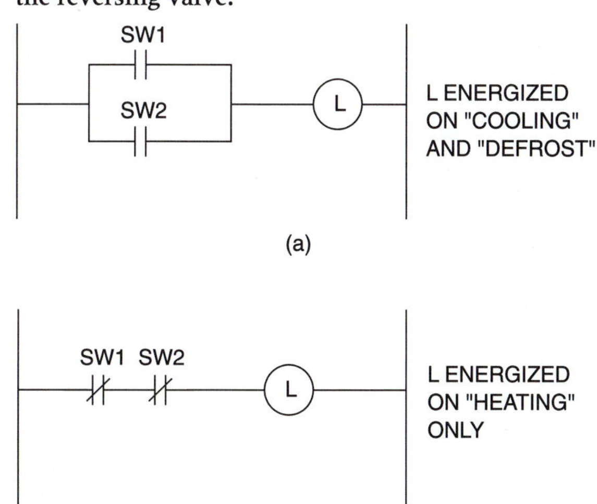

HEAT PUMP DEFROST CYCLE

Unlike a conventional heating/cooling system, the heat pump requires a defrost cycle. A standard air conditioner does not require a defrost because under normal conditions, the evaporator should never operate at a temperature below freezing. However, when the heat pump operates in the heating mode, it might be cooling outdoor air that is 40°F or less, and the outdoor coil that is cooling the outdoor air could be well below freezing. Moisture from the outside air condenses on the cold outdoor coil and freezes. Without a means to defrost the outdoor coil,

Figure 18–5 Reversing valve label.

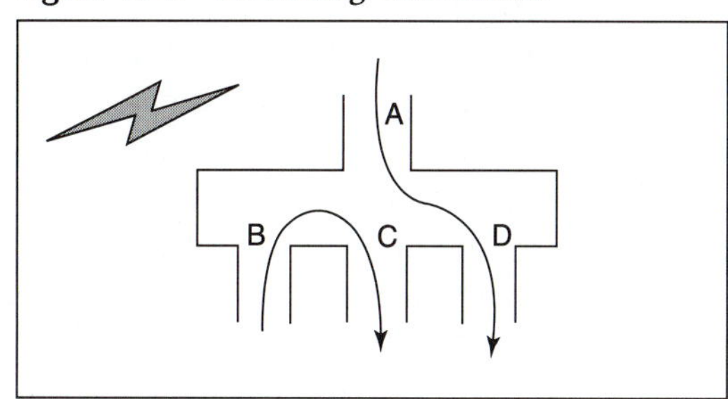

the airflow would be blocked within a few hours of operation on cool days.

During the defrost cycle, the following happens:

1. The reversing valve is energized or de-energized as required to return the system to the cooling mode of operation. The hot gas from the compressor will be directed to the outdoor coil, melting the ice from its fins.

2. The outdoor fan is switched off. This will allow the coil to get warmer, without losing heat to the cool outside air.

3. Supplemental electric heat will be energized in order to offset the room cooling that is being done by the cold indoor coil. (In exceptionally mild climates, where defrost seldom occurs, heat pumps are often supplied without supplemental heat.)

QUESTION:
During defrost, why not turn off the indoor fan also, thus preventing the cooling of the room and making it unnecessary to use the supplemental heat?

SOLUTION: The heat that is being supplied to melt the ice from the outdoor coil is actually heat that is coming from the room. If the indoor fan were shut off during defrost, there would not be sufficient heat available to defrost the outdoor coil. ■

HEAT PUMP CIRCUITS

Various manufacturers use different methods to accomplish control of the heat pump. However, they all accomplish essentially the same things. Some of the differences are as follows:

1. Some heat pumps energize the reversing valve on cooling and defrost; others energize the reversing valve on heating.

2. Some heat pumps use a heat pump thermostat; others use relays to accomplish the same thing with conventional thermostats.

3. Some heat pumps use the B terminal on a heat pump thermostat; others use the O terminal.

4. Some reversing valves are in the 24V circuit; others are in the 230V circuit.

5. Many different methods are used to sense when the heat pump requires a defrost and to put the system into the defrost mode of operation.

HEAT PUMP CIRCUIT NO. 1 (STANDARD THERMOSTAT, AIR SWITCH DEFROST)

Figure 18–6 shows an early design of heat pump circuit that was used before heat pump thermostats became popular. In order to understand the sequence for any heat pump system, your first analysis should be to determine whether the reversing valve solenoid is energized on heating or on cooling. In this case, the RVR switch and the DFR switch are in series with the RVS, so you suspect that the reversing valve will be energized on heating (see practice pointer above). To confirm your suspicion, you note that when the room thermostat makes R-W1 (heating), it energizes the RVR coil, closing the RVR switch, and energizing RVS. Your suspicion is confirmed.

SEQUENCE OF OPERATION

On a call for cooling, the room thermostat makes R-Y and R-G. The Y terminal energizes CR coil, operating two sets of contacts. One set of CR contacts energizes the compressor contactor, while another set of CR contacts brings on the outdoor fan motor. The G terminal energizes the indoor fan relay (IFR) coil, bringing on the indoor fan motor in the 230V circuit. The unit runs the same as if it were a conventional air conditioner.

On the heating cycle, the room thermostat only closes the R-W1 contact, which energizes the reversing valve relay (RVR). The RVR is simply a control relay with multiple sets of contacts. One set of RVR contacts is wired in parallel with the R-Y switch to energize the CR and operate the outdoor unit (compressor and outdoor fan). It also energizes the IFR. A second set of RVR contacts in the 230V circuit complete a circuit to energize the reversing valve solenoid (RVS). The unit operates the same as in the cooling mode, except that the reversing valve has operated to direct the hot gas to the indoor coil instead of the outdoor coil. If the heating capacity of the heat pump is insufficient to maintain room temperature, the room thermostat will make the second stage of heating, R-W2. This will energize the heating contactor (HC), which will, in turn, bring on the supplemental electric strip heater (SH).

The defrost cycle is initiated by an air pressure switch (AS) that senses when the airflow across the outdoor coil is becoming blocked due to ice. When the air pressure at the suction side of the outdoor fan decreases, it closes AS, energizing DFR, and de-energizing RVS. This places the reversing valve in the

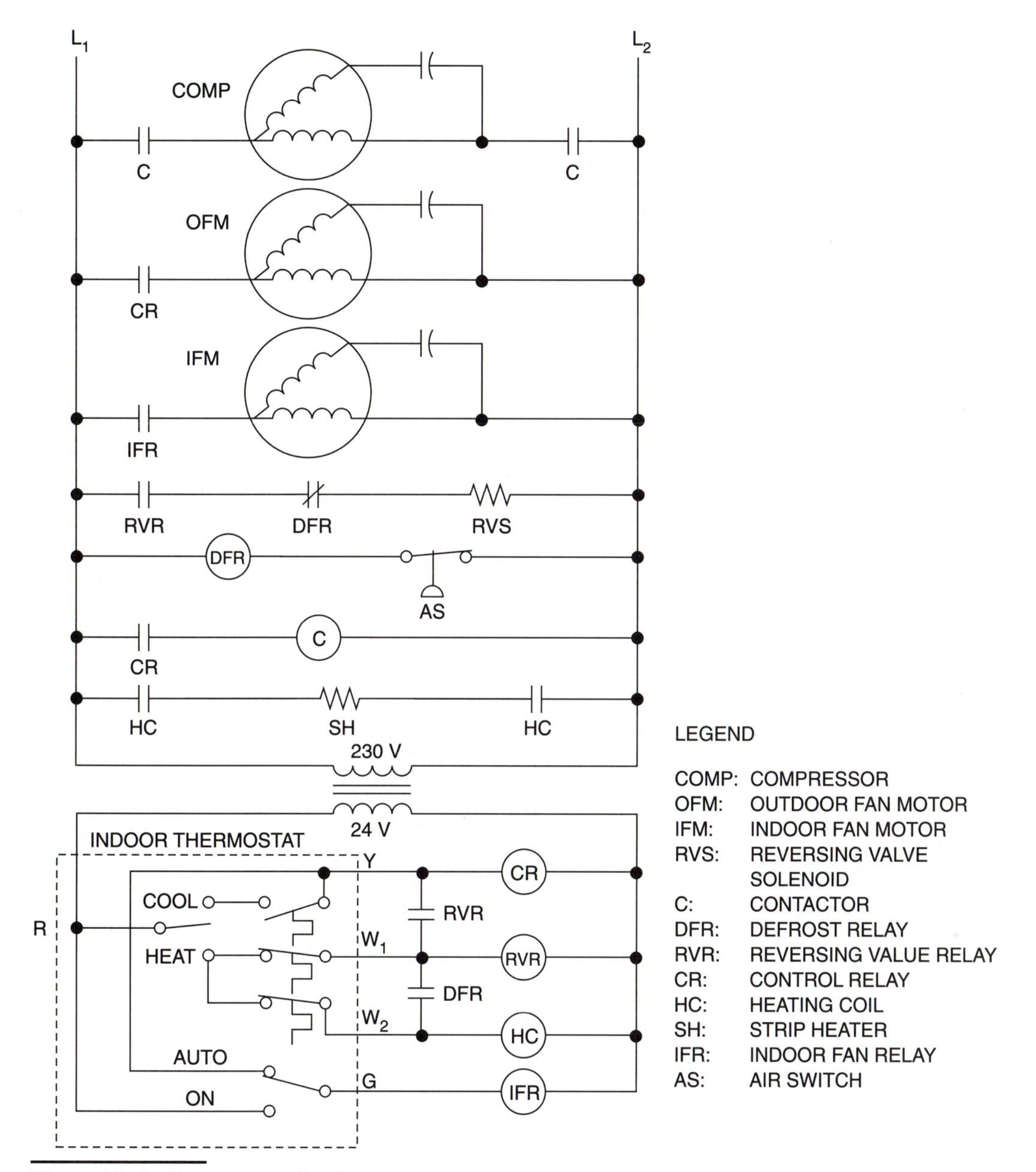

Figure 18–6 Heat pump circuit no. 1.

position that will direct hot gas to the outdoor coil. At the same time that the DFR energizes, a second set of DFR contacts in the 24V circuit closes, energizing the heating contactor, HC1. This will provide supplemental heat during the period of defrost. When the temperature sensing portion of the AS switch determines that the outdoor coil has been defrosted, it opens, de-energizing the defrost relay.

Outside Air Override

Figure 18–7 shows an almost identical circuit to the previous circuit, but with one important change. A thermostat sensing outside air temperature (OT) has been added in series with the contactor for the strip heater (HC1). The purpose of the outdoor air thermostat is to keep the strip heater off whenever the temperature outside is warm enough that the heat pump operation alone should be sufficient to maintain comfort conditions. Without the outdoor air thermostat, the following undesirable result can occur.

Suppose a homeowner turns the heater off at the thermostat upon leaving in the morning. It is 55°F outside, so the temperature in the house drifts down to 67°F over a period of several hours. When the homeowner returns, she sets the thermostat switch to Heat. The set point is 72°F. The thermostat is set up so that it will make R-W1 if the room temperature is below 71°F, and it will also make R-W2 if the room temperature is below 70°F. Therefore, with the room

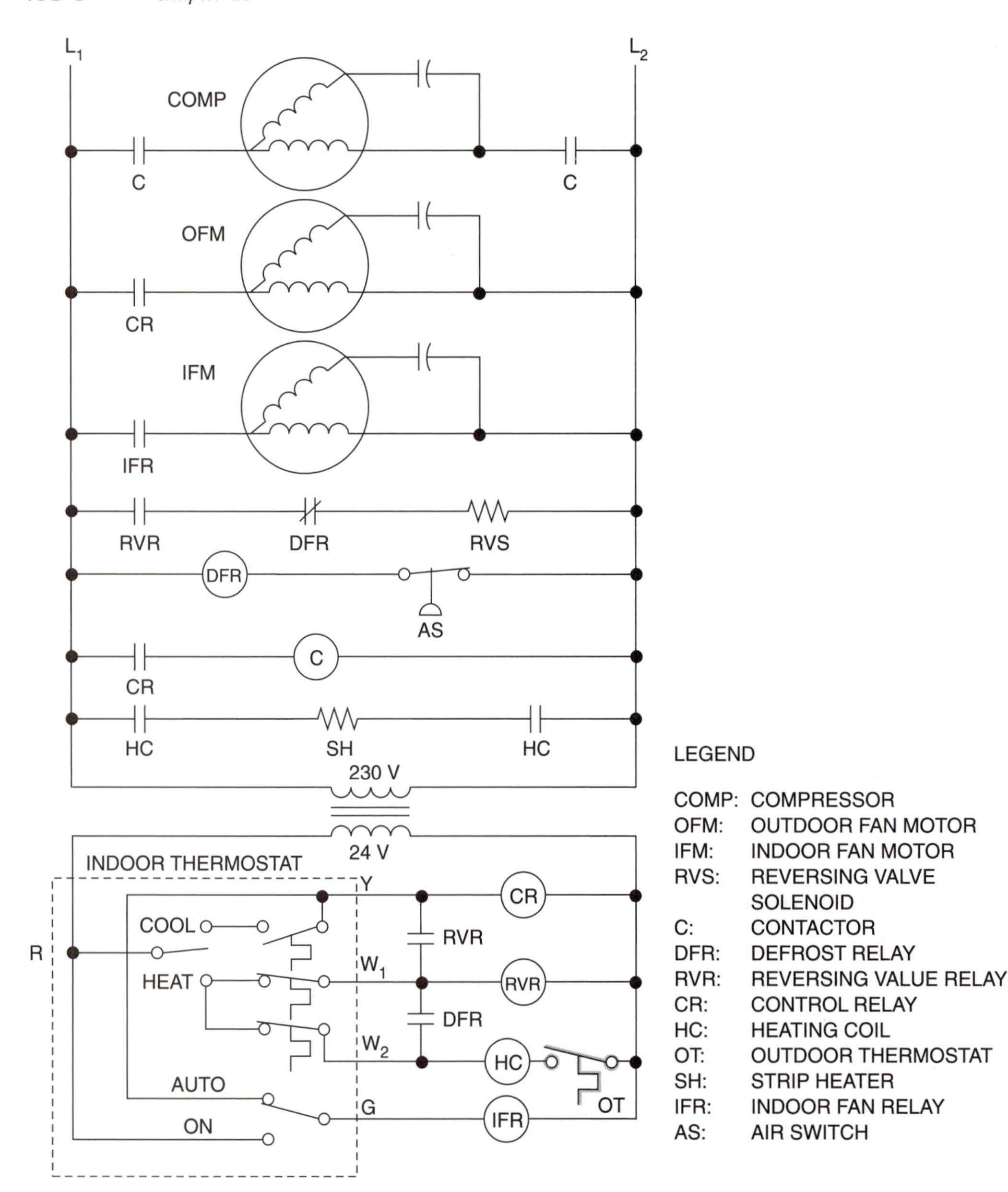

Figure 18–7 Circuit with outside air override.

at 67°F, the thermostat will bring on both the heat pump and the electric strip heater. It is far more expensive to bring the house up to 72°F using the strip heater, than to keep the strip heater off and simply let the heat pump bring the house temperature up to set point.

The outdoor air thermostat is typically set for 25°F to 45°F. So long as the temperature outside is above this set point, the electric heat is locked out. The heat pump, operating alone, should be able to bring the house temperature up to set point. It will take longer, but it will be much less expensive. When the outside air drops below the outdoor air thermostat set point, the heat pump might not be able to satisfy the room demand by itself, and the strip heater is allowed to operate.

HEAT PUMP CIRCUIT NO. 2 (STANDARD THERMOSTAT, OLD STYLE TIMER)

Figure 18–8 shows another heat pump circuit, similar to the previous circuit, except for the operation of the defrost cycle.

This defrost cycle is operated by a time clock, DTM (defrost timer motor). It has two sets of contacts (DT) that are operated off a cam that is driven by the DTM. Every few hours the DTM will cause the normally open contact to close momentarily, and thirty minutes later, the DTM will cause the normally closed contact to open momentarily. The defrost thermostat (DFT) senses whether or not a defrost is

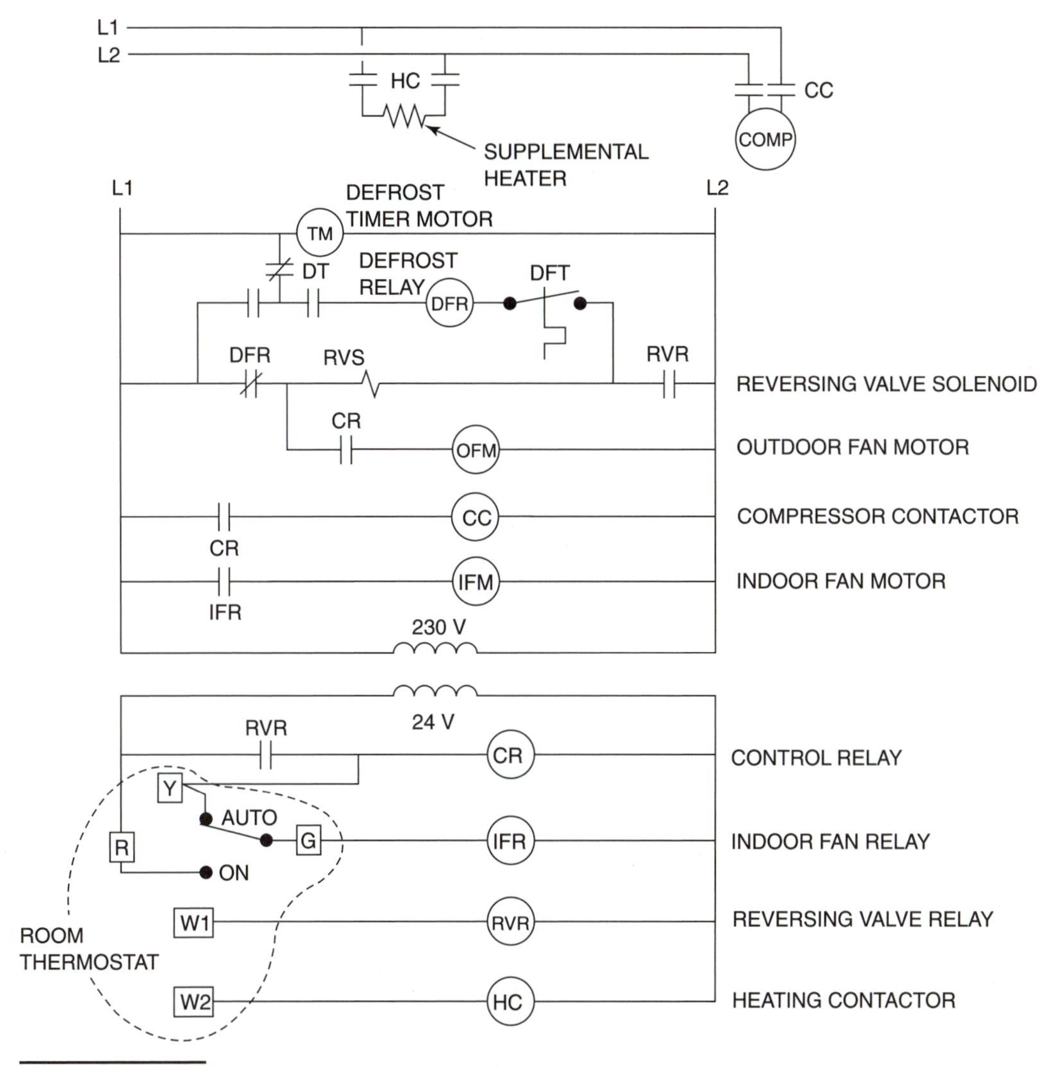

Figure 18–8 Heat pump with standard thermostat and old style timer.

needed by sensing the temperature of the outdoor coil. If the DFT is open, indicating that a defrost is not required, then the operation of the DTM and its switches has no effect. However, when the DFT is closed, the defrost relay (DFR) becomes energized as soon as the normally open DT switch closes. The DFR is simply another control relay with two sets of contacts. The DFR remains energized, even after the normally open DT contact reopens, because it gets "sealed-in" by the normally open DFR contacts.

With the DFR energized, the normally closed DFR contacts open, de-energizing both the RVS and the OFM. Hot gas is directed to the outdoor coil, accomplishing a defrost. At the same time, the indoor coil is receiving cold refrigerant, so the room temperature drops. The second stage heat on the room thermostat makes R-W2, energizing a heating contactor (HC) which energizes the supplemental heater. When the outdoor coil has been defrosted, it is sensed by the DFT which opens, de-energizing the DFR, and the system returns to normal operation. If, for some reason, the DFT does not open, the defrost will terminate anyway when the DTM causes the momentary opening of the normally closed DT contact.

TYPES OF MODERN DEFROST CONTROLS

Generally, the modern controls that place the heat pump into the defrost mode fall into two categories. The mechanical type (figure 18–9) consists of a clock motor that operates an SPDT switch that toggles between outdoor fan operation and placing the reversing valve in the cooling mode so that it will send hot gas to the outdoor coil for defrost. An alternative mechanical type of defrost control (figure 18–10) can use the air pressure difference across the outdoor coil to initiate the defrost, instead of a clock.

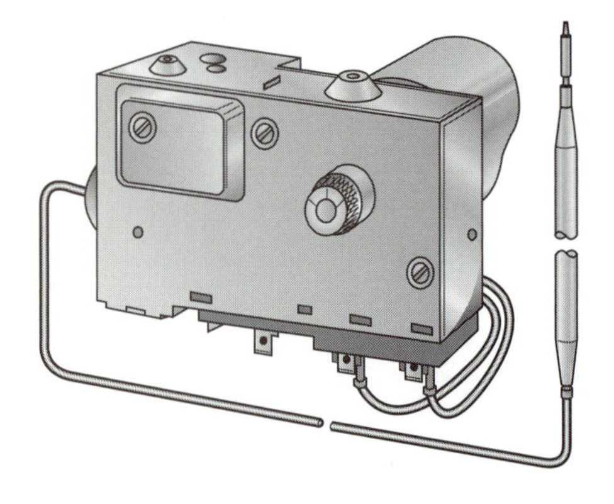

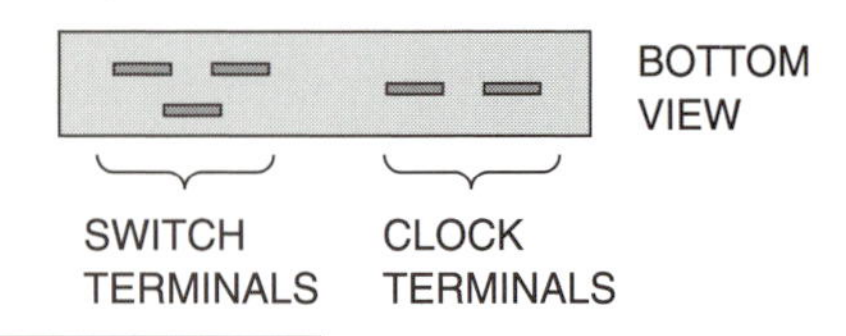

Figure 18–9 Mechanical defrost control.

The electronic type of defrost control (figure 18–11) consists of solid-state controls on a printed circuit board (a "black box" control). The printed circuit board logic may be one of two types:

1. Time-initiated, temperature-terminated: The board receives 24V power. There is a fixed time-interval between defrost cycles (there are usually several different time intervals available, one of which is selected). There is a defrost temperature sensor on the outdoor coil. When sufficient time has elapsed for the circuit board to place the heat pump into defrost, it energizes a defrost relay. When the defrost temperature sensor determines that the outdoor coil is defrosted, it interrupts the 24V circuit to the printed circuit board, and the defrost relay switches return to their normal position (heating).

2. Demand defrost: The printed circuit board for the demand-defrost system has two permanently attached thermistors which sense temperature on the outdoor coil and the outdoor ambient temperature. The board does not put the system into defrost on any regular schedule. It energizes a defrost relay (mounted on the board) whenever it is required.

Figure 18–10 Air temperature/pressure switch.

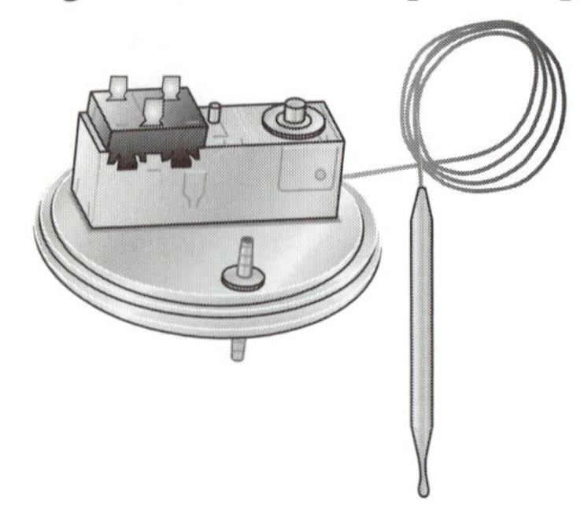

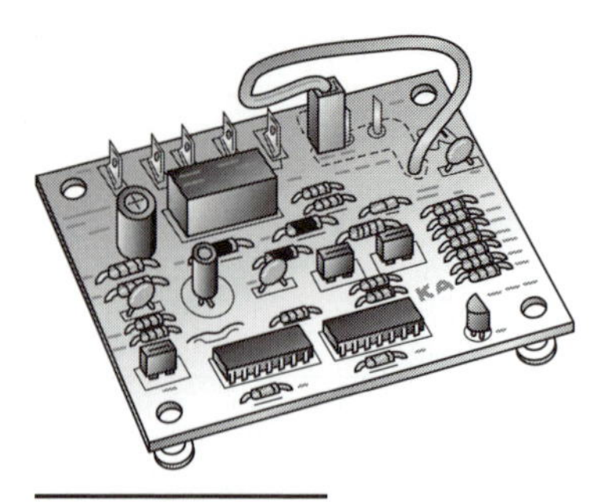

Figure 18–11 Electronic defrost control.

HEAT PUMP CIRCUIT NO. 3 (SPDT MECHANICAL DEFROST CONTROL)

The reversing valve for this system (figure 18–12) is de-energized in the heating mode, and energized in the cooling mode and the defrost mode.

The mechanical defrost control switch is normally in the 2-1 position, allowing the outdoor fan to operate whenever the compressor contactor is energized. Every few hours of time-clock operation, the defrost control will analyze the temperature signal being received from the thermal bulb (mounted on the outdoor coil) to determine if a defrost is required. If the sensing bulb is above freezing, nothing happens, at least for another few hours. If the sensing bulb indicates that a defrost is needed, the switch will move from 2-3 to 2-1. This energizes the defrost relay (DR), energizing the reversing valve to send hot gas to the outdoor coil. It also shuts off the outdoor fan. When defrost is complete, the switch in the defrost control returns to the 2-3 position, and normal heating operation resumes.

HEAT PUMP CIRCUIT NO. 4 (ELECTRONIC DEFROST CONTROL)

Figure 18–13 shows an interconnection diagram for a packaged heat pump with a heat-pump thermostat and electronic defrost control. It provides one stage of cooling and two stages of heating (as an optional accessory). This diagram can be intimidating to the new service technician, so it has been redrawn in a somewhat simplified ladder diagram format in figure 18–14. In many cases, you can troubleshoot the circuit and determine the problem without attempting to understand the entire se-

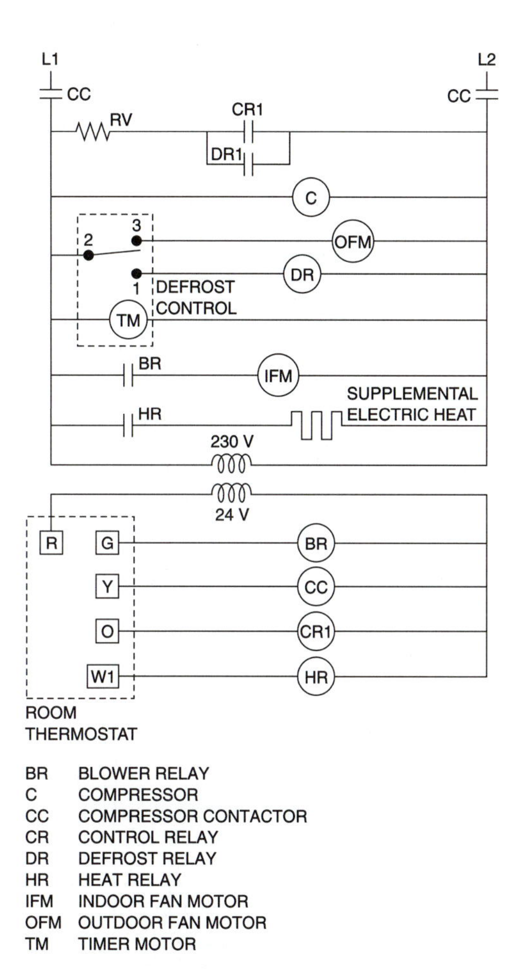

Figure 18–12 Heat pump circuit no. 3.

quence of operation. For example, if the compressor and outdoor fan won't run, you can always go to the compressor contactor to determine if the coil is energized. If it is not, it is simple to determine from the wiring diagram that the contactor coil is in series with the high pressure switch and the room thermostat. Those two switches can be checked for voltage drop to determine which is interrupting the power to the contactor coil.

But sometimes you must determine the whole sequence of operation in order to diagnose the problem. The cooling portion of the circuit is similar to any air conditioner. When the thermostat makes R-Y and R-G, the compressor contactor and blower relay coil are energized, starting all of the cooling components. One question that always arises on a heat pump circuit is whether the reversing valve is energized to place the system in cooling mode or in heating mode. There are two independent ways of arriving at that answer:

1. The reversing valve relay (RVR) is energized when the room thermostat makes R to B, which happens when the room thermostat system switch is set to Heat. When the RVR is energized, it closes contacts to energize the reversing valve (RV). Therefore, the reversing valve is energized during the heating mode.

2. During defrost, the reversing valve goes to the position that will produce cooling (so that hot gas is directed to the outdoor coil, defrosting it). When the defrost relay (DFR) coil is energized, calling for defrost, it opens the normally closed contacts between terminals 1 and 2, de-energizing the reversing valve. Therefore, during defrost or cooling, the reversing valve is de-energized, so it must be energized during the heating mode.

Defrost Control

The defrost control is an electronic "black box." If you are not familiar with the particular model being used (of which there are many), you have to figure out its operation based upon what you know it should do. The box has 24V applied to it continuously between the lower 24V terminal and the COMM terminal. The defrost sensor switch senses when a defrost is needed (it might be a thermostat that senses when the outdoor coil is excessively cold, or it might be a pressure switch sensing when there is excessive pressure drop across the outdoor coil). When that happens, the box should provide 24V to the defrost relay coil. This should de-energize the reversing valve, causing it to switch to cooling mode. If you come upon this unit, and

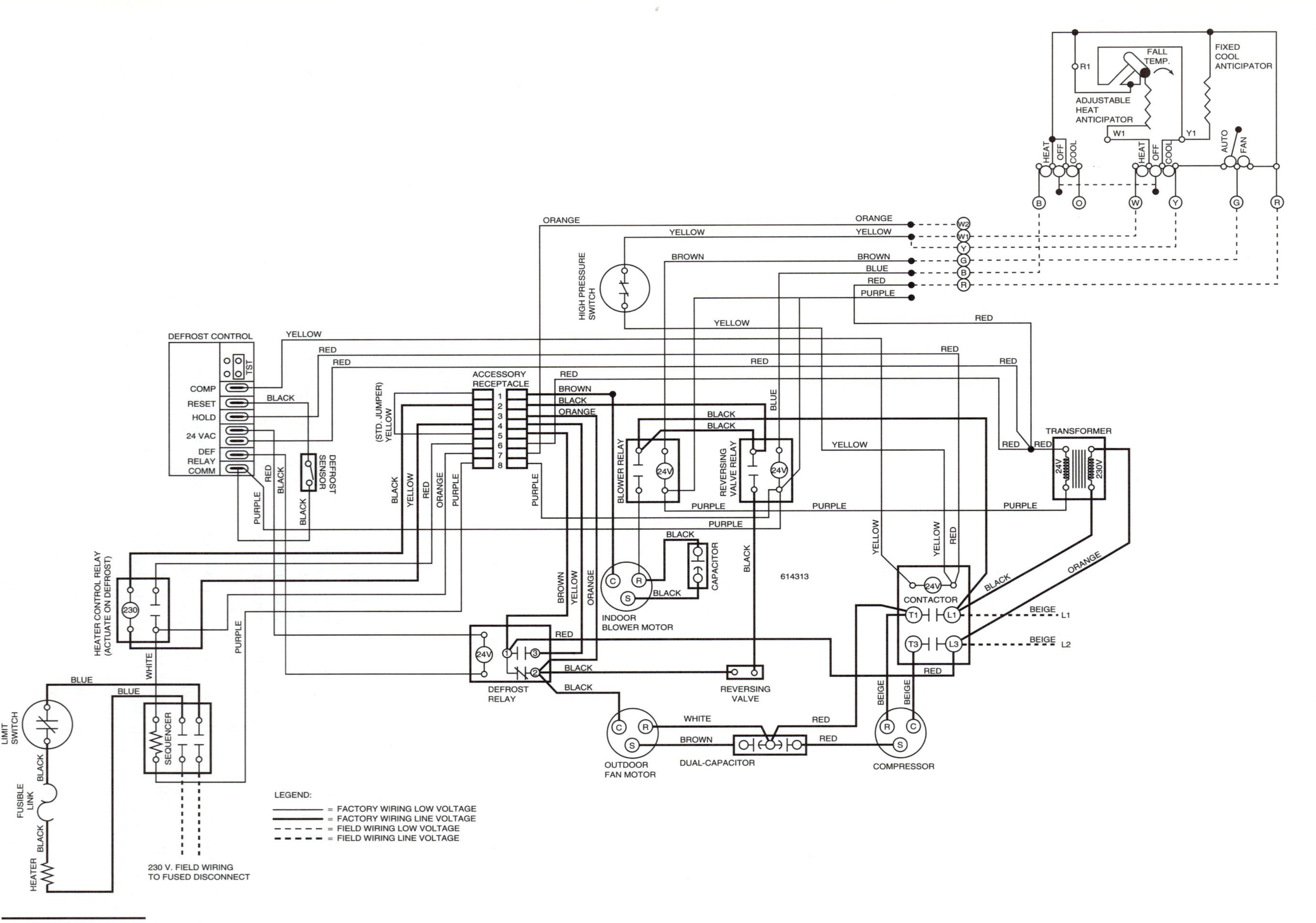

Figure 18–13 Packaged heat pump.

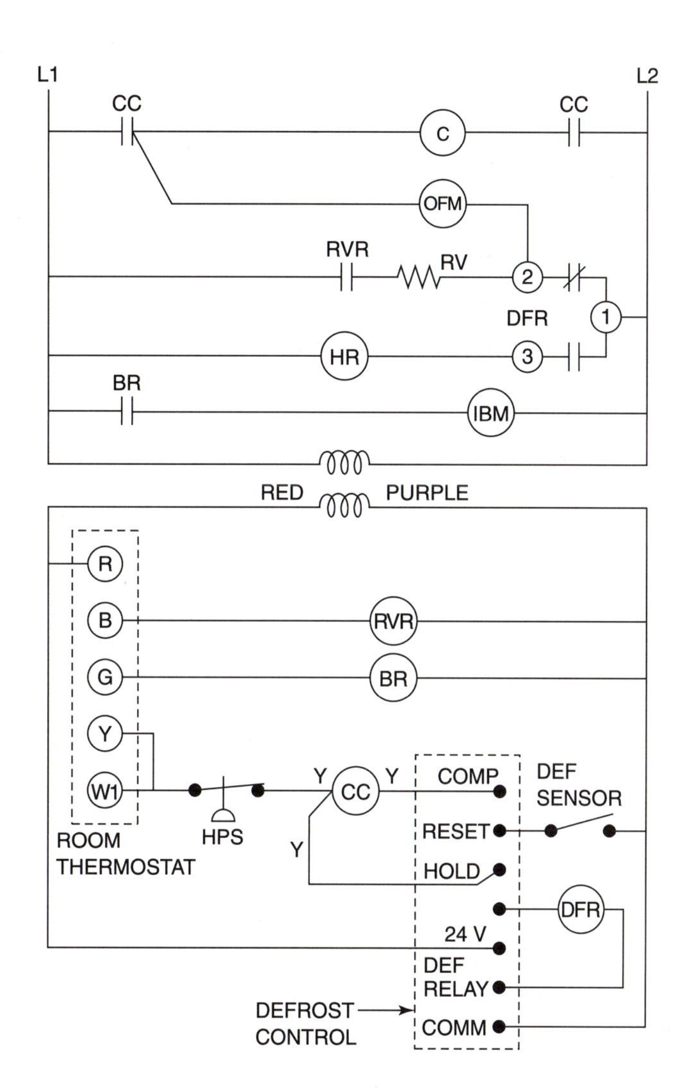

Figure 18–14 Heat pump circuit no. 4 (supplemental heat not shown).

the outdoor coil is covered with ice, you would check the defrost control as follows:

1. Check to see that you have 24V to the defrost control between the lower 24V terminal and the COMM terminal. If you do . . .

2. Check to see that the defrost sensor is closed (zero voltage across it). If it is . . .

3. Check to see if the defrost control is putting out 24V to the defrost relay (between the upper 24V terminal and the DEF RELAY terminal). If it is, then the defrost control is doing its job, and you will need to check the defrost relay coil, contacts, and reversing valve. If there is not 24V to the DFR coil from the defrost control, then the defrost control has failed because all of the required inputs to the defrost control are present, and you are not getting the proper output.

Supplemental Heat

When the unit goes into defrost, it will deliver cold air off the indoor coil. To offset the cooling effect in the room, the 230V heat relay coil becomes energized through the normally open contacts of the defrost relay. This closes the normally open heat relay switch, energizing the sequencer heater element. In this circuit, the supplemental heat is not used, except during defrost.

QUESTION:

For each of the following loads, describe the circuit from L1 to L2 by identifying each terminal and wire color. For example, the circuit through the compressor run winding can be described as L1, beige wire, terminal L1 on the contactor, terminal T1 on the contactor, beige wire, Run terminal on the compressor, Run winding, Common terminal on the compressor, beige wire, terminal T3 on the contactor, terminal L3 on the contactor, beige wire, L2.

1. Outdoor fan motor.
2. Heat relay coil.
3. Indoor blower motor.
4. Reversing valve relay coil.

QUESTION:
Is the room thermostat a heat-pump thermostat or a standard heating/cooling thermostat? How do you know?

If you have difficulty redrawing an interconnection diagram as a ladder diagram, try this. With one color, mark all of the wires on the interconnection diagram that are connected to what would be the left side of the ladder diagram, e.g., L1. Using a different color, mark all of the wires connected to what would be the right side of the ladder diagram, e.g., L2. If you are interested in both the power wiring and the control wiring, make a copy of the wiring diagram. Use one to color the power wiring, and the other to color the control wiring. From this, it should be much easier to construct the ladder diagram. Also, when constructing the ladder diagram, don't try to show all the detail. For example, if the interconnection diagram shows compressor windings, start relay, start and run capacitors, and an overload, you can simply use a circle labeled COMP on the ladder diagram to represent the compressor and all of its associated components.

EXERCISE:
Add the missing heat relay contacts, heat sequencer, heater, fusible link, and limit switch to the simplified heat pump ladder diagram in figure 18–14.

HEAT PUMP CIRCUIT NO. 5

Figure 18–15 shows the same circuit as figure 18–13, with the following additions:

1. The room thermostat has two bulbs to provide two-stage heating. If the room temperature continues to drop after the room thermostat has made R-W1, the thermostat will then make R-W2, energizing the small heater element inside the heat sequencer. The switches inside the sequencer will then close, bringing on the heater (through the safety devices).

2. A PTC has been added around the compressor run capacitor for added starting torque (refer to Chapter nine for a discussion on the PTC device).

3. There is a five-wire control power transformer. The extra tap is provided in the event that the line voltage is 208V instead of 230V.

4. A crankcase heater is provided. It is energized whenever there is power available to the unit.

HEAT PUMP CIRCUIT NO. 6—SPLIT SYSTEM HEAT PUMP

Figure 18–16 is the wiring diagram that applies for a condensing unit only. It is similar to an air conditioning condensing unit, except that it also contains a black box defrost controller, a defrost relay, and a reversing valve.

SEQUENCE OF OPERATION

When the room thermostat makes R-Y (on a call for heating or cooling), it energizes the R1-R2 terminals of the time delay relay (assuming that the HPC and LPCs are satisfied), energizing the compressor contactor. If the thermostat is set for Heat, R-B is also made, energizing the reversing valve coil through the normally closed contacts of the defrost relay.

The defrost control (DFC) gets input signals from a defrost coil sensor (DS) and a defrost ambient sensor (DAS). These are thermistors that sense the temperatures of the outdoor coil and the outdoor air. When the DFC decides to initiate a defrost, it energizes the defrost relay (DR), de-energizing the reversing valve from its terminal 5, and energizing an auxiliary heater from its terminal 6.

QUESTION:
On a no-cooling call, you find the compressor and outdoor fan running, but the unit is actually providing heat. You take the following measurements:
a. R-B = 24V
b. Defrost relay coil = 0V
c. Defrost relay switch 4-5 = 0V
d. Across RV = 0V
What do you suspect is the problem?

QUESTION:
On a no-heat call, the compressor and outdoor fan are running. The voltage measurements are the same as given above. What do you suspect is the problem?

WATER-SOURCE HEAT PUMP

Figure 18–17 is a ladder diagram for a packaged water-source heat pump. Instead of using outside

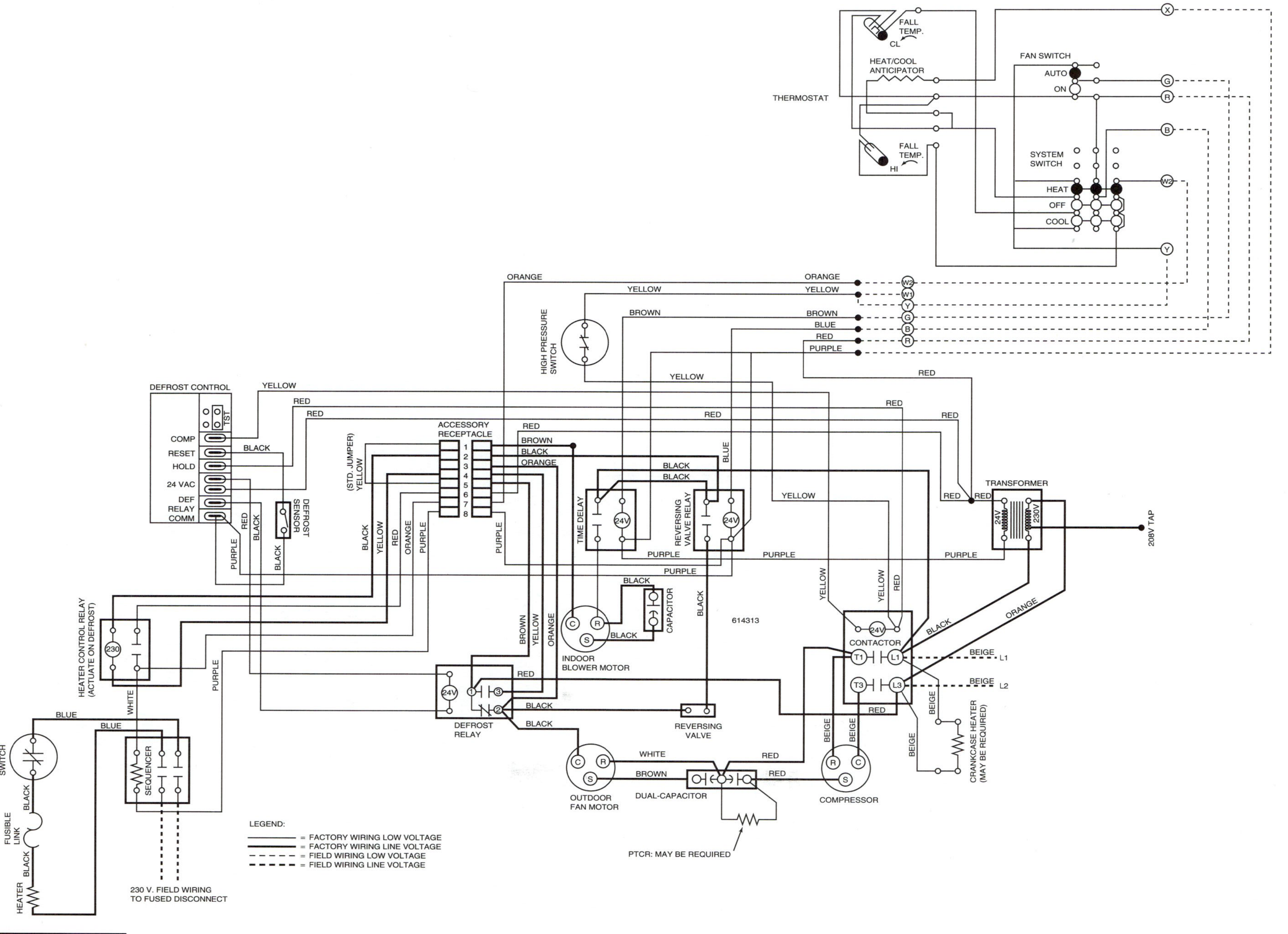

Figure 18–15 Heat pump circuit no. 5.

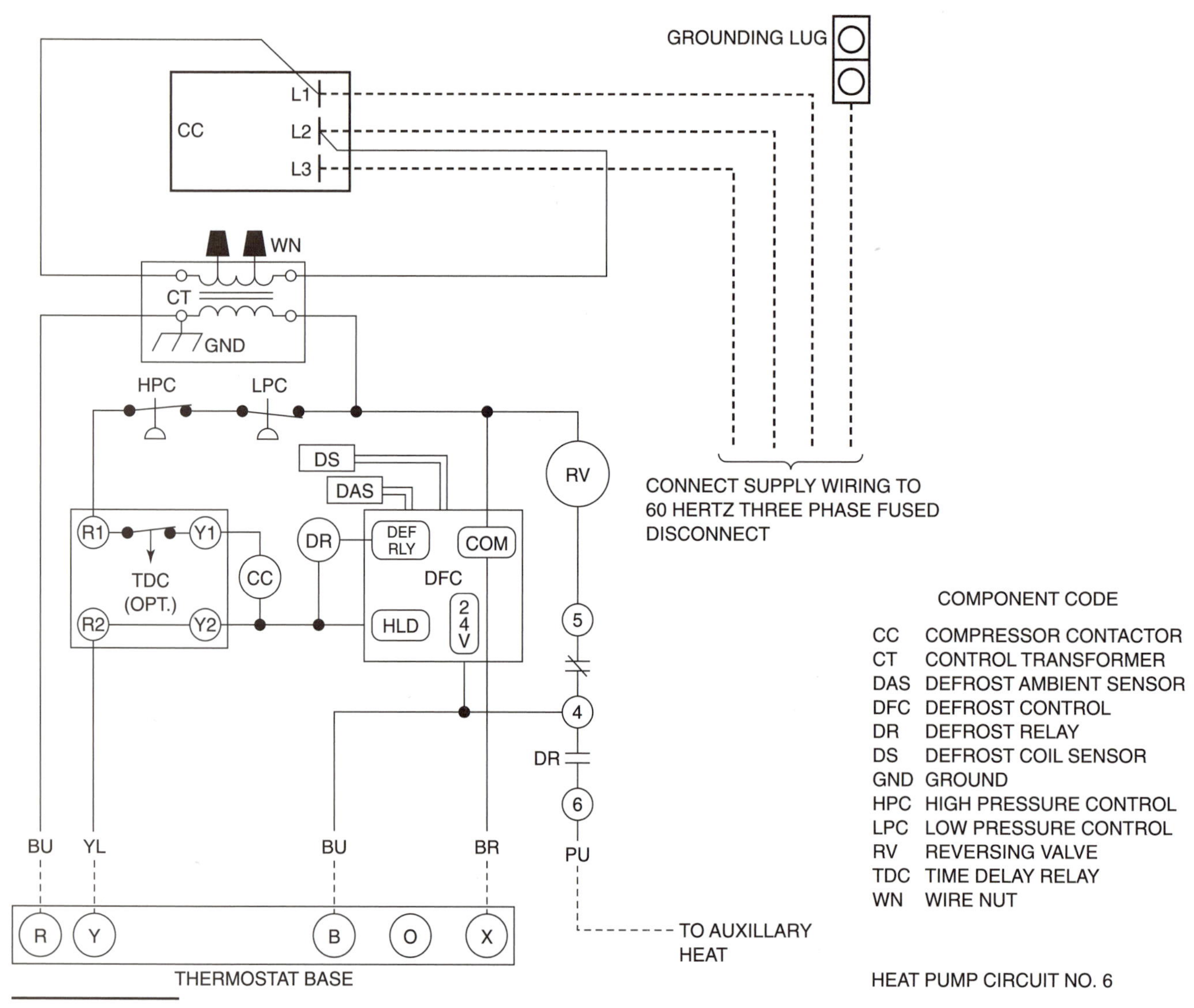

Figure 18–16 Heat pump circuit no. 6.

air as the medium to reject heat during the summer and draw heat from during the winter, this system uses well water. When the room thermostat calls for heating or cooling, it makes R-Y1, energizing the pump relay and energizing the pump motor. R-Y1 on the thermostat also brings 24V to the W1 terminal of the heat pump, energizing the compressor contactor through the time delay, the high pressure cut-out, the low pressure cut-out, and the normally closed contacts of the lock-out relay. The time delay in this circuit is really remarkable, as it is only a two-wire device, as compared to the previously discussed four-wire time delays. You can visualize the operation of the two-wire time delay as follows: when 24V are applied to the one terminal, some time later, it comes out from the second terminal.

The lock-out relay will become energized if either the HPC or LPC opens (see Chapter two to review the operation of the lock-out relay). In addition to locking out the compressor by opening the switch between terminals 4-5, the lock-out relay also closes a switch between terminals 2-4. This completes a circuit through the X terminal of the room thermostat, which will cause a trouble light to illuminate.

QUESTION:
What is the function of the slave relay?

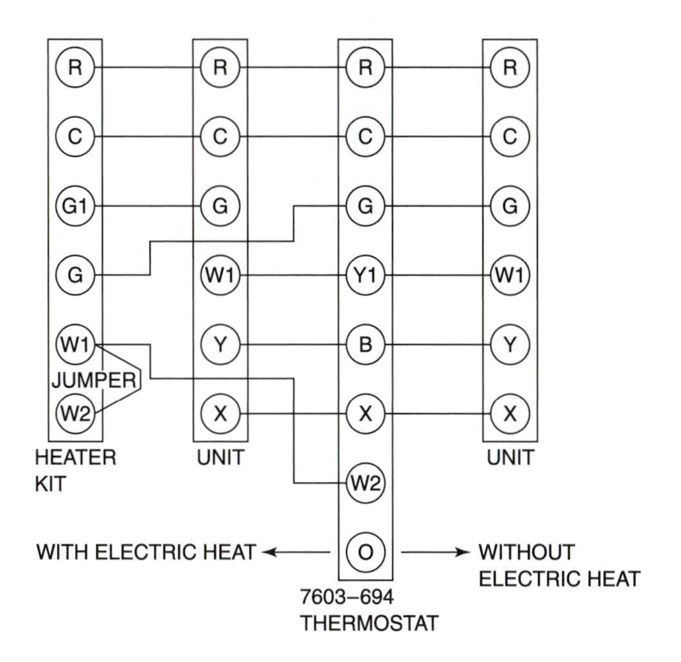

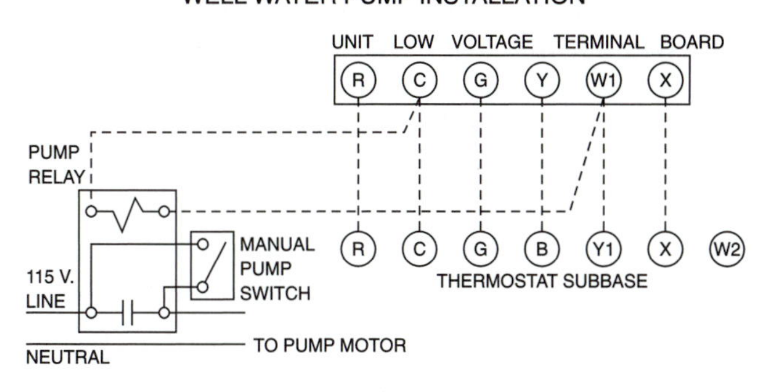

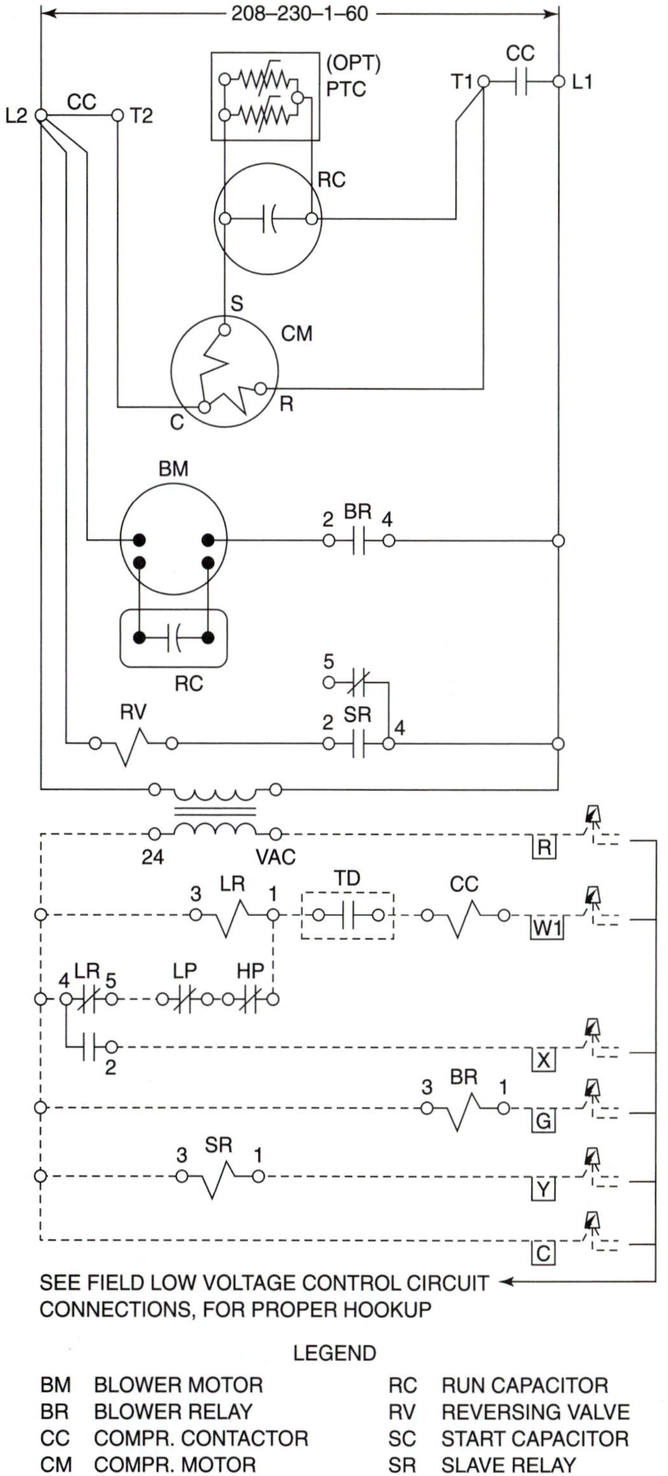

Figure 18–17 Water-source heat pump.

CHAPTER NINETEEN

Ice Makers

There are a great number of ice makers and ice maker manufacturers. There is much less uniformity among those manufacturers about how they control operation, as compared to air conditioners or refrigeration units. Understanding and troubleshooting ice machines is probably the most complex challenge that will be encountered by many technicians. But the task can be simplified if you know what tasks must be accomplished by the ice machine controls.

There are two general types of ice makers, **ice flakers** and **ice cubers**. An ice flaker (figure 19–1) consists of an evaporator formed into the shape of a container in which a level of water is maintained inside by the use of a mechanical float control. The evaporator temperature is 30–40°F below freezing, so ice forms on the inside surface of the evaporator. An auger inside the evaporator rotates slowly and continuously, so that as the ice forms, it is scraped off the inside surface, and pushed up and out the evaporator to a storage bin.

The ice flaker is a continuous operation. It continues until the bin is full, at which point a **bin thermostat** or **bin switch** turns off the refrigeration system. The bin switch usually is a thermostat that opens on a drop in temperature. The sensing bulb is located near the top of the bin. When the ice touches the probe, the switch opens.

The most common complexity that is added to the ice flaker circuit is a safety circuit that prevents the accumulation of ice in the evaporator. This is important, because if the auger becomes frozen in place, the auger motor will exert enough force on the output shaft of the gear box to snap it, or break the internal gears. If the compressor is allowed to run without the auger, the entire water contents of the evaporator may become frozen solid. When water freezes, it expands, and the result will be a cracked evaporator. The protection of the ice flaker generally takes two forms:

1. Auger delay—When the bin switch signals that the bin is full, it turns off the compressor. But the auger motor is allowed to run for a short while longer to clear all of the ice that has already been formed in the evaporator.

2. Proof of auger rotation—Two methods are used that check for proper auger operation before the compressor is allowed to start. One uses a centrifugal switch on the auger motor. This switch is wired into the compressor circuit, and if the auger motor doesn't run, the compressor won't start. The second method is to sense the auger motor amps. If the amps get too high, a switch will open, de-energizing the compressor motor circuit.

The ice cuber is far more complex than the ice flaker. Instead of being a continuous process like the ice flaker, it is a batch process. That means that it goes through different cycles of operation to make a batch of ice, and when a batch has been completed, the sequence of operation repeats. One cycle consists of a **"freeze" cycle** and a **"harvest" cycle.**

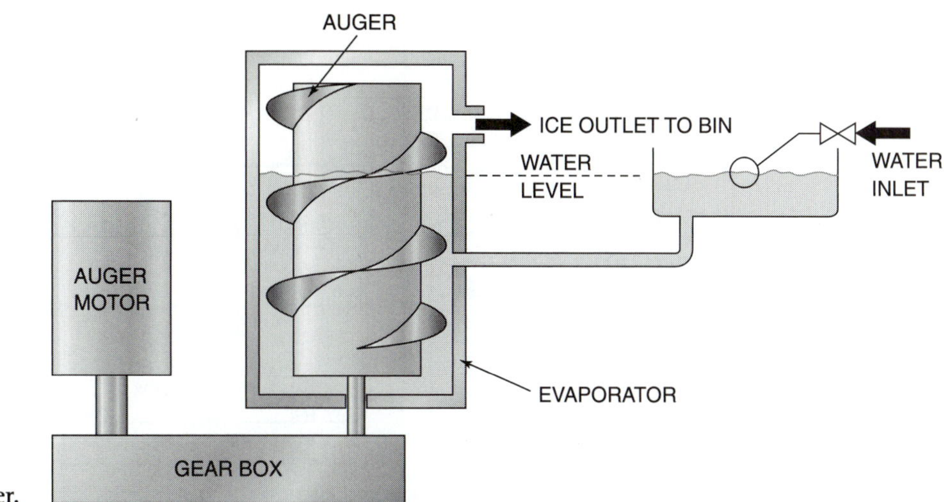

Figure 19–1 Ice flaker.

During the "freeze" cycle, a pump circulates water over a cold evaporator that has been formed into a shape that will make cubes (figure 19–2). As the water circulates over the evaporator, the ice cubes form. After the cubes are fully formed (usually 15 to 25 minutes), some method is used to sense that they are ready. That activates the "harvest" cycle in which the pump motor stops, the condenser fan motor stops, and a hot gas solenoid valve is energized, circulating hot gas through the evaporator. Some units energize a "dump valve" in which the unfrozen water is drained, and replaced with new water. Some units have an inlet water solenoid valve to accomplish the same purpose.

When the ice falls off the evaporator into the storage bin, the unit is returned to the "freeze" cycle. The operation of most modern cubers can incorporate many other features, and are controlled through a "black box" control board. There is no way for the service technician to know all of the functions of each manufacturer's black box controller, unless the service manual is available. Many of the features of these black boxes are explained later in this chapter.

Both ice flakers and ice cubers have a selector switch that has positions for Off, Ice, and Clean. In the Clean position, only the water pump is energized. This allows the technician to dump in some ice machine cleaner, and circulate it through the entire water system.

Figure 19–2 Ice cuber.

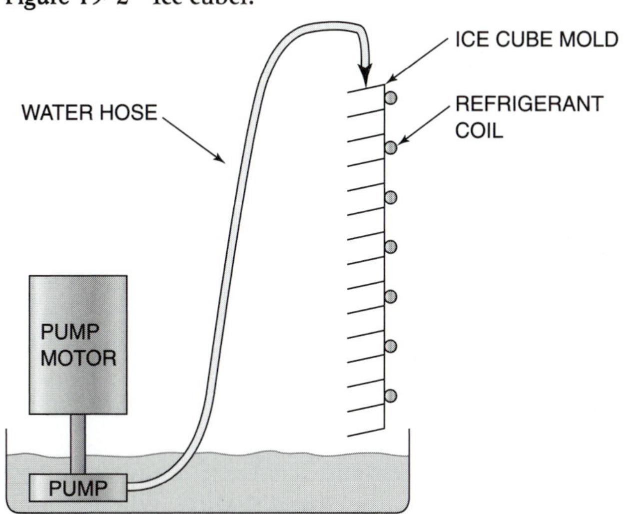

MANITOWOC J MODEL CUBER

Sequence of Operation

Figure 19–3 shows the ladder diagram. The control board has relays (switches labeled 3, 1, 2, 4), connections for an ice thickness probe, a discharge line thermistor, a low-voltage plug, and outputs to the hot gas solenoid, the dump solenoid, water pump, and contactor coil. When the bin switch first closes, relay switches 3 and 1 and 2 close, starting the water pump and the water dump solenoid for 45 seconds to purge old water from the ice machine. The hot gas valve is also energized during the water purge, but nothing flows through it because the compressor has not yet started. At the end of 45 seconds, switches 3 and 1 open, switch 2 remains closed, and 4 closes. The compressor and condenser fan motor start. The dump solenoid and water pump stop. Five seconds later, switch 2 opens and the hot gas solenoid is de-energized. For the next 30 seconds, the evaporator pre-chills. Then switch 1 closes again and the water pump starts. This is the "freeze" cycle.

The "freeze" cycle continues until the cubes are fully formed. This is sensed by a probe that consists

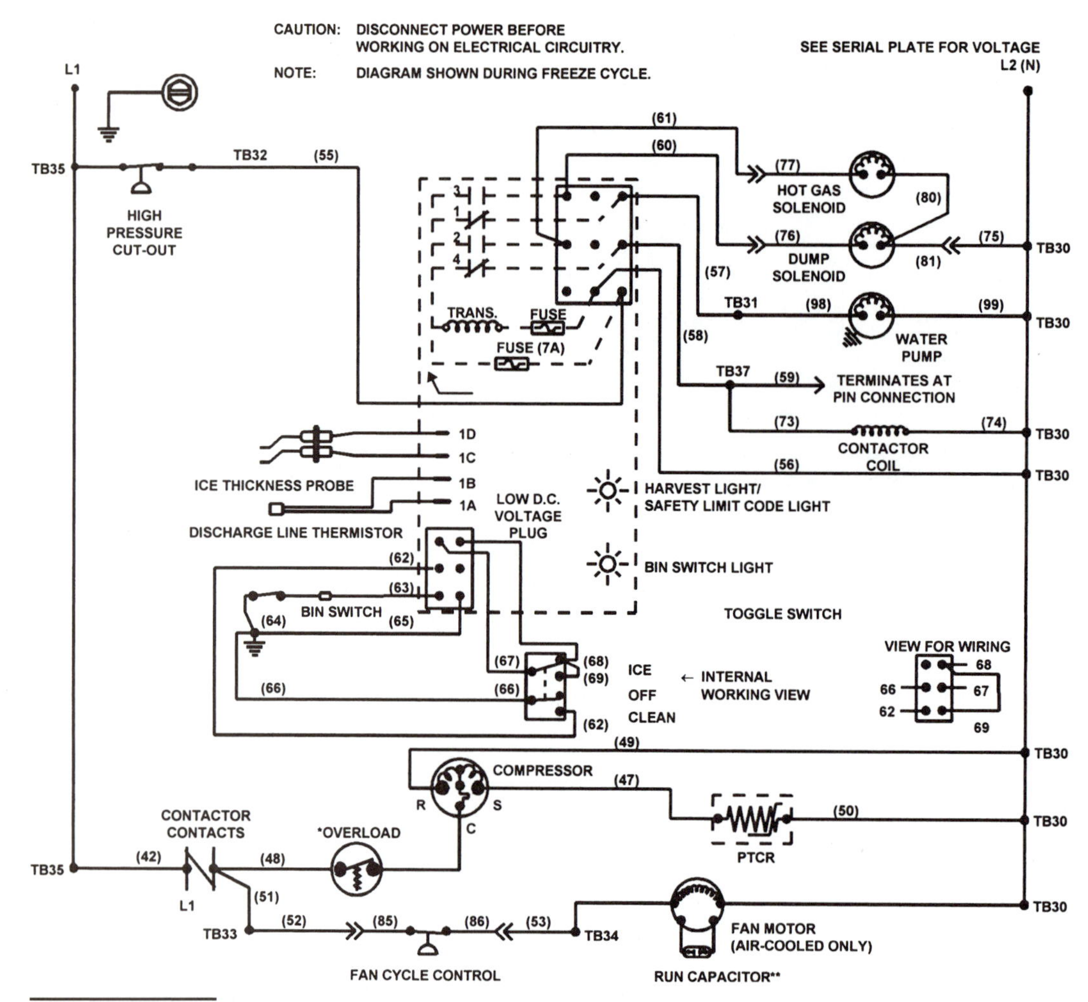

Figure 19-3 Manitowoc J model cuber. *(Courtesy of Manitowoc Ice, Inc.)*

of two metal fingers that hang in front of the evaporator. As the ice thickness builds, the circulating water moves closer and closer to the fingers until contact is made (the newer units use a one-finger probe). If the contact with water is maintained for seven seconds, the control board will initiate the "harvest" cycle. The water dump valve energizes for 45 seconds to purge the water from the trough, and the hot gas valve opens, diverting hot refrigerant gas into the evaporator. After 45 seconds, the water pump and dump valve de-energize, but the hot gas valve remains open (the newest control boards have an adjustable water purge time.) "Harvest" continues until the cubes slide out of the tray, in one sheet, moving a water curtain away from the evaporator which, in turn, trips a microswitch that acts as the bin switch as the cubes fall into the storage bin. This momentary opening and re-closing of the microswitch (bin switch) terminates the "harvest" cycle, and returns the machine to the "freeze" cycle. If the ice storage is full enough that the last ice made can't fall all the way into the bin, the bin switch will remain open and the ice machine will turn off until enough ice is removed to allow the sheet of new cubes to drop clear of the water curtain.

Safety Features

Once the unit has shut off, it will remain off for at least three minutes, even if the microswitch is closed. This is to allow the compressor PTC device to cool before a restart is attempted.

Once the unit has started, the control board locks the machine in the "freeze" cycle for six minutes. Even if water contacts the ice thickness probe during these six minutes, the ice machine will stay in the "freeze" cycle to prevent short cycling (earlier models did not have this feature). However, to allow the service technician to initiate a "harvest" cycle without delay, this feature is not used on the first cycle after moving the toggle switch to Off and then back to Ice.

If the water curtain has been removed during the "freeze" cycle, the "freeze" cycle will continue normally. However, if the water curtain remains off during the "harvest" cycle, there will be no way to trip the microswitch when the ice sheet falls. As a safety, any time the "harvest" cycle time reaches 3.5 minutes and the bin switch has not closed, the ice machine will stop as though the bin were full.

In addition to standard safety controls such as the high pressure cut-out, the control board has built-in safety limits. Safety limit #1 will shut down the system if the freeze time exceeds 60 minutes for three consecutive "freeze" cycles. Safety limit #2 will shut down the machine if the harvest time exceeds 3.5 minutes for three consecutive "harvest" cycles. Safety limit #3 will trip if the compressor discharge temperature falls below 85°F for three consecutive "harvest" cycles. Safety limit #4 will trip if the compressor discharge temperature exceeds 255°F for 15 continuous seconds.

Status Indicators

The control board has LEDs to indicate the status of the machine. The Bin Switch LED (green) is on when the bin switch (water curtain) is closed, and off when the bin switch is open.

The Harvest/Safety limit LED (red) primary function is to turn on as water contacts the ice thickness probe during the "freeze" cycle, and will remain on throughout the entire "harvest" cycle. Prior to the "harvest" cycle, the light will flicker as water splashes on the probes. The secondary function of this light is to continuously flash when the ice machine is shut off on a safety limit, and to indicate which safety limit shut off the machine.

The Clean LED (yellow, when provided) is on whenever a cleaning cycle is in progress.

Troubleshooting

Thermistors, which send temperature information to the control board by changing resistance, can fail. You can check its operation by removing it from the circuit and measuring its resistance at various temperatures. In an ice/water bath (32°F), the resistance should be 283K to 377K ohms. At 70–80°F, the resistance should be 93K to 119K ohms. In boiling water (212°F) the resistance should be 6.2K to 7.3K ohms. If the discharge line thermistor fails closed, the ice machine would stop on safety limit #4 15 seconds after contact #4 on the control board closes (15 seconds after the compressor starts). If the discharge line thermistor fails open, the ice machine would start and run through two normal "freeze" and "harvest" sequences. During the third "harvest" sequence, the ice machine would stop on safety limit #3.

The Ice-Off-Clean switch may fail. Because of the electronics involved, it is recommended by the manufacturer that a volt meter not be used to check the toggle switch. Instead, with the power off, isolate the toggle switch. Then, using ohms, check the switches for continuity in all three toggle switch positions. The switches between terminals 66–62 and between terminals 67–69 should be closed when the toggle switch is set to Clean. The switch between terminals 67–68 should be closed when the toggle switch is set for Ice.

The ice thickness probe (two-finger model) can fail if water hardness creates a bridge across the two probes, across the plastic insulator that separates them. On older model machines without the minimum freeze time feature, this would cause short cycling. The quickest check for this problem is to simply disconnect one of the probe wires from the control board. If the short-cycling stops, replace the probe. On the newer models, if the ice machine cycles into harvest before water contact with the probe, disconnect the ice thickness probe from the control board. Bypass the minimum freeze time feature by moving the Ice-Off-Clean switch to Off, and back to Ice. Wait about 1.5 minutes for water to begin flowing over the evaporator. If the harvest light stays off and the ice machine remains in the "freeze" sequence, the ice thickness probe was causing the malfunction. If the harvest light comes on, and 6–10 seconds later, the ice machine cycles from "freeze" to "harvest," the control board is causing the malfunction.

If the ice machine does not cycle into "harvest" when water contacts the ice thickness probe, place a jumper wire across the ice thickness probe terminals at the control board (figure 19–4(a)). If the "harvest" light comes on, and 6-10 seconds later harvest is initiated, replace the ice thickness probe. (On a single-finger probe, instead of jumping across the two thickness probe connections, jumper from the single thickness probe connection to ground as in figure 19–4(b).)

If the ice machine will not run at all, verify that voltage is available, the HPC is not open, the transformer and fuses at the control board are OK, the bin switch functions properly, the Ice-Off-Clean toggle switch is making contact from terminals 67–68, and that the DC voltage is properly grounded. If no problem is found, replace the control board.

Scotsman AC30 Cuber

Figure 19–5 shows the ladder diagram for a Scotsman AC30 model cuber. Ice maker wiring diagrams sometimes use a different convention. Instead of showing switches in their true "normal" position,

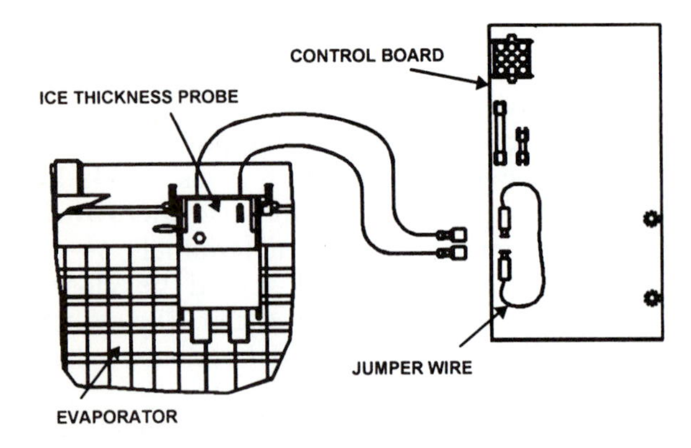

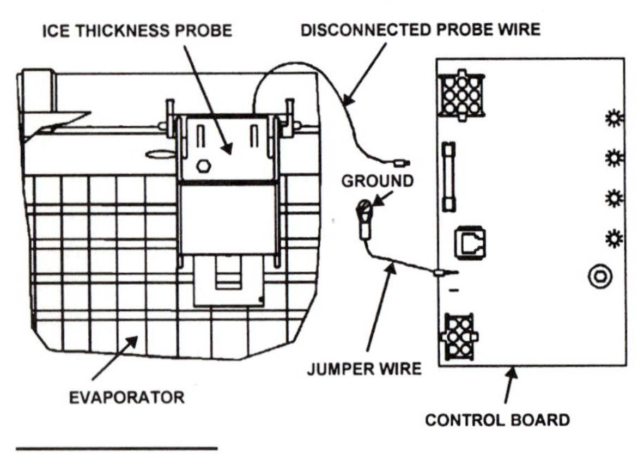

Figure 19–4 Bypassing the ice-thickness probe. *(Courtesy of Manitowoc Ice, Inc.)*

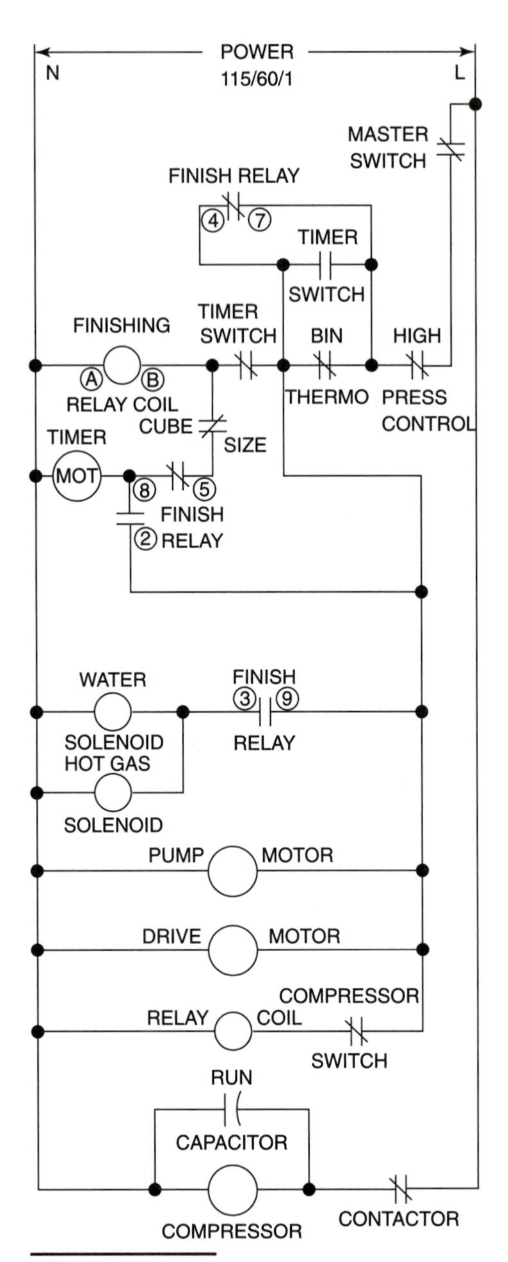

Figure 19–5 Scotsman AC30 cuber.

they will show the switches in the actual position during the "freeze" cycle. Figure 19–5 is actually shown in the "finishing freeze" portion of the "freeze" cycle, as explained below. The unique features of this ice machine are:

1. The **drive motor** operates a rotating water distribution tube similar to that which would be found in the bottom of a dishwasher. The water distribution tube sprays water upward, into an upside-down ice cube evaporator mold. During "harvest," the cubes drop down onto a table where the rotating water distribution tube sweeps them out into a chute that leads to the ice storage bin.

2. The "harvest" is controlled through the operation of a timer motor that operates an SPDT timer switch.

3. The "harvest" is initiated by a cube-size thermostat. It is actually a thermostat that senses evaporator temperature. As the cube size builds, the evaporator becomes more and more insulated from the 32°F circulating water. The reduced load on the evaporator results in a continuously dropping evaporator temperature as the cube size grows.

4. There is a **finishing relay**. Its function is to allow the completion of the batch of cubes that have been started and has entered into the "harvest" cycle, even if the bin switch should open during this time.

Sequence of Operation

During the "freeze" cycle, the manual master switch and compressor switch are both closed, as is the high pressure control and the bin thermostat. The normally open timer switch is actually closed, and the normally closed timer switch is open. The finishing relay is simply a three-pole double-throw control relay. The finishing relay coil is energized, causing the contacts to be closed between terminals 4–7 and 5–8. The contacts between terminals 2–8 and 3–9 are open. The master switch causes the

ADVANCED CONCEPTS

Instead of simply having the cube size thermostat begin the "harvest cycle," a timed "finish-freeze" cycle is used after the cube size thermostat closes. This is done because during the "freeze" cycle, the evaporator temperature (sensed by the cube-size thermostat) drops fairly rapidly through the first three-quarters of the cube formation. After that, the rate of temperature change flattens out considerably. If we attempted to use evaporator temperature to determine when the cubes were fully formed, we would find that small variations in the temperature at which the evaporator temperature started the "harvest" cycle would produce large variations in actual cube size.

pump motor and the drive motor to run. With the compressor switch Off, this is the cleaning mode of operation. With the compressor switch On, the compressor contactor coil is energized and the single pole contactor closes, bringing on the compressor.

The "freeze" cycle continues until the cube size is large enough to close the cube-size thermostat (not shown as a thermostat in the ladder diagram, but a thermostat none the less). This energizes the timer motor, beginning the "finish freeze" cycle. After six minutes, the timer switches operate. The normally open timer switch closes, sealing in around the bin thermostat. Now, it won't matter if the bin thermostat opens. The cycle will complete itself.

The normally closed timer switch opens, de-energizing the finishing relay coil, and starting the "harvest" by energizing the water and hot gas solenoid valves through finishing relay contacts 3–9. The timer motor continues to run, energized through finishing relay contacts 2–8. This "harvest" cycle continues for three minutes, until the timer motor returns its switches to the "freeze" cycle positions.

ICE-O-MATIC C SERIES CUBER (600 LB/DAY)

The four diagrams in figure 19–6(a)–(d) show the step-by-step sequence for an Ice-O-Matic C-61 cuber manufactured in the 1980s.

Sequence of Operation

During the first step of the "freeze" cycle, the bin switch is closed and the selector switch is in the Ice position. Power is supplied to the contactor coil, bringing on the compressor. The water pump is supplied power through the selector switch and the normally closed contacts of the cam switch. The fan motors are supplied power through the normally closed contacts of the control relay.

Step two begins when enough ice has formed on the evaporator to lower the suction pressure sufficiently to close the low-pressure control (note that this is a low-pressure cut-in, and not a low-pressure cut-out). This provides power to the Time Delay Module. By adjusting the set point of the time delay module to longer or shorter, the bridge thickness (the ice bridge connecting each cube to the adjacent cubes) may be adjusted. The proper bridge thickness is thick enough to hold the entire ice sheet together so that it releases in one piece, but thin enough so that the individual cubes break apart when they fall into the storage bin. Generally, a 1/8" bridge is about right.

After the power has been applied to the Time Delay Module for the set amount of time, it passes through (Step 3). This energizes the relay coil, the purge solenoid, the water pump (through the "A" normally open relay contacts), the hot gas solenoid and harvest motor (through the "B" normally open relay contacts). At the same time, the normally closed relay contacts "B" open, shutting off the condenser fan motors. The harvest motor has two functions. First, it pushes a probe (through a slip clutch) against the back of the ice sheet to help push the sheet of cubes out of the evaporator. Second, it drives a cam that operates an SPDT cam switch. Within a few seconds after the harvest motor is energized, it will rotate enough to operate the SPDT switch.

In Step 4, the harvest motor has operated the cam switch, the low-side pressure has built up due to the hot gas, opening the low-pressure control, and de-energizing the relay coil. The harvest motor and hot gas solenoid remain energized through the normally open contacts of the cam switch. Initially, the clutch between the harvest motor and the probe will be slipping, because the probe is unable to push the ice sheet out of the evaporator. This mode will continue, until the ice "releases" from the evaporator. Then, the probe will once again be able to move. As the probe pushes the ice out, and returns to its orig-

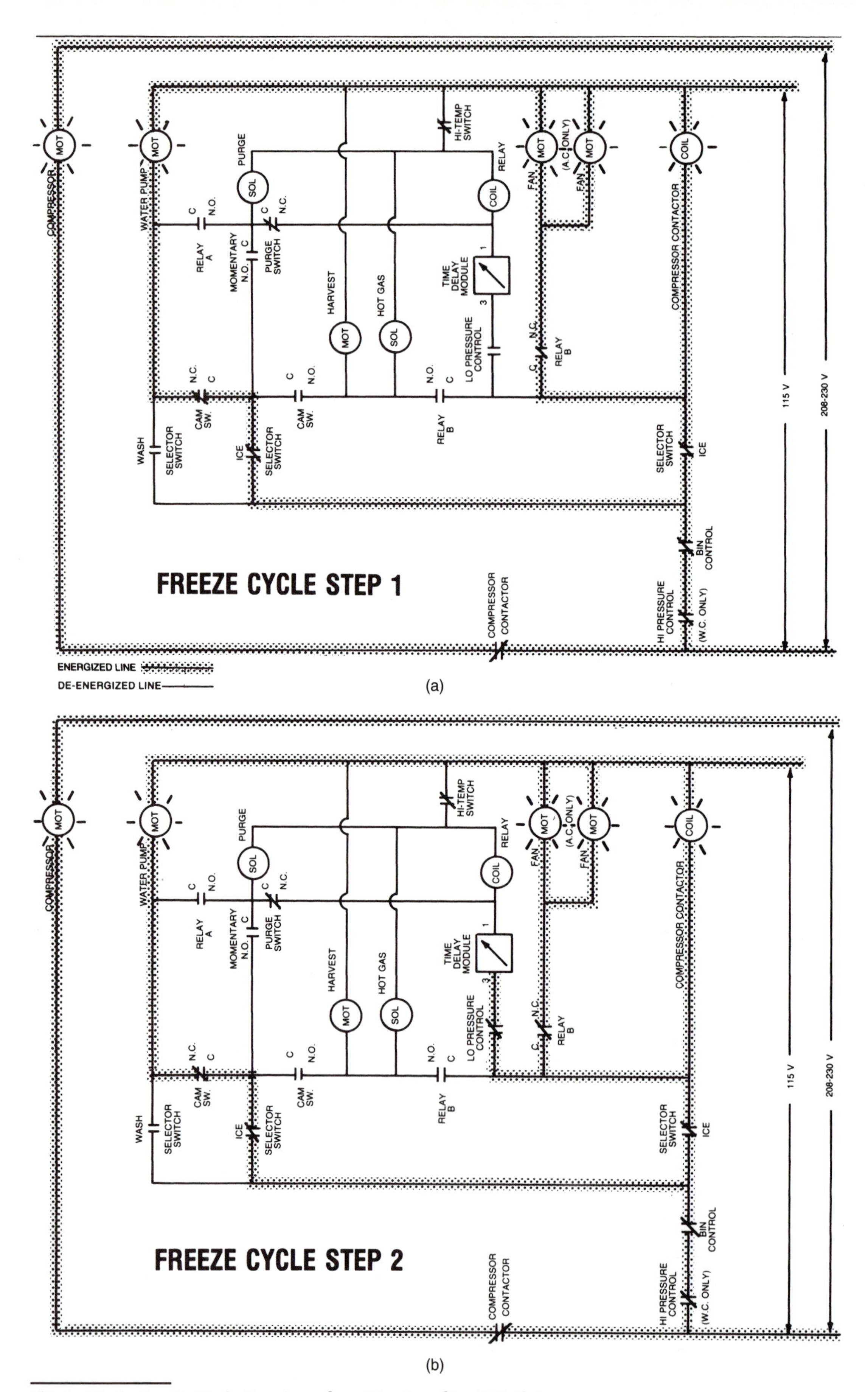

Figure 19–6 Ice-O-Matic C series cuber. *(Courtesy of Ice-O-Matic.)*

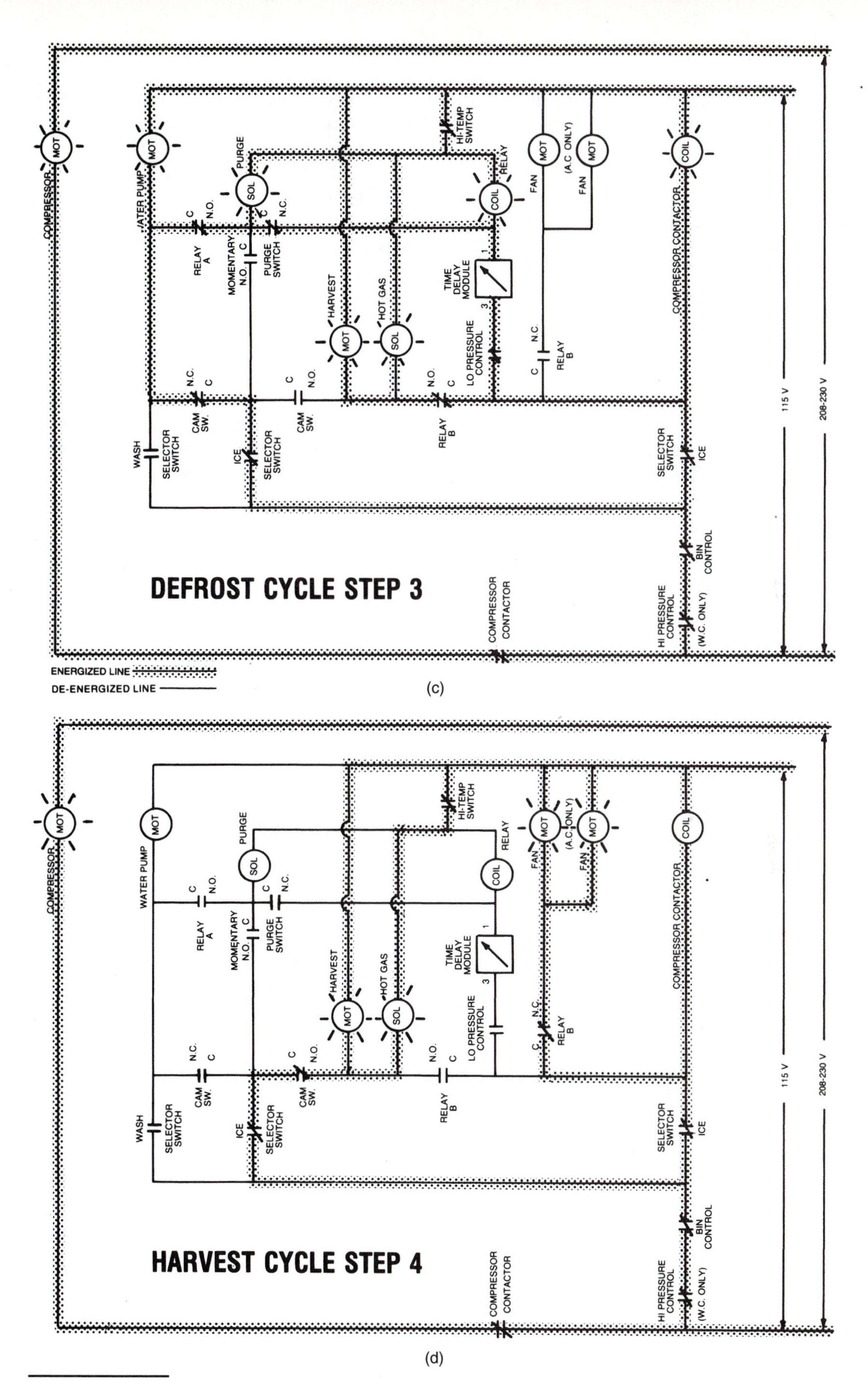

Figure 19-6 *(Continued)*

inal position, the cam switches return to their normal positions, and the "freeze" cycle begins anew.

EXERCISE:
Figure 19–7 shows an interconnection diagram for the C-61 ice maker. Use a highlighter or other method to indicate which wires are at L1 pressure, and a different color to indicate which wires are at L2 pressure during the "freeze" cycle.

ICE-O-MATIC C SERIES CUBER (1200 LB/DAY)

Figure 19–8(a)–(d) shows a diagram similar to the previous wiring, but with some added complexity. The differences are:

1. There are two evaporators, each with its own harvest motor. The two ice sheets will not release at the same identical moment. The unit must remain in the "harvest" cycle until both ice sheets have been harvested.

2. The compressor operates on 230V, while everything else operates on 115V.

3. There is a finishing relay, which does not allow the unit to shut down if the bin switch opens during the "harvest" cycle.

Sequence of Operation

In Step 1, the machine is in the "freeze" cycle. The compressor, condenser fans, and water pump are running. Relay 3 is energized, and will remain so as long as the high-temperature safety switch does not

Figure 19–7 Exercise. *(Courtesy of Ice-O-Matic.)*

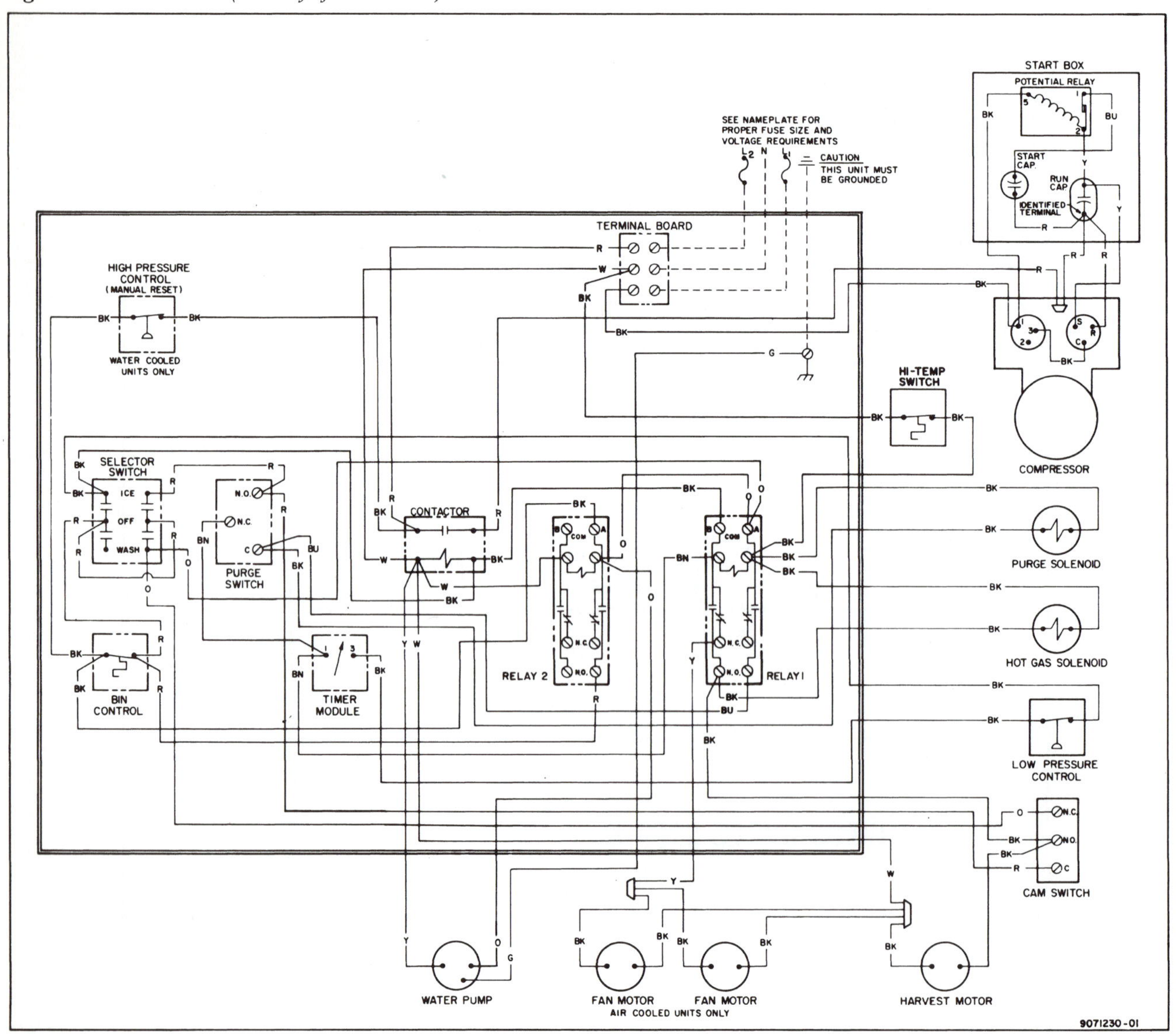

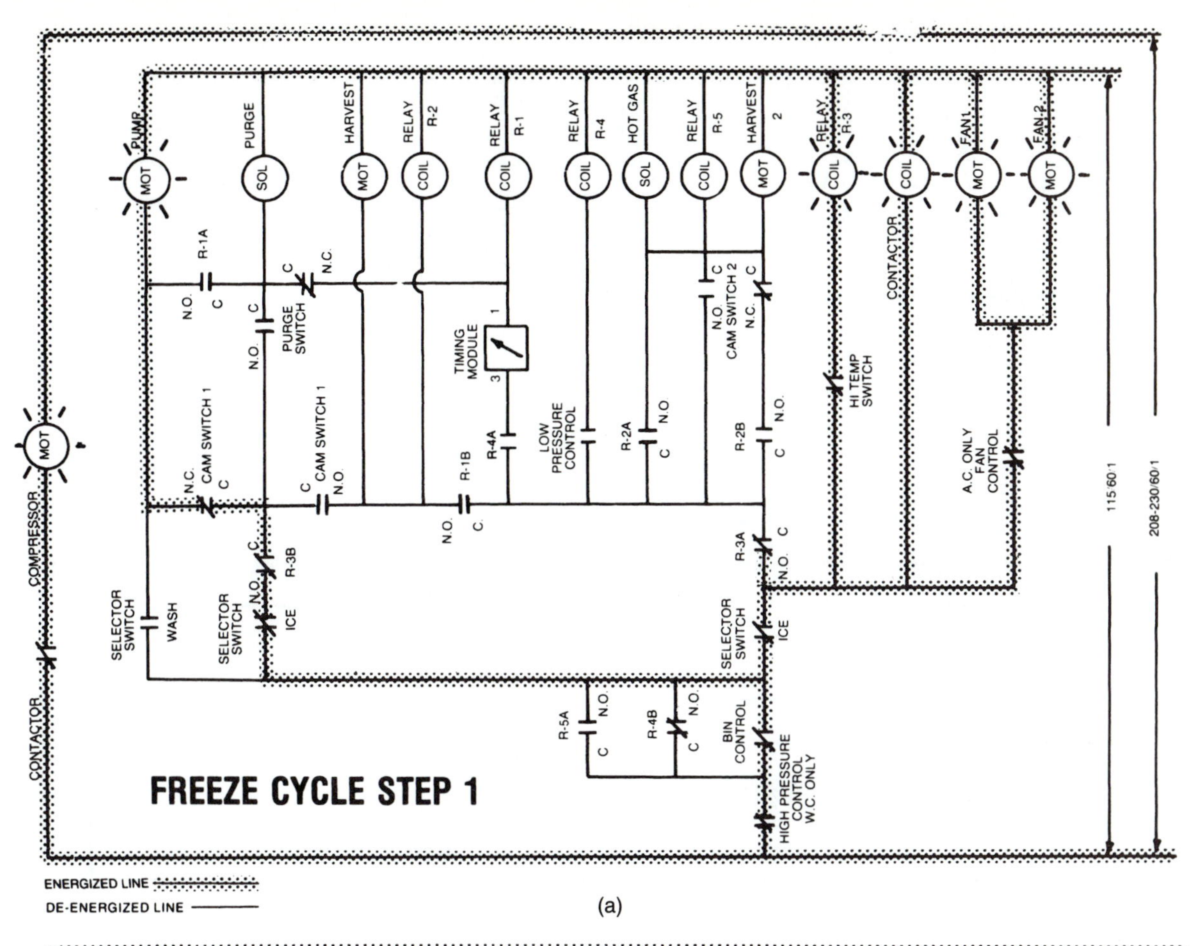

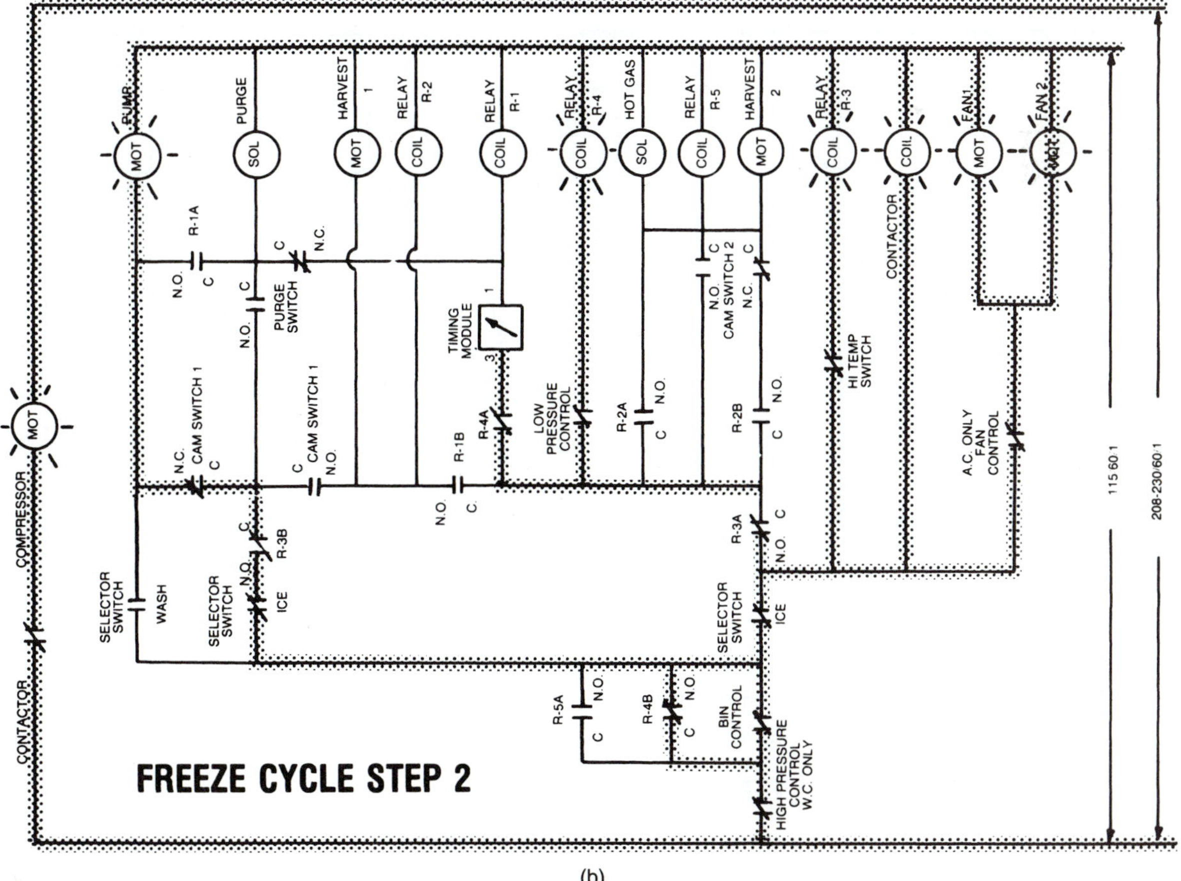

Figure 19–8 Ice Cuber with two evaporators. *(Courtesy of Ice-O-Matic.)*

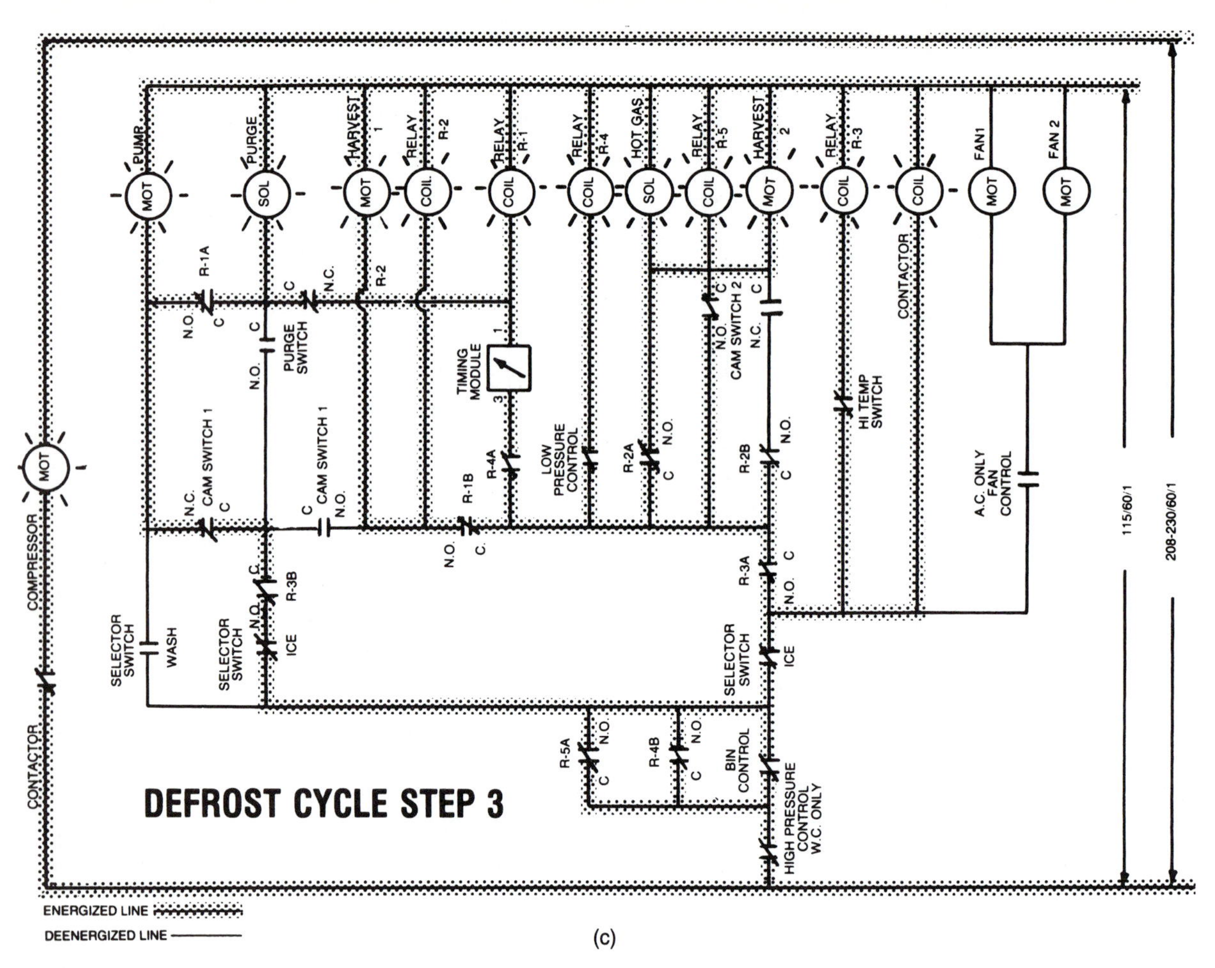

(c)

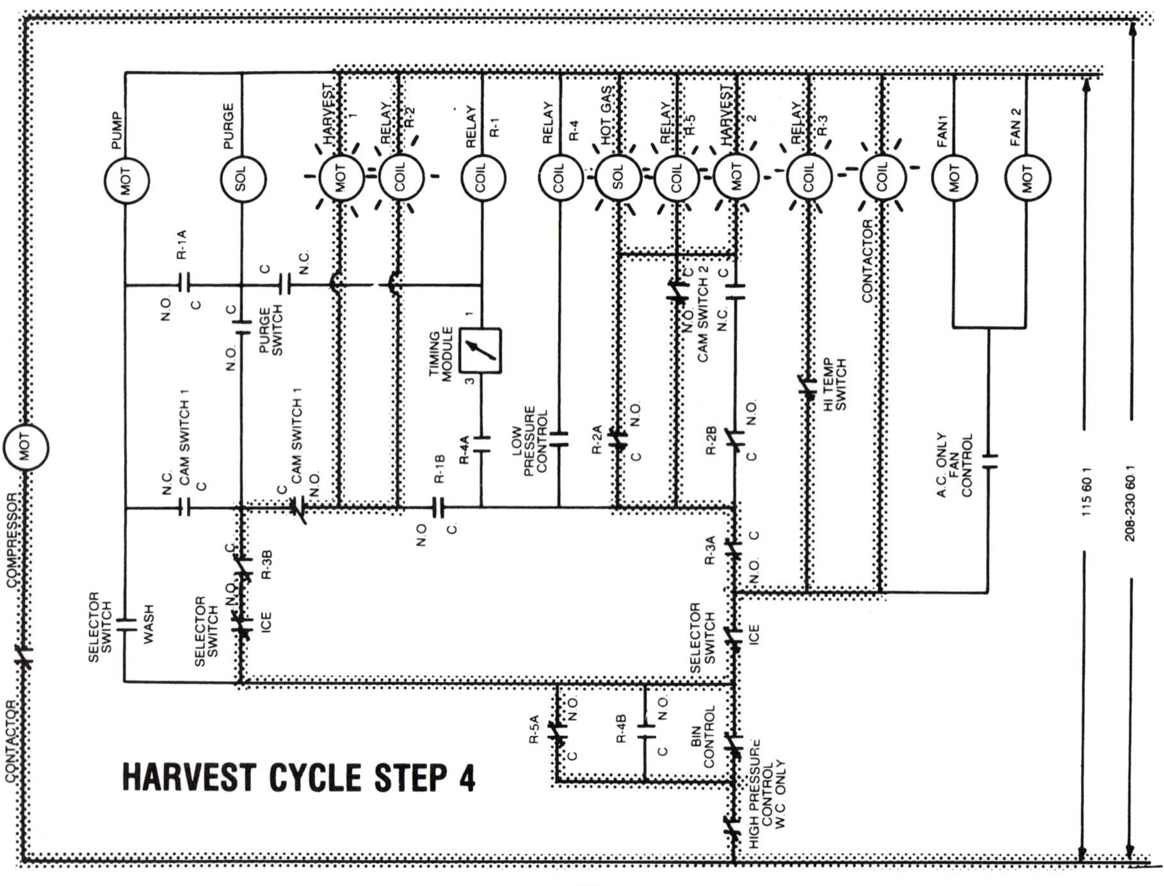

(d)

Figure 19-8 *(Continued)*

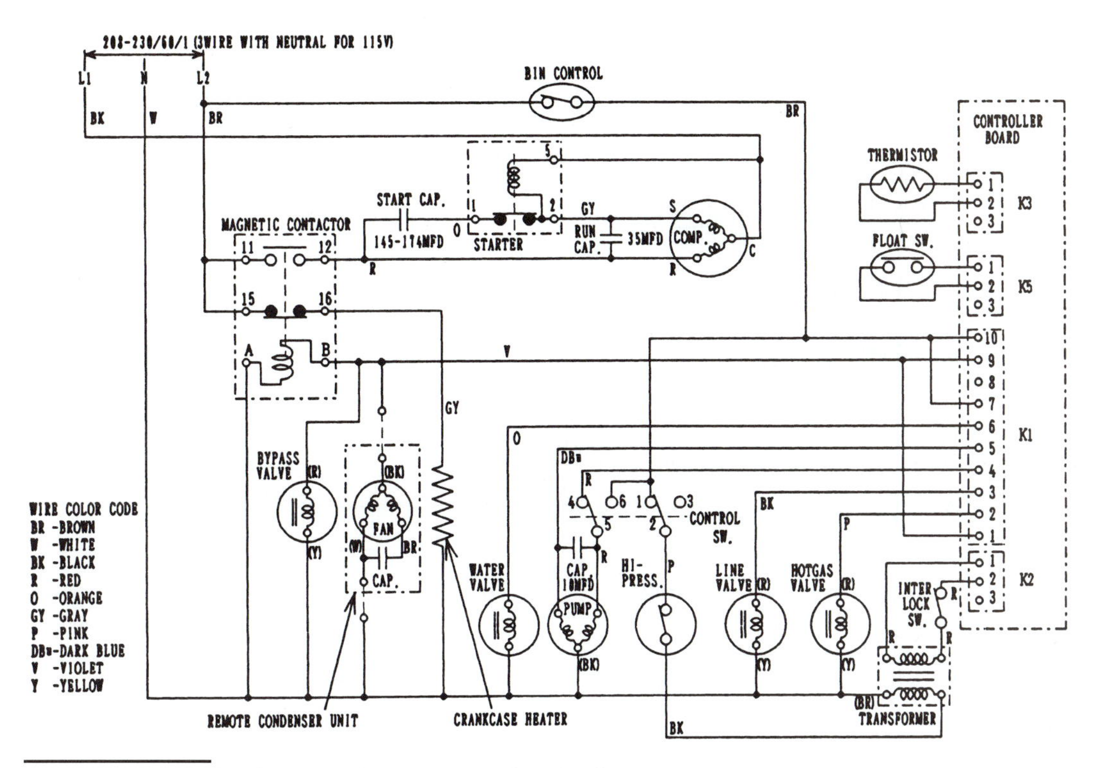

Figure 19–9 Hoshizaki KM cuber. *(Courtesy of Hoshizaki America, Inc.)*

trip. When a sufficient thickness of ice has been formed, the low-pressure cut-in closes.

In Step 2, instead of the low pressure cut-in energizing the timing module directly, it energizes relay R-4. Relay R-4A contacts then energize the timing module, and R-4B contacts close in parallel around the bin control. If the bin control opens after this point in time, it will have no effect so long as R-4B contacts are closed (seal-in circuit) through the end of the "harvest" cycle.

Step 3 begins when the timing module allows current to pass, starting the hot-gas cycle. This energizes relay R-1 and the water purge solenoid. The R-1 normally open "B" contacts close, energizing relay R-2 coil and harvest motor #1. The relay R-2 contacts energize both the hot-gas solenoid and harvest motor #2. When harvest motor #2 moves the cam slightly, the normally open contacts of cam switch 2 close, energizing relay R-5, whose contacts seal in again around the bin control.

Shortly after the hot gas starts, the low side pressure rises and the low-pressure cut-in opens, de-energizing relay R-4, the purge solenoid, and the water pump. Harvest motor #2 will continue to turn until harvest motor #1 returns its cam switch to the normal position. Harvest motor #1 will always stop turning before harvest motor #2. The hot gas valve and relay coil R-5 will remain energized as long as either cam switch is in the not-normal position. As soon as both cam switches return to their normal positions, the "freeze" cycle will start over, unless the bin switch has opened during "harvest." In that event, the machine will stop at the completion of the "harvest" cycle.

HOSHIZAKI MODEL KM CUBER

Figure 19–9 is typical of the Hoshizaki cubers. As with the other modern ice makers, a "black box" controller board is used to control the "freeze" and "harvest" cycles. Figure 19–10 shows the layout of the control box for this cuber. One unique feature of this sequence is that in between the "freeze" cycle and the "harvest" cycle, there is a pump-out cycle when the water pump is used to help clean out any deposits in the water system.

Sequence of Operation

The unit always starts in the 1-minute fill cycle shown in figure 19–11(a). When power is applied to the unit, the water valve is energized and the fill period begins. After one minute, the controller board checks for a closed float switch, indicating that sufficient water has been supplied to the unit water sump. If the float switch has not closed, the water valve will remain energized through additional 1-minute cycles until the float switch does close.

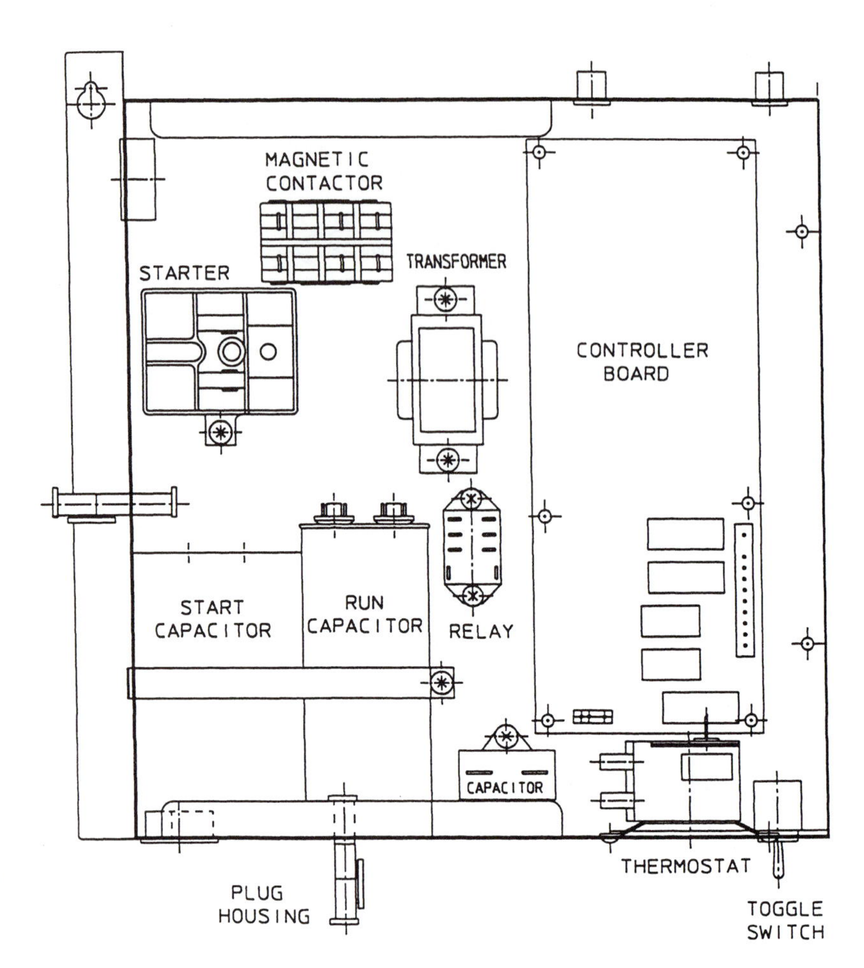

Figure 19–10 Hoshizaki KM cuber basic control box layout. *(Courtesy of Hoshizaki, America Inc.)*

In figure 19–11(b), the unit actually goes through a short "harvest" cycle prior to making any ice. The compressor starts, the hot gas valve opens, the water valve remains open and "harvest" begins. The evaporator will warm quickly, because there is no ice on it. When the suction line thermistor reaches 48°F (average 2 minutes), the "harvest" is turned over to the adjustable control board defrost timer. This adjustment can vary the defrost timer from 1 to 3 minutes.

Figure 19–11(c) shows the "freeze" cycle. After the timer terminates the "harvest" cycle, the hot gas and water valves close, and ice production starts. For the first 5 minutes, the controller board will not accept a signal from the float switch. This 5-minute minimum "freeze" cycle acts as a short cycle protection. As ice builds up on the evaporator, the water level in the sump lowers. When the float switch opens, the "freeze" cycle is over.

In figure 19–11(d), the float switch has opened, and the unit begins its "harvest" cycle. The hot gas valve opens and the compressor continues to run. The drain timer starts counting the 10/20 second pump out. The water pump stops for 2 seconds, and reverses, taking water from the bottom of the sump and forcing pressure against the check valve seat allowing water to go through the check valve and down the drain. At the same time, water flows through the small tube to power flush the float switch. When the drain timer stops counting, the pump-out is complete. Pump-out always occurs on the second harvest after start-up. Some control boards allow for adjustment for pump-out to occur every cycle, or every second, fifth, or tenth cycle from this point.

After the pump-out, the unit goes into its normal "harvest" cycle. The water valve opens to allow water to assist the "harvest." When the evaporator warms to 48°F, the control board receives the thermistor signal and starts the defrost timer. The water valve is open during "harvest" (defrost) for a maximum of 6 minutes, or the length of the "harvest," whichever is shorter. When the defrost timer completes its count down, the defrost cycle is complete and the next "freeze" cycle begins. The unit continues through the "freeze," pump-out, and normal "harvest" cycles until the bin control senses ice and shuts the unit down.

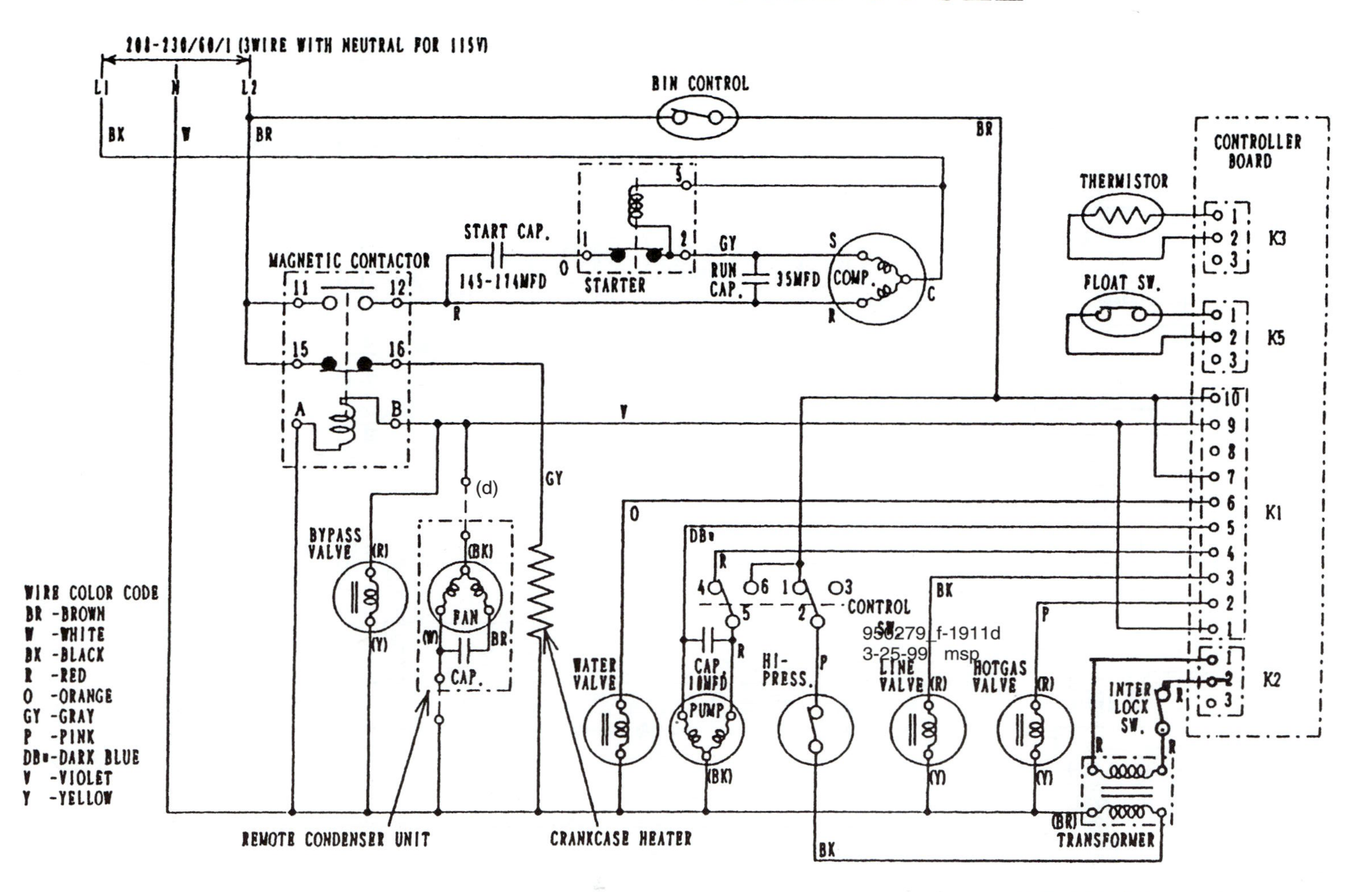

(a)

Figure 19–11 Hoshizaki sequence of operation. *(Courtesy of Hoshizaki America, Inc.)*

INITIAL HARVEST / NORMAL HARVEST

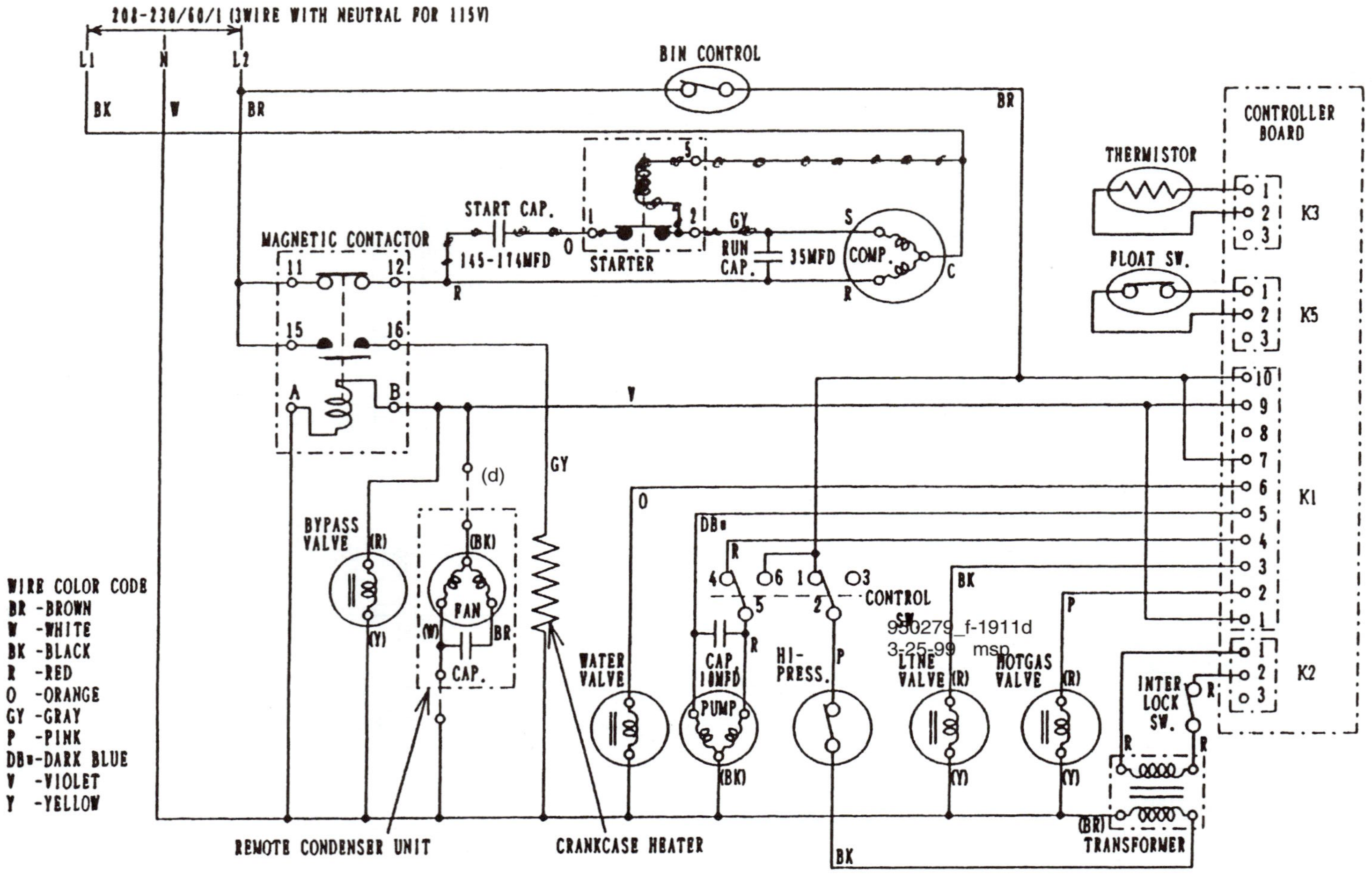

TO FREEZE CYCLE ↓

(b)

Figure 19–11 *(Continued)*

FREEZE CYCLE

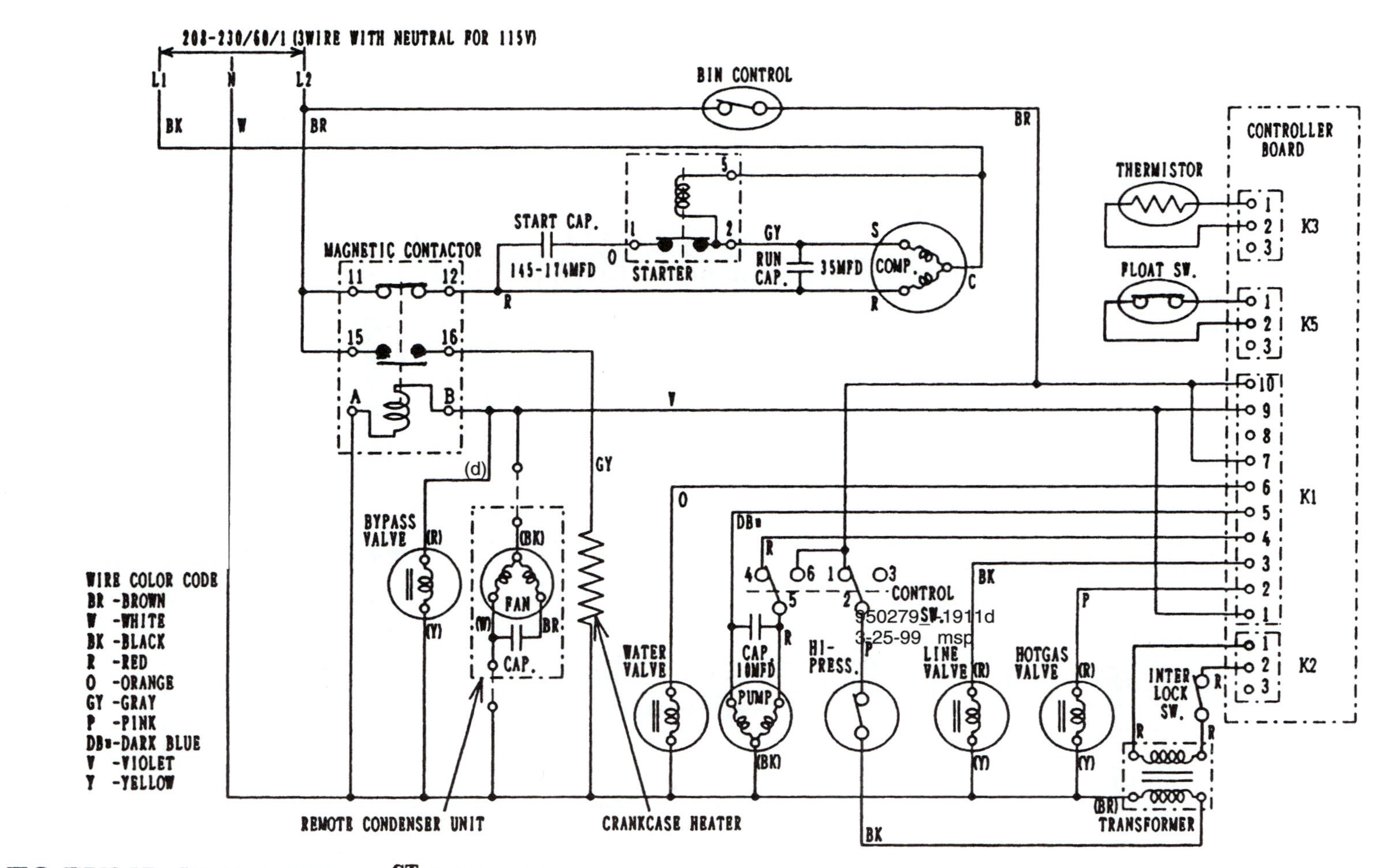

TO PUMP-OUT AFTER 1ST FREEZE (MAY SKIP TO NORMAL HARVEST DEPENDING ON BOARD ADJUSTMENT)

(c)

Figure 19–11 *(Continued)*

10 /20 SECOND PUMP-OUT CYCLE

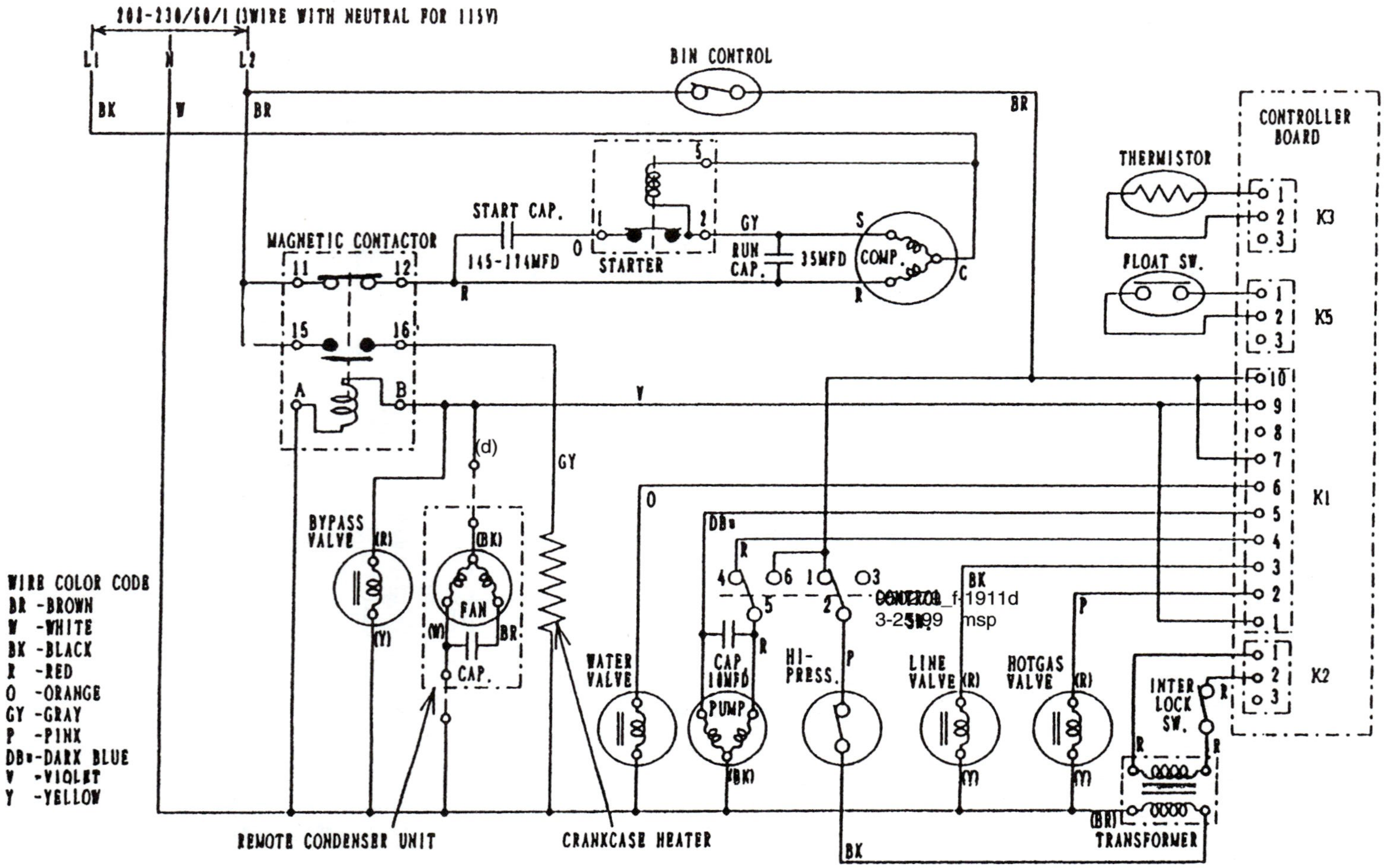

208-230/60/1 (3WIRE WITH NEUTRAL FOR 115V)
L1
N
L2
BK
W
BR
BIN CONTROL
BR
CONTROLLER BOARD
THERMISTOR
K3
FLOAT SW.
K5
K1
K2
MAGNETIC CONTACTOR
START CAP.
145-174MFD
STARTER
GY
RUN CAP.
35MFD
COMP.
S
C
R
Y
GY
O
DB
BK
P
BYPASS VALVE
FAN
CAP.
WATER VALVE
CAP. 10MFD
PUMP
HI-PRESS.
LINE VALVE
HOTGAS VALVE
INTER LOCK SW.
TRANSFORMER
REMOTE CONDENSER UNIT
CRANKCASE HEATER
WIRE COLOR CODE
BR -BROWN
W -WHITE
BK -BLACK
R -RED
O -ORANGE
GY -GRAY
P -PINK
DB=-DARK BLUE
V -VIOLET
Y -YELLOW

TO NORMAL HARVEST CYCLE →

(d)

Figure 19–11 *(Continued)*

Troubleshooting Tips

The thermistor can be checked using the following temperature/resistance table:

Sensor temp.(°F)	Resistance (K ohms)
0	14.4
10	10.6
32	6.0
50	3.9
70	2.5
90	1.6

If the thermistor is open, the unit will go through a 20-minute "harvest" cycle. If the thermistor is shorted (zero ohms), the unit locks out on manual reset high temperature safety. If the evaporator reaches 127°F, the thermistor resistance will drop to 500 ohms, shutting down the unit on manual reset. To reset, turn the power off and back on. This is the only manual reset safety on this unit.

KOLD-DRAFT CUBER

The Kold-Draft ice cuber is unique in many respects. The ice is formed in an evaporator that has the cube openings on the bottom side. There is a water distribution plate that fits up under the evaporator. Water is pumped through the distribution plate and out the top of the distribution plate through holes. Each hole squirts water into a different compartment of the ice cube mold.

As the ice cubes form, the space between the ice and the water plate diminishes. When the cubes are fully formed, the ice is sufficiently close to the water plate holes that it causes an increase in the output pressure of the pump that supplies the water to the water plate. This increased pressure causes some of the water to shoot over a weir, and into a drain. The level of water in the system falls quickly. A water level switch is used to sense when it is time to begin the "harvest" cycle.

During "harvest," an actuator motor pivots the water plate out of the way. Hot gas is introduced into the evaporator (in the usual fashion). The cubes fall out by gravity.

Sequence of Operation

The wiring diagrams are shown in Figures 19–12(a)–(g). The condensing unit runs continuously, so long as the bin is not full.

Figure 19–12(a). The sequence begins by filling the water sump. A water level controller energizes the water valve, and the pump & defrost snap action toggle switch operates the water pump. The pump & defrost snap action toggle switch is in the Up position when the water plate is up, next to the evaporator. Later, when the water plate moves down, away from the evaporator, it will kick the pump & defrost snap action toggle switch into the Down position. If the incoming water is below 40 to 45°F during this fill process, the cold water control will energize the defrost valve to prevent freezing until the fill is complete. When there is sufficient water in the sump, the water level control opens, de-energizing the water valve. The water level switch actually consists of a tube, with a thermistor probe near the top and another near the bottom. The thermistor probes are connected to a circuit board, which interprets the thermistor probe signals to determine the water level in the sump.

Figure 19–12(b). As the "freeze" cycle continues, the temperature of the evaporator drops. When it falls to 20°F, the actuator control (operated by another thermistor that senses evaporator temperature) moves to the Cold position. This does not affect the "freeze" cycle. It simply is preparing for the "harvest" cycle.

Figure 19–12(c). When the cubes are almost fully formed, the interference with the water supply holes in the water plate cause the pump discharge to increase. This pushes water over a weir to a drain. The water level in the sump drops quickly. When it reaches the lower thermistor probe in the level sensing assembly, the water level control switch switches to Low, energizing both the defrost valve and the actuator motor. The actuator motor is a reversible motor. In this mode, it lowers the water plate away from the evaporator.

Figure 19–12(d). As the water plate moves away from the evaporator, it trips the pump & defrost snap action toggle switch to the Down position. This provides a parallel switch to operate the defrost valve (in addition to the evaporator-temperature-sensing actuator control). It also de-energizes the water pump.

Figure 19–12(e). When the water plate opens fully, the actuator toggle lever on the actuator motor shaft pushes the actuator toggle switch (to the left on the wiring diagram), stopping the actuator motor, and completing a third circuit to the defrost valve. As this cycle continues, the evaporator warms, and the cubes fall out.

Figure 19–12(f). With the ice out of the evaporator, the evaporator thermistor warms, causing the actuator control to switch from 2–1 to 2–3. This causes the actuator motor to operate in the opposite direction, raising the water plate back towards the evaporator. The cold water control will warm up at about this time and switch to the warm side, keeping the defrost valve energized.

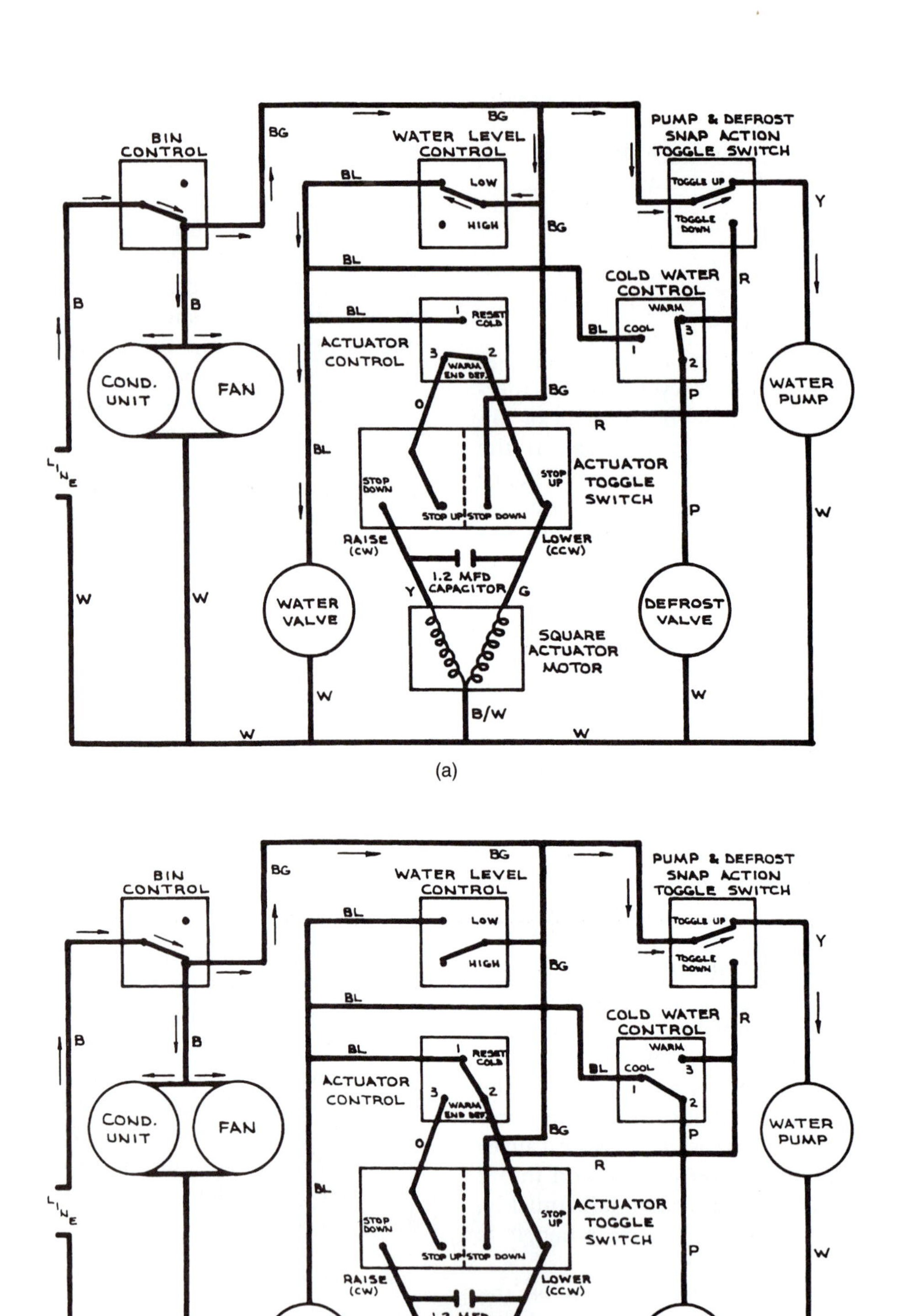

Figure 19–12 Kold-Draft sequence of operation. *(Courtesy of Kold-Draft®.)*

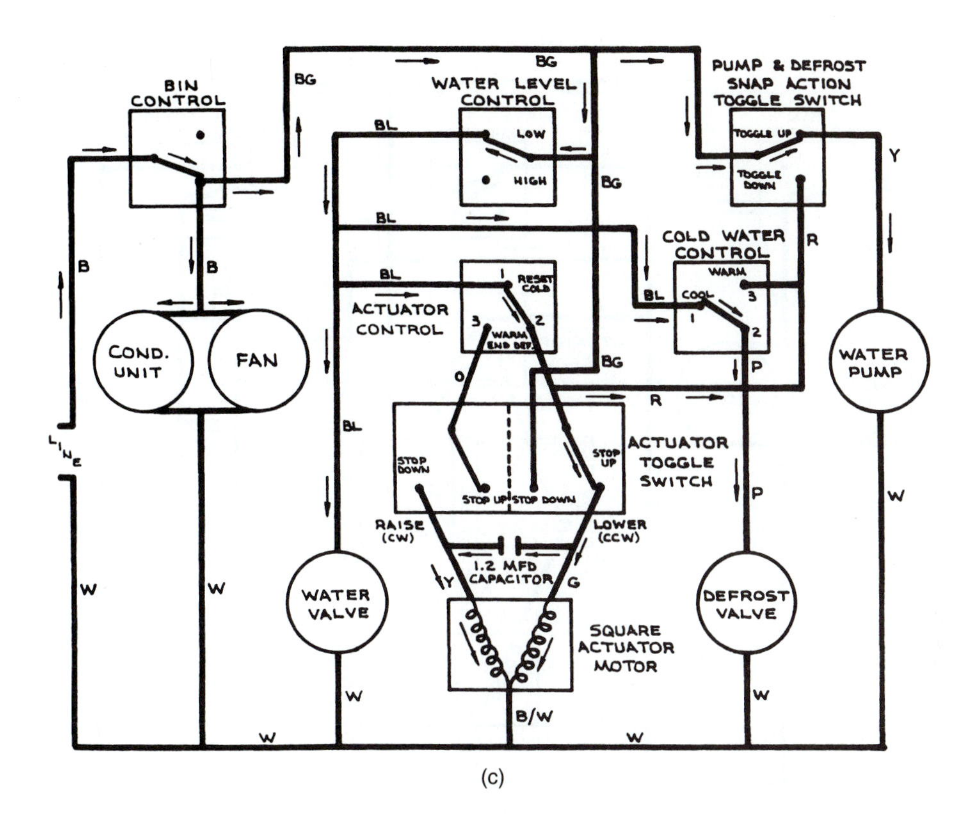

(c)

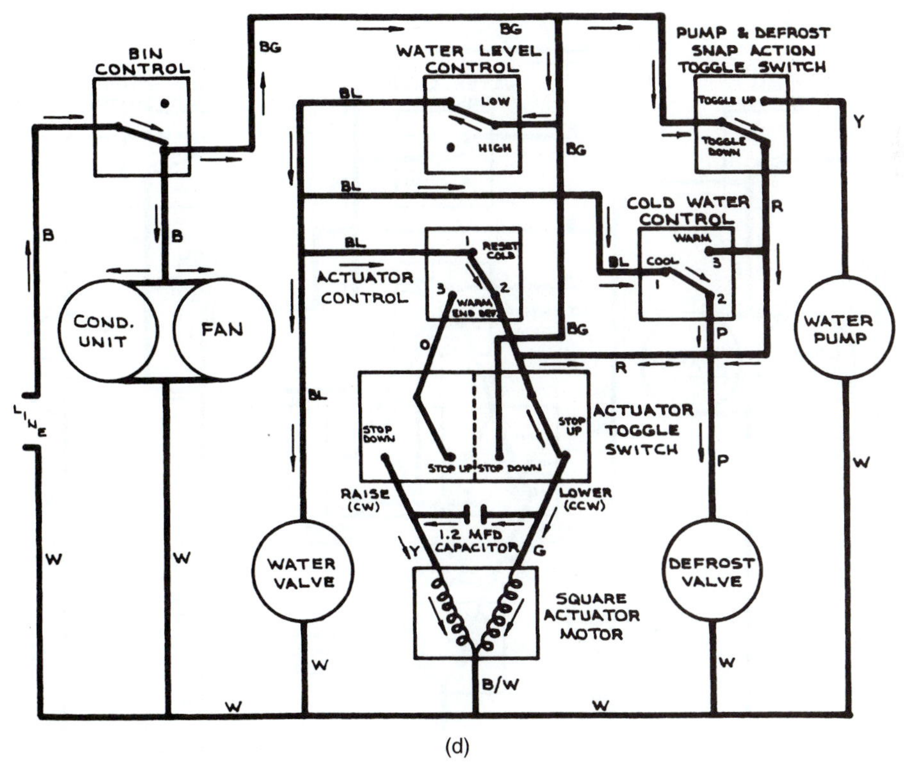

(d)

Figure 19–12 *(Continued)*

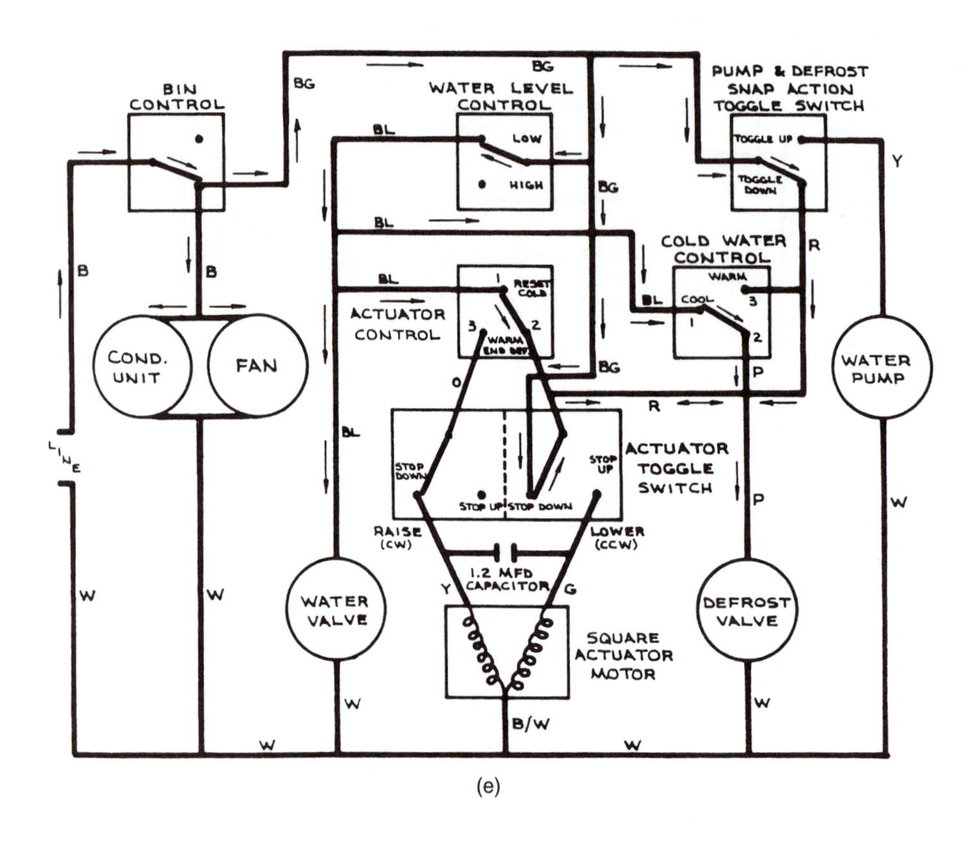

(e)

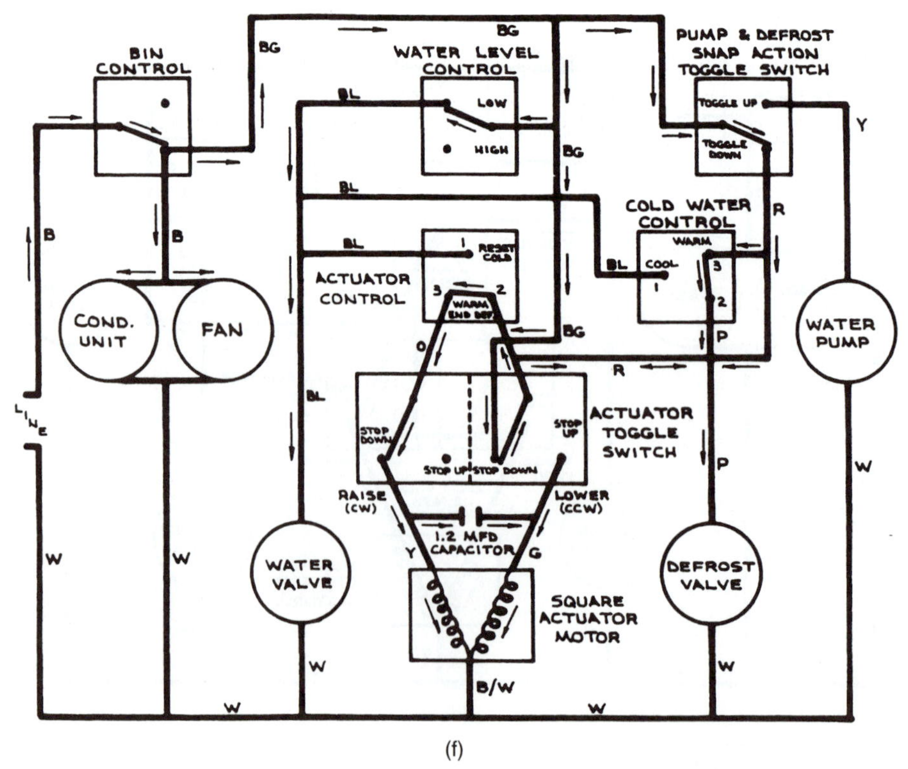

(f)

Figure 19–12 *(Continued)*

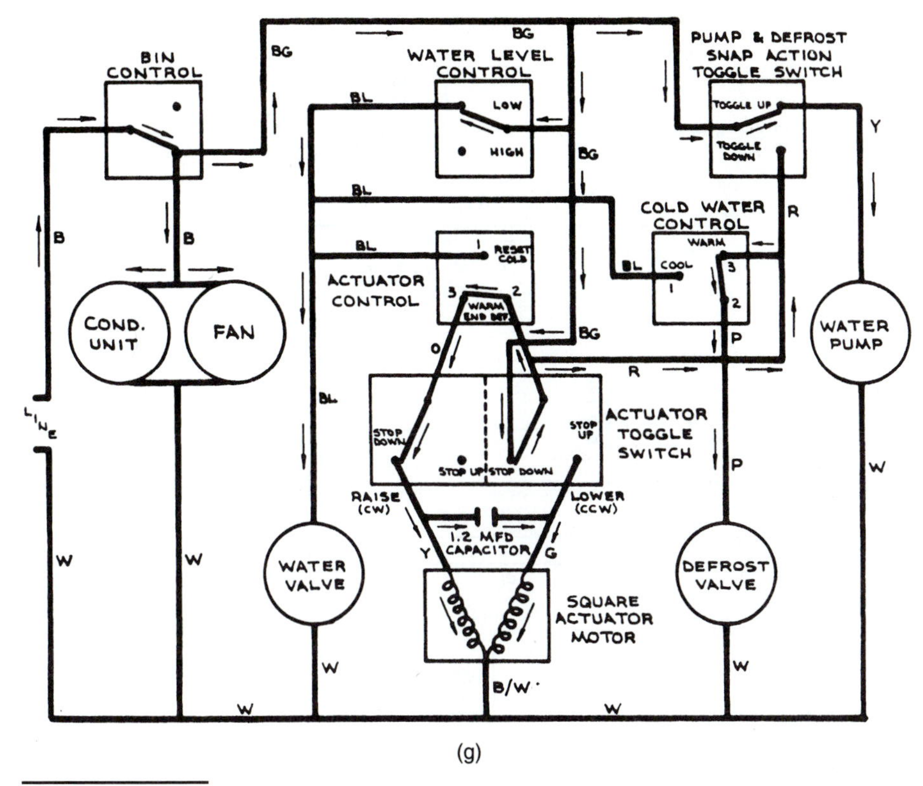

Figure 19–12 *(Continued)*

Figure 19–12(g). With the water plate almost closed, it pushes up the pump & defrost toggle switch, starting the water pump. When the water plate reaches the fully closed position, the actuator toggle switch is pushed to the right, stopping the actuator motor with the water plate up. This then is the same as figure 19–12(a). The cycle repeats until the bin control is satisfied, and then everything turns off.

HOSHIZAKI MODEL F OR DCM ICE FLAKER

As you will recall, flakers do not have separate "freeze" and "harvest" cycles. They make ice and harvest continuously, at the same time, until the bin is full. This flaker also incorporates, on some models, a periodic flush cycle.

SEQUENCE OF OPERATION

The interconnection diagram and control box layout are given in figures 19–13 and 19–14. On initial start-up, there will be a Fill cycle before the gear motor starts (figure 19–15(a)). With the Flush switch and Ice switch closed, power is supplied to the inlet water valve. The unit will not start until the reservoir is full and both floats on the dual float switch are closed (figure 19–15(b)). The operation is then turned over to the bin control. If the bin control is closed (calling for ice), the gear motor and condenser fan motor are energized. One minute later, the compressor starts (figure 19–15(c)). As the refrigeration system cools the water in the evaporator, ice will start to form within 2 to 5 minutes. Ice production will continue until the bin control opens.

When the bin switch opens, on some models, the entire unit shuts down within six seconds (figure 19–15(d)). On other models, 90 seconds after the bin control switch opens, the compressor stops. One minute later, the gear motor and condenser fan motor stop. This sequence is accomplished through a series of timers within the solid state timer board. Its purpose is to provide a period of time to clean the ice out of the evaporator to prevent the auger from freezing in place during the Off cycle.

PERIODIC FLUSH CYCLE

On some of the larger flakers, a periodic flush cycle is included to prevent the buildup of mineral deposits from the water system. A 12-hour timer will cycle the unit down and open the flush valve which allows the complete water system to drain (figure 19–15(e)). The unit will remain off for 15 minutes which allows any ice remaining in the evaporator to melt and flush the evaporator walls and mechanical

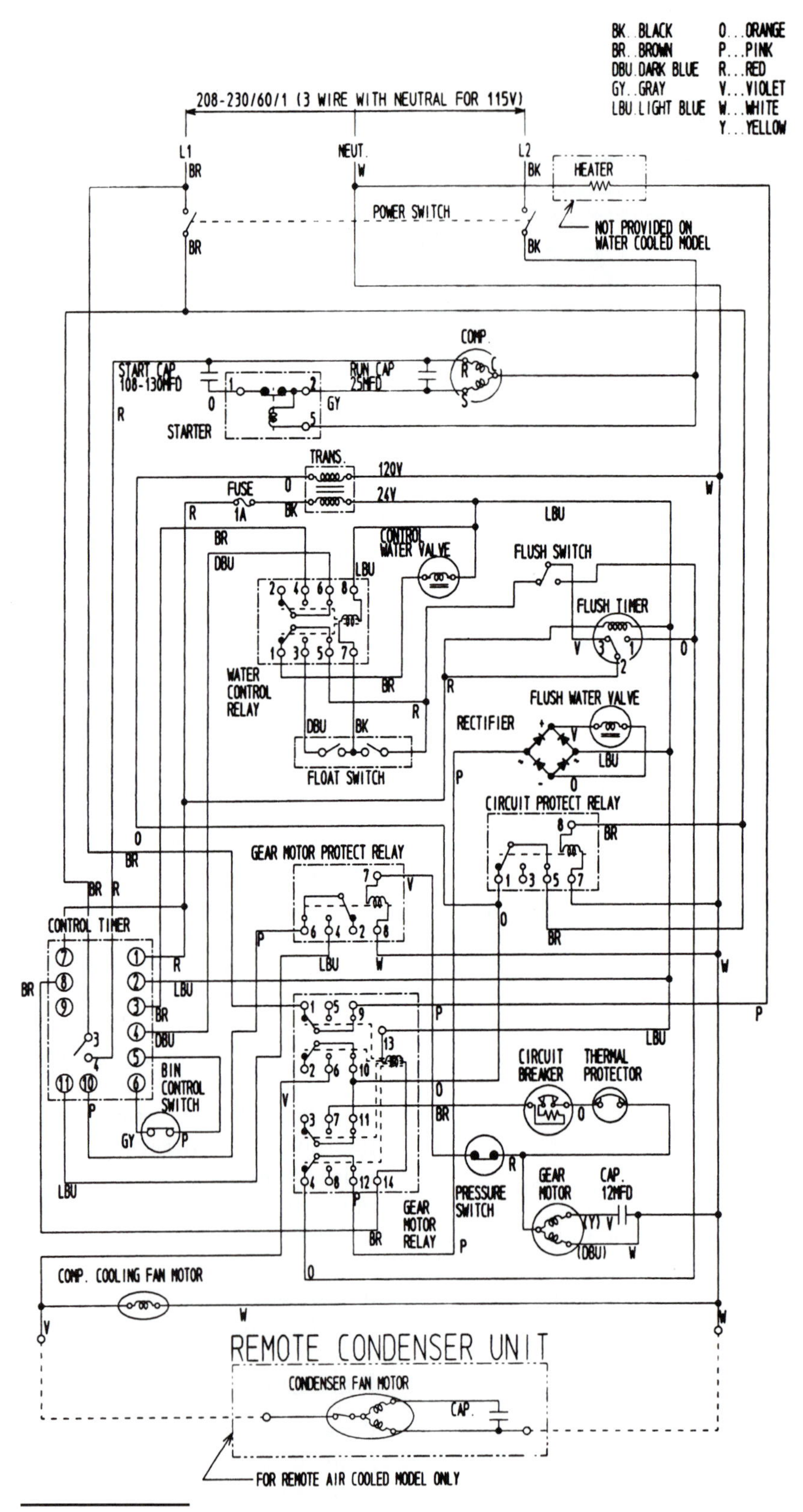

Figure 19–13 Hoshizaki interconnection diagram
(Courtesy of Hoshizaki America, Inc.)

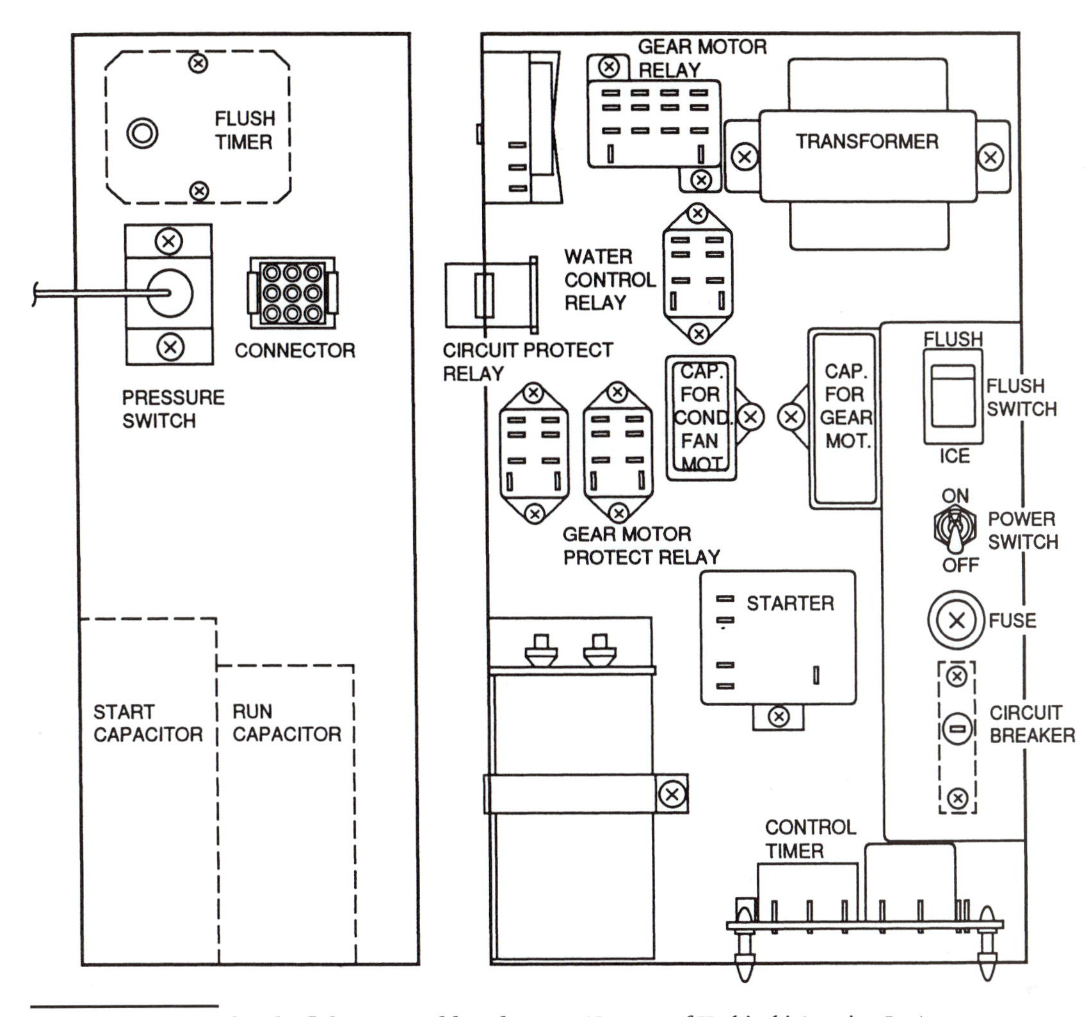

Figure 19–14 Hoshizaki flaker control box layout. *(Courtesy of Hoshizaki America, Inc.)*

seal. The inlet water valve is not energized during this flush period. The unit will automatically restart after 15 minutes on the flush timer.

Circuit Protect Relay

This flaker, as does many others, uses a 230V compressor, while all the other loads operate on 115V. However, this flaker uses an interesting circuit protection relay to protect the 115V circuit from a careless technician mis-wiring the power supply (figure 19–15(f)). If either of the power leads (L1 or L2) is mistakenly connected to the terminal that should have the neutral wire, 230V would be applied to the 115V portion of the circuit, causing significant damage. However, here, if 230V is applied to the coil in the circuit protect relay, the switch will move from terminal 1 to terminal 3. No part of the circuit is connected to terminal 3, so no part of the 115V circuit will "see" the improper 230V. When the proper 115V is applied to the circuit protect relay coil, the switch remains connected to terminal 1. Once the unit has been properly wired, the circuit protect relay has no further function.

Ross-Temp Model RF Flaker

The flaker in figure 19–16 operates as follows. When the bin switch closes, the power relay and the gear motor are energized, and the compressor runs. After the delay thermostat senses a cold evaporator, it switches from the RD-BK position to the BL-BK position. This bypasses the bin switch, but it has no effect until the end of the cycle when the bin is full. The gear motor and the compressor motor are both started through current relays.

When the bin is full, the bin thermostat opens, de-energizing the power relay. But the gear motor continues to run through the delay thermostat. This allows the auger to clear out any accumulated ice in the evaporator. After a few seconds, the evaporator

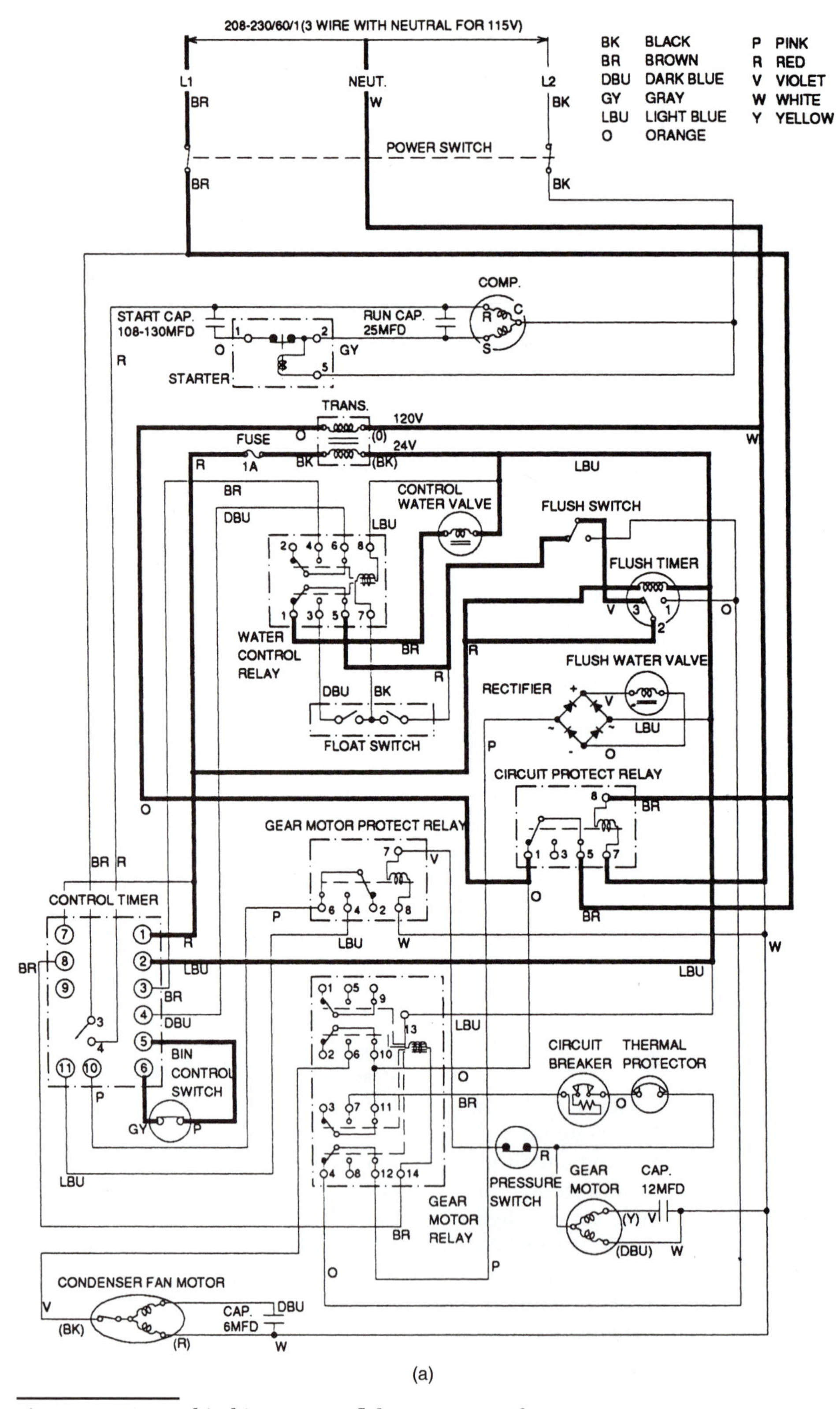

(a)

Figure 19–15 Hoshizaki F or DCM flaker sequence of operation. *(Courtesy of Hoshizaki America, Inc.)*

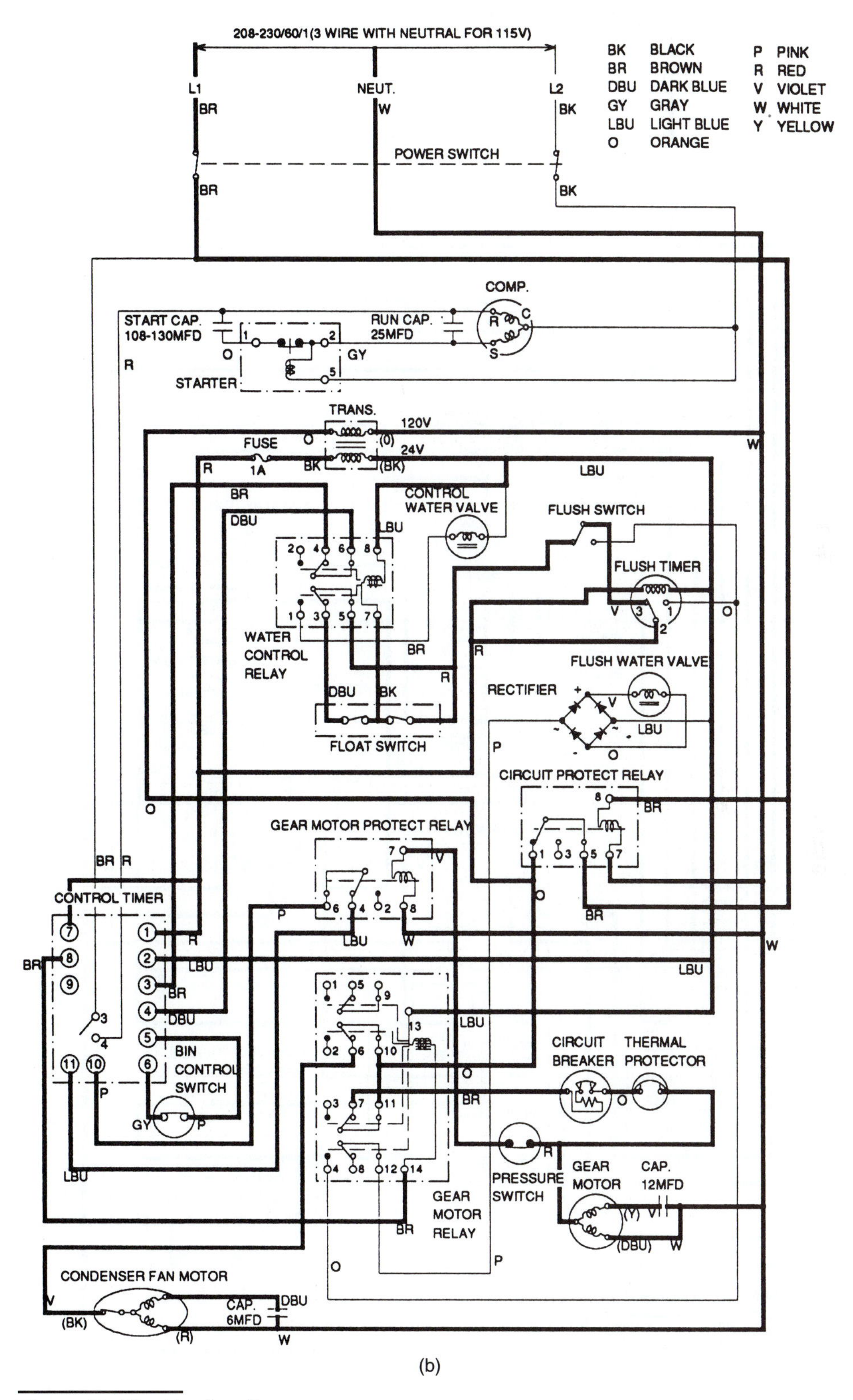

Figure 19–15 *(Continued)*

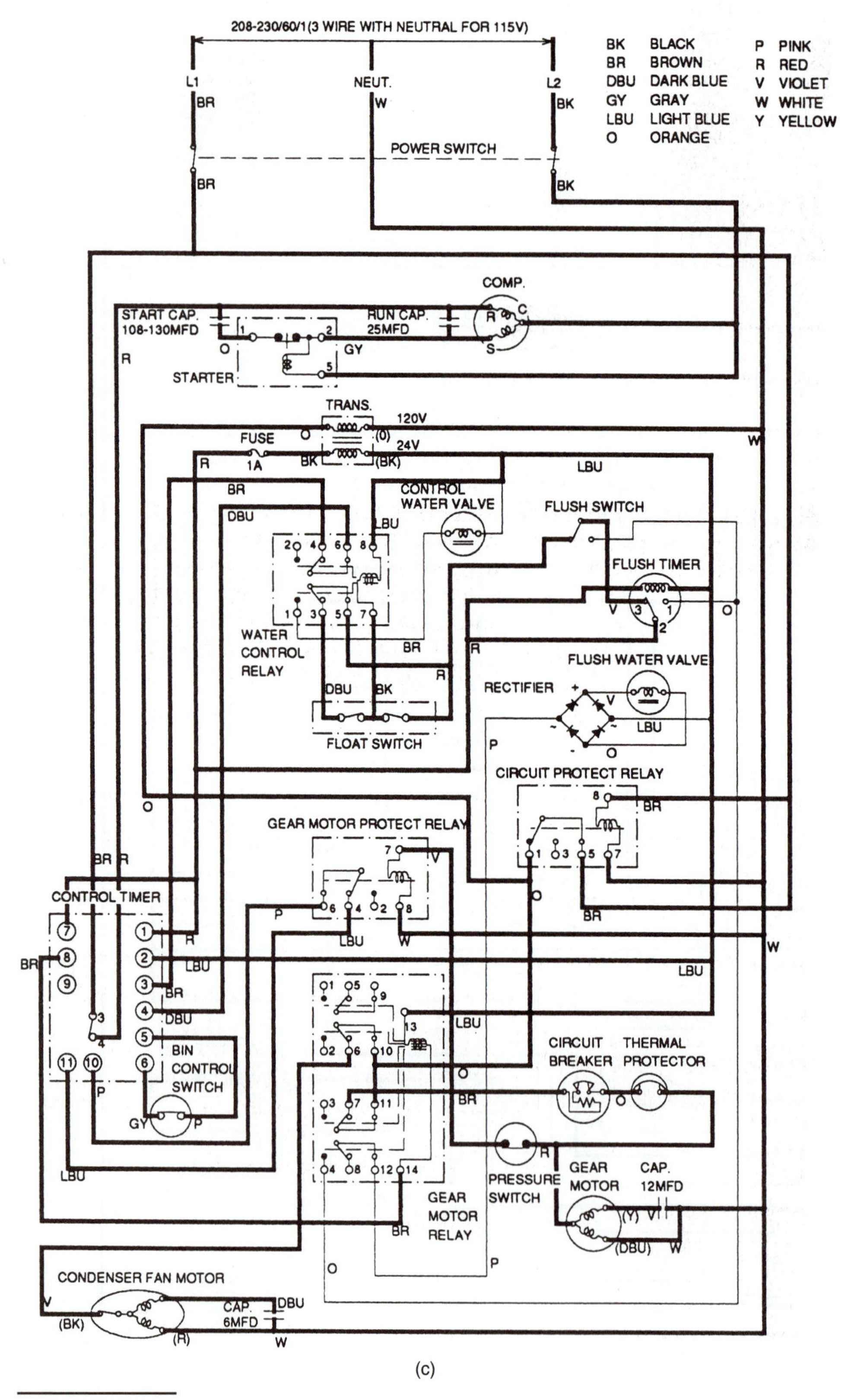

(c)

Figure 19–15 *(Continued)*

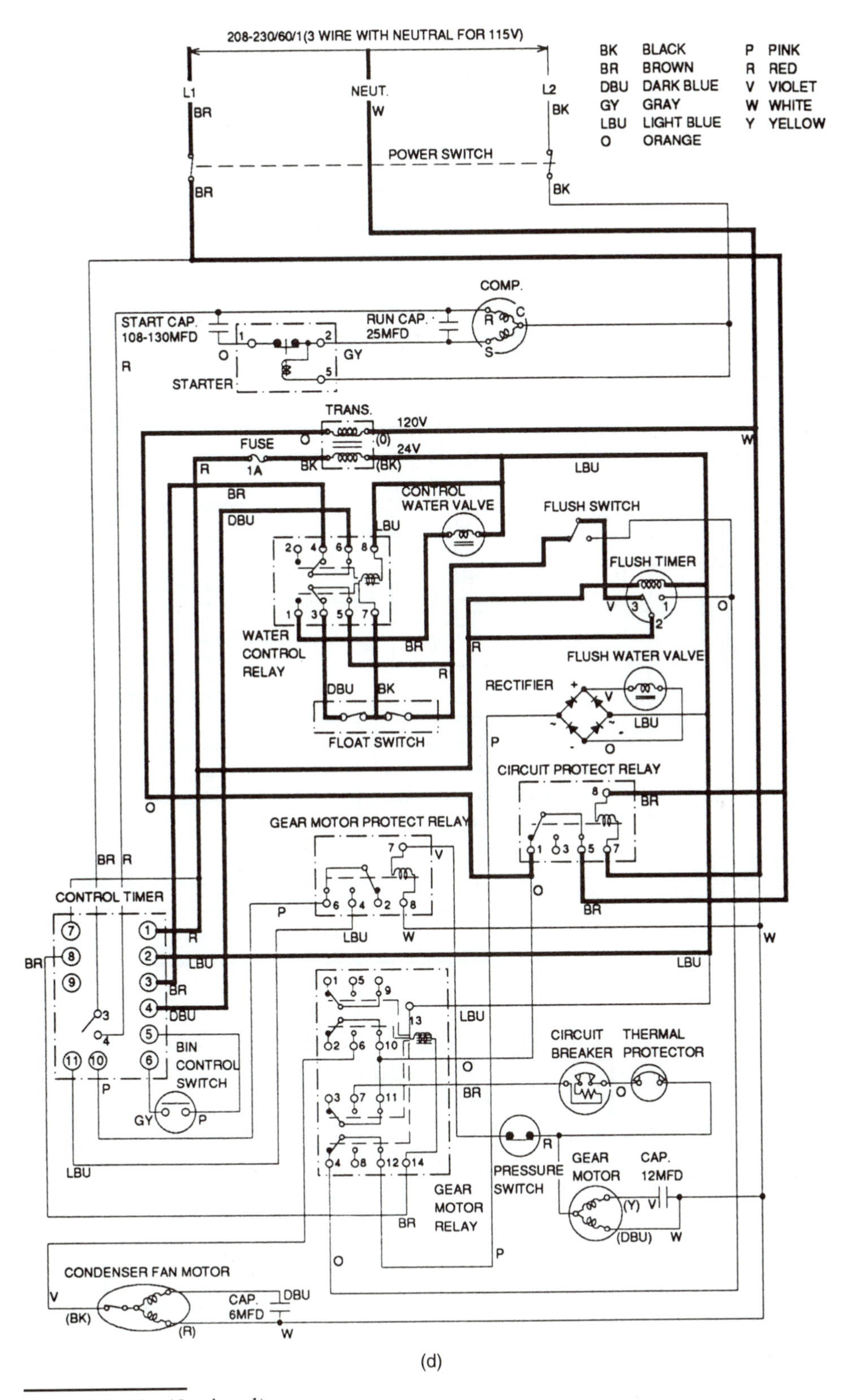

(d)

Figure 19–15 *(Continued)*

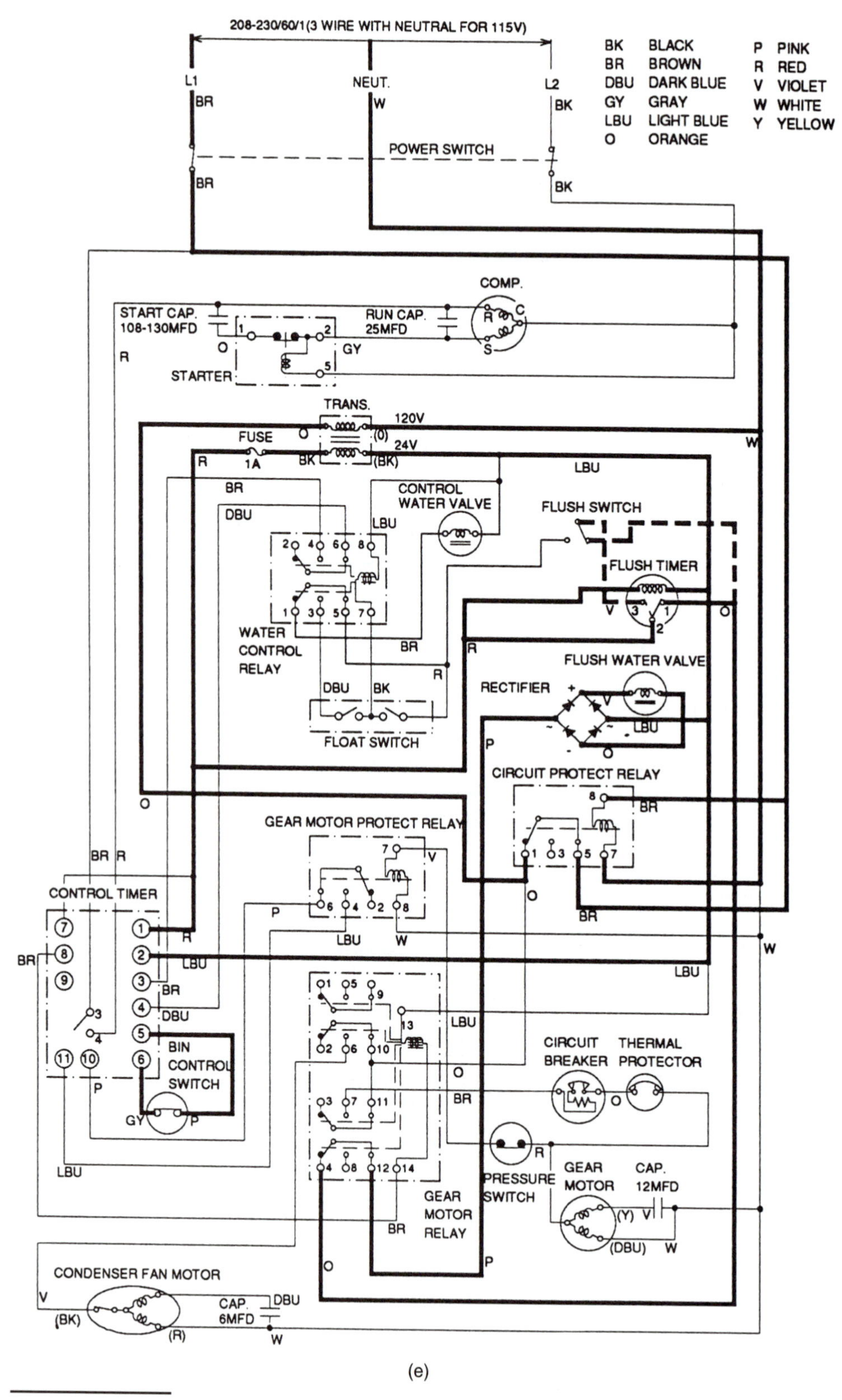

(e)

Figure 19–15 *(Continued)*

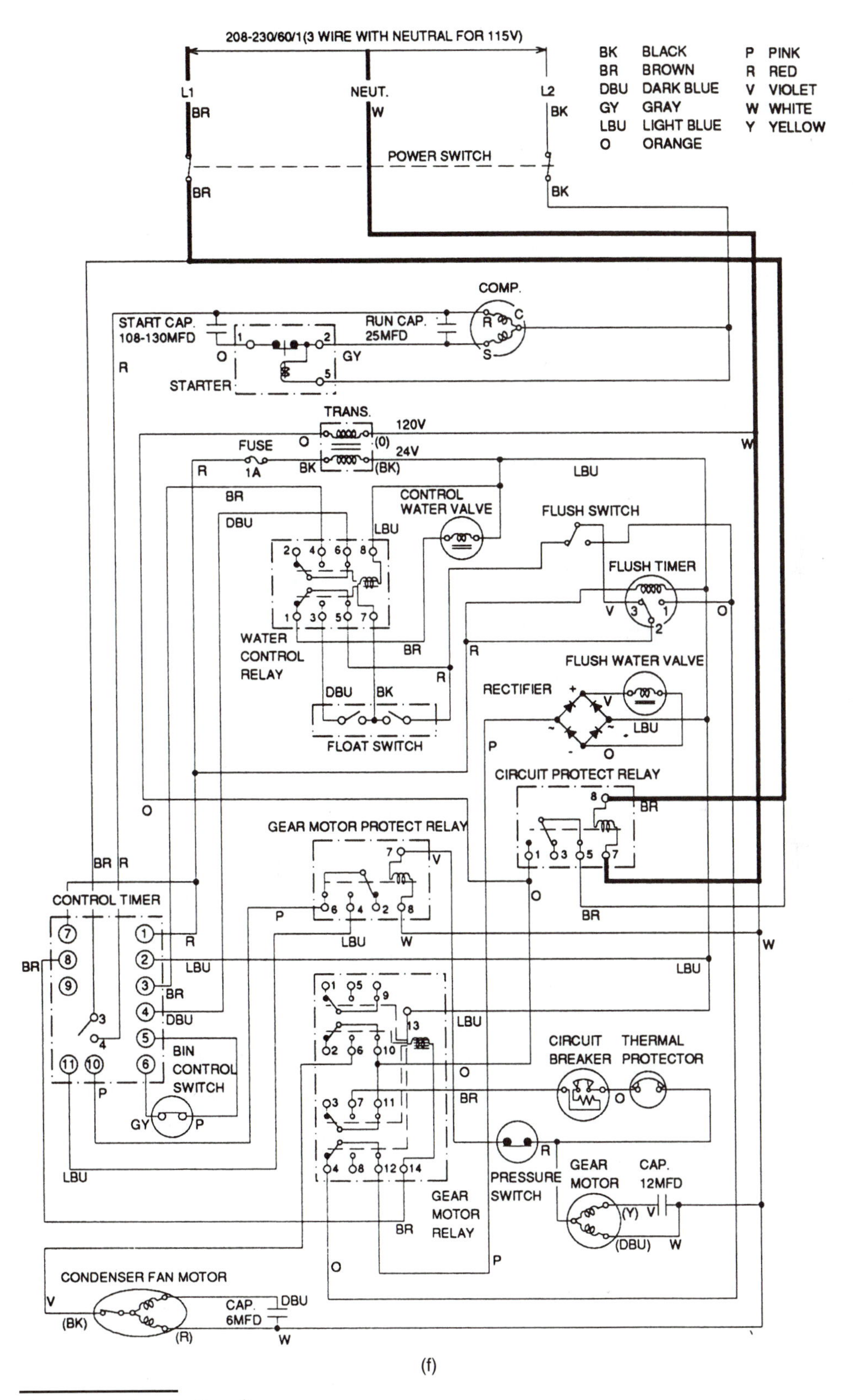

(f)

Figure 19–15 *(Continued)*

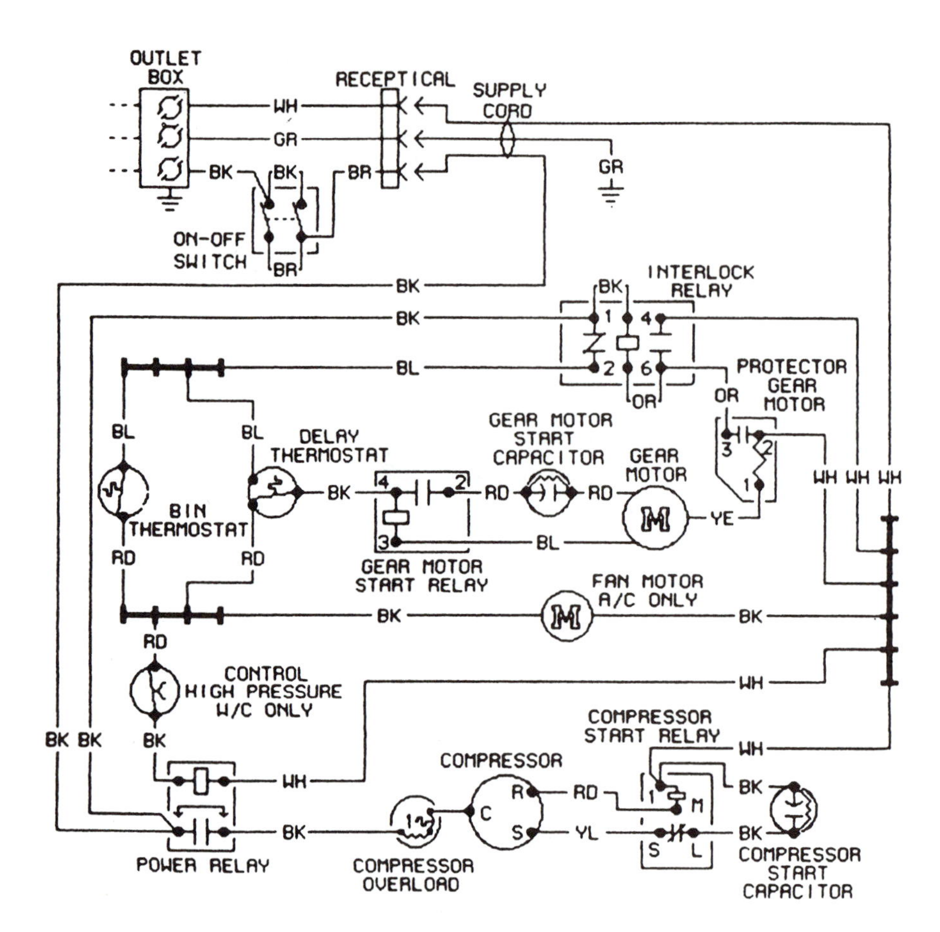

Figure 19–16 Ross Temp RF flaker. *(Courtesy of Ross Temp, a division of IMI Cornelius, Inc.)*

warms, and the delay thermostat switches back to the RD-BK position, and the gear motor stops.

The interlock relay has no function, until something goes wrong. The current through the gear motor is also passed through a current-sensing element in the gear motor protector (between terminals 1–2). If the current through the gear motor becomes higher than normal, the current-sensing element in the gear motor protector will close the switch between terminals 2–3, energizing the coil in the interlock relay. (This high current might be caused by binding of the auger bearings or lack of lubrication in the gear box.) This operates two switches in the interlock relay. The normally closed switch (terminals 1–2) open, de-energizing the power relay and the gear motor. The normally open switch (terminals 4–6) closes, locking in the interlock relay coil. Even if the gear motor protector now returns to its normal position, the interlock relay will remain energized, and the power relay will remain de-energized. The only way to return from this lock-out position is to turn the unit off at the On-Off switch.

EXERCISE:

Redraw the diagram in figure 19–16 as a ladder diagram. Show all the wire colors. You may combine the gear motor, its start relay and capacitor into a single circle in your diagram. The same applies for the compressor motor.

CHAPTER TWENTY

Miscellaneous Devices and Accessories

TIMERS AND TIME DELAYS

Figure 20–1 shows a mechanical timer that has been used for many years to turn off heating or air conditioning in applications where the space would be unoccupied for many hours at a time (such as in an office building). The normally open contacts can be wired in series with the low voltage side of the transformer (or the line voltage side for that matter) so that whenever the timer switch opens, the heating or air conditioning system will be inoperative. It is a seven-day timer, so the large wheel rotates once each week. The trippers on the outside of the wheel may be placed by the service technician for any On time or Off time each day. There are 24-hour timers available as well, but they are only useful if the same schedule is to be used every day. The seven-day timer can provide different schedules for each day, which is important if the space will be unoccupied during part of the day.

There is a fixed pointer, which should be pointing to the actual time of day when you are setting the trippers. Then, as each On tripper passes a trip point, the normally open switch closes. At the end of the occupied period, the Off tripper will pass the same trip point, returning the normally open switch to its open position. There is a manual override lever so that the unit may be turned on or off, regardless of the position of the trippers.

One of the problems with these timers is that they accumulate power interruption time. Suppose an office building experiences six 15-minute power outages over a two-year period. At the end of two years, the timer clock is an hour and a half behind the actual time. Your customer will call you in the morning, and by the time you get there, the unit will be running fine. Go check to see if there is a time clock tucked away someplace.

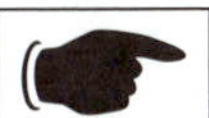

Suppose the customer tells you that his building will be occupied from 8:00 A.M. to 5:00 P.M. each day. Do not set the timer trippers for those times. If you do, and it happens to be summertime when you are setting the clock, then when daylight savings time ends and nobody bothers to reset the time clock, the heating will then shut off at 4:00 P.M. just as winter is approaching. If it happens to be wintertime when you are setting the clock, when everybody "springs ahead" for daylight savings time, they will be getting to the office an hour before the air conditioning becomes operable.

QUESTION:
Suppose the owner told you that the office was going to operate from 7:00 A.M. to 6:00 P.M. It's daylight savings time. At what hour should you

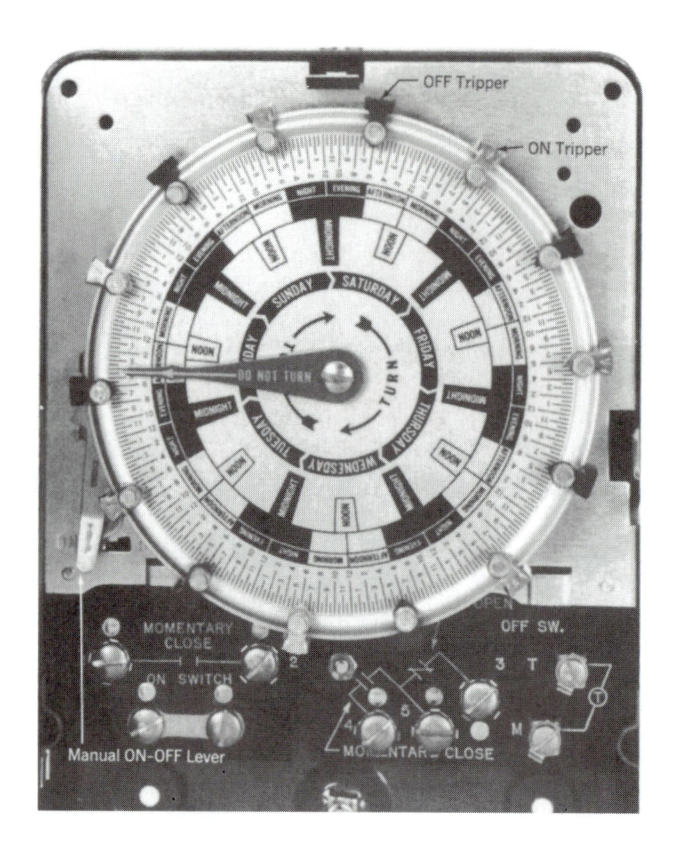

Figure 20–1 Mechanical timer. *(Courtesy of Maple Chase Co.)*

place the On tripper and the Off tripper so that the heating and air conditioning will be operable for the least number of hours, and yet cover the occupied hours year round?

Electronic Delay-On-Make, Delay-On-Break Timers

Figure 20–2 shows a "black box" timer. It is a significant departure from the electro-mechanical timers that use a clock motor to drive a series of gears or cams to operate a switch. It is simply a two-wire device. When a voltage is applied to one terminal, some time later, that same voltage comes out the other terminal as if an internal switch closed. Remarkably, some of these timers will operate over a dramatic range of input voltage (18–240V), and the input voltage may be AC or DC. Some models have a fixed time delay, while others have a knob that can be used to adjust the length of the time delay.

A timer may be a Delay-On-Make timer or a Delay-On-Break timer. For the circuit shown, when the thermostat closes, a Delay-On-Make timer will allow 24V to the control relay coil five minutes (for example) later. If a Delay-On-Break timer is used, when the thermostat closes, it will provide 24V to the control relay coil immediately, except if the thermostat has been open for less than five minutes. If the thermostat has only been open for two minutes, then the Delay-On-Break timer will time out another three minutes before it allows the voltage to pass to the control relay coil. The Delay-On-Break timer is actually superior for this application (anti-recycle timer) because it doesn't delay the operation of the condensing unit unless there has been an off-cycle of less than five minutes. The Delay-On-Make will provide a time delay even if the condensing unit has not run for two hours. The Delay-On-Make timer would be more appropriate for staging the starting of multiple units to reduce peak power requirements.

A less sophisticated model of this electronic timer uses four terminals. Voltage is applied to the two input terminals, completing a circuit. Minutes later, the voltage is available at the two output terminals to supply voltage to the contactor or relay coil.

Electronic Post-Purge Fan Delay Timer

Figure 20–3 shows a timer that provides an OFF delay, rather than the ON delay provided by the timers described above. When the room thermostat calls for cooling, it makes R-G, providing power to

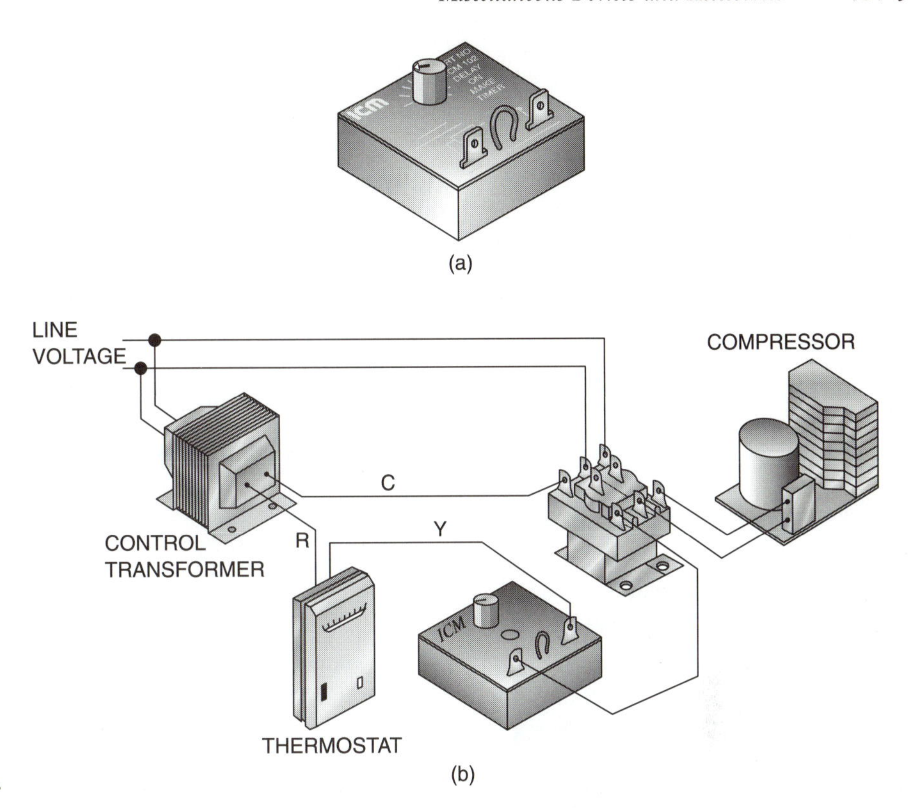

Figure 20–2 Delay-on-make timer.

terminal 4 of the timer. As soon as power is applied to terminal 4, the power on terminal 2 (which is always available) will be applied to the fan relay coil. There is no delay during start-up. However, when the thermostat is satisfied, and it opens R-G, the power on terminal 2 will continue to be applied to the fan relay coil. The fan will run during the time delay period, and then it will shut down when power is no longer provided from terminal 1.

ELECTRONIC MONITORS

An electronic monitor can be used for a number of different functions:

1. If the available line voltage is not within acceptable limits, the monitor will prevent the compressor contactor from becoming energized.
2. If, on a three-phase system, there is an unacceptable voltage imbalance between the phases, the compressor contactor will not be energized.
3. If the contactor wants to short cycle based upon input from the operating or safety controls, the monitor will provide a delay on break period.

Figure 20–4 shows a line monitor for a single-phase system. Figure 20–5 shows a monitor for a three-phase system.

COLD CONTROLS

The term **cold control** (figure 20–6) is jargon in the refrigeration industry for line-voltage thermostat in a small refrigerator. There are two general types:

1. Cold controls that sense coil temperature.
2. Cold controls that sense box temperature.

The difference is important, because although the two types may look identical, they are not interchangeable. The difference lies in their cut-in and cut-out temperatures. For example, a cold control that senses box temperature for a 38°F box might have a cut-in temperature of 39°F, and a cut-out of 37°F. However, for a cold control that senses coil temperature, when the compressor is running, the coil temperature will be much colder than the box. This type of cold control for the same 38°F box might have a cut-in setting of 39°F, but a cut-out of 0 to −15°F, depending upon the design of the coil operating temperature. A cold control that senses coil temperature may be replaced with a box-temperature-sensing cold control if the sensing element can be placed where it senses box temperature. However, the reverse is not true, because you would not know the actual design coil temperature of the unit you are servicing.

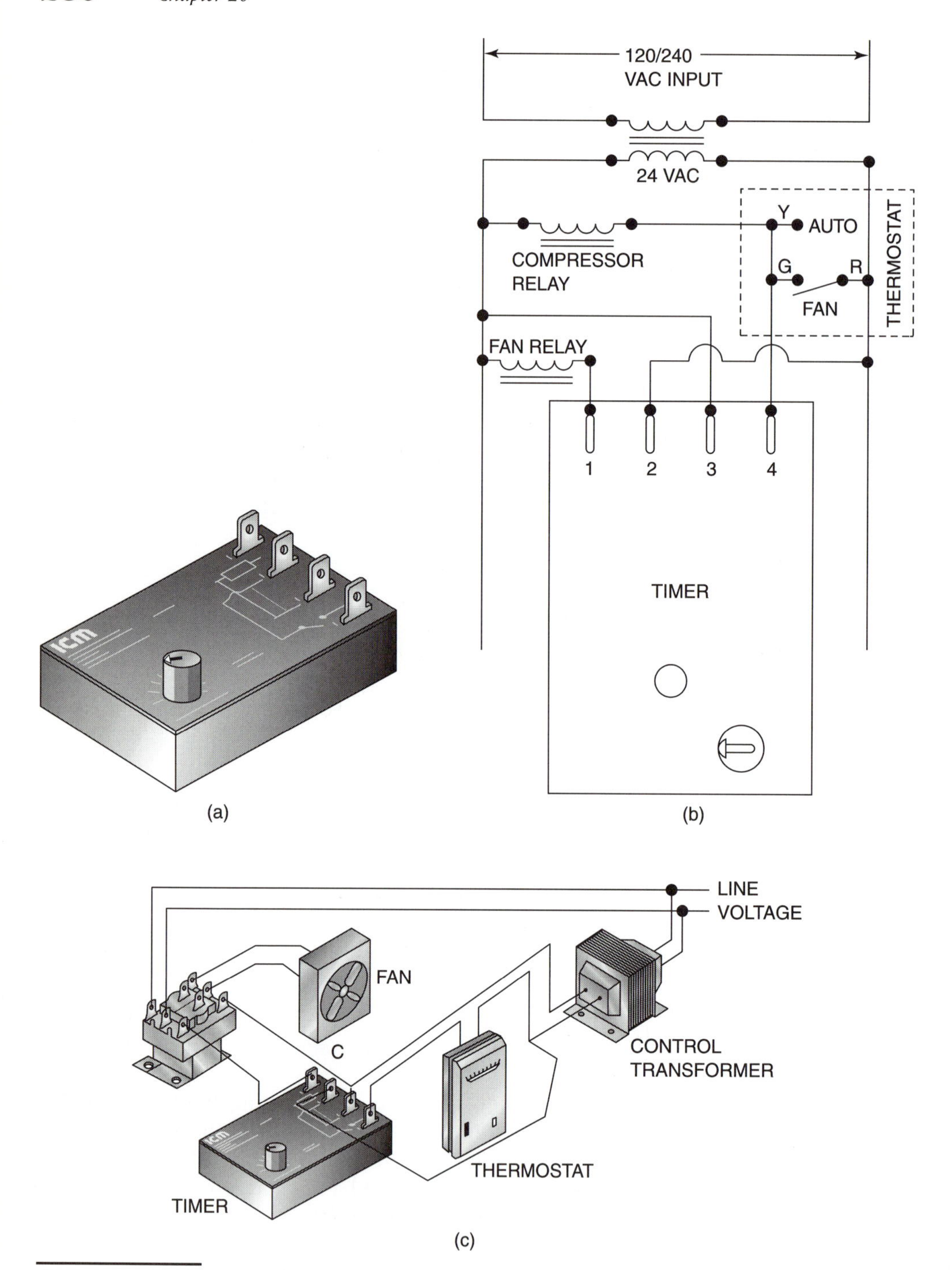

Figure 20–3 Post-purge fan delay timer.

When either of the above two types of cold control is adjusted by the customer, it lowers both the cut-in and cut-out temperature. This presents a risk of the customer setting the thermostat for a too-low temperature . . . so low that the ice that forms on the evaporator coil during the "freeze" cycle does not completely melt during the "off" cycle. There is another type of cold control for refrigerators called a constant cut-in that remedies this problem. Regardless of how the customer adjusts the control, the cut-in setting remains fixed at 38 to 39°F. This assures that all ice will be melted off the coil before the compressor is allowed to restart.

EVAPORATIVE COOLERS

Evaporative cooling is a low-operating-cost alternative to air conditioning. It works by passing outside air over a wet pad. As the air passes through the

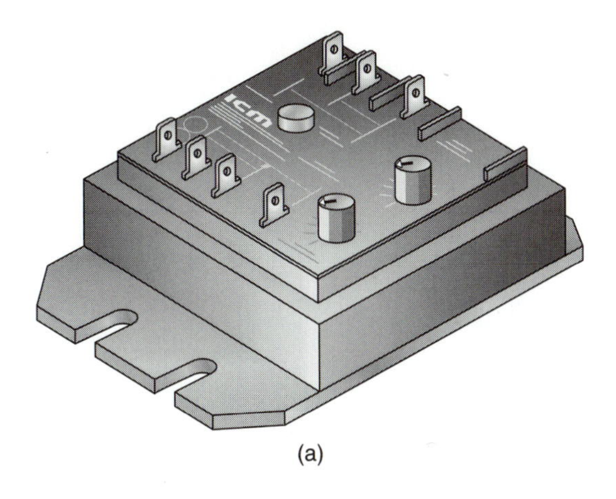

(a)

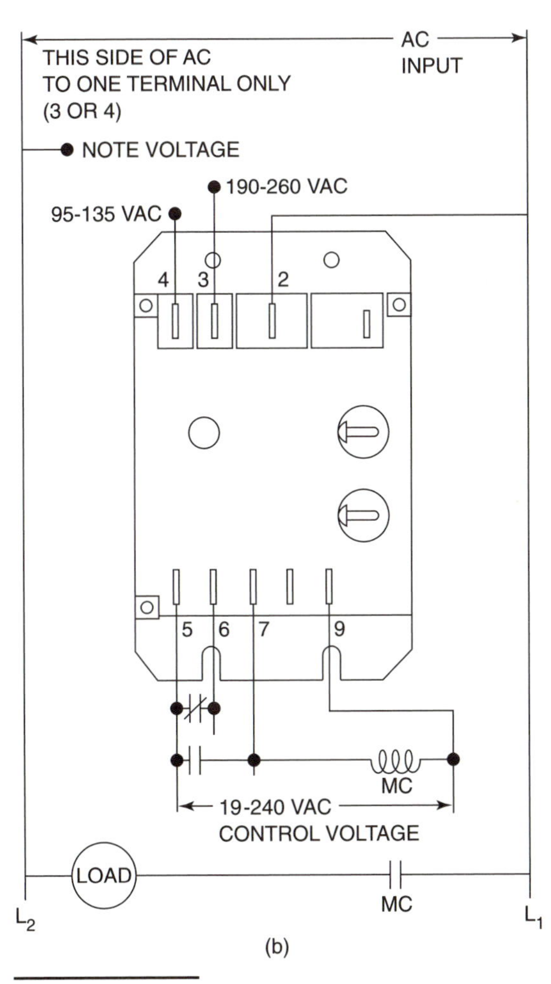

(b)

Figure 20–4 Electronic monitor-single-phase.

pad, it is cooled (but also humidified) by the evaporation of the water from the pad. The cool, moist air is then supplied to the occupied space. Figure 20–7 shows a wiring diagram for an evaporative cooler. The switch on the 115V line (unlabeled Hot) supplies power to either the black or the red wire on the motor that drives the fan that draws the outside air through the pad and delivers it to the room. The common side of the fan motor is attached, through the junction box, to the other side of the power line (the unlabeled neutral line). A separate 115V supply comes into a different switch, and supplies power to the recirculating pump motor. The other side of the pump motor is connected to the neutral through the junction box.

EXERCISE:
Redraw the above wiring diagram as a ladder diagram. Label one of the switches Fan, and the other switch Pump.

The Honeywell R8183 control panel(figure 20–8) is a packaged control scheme that can be used to control the operation of an evaporative cooler. It contains 3 relays (fan on/off, fan high/low, and pump on/off) and a 20VA transformer. The pump relay operates the water pump and provides wetting of the evaporator pad. Two fan relays control fan operation and fan speed. All R8183 functions are remotely controlled through the thermostat subbase.

HUMIDIFIERS

Figure 20–9 shows a typical residential humidifier. On this type of system, the furnace motor pushes air through the furnace heat exchanger where it is warmed and delivered to the room. But a small portion of the supply air is diverted to flow to the humidifier, where it passes over a wetted pad, and then back to the return air side of the furnace. A small motor in the humidifier causes the wetted pad to rotate through a pan of water, where the pad is continuously wetted (the level of water is maintained by means of a water supply through a mechanical float valve). Figure 20–10 shows the wiring diagram to control this system. The furnace fan runs under two different circumstances:

1. The normally open switch on the blower relay is closed. The furnace fan runs on high speed. Or,
2. The blower relay is not energized, but the furnace is running **and** the bonnet switch (labeled as fan switch) is closed.

There is a humidistat in the room that only closes when the room humidity is lower than the set point of the humidistat. When the humidistat closes, it causes the humidifier motor to operate (in this case, it is a 24V motor.) When the humidistat is satisfied, the wetted pad stops rotating, and it dries out, causing the humidification to stop.

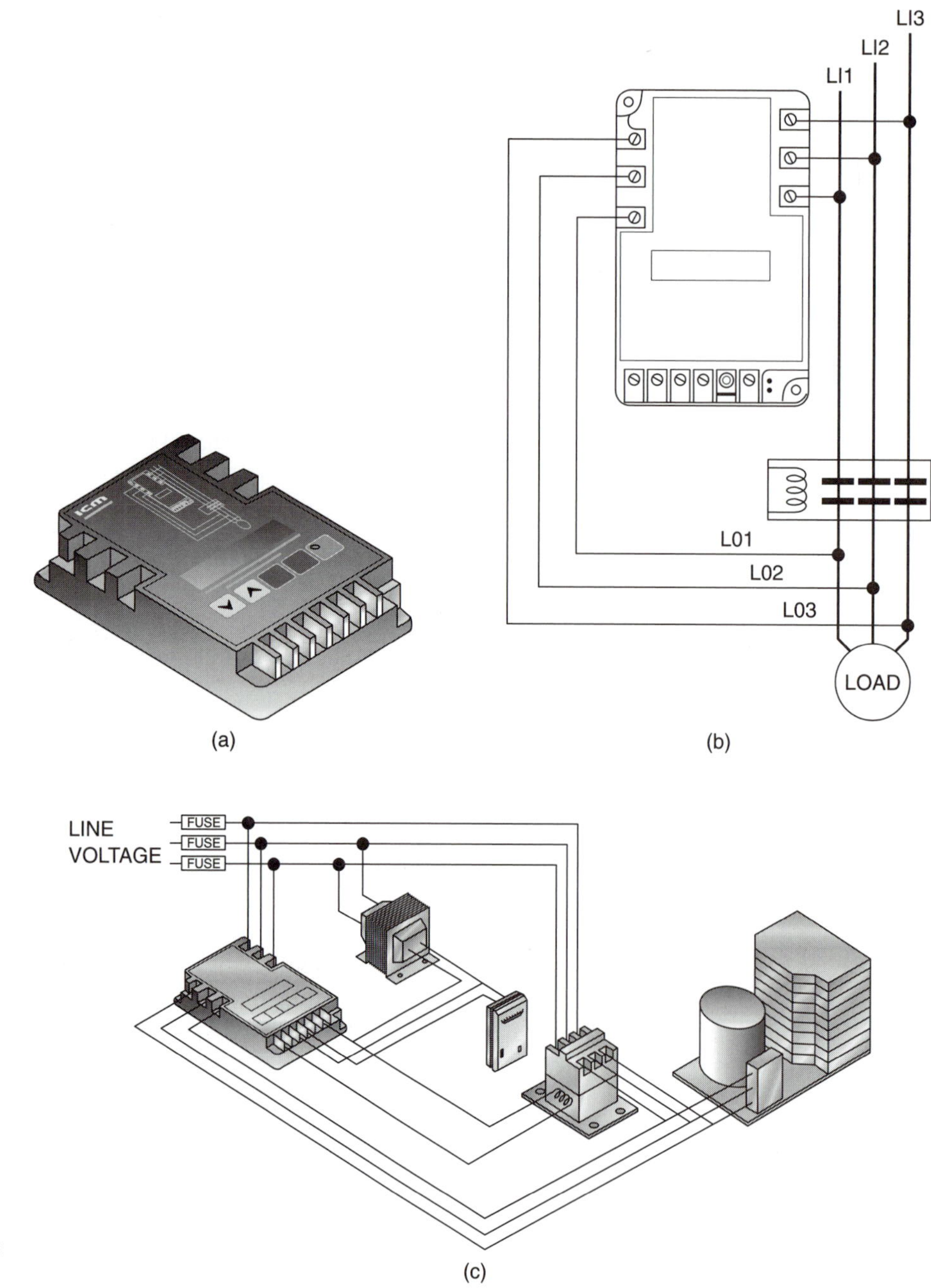

Figure 20–5 Electronic monitor-three-phase.

Figure 20–6 Cold control.

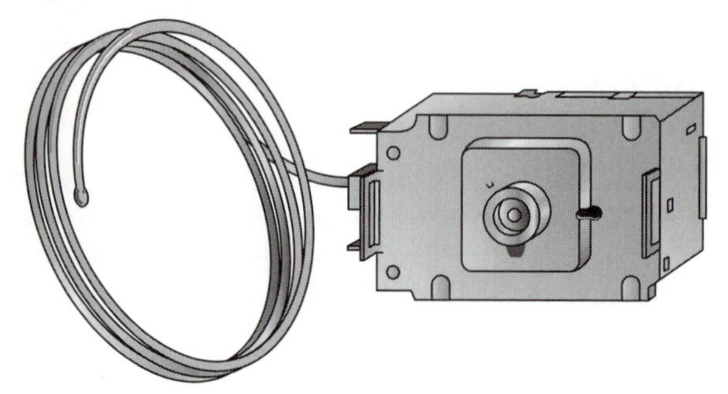

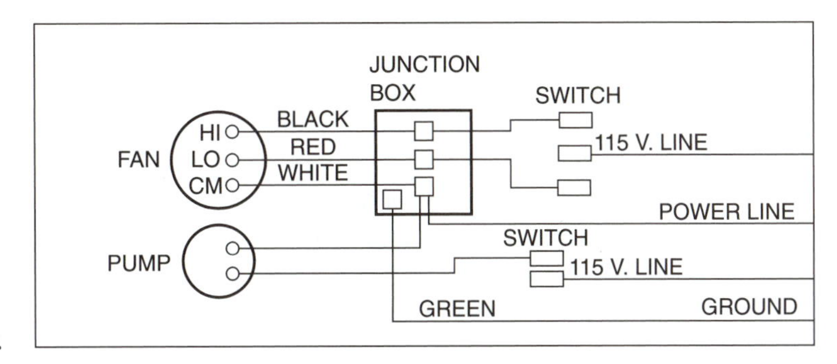

Figure 20–7 Evaporative cooler.

> **EXERCISE:**
> Redraw the humidifier wiring as a ladder diagram. Include the following items that are not shown on the interconnection diagram: Room thermostat (including terminal colors), gas valve, furnace transformer, relay coil.

CONDENSATE PUMPS

Normally, the condensate that forms on the evaporator coil of an air conditioner is allowed to drain by gravity. However, sometimes, a gravity drain is not possible. For example, if air conditioning is being added to a central heating system, and the furnace is located in a closet area in the middle of the house, you cannot install a gravity drain to the outside. In that event, the condensate is allowed to drain into a condensate pumping unit (figure 20–11). This unit has a level control that operates a pump motor. When the tank is almost full, the level switch closes, the pump motor starts, and the unit pumps the water out of the tank. When the tank is emptied, the level switch opens and turns off the pump.

In addition to the above, there is another level switch that is used as a safety switch. While the main level switch closes on a rise in level, the safety switch opens on a rise in level. The set point of the safety switch is slightly higher than the main level switch, and if everything works normally, the safety switch never operates. However, if the tank level rises above the set point of the main level switch due to a failed pump or a clogged discharge line, the safety switch will open. The safety switch is not wired to anything in the condensate pump unit (dry contacts). The technician wires the safety switch in series with the compressor contactor. So, if the level gets too high, the air conditioning will not be allowed to operate, thus preventing a flood below the evaporator.

MODULATING DAMPER MOTORS

Figure 20–12 shows a modulating motor that can be used to control damper positions, valves, or a series of switches. Unlike a conventional motor that rotates to turn a fan blade or a compressor, this motor shaft only rotates through a limited arc (one quarter to less than one half of a turn). The electronic damper motor operates on a principle of a circuit called a **Wheatstone bridge** (figure 20–13). In this circuit, when the switch is closed, if the resistances are all equal, the potential at C equals the potential at D. The net potential difference between C-D is zero, and there is no flow of electricity through the galvanometer (G). However, if the resistance of any one leg is changed, the bridge will become unbalanced, and there will be a current flow between C-D. This current flow is used to drive the damper motor.

Typically, the resistors that form one part of the bridge are located in a temperature sensor, and the resistance changes in response to temperature. The rotational movement of the damper motor operates a device in response to this temperature change. Internally, the rotational movement also

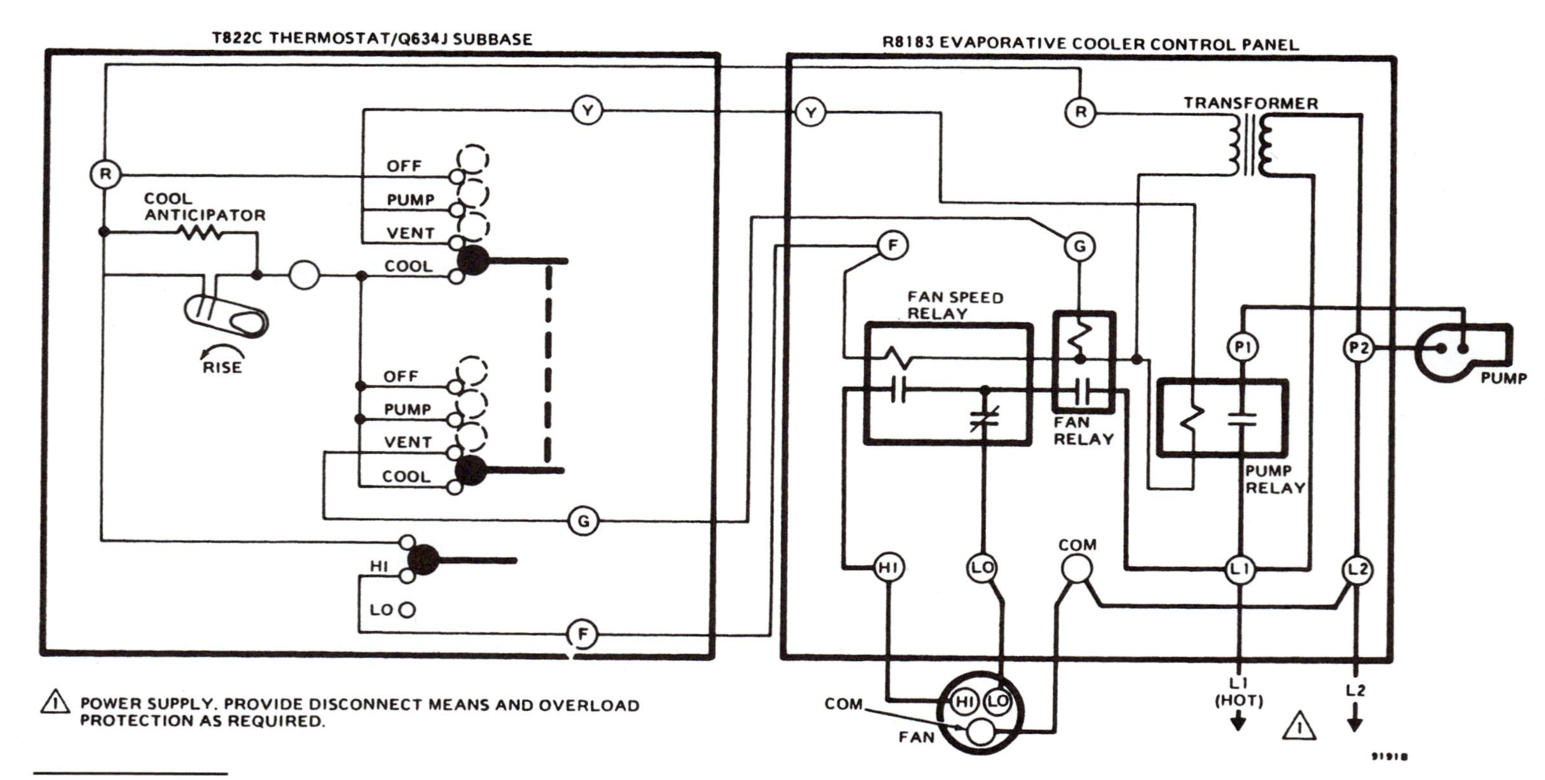

Figure 20–8 Evaporative cooler control panel. *(Courtesy of Honeywell, Inc.)*

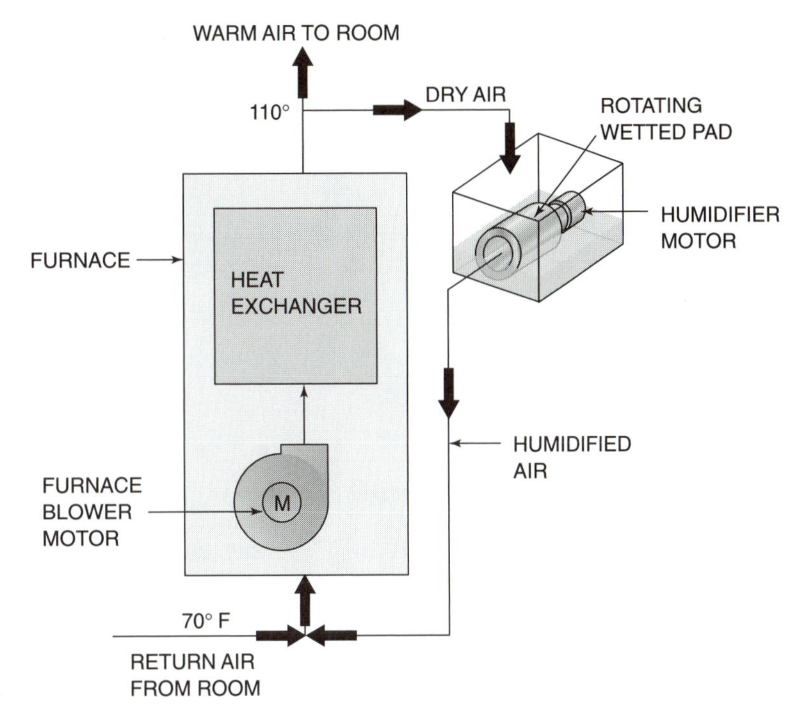

Figure 20–9 Residential humidifier.

changes the resistances internal to the damper motor, so that the motor reaches a new electrical balance point, but with the damper motor shaft in a new position. In this way, the damper position is proportional to the temperature input from the sensor. Figure 20–14 shows a typical wiring connection diagram for a temperature sensor and an electronic damper motor. As the sensed temperature rises, the damper motor shaft rotates in one direction. As the sensed temperature falls, the damper motor moves in the other direction.

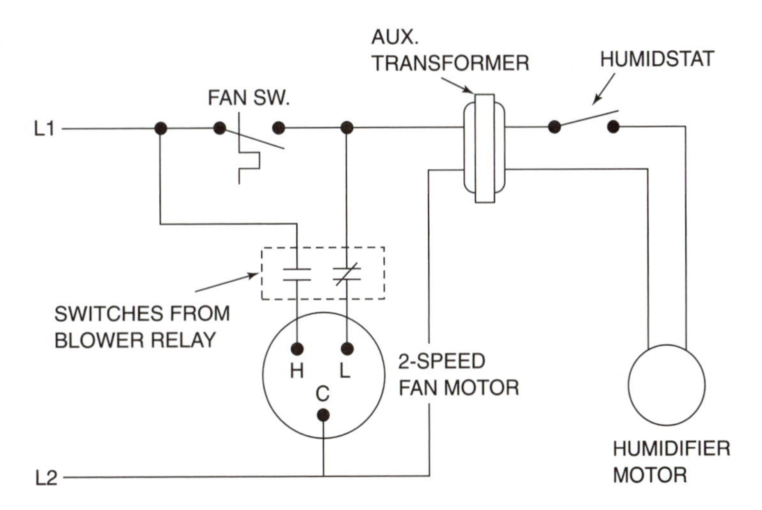

Figure 20–10 Humidifier wiring with two-speed furnace fan.

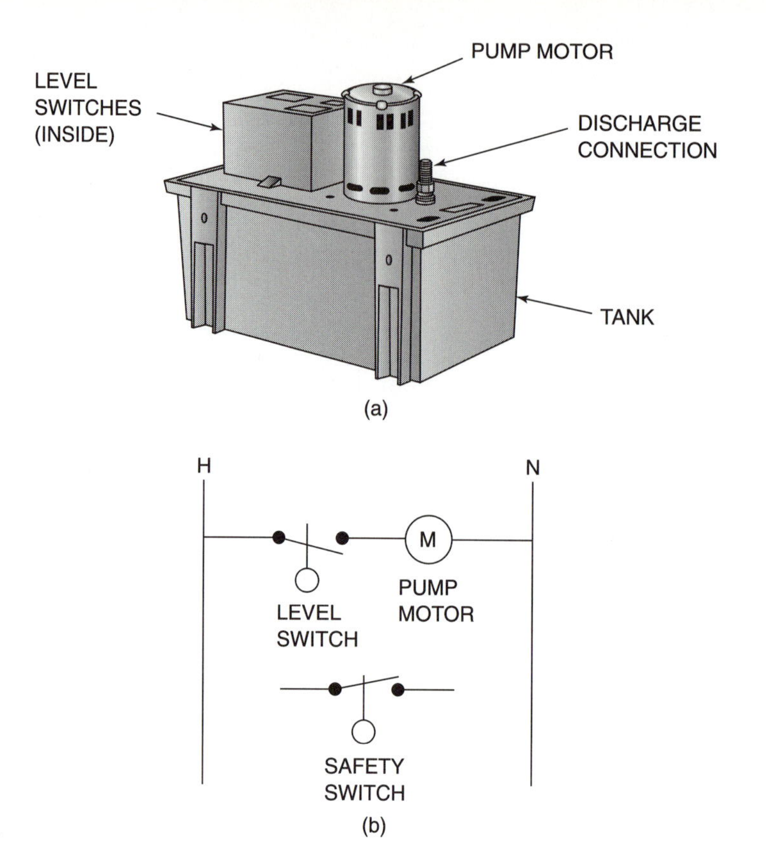

Figure 20–11 Condensate pump.

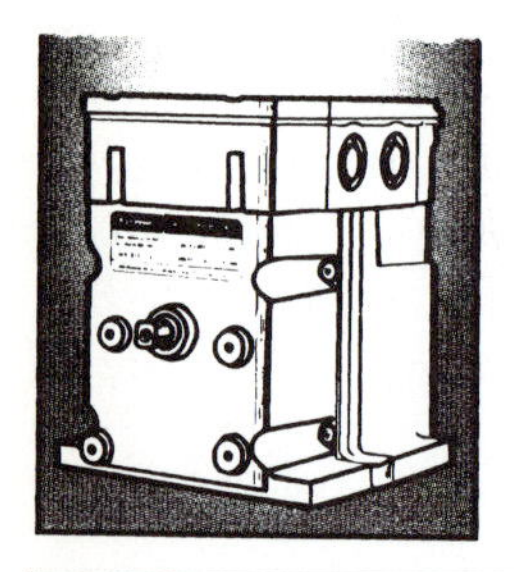

Figure 20–12 Modulating damper motor.

Figure 20–13 Wheatstone bridge circuit. *(Reprinted from Refrigeration and Air Conditioning, 3rd Ed. by Air Conditioning and Refrigeration Institute, copyright © 1998, Prentice Hall, Inc. Reprinted by permission.*

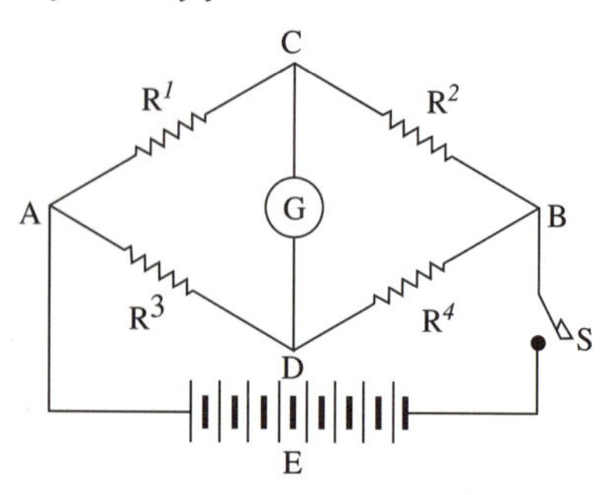

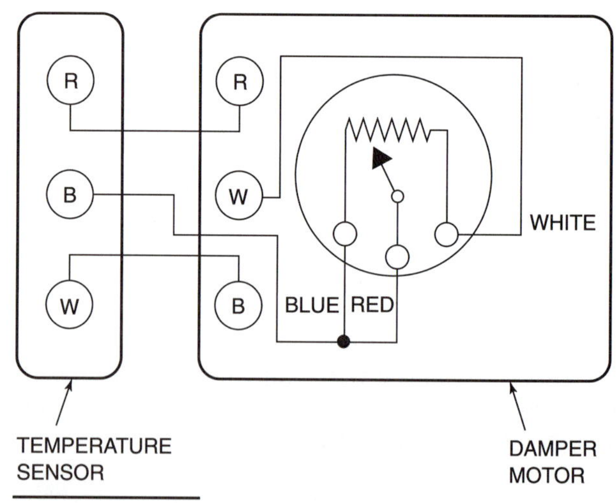

Figure 20–14 Damper motor responds to temperature change.

TWENTY-ONE

WIRING TECHNIQUES

SOLDERLESS CONNECTORS

The electrical connections made by the service technician will almost always be non-soldered mechanical connections. The most common type of connector is the wire nut (figure 21–1). The wire nut has a coil of wire inside. When bare conductors are placed inside the wire nut, the wire nut may be "screwed" onto the wires, connecting the wires together and to the internal coil of wire. If the conductor has been stripped of insulation to the correct length, the wire nut will extend past the end of the insulation so that no bare conductor is visible.

After the connection has been made, pull hard on each individual wire. If the connection has been properly made, it will not come loose.

Do not twist the wires together prior to inserting them into the wire nut. They will become twisted by the wire nut. If you are connecting a solid wire to a stranded wire (figure 21–2), leave the stranded wire slightly longer than the solid wire. Otherwise, the wire nut may grab securely only onto the solid wire, with the stranded wire wrapped around the solid wire. Over time, this may cause the connection between the two wires to deteriorate, creating a resistance, then heat (due to the $I \times R$ drop through the resistance), and eventually a burned connection that fails.

POWER CORDS

A power cord for an appliance sometimes is not supplied with the appliance (such as a residential furnace). It must be connected by the service technician. Use a 14–3 cord of the required length. 14–3 means that the cord contains three 14ga wires. At one end, you must attach the plug. A 115V plug is shown in figure 21–3b. The U-shaped connection is the ground. It is longer than the other connectors so that it will make contact first when the plug is inserted into an outlet (and, when removing the plug from the outlet, the ground wire breaks contact last).

The terminals on the attachment plug are:

1. Brass for the Hot blade.
2. Silver for the Neutral blade.
3. Green hex screw for the ground.

The power cord has a black, a white, and a green wire inside the sheath (figure 21–3a). Remove 1 inch of the outer sheath. Strip 1/2 inch of insulation from each conductor. Insert the conductor through the center hole of the plug. Twist the ends of each conductor (to prevent them from spreading), and bend each around the appropriate terminal (black on brass, white on silver, green on green). It is important that you bend the wires around the terminals in a clockwise direction. Tighten the terminals, taking care to keep all strands of the conductor under the screw head. Inspect for loose strands which can cause shorts.

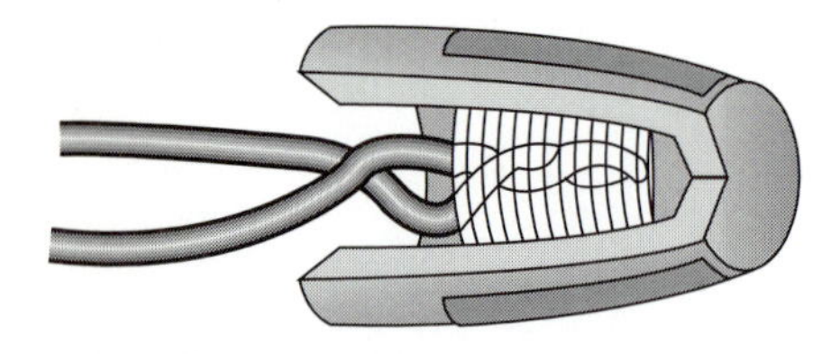

Figure 21–1 Wire nut.

After the wires have been attached to the plug, place the insulating disc over the blades, and tighten the clamp around the cord where the cord enters the plug. If the plug does not have a clamp to relieve strain on the terminals when the cord is pulled, you will have to make your own as shown in figure 21–3c. The black and white wire can be tied as shown prior to making the connection to the terminals. Then the knot may be pulled into the plug body. It will take the strain off the terminals.

At the other end of the cord, connect the conductors to the unit wiring using wire nuts or a terminal strip if provided (black to black, white to white, and green to green or if no green is provided, to the casing).

TERMINAL STRIPS

Sometimes, a manufacturer provides terminal strips for the wiring connections made by the manufacturer and for the field wiring. It serves two purposes:

1. It eliminates the need for wire nuts, making for a neater installation.

Figure 21–2 Connecting solid to stranded.

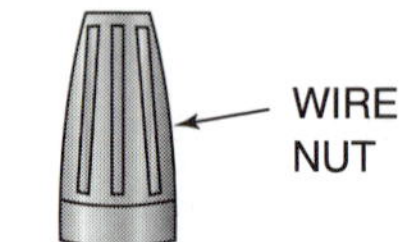

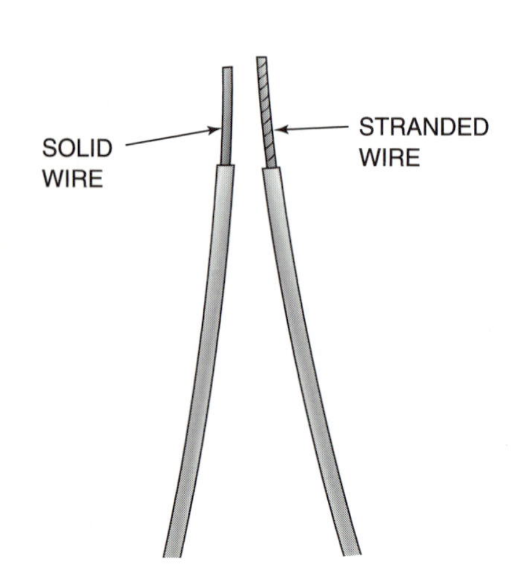

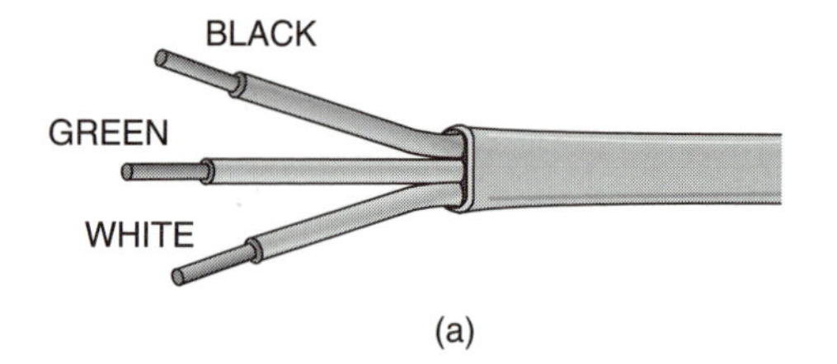

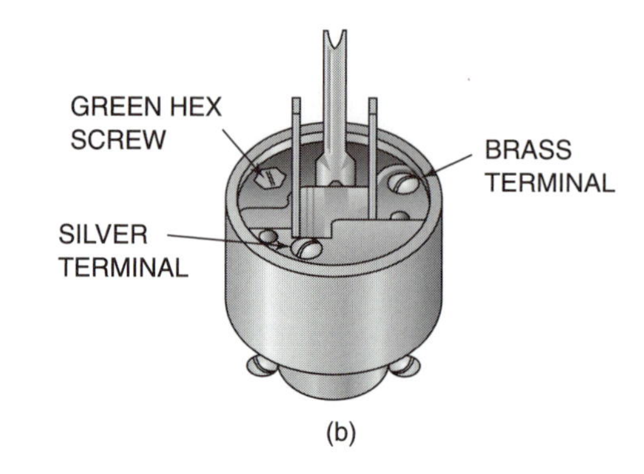

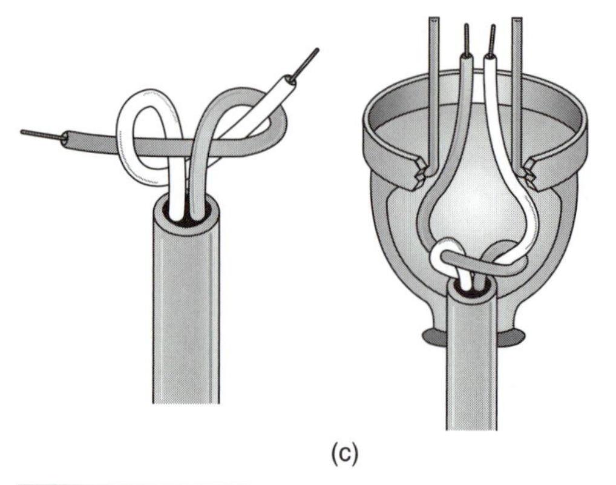

Figure 21–3 Wiring a plug.

2. The terminals can be labeled to correspond with terminal identifications on a wiring diagram. This makes it easy for the technician to troubleshoot. All you need to do is find the terminal numbers on the wiring diagram where you want to measure voltage, and go to the appropriate terminal numbers on the strip.

STRAIN RELIEF

Figure 21–4 shows several types of **strain relief** that are used where a power cord or a thermostat wire passes through a hole in the metal casing of a unit. The strain relief serves two functions:

1. It prevents the insulation from being cut where it rubs on the sharp metal edge of the unit.
2. It prevents strain on the wiring connections when the cord is pulled.

A wire should never be routed through a metal hole without using a strain relief.

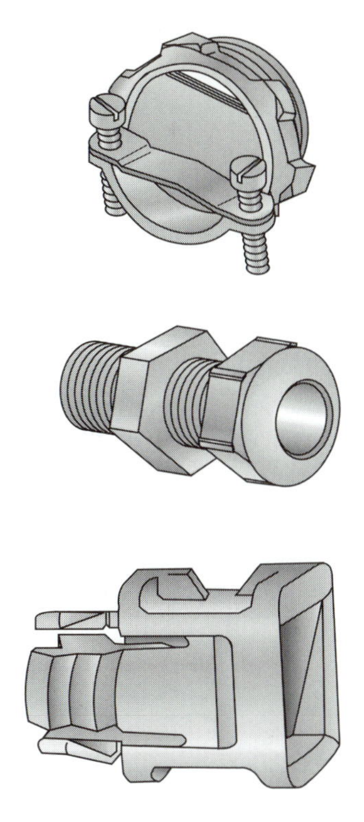

Figure 21–4 Strain relief.

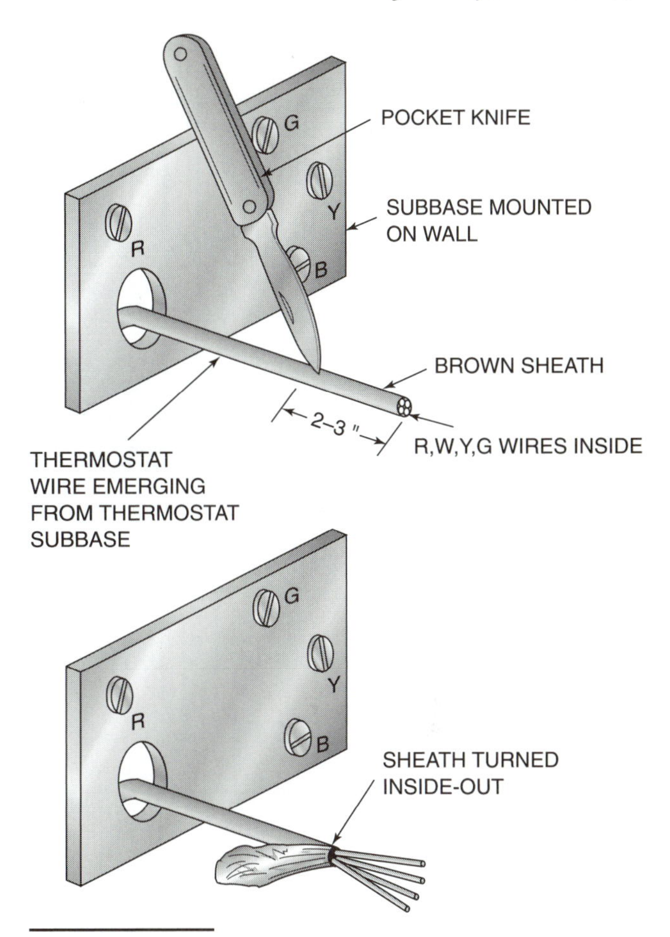

Figure 21–5 Stripping thermostat wire.

WORKING WITH THERMOSTAT WIRE

Connecting thermostat wire to a room thermostat subbase is one of the most-watched tasks you will perform. A customer's opinion of your competence will very likely be affected by how professionally you can do this task. That means it must be done quickly, and when you're done, it should look good.

Start by stripping 2–3in. of sheathing from the thermostat wires. Manipulate the wire to flatten the thermostat wires inside, so they are laying side by side. This will allow you to insert the blade of a knife between the thermostat wires, without cutting into the individual wire insulation. Figure 21–5 shows a technique that your mother warned you against, but it is very effective if you are very careful. Push the blade through one side of the sheath, and then cut the sheath out to the end (make sure you don't put your fingers between the blade and the handle if you are using a folding pocket knife). Then you can pull the cut portion of sheath back, turning it inside-out and cutting off the excess. If there are more wires inside the sheath than you plan to use, don't cut them off. Instead, fold them back and wrap them around the remaining sheath. That way, they will be available for future use, if needed.

Push the sheathing and any excess thermostat wire back into the wall. Lay out the individual wires to the individual terminals on the subbase, and strip the insulation as shown in figure 21–6. Note that you are making each wire a different length to match the distance to the terminal. Loosen the terminal screw. Grab the end of the exposed thermostat wire with a needle-nose pliers, and pull it around the terminal in a clockwise direction. Tighten the screw terminal to secure the wire. Then, still holding the end of the wire in the pliers, rock the end of the wire back and forth until it breaks off close to the terminal. Repeat for each terminal. The result will be a connection where the insulation goes all the way up to the screw terminal, the wire is the exact right length, and the tag end of the wire is as short as possible.

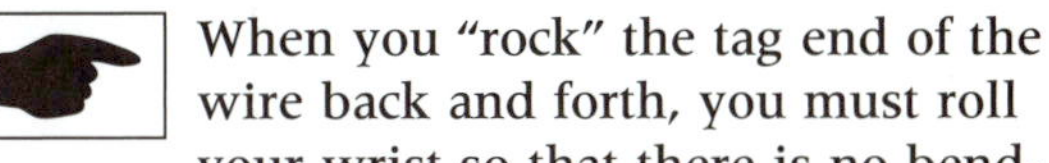

When you "rock" the tag end of the wire back and forth, you must roll your wrist so that there is no bending of the wire at the pliers. If you do it wrong, the wire will break at the pliers. If you do it right, the wire will break at the terminal. With practice, you can make it break at the terminal every time.

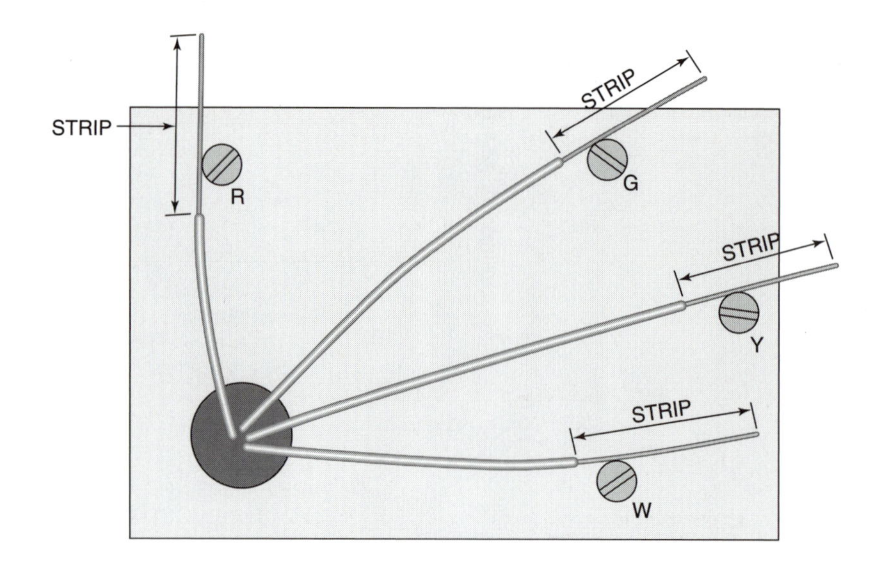

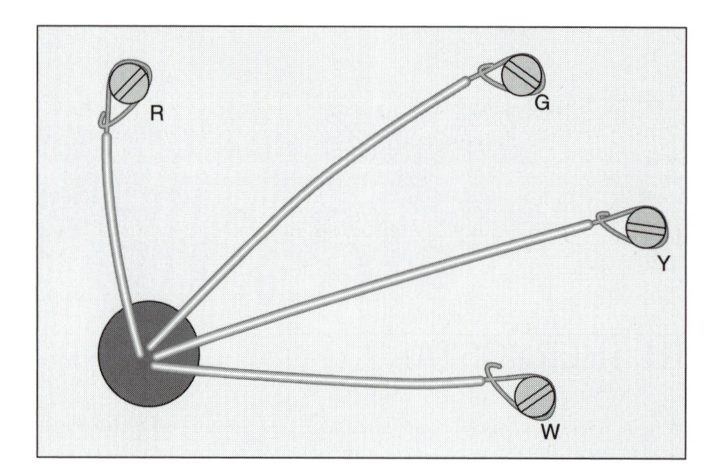

Figure 21–6 Connecting thermostat wire to subbase.

Figure 21–7 EMT bender.

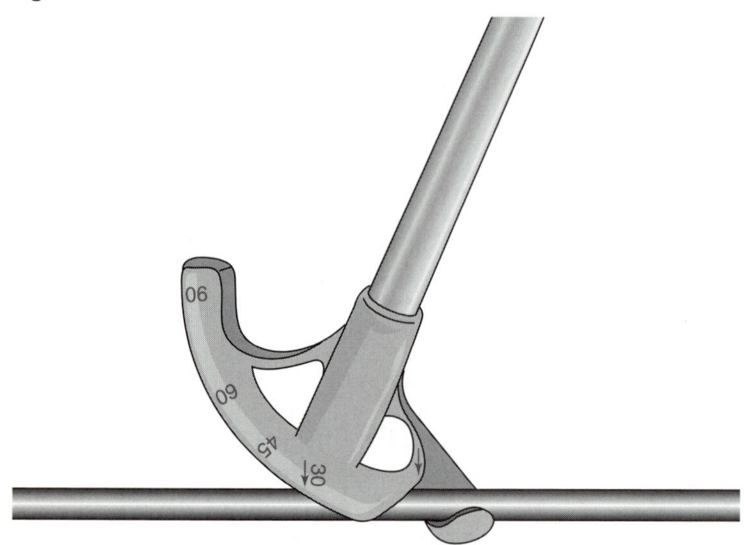

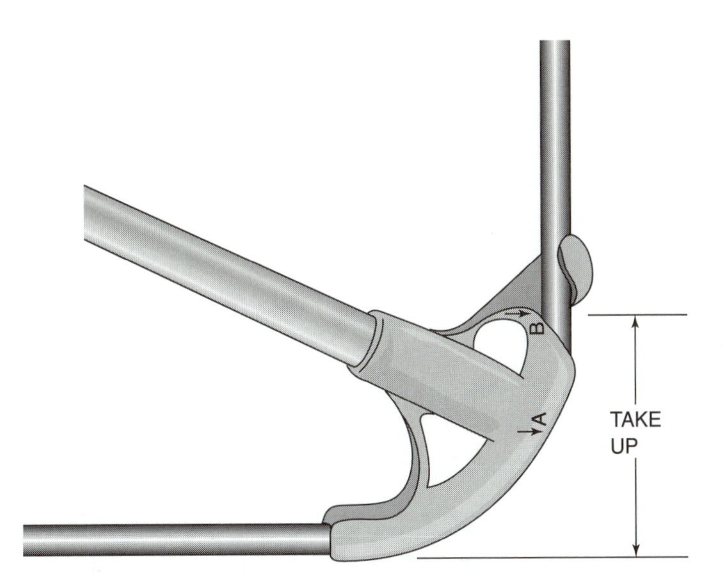

Figure 21–8 Take-up.

> **EXERCISE:**
> Using a thermostat subbase mounted on a board, practice the above technique until you can do it quickly and easily.

CONDUIT BENDING

Installation of conduit is usually done by the electrician. But there may be times when the air conditioning technician may have to install incidental runs of conduit. EMT (thinwall conduit) can be bent using an EMT bender, as shown in figure 21–7. They come in 1/2″, 3/4″, and 1″ sizes. The difficulty in using a conduit bender is in predicting the length of the finished bend. Each bender (depending upon the manufacturer) has a certain amount of **take-up** (figure 21–8). For example, if the bender you are using has a 6-in. take-up, and you want to make a bend that will produce a stub that is 16″ from the outside of the main leg, begin by subtracting the take-up dimension from the desired 16″ dimension (10″). Place the bender with mark "B" positioned 10″ from the end of the tube (figure 21–9). With the tube on the floor, pull on the tube bender until the stub rises to vertical. The resulting stub will terminate 16″ from the floor.

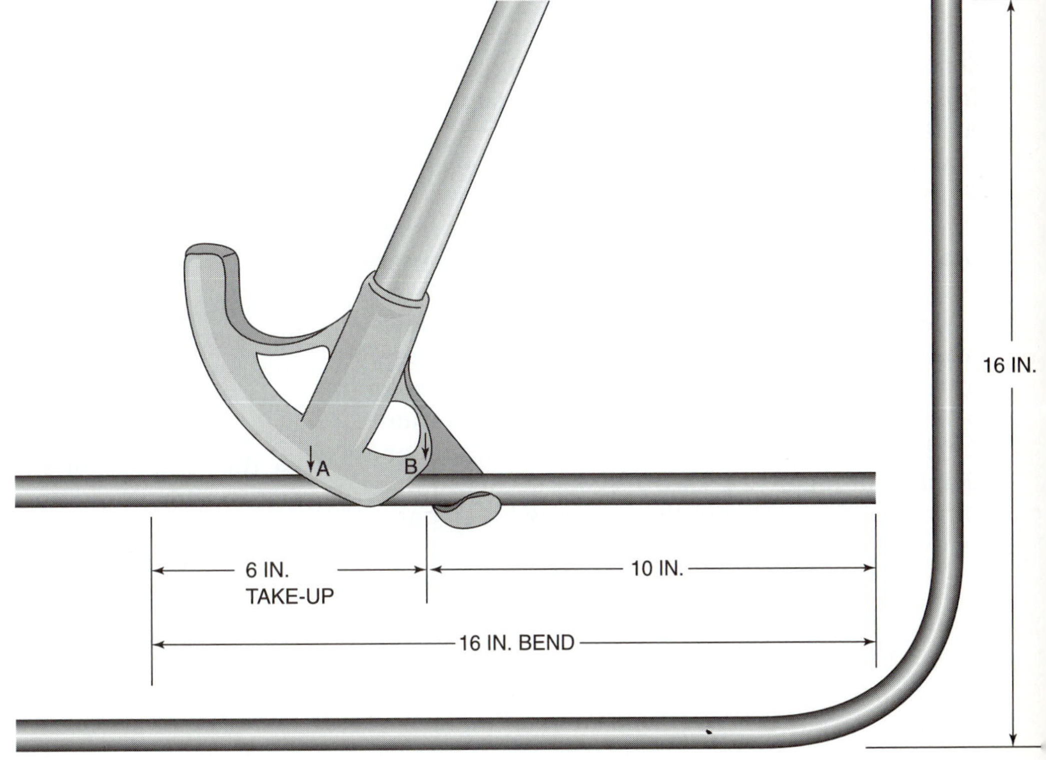

Figure 21–9 Positioning the bender.

CHAPTER TWENTY-TWO

DDC CONTROLS

Direct digital controls (DDC) are a part of computer-based building automation systems. A computer is integrated into the control system as shown in figure 22–1. A control valve is being used to control the supply air temperature from an air handler. Information from a sensor in the supply air stream is converted to a digital form and fed to the computer. Based on the information from the sensor (and possibly other sensors as well), the computer sends an output signal that is converted to an electric signal that will modulate the electric water valve. The relationship between the input signal from the sensor(s) and the output signal from the computer follows a program that is stored in the computer. From the computer terminal, an engineer or technician can reprogram the operation of the control system.

A variation of the DDC control is shown in figure 22–2. Here, the output signal from the computer is eventually converted to an air pressure signal that varies from 3–15 psi, and is used to operate pneumatic controls. Pneumatic controls have been used for many years. The operators (valve actuators or damper motors) can modulate, depending on how far the incoming air pressure can move a piston or a diaphragm (figure 22–3). DDC controls are in their infancy in the 1990s but they are growing in popularity for the following reasons:

1. They are more reliable and accurate than pneumatic controls, which require a well-maintained clean, dry, compressed air source.
2. The calibration of the sensors and the set point of the controller may each be adjusted from the computer terminal.
3. Very complex control logic schemes can be programmed into the computer software.
4. The controls can react very quickly, without "**hunting.**" Hunting is a term used to describe a control signal that oscillates wildly around a setpoint before settling down. This happens with pneumatic controls if you try to set them up to respond very fast.
5. The cost over the lifetime of the control system is low, compared with pneumatics.
6. The control scheme and logic is easily changed. It can be accomplished by merely reprogramming the software instead of changing hardware.

In addition to the normal functions of controlling temperature within a building, the DDC control system can be integrated into a total building automation system. It can sense the indoor and outdoor temperatures each morning, and compare it with past building operating history in order to determine the latest possible starting time for the heating and air conditioning equipment so that it achieves comfort by the time the occupants arrive.

At the end of the day, the building automation system can determine what is the most energy efficient time to shut down the heating and cooling systems. They can be turned off before quitting

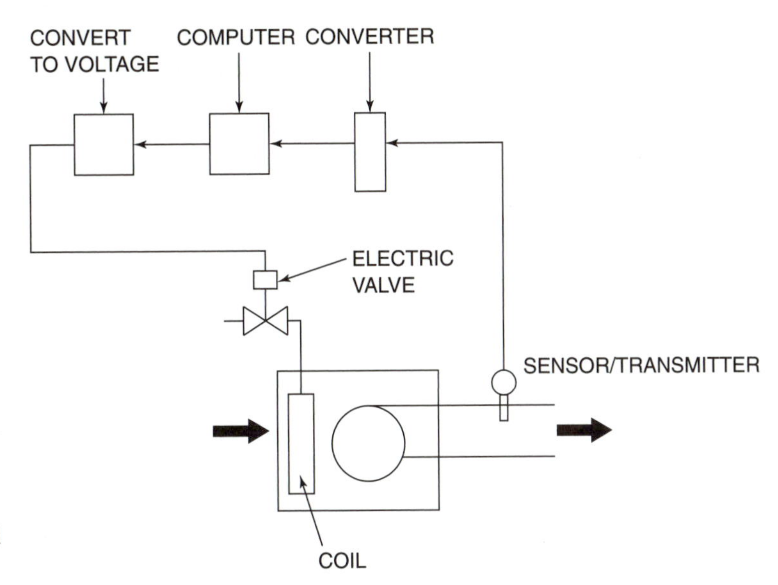

Figure 22–1 Computer-based control system.

time, and the building will "coast" down to its nighttime conditions while still maintaining comfort for the last occupants of the building. This is commonly referred to as providing the optimum start/stop times.

The building manager can get almost unlimited reports from the computer, including monthly energy use. These reports, coupled with other reports on where the energy is being used, and other energy-saving measures described later, can save 50% or more of the energy usually consumed by a similar building without computerized controls.

BUILDING CONTROL STRATEGIES

MODULATING CONTROLS

All of the controls described prior to this point in the text have been on-off controls. For example, a room thermostat has a set point of 74°F. If the room temperature rises to 75°F, the air conditioning system turns on. When the room temperature falls to 73°F, the air conditioning system turns off. Whenever the air conditioning system operates, it operates at 100% of its capacity, and the control scheme

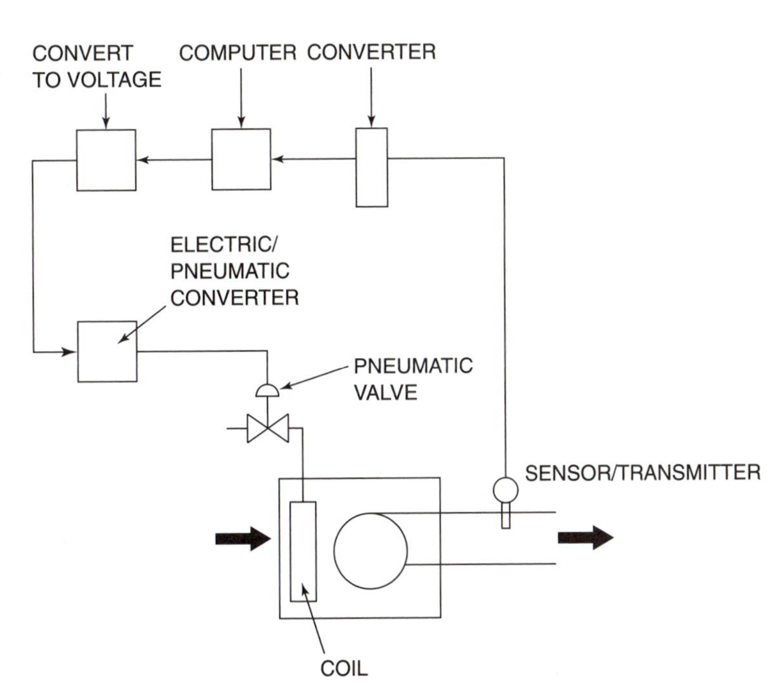

Figure 22–2 DDC control with pneumatic valve.

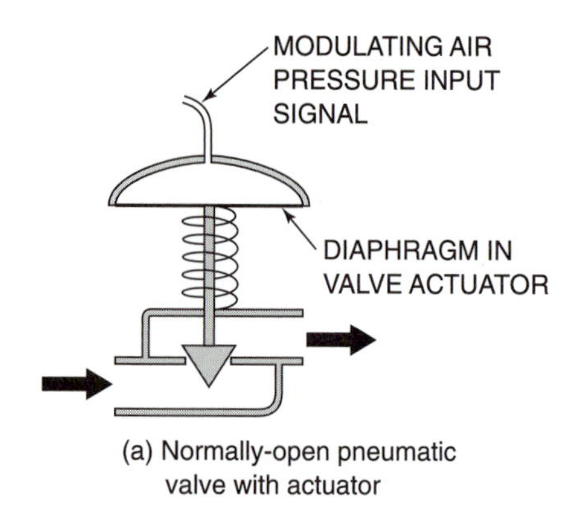

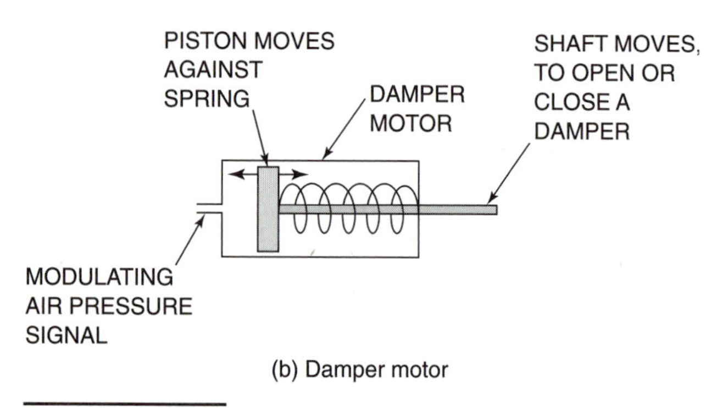

Figure 22–3 Pneumatic devices.

maintains room temperature by only operating the cooling for as much time as necessary.

The first control scheme shown in this chapter (figure 22–1) would be difficult to do with on-off controls. Suppose we were trying to maintain a supply air temperature of 60°F from a cooling system. If we tried to do this with a conventional rooftop air conditioner, when the air conditioner was off, the supply air temperature would be the same temperature as the return air (i.e., 74°F). The thermostat would turn on the compressor, and the supply air temperature would fall to 54°F, turning the compressor off. The temperature would then rise once again, very quickly, resulting is unacceptable cycling of the compressor. However, with the controls shown in figure 22–1, the cooling is being

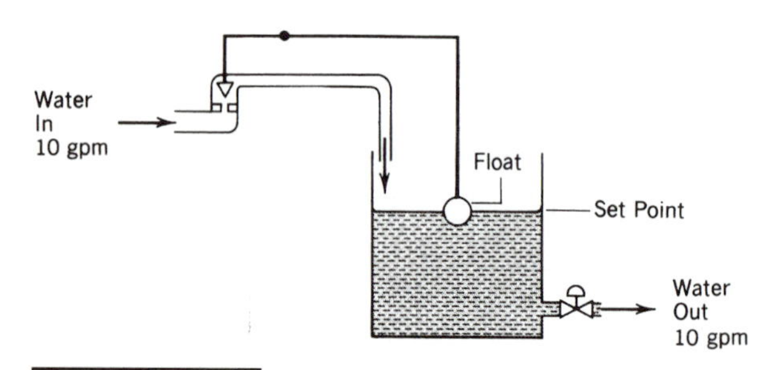

Figure 22–4 Level control.

provided by chilled water, and the amount of cooling that is being provided to the air can be modulated (adjusted) by modulating the amount of water that is allowed to flow through the coil.

While most small commercial and all residential applications will use On-Off control, most larger installations will use hot and chilled water for building heating and cooling, and will control the building temperatures with modulating controls.

Economizer Control

Figure 22–6 shows an air conditioning system with an **economizer.** In many large buildings, there are interior areas that have no outside walls. There is heat generated in these areas from lights, people, and equipment that must be removed, even when it is cold outside. Some air conditioning systems must be available year-round to cool these interior zones. But systems with economizers can take advantage of the cold outside air for "free cooling."

When it is hot outside, the dampers are positioned so that the outside-air and exhaust-air dampers are mostly closed, while the return-air damper is wide open. This allows only the minimum required amount of outside air into the building (too much outside air would dramatically increase the air conditioning load).

When outside-air temperature drops, at some point it reaches a condition where its **enthalpy** (heat content) is lower than the enthalpy of the return air. This is sensed by enthalpy sensors in both the outside-air duct and the return air duct, and their signals are sent to the computer. When the outside-air temperature is sufficiently low, the computer sends an output signal to the damper motors. The return-air damper will close, while the exhaust-air and outside-air dampers open wide. This allows the use of 100% outside air.

As the outside-air temperature continues to fall, the amount of cooling required from the chilled water system also is reduced, until there is no need for mechanical cooling (all the cooling is being provided by the outside air). If the temperature of the outside air continues to fall, the computer will signal the outside-air and return-air dampers to begin closing, while the return-air damper opens slightly. As the outside-air temperature continues to fall, the quantity of outside air continues to be reduced, using only as much outside air as is necessary to satisfy the room cooling demand.

Load Shedding

The largest single component of cost for an electric utility is not the cost of fuel to generate the electricity. In most cases, the largest single cost is the debt

ADVANCED CONCEPTS

Modulating controls have several different modes of operation. In order to explain them, we will describe the liquid level control system shown in figure 22–4. The level in the bucket (similar to temperature in a room) is being controlled at a relatively constant set point. There is a loss of water from the bucket (similar to heat loss from a room), and there is a source of make-up water (similar to heat input from a furnace) to keep the level constant. A modulating control valve gets its input signal from a control arm that moves up or down in response to the position of a float. The water flow out of the bucket is the load. The system as shown is in equilibrium. That is, the flow of water out of the bucket is 10 gpm, and the input from the water valve is also 10 gpm. Because the flow in exactly matches the flow out, the level is remaining constant. If the load changes to 8 gpm output (figure 22–5) the level in the bucket will begin to rise. As it rises, it will move the control arm, closing off the control valve. The level will continue to rise until the input has been reduced to exactly 8 gpm, once again matching the load. In this case, the control mode is called **proportional.** That simply means that the corrective signal (movement of the control arm) is proportional to how different the actual level is from the desired level (set point). If the level is 1-in. higher than the desired level, the control arm moves a certain amount. If it is 2-in. higher than the desired level, the control arm moves twice as much. In other words, the corrective signal to the control valve is proportional to the error signal. Prior to DDC controls, pneumatic controls were simply proportional controls. But with DDC, more sophistication is available, and is commonly used.

The problem with proportional control in the example above is that if the load decreased from 10 gpm to 8 gpm, we were stuck with a new equilibrium level that was slightly higher than our set point. This is called **offset.** DDC controls commonly add an additional mode to correct this offset. It is called **proportional plus integral.** After a new equilibrium has been reached, the integral mode looks at the level, compares it with the set point, and adjusts the output to eliminate the offset. This would be analogous to using a turnbuckle to adjust the length of the vertical part of the control linkage to restore the original level, even though the load is now 8 gpm instead of 10 gpm. Still another refinement in control modes available with DDC controls is called **proportional plus derivative plus integral** control, sometimes called **PID** control. With PID controls on our level control example, the turnbuckle adjustment to eliminate the offset would not be made at a constant speed. Rather, the turnbuckle would initially be adjusted quickly. But as the actual level approached the desired level, the rate of adjustment of the turnbuckle would slow down. This may seem like an absurd level of overkill, but with DDC controls, there is little or no additional cost because the control modes are determined by the software in the computer.

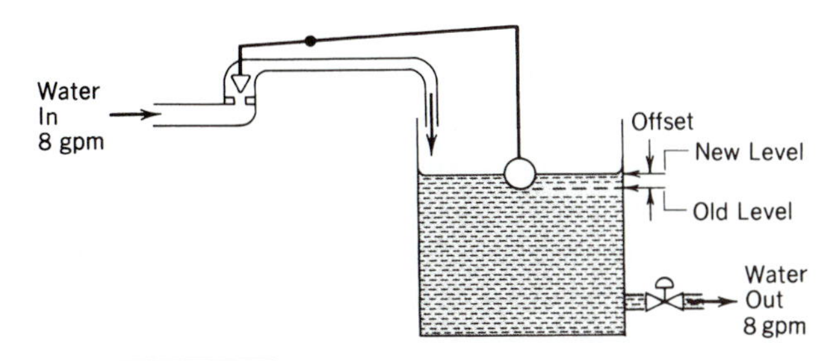

Figure 22–5 Level control-change in load.

repayment on the loan that was taken by the utility company to finance the construction of the electrical generating station. Unfortunately for the electric company, they must build (and finance) enough generating capacity to satisfy the peak electrical demand, but most of the time (when it is not exceptionally hot outside), some of the electrical generating capacity sits idle because it is unneeded.

Figure 22–6 Economizer.

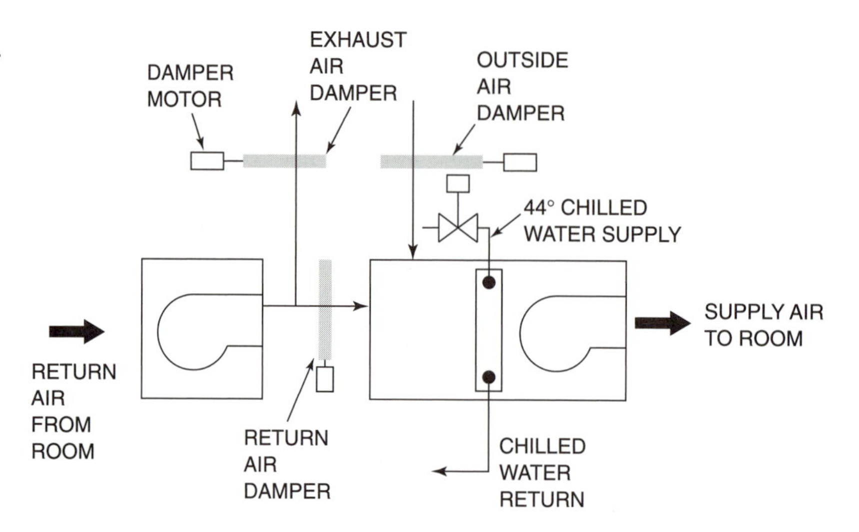

Therefore, electric utilities have devised two methods of attempting to level their load:

1. **TOU rates** Time-Of-Use rates charge customers different rates per kilowatt of electricity consumed, based upon the time of day when it was consumed. Some rate schedules provide electricity at 3:00 a.m. (when demand is typically very low) for 10 to 30% of the price of the same quantity of electricity consumed at 3:00 p.m.
2. **Demand charges** These are imposed on customers *in addition to* the electrical consumption charges. For example, the utility company might charge $5.00 per kilowatt based on the highest rate of kilowatt usage by the customer during the previous month.

Load shedding is a strategy whereby the building automation senses the total electrical consumption at each moment, and compares it with the previous highest rate of consumption for the month. If it appears that the building is on its way towards setting a new peak, certain non-essential equipment will be turned off automatically to avoid setting a new peak demand. The computer in the building automation system has been previously programmed with all the decisions about which equipment will be shut off, when, and for how long. With the computer control, the load shedding strategy can actually become quite sophisticated. For example, instead of a simple "rolling shutdown" of preselected equipment, the computer can be programmed to change the room temperature set points for the period where there is risk of setting a new demand.

Night Purge

In the typical system without night purge, the air conditioning equipment is started an hour or two before the occupants arrive. At the time when the occupants are expected to be gone, the night setback temperature begins. Typically, the temperature of the building will rise during the night, so it might be 80°F when the equipment is started again the next morning. This is because much of the heat that started into the building during the afternoon takes that long to actually make its way into the occupied space.

With a night purge system, outside air is introduced into the building during the evening. This has two beneficial effects:

1. The building will be cooler just before the cooling equipment must be started. This will allow for a later equipment start time, and will require removal of less heat from the building before the occupants arrive (the "pull-down" period).
2. The quality of the indoor air is improved because the air contaminants that collected in the building during the day will be purged to outside.

Of course, the building automation system senses if the outside air is pre-cooling the building too much. This would be undesirable, because then the building owner would incur additional heating costs. However, a properly set-up building automation will allow the introduction of only just the right amount of air.

Chilled Water Reset

The typical air conditioning system in a large building consists of a chiller that produces water at 44°F, and air handlers that use that 44°F water to cool the room air (figure 22–7). However, during many times of the year when peak cooling is not required by the air handlers, it would be desirable to reset the chilled water supply temperature to higher than

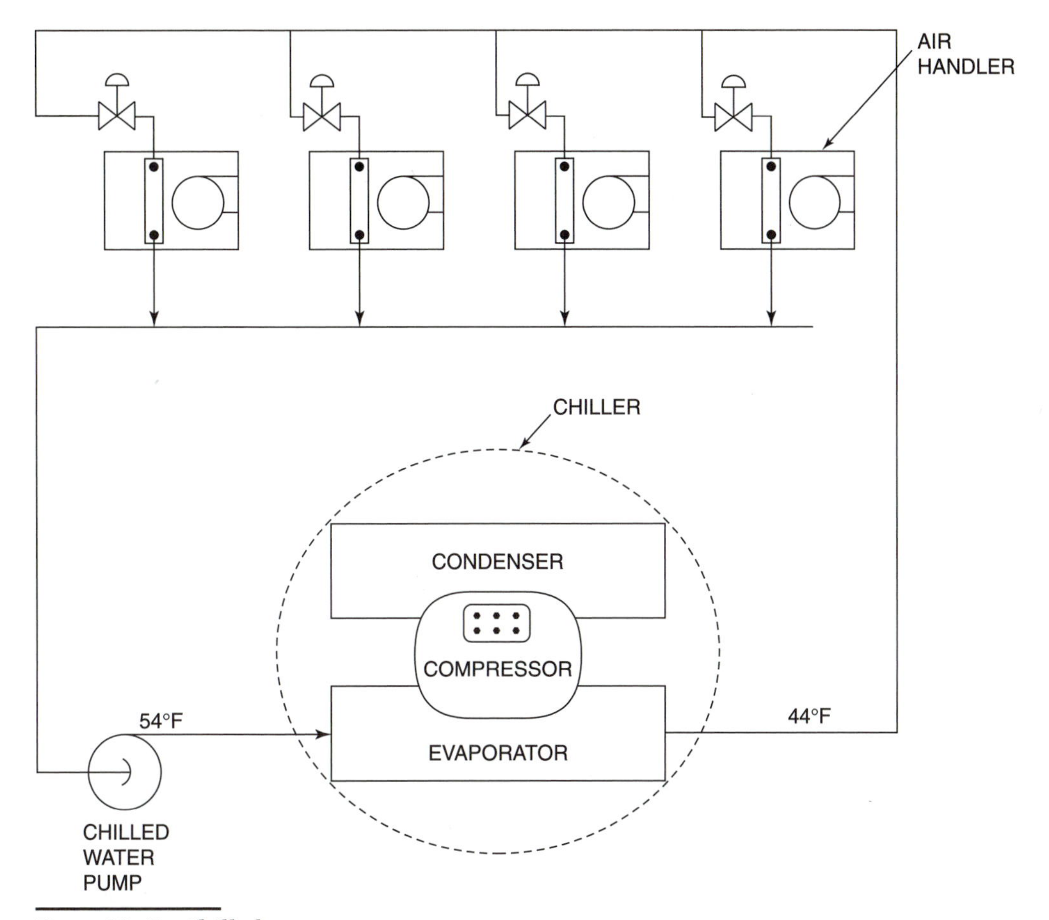

Figure 22–7 Chilled water system.

44°F. A chiller can produce a ton of cooling with 48°F leaving water temperature with much less electricity than a ton of cooling with 44°F leaving chilled water. Each degree of chilled water reset reduces the chiller energy costs by about 1 to $1\frac{1}{2}$%. This may not sound like much, but it represents a potentially major reduction in operating costs.

The building automation computer can be programmed to look at all of the air handlers, and select the one that, at that instant, requires the coolest chilled water to satisfy the room cooling demand. The computer will then send a signal to the chiller, resetting the chilled water supply temperature as high as possible, but not so high that it would create a problem for the zone that is demanding the lowest chilled water temperature.

Condenser Water Reset

At design conditions, condenser water will usually be supplied to a water chiller at about 85°F. Some systems maintain that temperature constant by either cycling the fan on the cooling tower, or using a temperature control valve to bypass water around the cooling tower (figure 22–8). Other systems will allow the condenser water temperature to drift down as it becomes cooler outside. This results in lower operating costs for the chiller due to lower head pressure. However, the actual savings due to lower condenser water temperatures are limited. Chiller efficiency does not continue to improve with each degree of reduction in condenser water temperature. At some temperature, as condenser water temperature (and head pressure) continues to fall, the chiller efficiency actually begins to deteriorate and energy consumption *increases.*

With a DDC computerized control system, it can be programmed to continuously monitor the condenser water temperature and the electrical consumption of the chiller. The control system can then allow the condenser water temperature to drift only to the point where the best chiller efficiency is achieved.

Fan Cycling

Historically, chilled water/air handler systems have been controlled so that the air handlers operated continuously, while the temperature of the air being supplied to each zone was modulated by modulating the flow of hot or cold water through a heating or

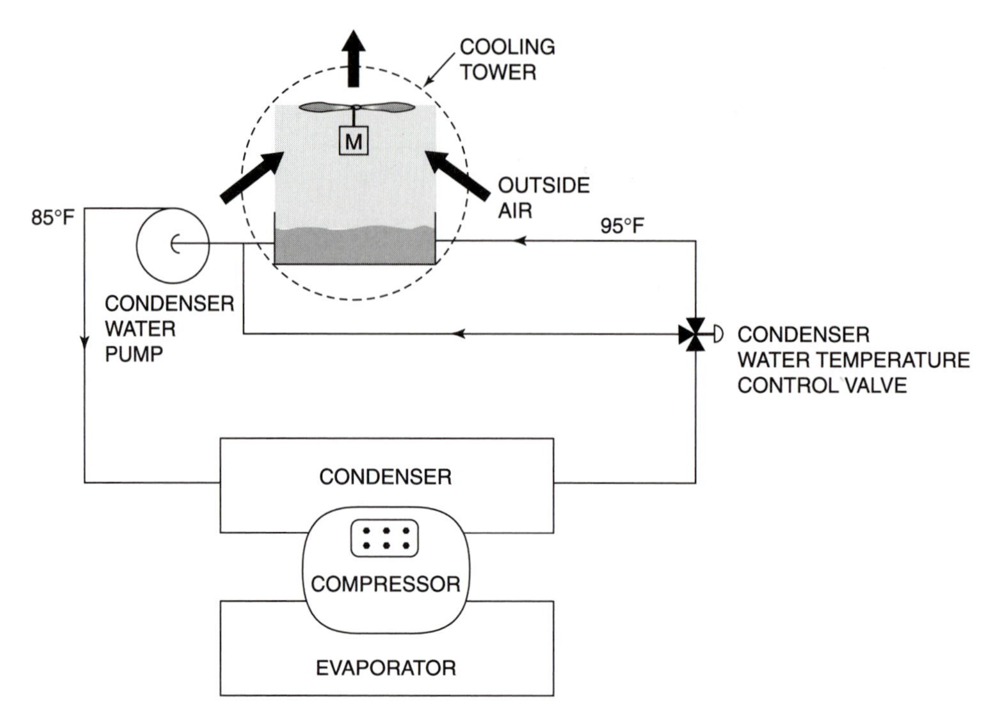

Figure 22-8 Condenser water reset.

cooling coil. However, the amount of air supplied by the air handlers was determined by the design engineers to be the quantity of air needed to maintain comfort conditions when it is approximately 95°F outside. Whenever it is cooler than 95°F outside, there is actually more air being supplied than is necessary. This has led to the development of **variable air volume** (VAV) systems, in which terminal air units actually modulate the quantity of air being supplied to each zone. They are more energy efficient than constant volume systems. But some of the benefits of variable volume systems can be achieved with constant volume systems using fan cycling.

By continuously measuring room temperature and other variables, it is possible, at times, for the computer to turn off some of the air handler fans for short periods of time, without adversely affecting the comfort of the occupants. Fans may be cycled off for five to 15 minutes, at a frequency of every 20 minutes to every hour.

Another use of fan cycling is to avoid setting a new electrical demand for the electrical billing period (see "Load Shedding" above). Some applications only initiate fan cycling as part of the load shedding strategy.

Fire and Smoke Alarm Systems

The fire and smoke control systems in large buildings have two functions:

1. detecting and reporting alarms
2. taking action in response to alarms.

The fire alarms can include smoke detectors, heat detectors, manual alarms, and sensors to indicate that water is being released from a fire sprinkler system.

Upon activation of any of the above alarms, a signal is sent to the computer console. It will respond, as programmed for each situation, to sound audible alarms, display graphically the location of the sensor that has detected a problem, and automatically take over the control of the air systems (to prevent the spread of fire), and even fire doors and elevators.

Lighting Control

Lighting control is usually not within the realm of duties of the heating/air conditioning/refrigeration person in the maintenance department of a large building. But lighting comprises a significant portion of the heat load in a building. Any steps that can be taken to operate the lights for no longer than required by the occupants will yield energy savings for the cooling system. Lighting is easily controlled through the same building automation system that controls the building environmental systems.

The computer program that operates the light switches has several functions, in addition to turning the lights on and off at preprogrammed times:

1. The occupants (if there are any) must be warned before the lights automatically go out. This is an obvious safety issue. Commonly, the computer controls the lighting control panels to momentarily

shut off the lights 10 minutes prior to actually shutting off the lights for the evening. This will give any remaining occupants a chance to leave the building while the lights are still on.

2. A means may be provided for occupant override of the programmed operation. If somebody is working later than usual, and doesn't want to leave the building when the warning of shutdown comes, there must be a means for the occupant to have the lights left on. There are three ways that this is usually accomplished. The simplest is to merely provide switches that the occupant can operate to tell the computer not to turn off the lights for another hour. A second method is to have motion detectors that sense whether there are occupants remaining in the area where the lights are scheduled to be turned off. The computer can then automatically skip that area until the occupants have left. A third method of the occupant advising the computer that the lights must be left on is through the use of the occupant's telephone that is tied into the computer. The computer can also keep a record of which occupants are staying late (for security purposes) by requiring the input of a security code that is unique for each occupant.

Glossary and Abbreviations

Air Over (AO) Motor A motor designed to drive a fan, and be cooled by the stream of moving air.

Alternating (AC) Current An electric current that reverses its direction at a fixed frequency.

Alternator Alternating current generator.

Ammeter A meter used to measure the flow of amps.

Ampacity The current-carrying capacity of a wire.

Ampere (amp) The unit of measurement of electrical current flow.

Analog Meter A meter that uses a moving pointer instead of a digital display.

Anti-Recycle Timer A clock/switch combination that prevents a unit from restarting until after it has been off for a predetermined time.

Anticipator A small resistor in a room thermostat that heats up the bimetal element slightly faster than it would be heated by the rising room temperature alone.

Aquastat Honeywell brand name that is used generally to denote an immersion-type water-temperature-sensing thermostat.

Armature The part of an electrical motor or contactor that moves.

Armored Cable Flexible metal conduit.

Automatic reset An automatic safety control that will reclose after the abnormal condition that caused it to open returns to a normal value.

Autotransformer A means of reducing motor speed by using different transformer taps to supply different reduced voltages to the motor.

AWG American Wire Gauge. A system of designating wire size (diameters). The smaller the number, the larger the gauge of the wire.

Back EMF The voltage produced by an inductive load that opposes the applied voltage.

BHP Brake horsepower. A measure of the output power of a motor.

Bimetal A temperature-sensing element that bends when its temperature changes.

Bin Switch Senses when the ice maker bin is full, and turns off the refrigeration.

Bin Thermostat A thermostatically controlled bin switch.

Bleed Resistor A resistor connected across the terminals of a capacitor to slowly dissipate the stored electrical charge.

Blower Relay A control relay in a furnace that controls the operation of the furnace fan.

Bonnet The area of a heat exchanger in a furnace through which the room air passes.

Bonnet Switch A thermostat that senses bonnet temperature, and closes when the temperature climbs to approximately 130°F.

Breaker A circuit breaker.

Brushes The stationary contact for bringing power to the rotating contact surface on the rotor of a brush-type motor.

BX Type of flexible metal conduit. Also called armored cable.

Cad Cell A device whose resistance decreases when it senses light.

Calibrate The adjustment of a device so that it actually operates at the temperature that is indicated on the device.

Capacitance The electrical storage capacity of a capacitor.

Capacitor A device that can store electrical energy and is used in motor circuits to increase starting torque or improve motor efficiency.
Casing Ground A ground wire attached to the casing of equipment to prevent the possibility of a shock resulting from a ground fault.
Circuit Any combination of wires, switches, and loads that permit an electric current to flow from a source of electricity at a higher potential to another source at a lower potential.
Circuit Breaker An automatic switch that senses current, and will open if the sensed current exceeds the circuit breaker rating.
Cold Control Jargon for the line-voltage thermostat in a small refrigeration system.
Combination Gas Valve Includes main valve, pilot valve, and pressure regulator in one unit.
Common (a) A terminal that is a part of two or more circuits; (b) The neutral side of a 115V system.
Commutator The rotating contact surface on the rotor of a brush-type motor.
Complete Circuit A continuous path for electrons to flow from a higher potential to a lower potential.
Conductor A material with very low electrical resistance.
Conduit Tubing that is used to enclose and protect electrical wiring.
Contactor Heavy-duty switch or switches closed by a coil in a different circuit.
Continuity There is a path without a break for electrons to flow.
Control Circuit The low-amperage circuit that includes the switches and coils for contactors and relays, but not the large motors that they may control.
Control Relay Light-duty switch that is operated by a coil in a different circuit.
Cooling Anticipator A resistor in a cooling thermostat that causes the compressor to be energized sooner.
Current The flow of electrons.
Current Relay A start relay used on compressors that senses the motor speed by sensing the current through the Run winding.
Cut-in The pressure or temperature at which the switch in an automatic pressure or temperature control will close.
Cut-out The pressure or temperature at which the switch in an automatic pressure or temperature control will open.
Cycling A motor repeatedly turning on and off.
DDC Direct Digital Control; a computer-based means of programming the operation of operating controls.
Defrost Termination Switch A thermostat that senses evaporator temperature and opens when the temperature rises to approximately 45°F.
Defrost Timer A clock/switch combination that initiates and terminates the defrost cycle in a freezer.
Demand Charge A charge made by the electric utility based upon the peak usage of electricity.
Dielectric The high-resistance material between the storage plates of a capacitor.
Differential The difference between the cut-in and cut-out settings on a controller.
Diode The electronic equivalent of a check valve. It only allows current flow in one direction.
Direct Current(DC) An electric current in which the electrons flow in one direction only.
Domestic Hot Water Hot water for use in sinks, baths, and showers.
Door Switch A momentary switch that is operated by the opening or closing of a door.
Double-shaft Motor A motor with a long shaft that extends out of both ends of the motor.
DPDT Double pole double throw (switch).
DPST Double pole single throw (switch).
Dripproof Motor A motor constructed with ventilation openings at the bottom so that falling liquid at an angle of 15 degrees from vertical will not enter the motor.
Dry Switch A switch that is operated by a device, but is not electrically connected to that device, such as an auxiliary contact on a contactor.
DTS Defrost termination switch.
Duct Heater An electric heating element mounted inside a duct.
Economizer A mode of operation of an air conditioning system where cool outside air is used to supplement the mechanical cooling.
Efficiency The ratio of useful energy output, divided by the energy input.
Electrical Interlock A scheme where a control relay is energized at the same time as a load. The control relay contacts energize a second load.
Electric Potential Electrical pressure (voltage).
Electromagnet A magnet that is created by current flowing through a coil of wire wound around an iron core.
EMF Electromotive force; a scientific term for voltage.
EMT Galvanized, light-weight conduit.
Enthalpy Heat content of air.
Fan-Limit Switch A fan switch and a limit switch in a common enclosure that use a common sensing element to sense bonnet temperature.
Farad the unit of measurement of electrical capacitance.
Field Wiring Wiring that is done by the installing technician, as compared to factory wiring.
Finishing Relay A control relay that keeps an ice maker operating to the end of the cycle, even after the bin switch has opened.

Fish Wire Tape made of tempered spring steel used to pull wires through conduits and wall openings.
FLA Full Load Amps.
Float Switch A switch that opens or closes in response to the liquid level in a tank.
Flow switch An automatic switch that senses flow (usually of a liquid).
Frame Size A set of physical dimensions of motors, as established by NEMA.
Freezestat Jargon for a thermostat that opens when the sensed temperature drops below 32°F.
Frequency The rate at which alternating current changes its direction of flow.
Fused Disconnect A main disconnect switch plus fuses in each leg, all in one box.
Ganged Switches Two or more switches that are mechanically linked to operate together.
Generator A device that converts mechanical energy into electrical energy.
GFCI Ground Fault Current Interrupter. An outlet that disconnects the power supply if it detects an unequal amperage flow in the two legs of a single-phase supply.
Grounded (Ground Fault) A failure mode wherein a device has allowed unintended contact from current-carrying wiring to ground.
Ground Wire A connection to a conducting material that is at zero electrical potential.
Hard Start Kit Contains a start capacitor and solid-state start relay to give added starting torque to a PSC motor.
Harvest Cycle The part of an ice cuber operation where the cubes are melted out of the mold.
Head Pressure Control A control scheme to reduce condenser capacity to prevent the high-side refrigerant pressure from dropping too low.
Heat Anticipator An adjustable resistance in a heating thermostat that causes the switch contacts to open sooner.
Heat Sequencer A time delay relay used to sequence the operation of several electrical strip heaters.
Henry The unit of measurement of electrical inductance.
Hertz The unit of measurement of AC frequency, equal to one cycle per second.
High Leg The leg connected to the terminal of a delta transformer that is opposite the center-grounded leg.
High Limit A safety thermostat that opens in the event that a sensed temperature becomes abnormally high.
High Pressure Cut-out A pressure switch that opens on a rise in pressure.
Holding Contacts *See* Seal-in Circuit.
Hot Jargon for the electrically pressurized leg of a 115V power supply.
HPC High Pressure Cut-out.
Humidistat A switch that closes to turn on a humidifier when the sensed humidity becomes too low.
Hunting A control problem characterized by repeated wide deviations above and below setpoint of a controlled device.
Identified Terminal The terminal on a capacitor that is marked to show that it is the terminal that is connected to the plate nearer to the outside casing.
Impedence Relay Lock-out relay.
Inducer Fan A small fan that helps pull the products of combustion through the heat exchanger in a furnace.
Induction The production of an electric current in a conductor by a magnetic field in close proximity to it.
Induction Motor A motor in which a current is induced in the rotor.
Insulator A material with a very high resistance.
Interlock A scheme where when one device is energized, it energizes a second device.
J-Box Jargon for junction box. An enclosure where mechanical connections of wiring is made.
Kilowatt-hour A quantity of energy equal to a rate of one kW being consumed for one hour.
Klixon A trademark brand name, commonly used as a generic term for a line-voltage overload.
kWh Kilowatt-hour.
LED Light emitting diode.
Level Switch *See* Float Switch.
Limit Switch A safety switch (usually a thermostat) that will open if the sensed temperature or pressure rises above a maximum safe value, or drops below a minimum safe value.
Line Drop The voltage loss in conductors in a circuit due to their resistance.
Line Duty A switch designed to carry the heavy load of a compressor or other load.
Load The resistance connected across a circuit that determines the amount of current flow.
Lock-out A condition that occurs when an electronic ignition system fails to sense a flame, and the control box will not energize anything until it is reset.
Lock-out Relay A control scheme using a control relay that locks out the operation of a device until the circuit is de-energized to reset itself.
Low Limit A safety thermostat that opens if the sensed temperature becomes too cold.
Low Pressure Cut-out: An automatic switch that opens on a decrease in pressure.
Low-voltage Thermostat A thermostat whose switch is designed to operate a 24V load.

LPC Low Pressure Cut-out.
LRA Locked Rotor Amps.
LSV Liquid Solenoid Valve.
Main Winding Same as Run winding.
Manual Reset An automatic safety control that will remain open even after the abnormal condition that caused it to open returns to a normal value.
Mechanical Interlock A scheme where a device is energized, and the operation of that device mechanically operates a switch that energizes a second device.
Megohm Meter An ohm meter with a very high scale, used to sense the impending breakdown of insulation. One megohm equals one million ohms.
Microfarad The commonly used measure of capacitance, equal to a millionth of a farad.
Microprocessor A small computer using an integrated circuit.
Micro-switch A small momentary switch used to sense the physical position of a device.
Momentary Switch A switch that opens (or closes) when it is pressed, but then returns to its normal position when it is released.
NEC National Electric Code.
NEMA National Electrical Manufacturers Association.
Neutral The part of a 115V circuit that carries current, but is at zero electrical pressure.
Nichrome A type of wire that is used as the element in a resistance heater.
Nichrome Wire A wire that has a high resistance. When a current passes through it, it becomes hot.
Normally Closed A coil-operated switch (or valve) that is closed when the coil is not energized.
Normally Open A coil-operated switch (or valve) that is open when the coil is not energized.
Ohms The measurement of resistance to the flow of electricity.
OPC Oil pressure control.
Open Circuit A circuit that does not provide a complete path for the flow of electricity.
Operating Control The thermostat or pressurestat that normally turns a compressor or other device on and off to maintain the setpoint.
Overload 1. A condition where a device carries more current than it is rated to carry; 2. An automatic switch that senses current and opens if the current exceeds a preset rating.
Parallel Circuit A circuit that contains two or more paths for electrons to flow.
Phase 1. One of the legs of a power supply. 2. The alignment (in time) of the voltage and the amperage in a wire.
PID Proportional plus integral plus derivitive control.
Pilot Duty Switch A light-duty switch that operates a low amp draw device such as a contactor.
Pilot Generator A series of thermocouples that is heated by a pilot flame, and produces 250–1000mV to operate the gas valve.
Pilot Light A light bulb that is wired in parallel with a load to indicate when that load is energized.
Pilot Safety Switch A switch that senses the high temperature produced by a pilot flame in a furnace. If there is no flame, the switch opens.
Poles The number of moving electrical switches that are operated by a coil in a contactor or control relay.
Potential Electrical pressure (voltage).
Potential Relay A start relay used on compressors that senses the motor speed by sensing the amount of back EMF being generated by the Start winding.
Potentiometer A variable resistor that may be adjusted manually.
Power Work per unit of time.
Power Factor A measure of lag between the time when peak voltage is reached and the time when peak current flow occurs.
Powerpile *See* Pilot Generator.
Pressurestat An automatic switch that opens or closes in response to a change in pressure.
Primary Control The main controller on an oil burner.
Primary Winding The input side of a transformer.
Printed Circuit A circuit on which the wiring has been etched into a board.
Proportional Control The modulating control mode where the corrective output signal is proportional to how far the actual sensed temperature (or pressure) is different from the setpoint value.
PSC Motor Permanent split capacitor; a type of split-phase motor that uses a run capacitor, but no start relay.
PTC Device Positive temperature coefficient device, whose resistance changes with its temperature. It is used in the starting circuits of PSC motors.
Pumpdown A refrigeration control scheme wherein all refrigerant is pumped out of the evaporator prior to allowing the compressor to shut down.
Purge Blower A fan that blows clean air through the heat exchanger of a furnace.
Push Button A momentary switch.
Range The bounds of the available setpoints of a controller.
Rectifier A circuit that converts an AC input voltage to a DC (or pulse DC) output voltage.
Reset Relay *See* Lock-out Relay.
Resistance A measurement of a material's tendency to allow or restrict the flow of electrons.
Resistance Heating Producing heat by passing a current through a resistance wire.
Resistor An electronic device used to create a fixed resistance.

Rigid Conduit Steel conduit made to the same dimensions as standard pipe.
RLA Running Load Amps.
Roll-out Switch A switch that opens if the flame in a furnace escapes from inside the heat exchanger.
Run Winding The winding in a split-phase motor that is energized continuously when the motor is running. Also called Main Winding.
RVS Reversing Valve Solenoid.
Safety Control Any switch that does not operate during normal operation of the system, but shuts the system down if some variable gets outside of acceptable limits.
Sail Switch A flow switch that senses the flow of air.
SCR Silicon-controlled rectifier. An electronic device that opens to allow flow through an anode-cathode when a predetermined current flows through the gate and then shuts off when the flow through the anode-cathode falls.
Seal-in Circuit A circuit that continues to energize a relay coil, even after the switch that initially energized the coil opens.
SealTite A brand of weatherproof flexible conduit.
Secondary Winding The output side of a transformer.
Series Circuit A circuit in which the same electrons flow through all of the devices.
Service Factor The multiple of full-load amps that a motor can draw for short periods of time without damaging the motor.
SF Service Factor.
SFA Service Factor Amps.
Shaded Pole An extra pole on a stator winding that is wrapped with a copper ring, causing it to be out of phase with the stator winding.
Shaded Pole Motor A motor type characterized by no start winding, and used on very small loads.
Short Circuit A path for electron flow between two different potentials that does not contain a load.
Short Cycling Rapid repeated starting and stopping of a motor or other load.
Silicon-Controlled Rectifier An electronic device used to control the flow of electricity to a high-amp load.
Slip The difference between the synchronous speed and the actual speed of a motor.
Slow Blow Fuse A fuse that can carry four to six times its rated current for a few seconds.
Solenoid A coil of wire that operates a switch or a valve.
Solenoid Valve A valve that is operated by a solenoid coil.
Solid State Device An electronic component that is built from semiconductor materials.
Solid State Relay A start relay used on compressors that allows the Start winding to remain in the circuit for a predetermined time.
SPDT Single pole double throw (switch).
Split Phase Motor A motor that has a start winding and a run winding.
SPST Single pole single throw (switch).
Stack Control A primary control that incorporates an integral stack switch.
Stack Switch A switch that senses flue stack temperature in an oil-fired system to determine if flame has been established.
Starter A combination of a contactor and an overload for starting a motor.
Start Winding The winding in a split-phase motor that is energized for only a few seconds on start-up.
Starting Torque A measure of the ability of a motor to start up against a mechanical resistance to the rotation of the motor.
Stator The non-rotating windings in a motor.
Step-down Transformer A transformer whose output voltage is lower than the input voltage.
Step-up Transformer A transformer whose output voltage is higher than the input voltage.
Stinger Leg *See* High Leg.
Stratification The ``layering" of air caused by buoyant warm air collecting near the top of a room.
Strip Heater Jargon for an electric resistance heater.
Subbase A wallplate that contains the terminals for thermostat wires and accepts the mounting of a thermostat.
Synchronous Motor A motor that rotates at the same speed as the rotating magnetic field in a motor.
Synchronous Speed The rotational speed of the magnetic field in a motor.
Thermal Fuse A one-time thermostat that opens when it is subjected to a temperature higher than its temperature rating.
Thermal Overload A thermostat that senses motor temperature and opens if it gets too hot.
Thermistor An electronic device whose resistance is proportional to its temperature.
Thermocouple A junction of dissimilar metals that produces a voltage that is proportional to temperature.
Thermopilot Relay A brand name to describe a pilot safety switch that is held closed by a hot thermocouple.
Thermostat An automatic switch that opens or closes in response to a change in temperature.
Thinwall Conduit Galvanized, lightweight conduit. Also called electro-mechanical tubing, abbreviated EMT.

Throws The number of electrical positions available to a pole on a switch.
Time-delay Fan Switch An automatic switch that closes (to turn on a furnace blower) motor one minute after a 24V heater is energized.
Time-delay Fuse A fuse that can carry four to six times its rated current for a few seconds.
Torque The force that produces rotation. Common units are foot-pounds and inch-pounds.
Totally Enclosed Air Over (TEAO) Motor A motor constructed with no ventilation openings that requires external airflow over the motor enclosure for cooling.
Totally Enclosed Fan Cooled (TEFC) Motor A motor constructed with no ventilation openings that includes an external fan with a shroud to move cooling air over the motor.
Totally Enclosed Non-Ventilated (TENV) Motor A motor constructed with no ventilation openings.
TOU Rate Time of use. Electrical consumption prices are based upon the time of day when it is used.
Transducer Any device that converts energy of one form into energy of a different form.
Transformer A device that uses an AC source of current to produce an AC output current at a different voltage.
Transistor The electronic equivalent of an electric control valve. When a small voltage is applied, a gate is opened, and a larger current flow is allowed.
Triac An electronic component that controls the AC voltage that is allowed to pass to a load.
Trickle Heat Circuit A small current that is allowed to pass through compressor motor windings to act as a crankcase heater during the off-cycle.
Unloader Solenoid A solenoid on a compressor that operates to disable one or more cylinders.
VA Rating For transformers, the product of the output voltage times the maximum amps that can be supplied by the secondary winding.
VAV Variable Air Volume.
Voltage Electrical pressure.
Volt Meter An instrument used to measure a difference in electrical potentials between two points.
VOM Volt-Ohm meter.
Watt The unit of measurement of electrical power, equivalent to 3.4 Btu/hr.
Watt-hour The amount of energy consumed at a rate of one watt for one hour.
Wild Leg *See* High Leg.
Zone An area of a building that is controlled by its own thermostat.
Zone Valve A hot-water valve on a boiler system that controls the flow of water to a particular zone.

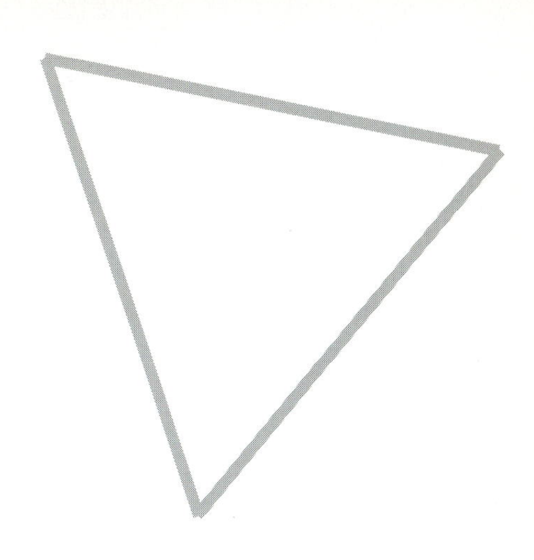

INDEX